Diagnostic Molecular Biology

Diagnostic Molecular Biology

Chang-Hui Shen

ACADEMIC PRESS

An imprint of Elsevier

Academic Press is an imprint of Elsevier
125 London Wall, London EC2Y 5AS, United Kingdom
525 B Street, Suite 1650, San Diego, CA 92101, United States
50 Hampshire Street, 5th Floor, Cambridge, MA 02139, United States
The Boulevard, Langford Lane, Kidlington, Oxford OX5 1GB, United Kingdom

Notices
Knowledge and best practice in this field are constantly changing. As new research and experience broaden our understanding, changes in research methods, professional practices, or medical treatment may become necessary.

Practitioners and researchers must always rely on their own experience and knowledge in evaluating and using any information, methods, compounds, or experiments described herein. In using such information or methods they should be mindful of their own safety and the safety of others, including parties for whom they have a professional responsibility.

To the fullest extent of the law, neither the Publisher nor the authors, contributors, or editors, assume any liability for any injury and/or damage to persons or property as a matter of products liability, negligence or otherwise, or from any use or operation of any methods, products, instructions, or ideas contained in the material herein.

Library of Congress Cataloging-in-Publication Data
A catalog record for this book is available from the Library of Congress

British Library Cataloguing-in-Publication Data
A catalogue record for this book is available from the British Library

ISBN 978-0-12-802823-0

For information on all Academic Press publications visit our
website at https://www.elsevier.com/books-and-journals

Working together
to grow libraries in
developing countries

www.elsevier.com • www.bookaid.org

Publisher: Andre Gerhard Wolff
Acquisition Editor: Mary Preap
Editorial Project Manager: Carlos Rodriguez
Production Project Manager: Punithavathy Govindaradjane
Cover Designer: Greg Harris

Typeset by SPi Global, India

Dedication

To my parents, Ching-Piao and Chen-Mei, who encouraged me to chase my dreams; to my wife, Tanya (Mi E), and my children, Andrew, Owen, and Natalie, whose patience, understanding, and support made it possible for me to undertake this project; and to my Ph.D. mentor, Dr. James Allan, and my postdoctoral mentor, Dr. David Clark, who have taught me as well as given me guidance throughout my education and career.

Contents

Preface

Diagnostic molecular biology is the integration of the knowledge and technology in molecular biology with clinical laboratory techniques. It is a clinical reality with its roots deep inside the fundamental understanding of genome structure and cellular activities at the molecular level. In the past few years, the field of molecular diagnosis has grown rapidly due to the revolutionized developments and discoveries in the fields of molecular genetics and biotechnology, and the need to provide diagnosis at the molecular level for all kinds of diseases. As the study of diagnostic molecular biology has increasingly gained attention, more and more clinical laboratories are preparing to deal with its demands and challenges, and molecular diagnosis has now become an important part of healthcare. As such, this book aims to deal with all aspects of diagnostic molecular biology from its principles and techniques to application of molecular diagnostics in a variety of fields.

This textbook is aimed at advanced students, entry-level scientists, and technicians who intend to enter the clinical laboratory professions with knowledge in general biology, genetics, and microbiology. In order to provide the most current understanding of the rapidly improving techniques in molecular diagnosis, this book is organized into three parts. Part 1 provides the fundamentals of molecular biology and comprises Chapters 1 through 5. Chapter 1 serves as the introduction to the structure and function of nucleic acids. Chapters 2–4 contain basic coverage of nucleic acid-based cellular activities including replication, repair, and gene expression. Chapter 5 discusses genome structure and evolution.

Part 2 describes commonly used molecular biology techniques and newly developed techniques in molecular diagnosis, including real-time PCR and NGS. It comprises Chapters 6 through 12. Chapters 6 to 8 cover the techniques commonly used in nucleic acid and protein preparation, analysis, and quantitation. Chapter 9 discusses amplification techniques for nucleic acids. Chapter 10 contains the techniques to characterize nucleic acids and proteins. Chapters 11 and 12 provide detailed information on the principles and applications of sequencing techniques.

Part 3 includes Chapters 13 through 16 and contains current molecular diagnostic techniques in identifying diseases. Chapter 13 covers molecular cytogenetic diagnosis in identifying chromosomal disorders. Chapters 14 and 15 discuss molecular diagnosis in genetic diseases and infectious diseases. Finally, Chapter 16 provides guidance for operating a molecular clinical laboratory.

This book gives an overview of important topics of interest within molecular diagnosis and shows the remarkable recent progress of molecular techniques in clinical analysis. The length is intended to provide instructors with a choice of favorite themes without sacrificing the important topics and without providing an overwhelming curriculum for the limited amount of time available in one semester. As such, this book can lead students to understand the essentials of molecular biology and to know how molecular techniques are used in clinical diagnosis.

Acknowledgments

The help of many made this textbook possible. I would like to thank Dr. Michelle Esposito, Jaclyn Dibello, Fina Vitale, and Giuseppe Minniti for their help in designing beautiful illustrations and providing comments on the text. I also would like to acknowledge colleagues and reviewers who contributed their ideas, advice, and critiques to this book. Furthermore, I would like to thank the editorial, production, marketing, and sales team at Elsevier. They have been exemplary in all aspects during the development of this book: Jill Leonard, Acquisitions Editor; Mary Preap, Acquisitions Editor; Tari K. Broderick, Senior Acquisitions Editor; Halima Williams, Editorial Project Manager; Fenton Coulthurst, Editorial Project Manager; Carlos Rodriguez, Editorial Project Manager; Sandhya Narayanan, Copyrights Coordinator; and Punithavathy Govindaradjane, Project Manager—Book Publishing Division. I am grateful for their efforts in helping me bring this project to completion.

Chapter 1

Nucleic Acids: DNA and RNA

Chapter Outline

DNA/RNA IS GENETIC MATERIAL

Transformation in Bacteria

In 1928, in an attempt to develop a vaccine against pneumonia, Frederick Griffith became the first to identify bacterial transformation, in which the form and function of a bacterium changes. Both virulent and avirulent *Streptococcus pneumoniae* were under his study. The virulence of the bacterium is determined by its capsular polysaccharide. Virulent strains have a capsule which is enclosed in a capsular polysaccharide, whereas avirulent strains do not. The nonencapsulated bacteria are readily engulfed and destroyed by phagocytic cells in the host animal's circulatory system. However, due to their protective outer polysaccharide capsule, virulent strains are not easily engulfed by the host's immune system, so they can multiply and cause pneumonia.

The presence or absence of the capsule also causes a visible difference between colonies of virulent and avirulent strains. Encapsulated bacteria form smooth, shiny-surfaced colonies (S) when grown on an agar culture plate. On the other hand, nonencapsulated strains produce rough colonies (R). As such, it is easy to identify the difference between these two strains through standard microbiological culture technique.

In Griffith's experiment, the virulent *S. pneumoniae* that has a smooth (S) capsule in its appearance was capable of causing lethal infections upon injection into mice (Fig. 1.1). Because of their lack of a protective coat, the R-type bacteria are destroyed by the animal after the injection, as previously described. As such, the mice are still alive after the injection of R-type bacteria. When S-type bacteria were killed by the heat, they were no longer able to cause a lethal infection upon injection into mice alone. However, when the heat-killed S-type bacteria and live R-type bacteria were injected together, neither of which causes lethal infection alone, the mice died as a result of pneumonia infection. It was found that the virulent trait that was responsible for production of the polysaccharide capsule was passed from the heat-killed S-type cells into the live R-type cells, thus converting the R-type bacteria into S-type bacteria, allowing it to become virulent and lethal by evading the host's immune response. Griffith concluded that the heat-killed bacteria somehow converted live avirulent cells to virulent cells, and he called the component of the dead S-type bacteria the "transforming principle."

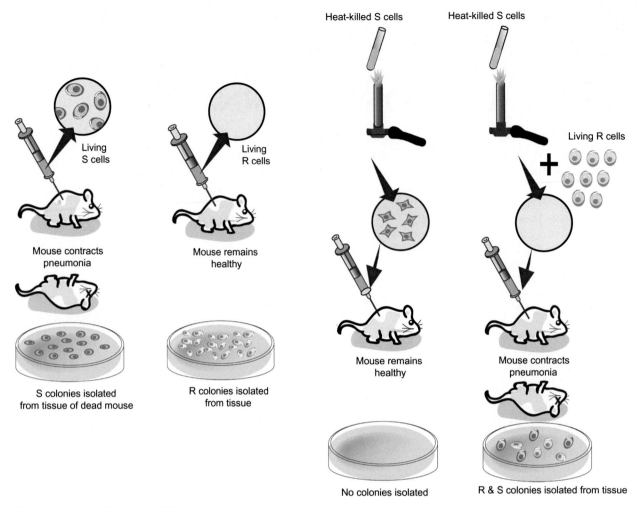

FIG. 1.1 Schematic diagram of Griffith's experiment which demonstrates bacterial transformation.

DNA Is the Genetic Material for Bacteria

Griffith's work led to further research of the transformation phenomenon. In 1944, Oswald Avery, Colin MacLeod, and Maclyn McCarty published what is now considered a classic paper in the field of molecular genetics. In this work, they demonstrated that DNA is the transforming principle. The schematic diagram of their experiment is shown in Fig. 1.2. In their experiments, they removed the protein from the transforming extract through organic solvent extraction. After this treatment, proteins were absent from the transforming extract. They found that the transforming principle was still active, which meant the heat-killed bacteria were still able to convert live avirulent cells to virulent cells. They also performed chemical, enzymatic, and serological analysis, together with the results from electrophoresis, ultracentrifugation, and ultraviolet spectroscopy. These treatments can remove carbohydrates, lipids, protein, or RNA from the extract. They found that carbohydrates, lipids, protein, and RNA were also not the transforming substance. Chemical testing of the final product gave a strong positive reaction for DNA. The final confirmation came with experiments using crude samples of the DNA-digesting enzyme deoxyribonuclease (DNase), which can degrade DNA, specifically. They demonstrated that the transforming principle can be destroyed by this enzyme. There was no loss of transforming activity after heat inactivated this enzyme. As such, their observations confirmed that DNA is the transforming substance.

DNA Is the Genetic Material for Bacteriophage

The second major piece of evidence supporting DNA as the genetic material was through experiments conducted by A. D. Hershey and Martha Chase in 1952. Hershey and Chase used T2 bacteriophage in their experiment to identify whether DNA

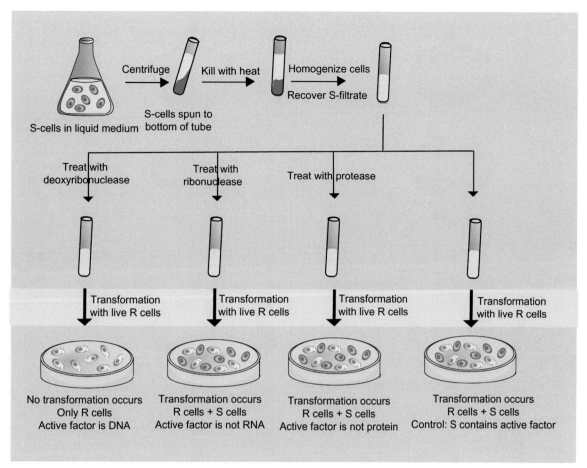

FIG. 1.2   Schematic diagram of Avery, MacLeod, and McCarty's experiment which demonstrates that DNA is the transforming principle.

or protein is the genetic material. Bacteriophage can infect *E. coli* and use the host to synthesize new phage particle. The phage consists of a protein coat surrounding a core of DNA. The phage attaches to the bacterial cell, and the genetic component of the phage enters the bacterial cells. Following infection, the viral genetic component dominates the cellular machinery of the host cells and leads to viral reproduction. Subsequently, many new phages are constructed, and the bacterial cell is lysed, releasing the progeny viruses. This process is normally referred to as the lytic cycle.

To define the function of the protein coat and nucleic acid in the viral reproduction process, Hersey and Chase radioactively labeled phage DNA with phosphorus-32 (^{32}P) and labeled phage protein with sulfur-35 (^{35}S). This is because DNA has phosphorus but not sulfur, whereas protein contains sulfur, but not phosphorus. Hershey and Chase let the labeled T2 bacteriophages infect the unlabeled bacteria and inject their genetic material into the cells (Fig. 1.3). After the attachment and genetic material entry, the empty phage coats were removed through high shear force in a blender. The force stripped off the attached phages so that the phage and the bacteria could be analyzed separately. Centrifugation separated the lighter phage particles from the heavier bacterial cells. Following this separation, the bacterial cells, which now contained viral-labeled DNA, were eventually lysed as the new phages were produced. These progeny phages contained ^{32}P but not ^{35}S. These results suggested that the protein of the phage coat remains outside the host cells and is not involved in directing the production of new phages. On the other hand, phage DNA enters the host cells and is directly involved in phage reproduction. Because the genetic material must first enter the infected cells, they concluded that DNA is the genetic material, and that it contains genes passed along through generations.

Taken together with work that had been done before, Hershey and Chase's work provided final, strong evidence to prove that DNA is the genetic material. Although these experiments demonstrated that DNA is the genetic material in bacteria and viruses, it was generally accepted that DNA is a universal substance as the genetic material in eukaryotes. This is because some indirect evidence has indicated that DNA is the genetic material in eukaryotes. For example, the genetic material

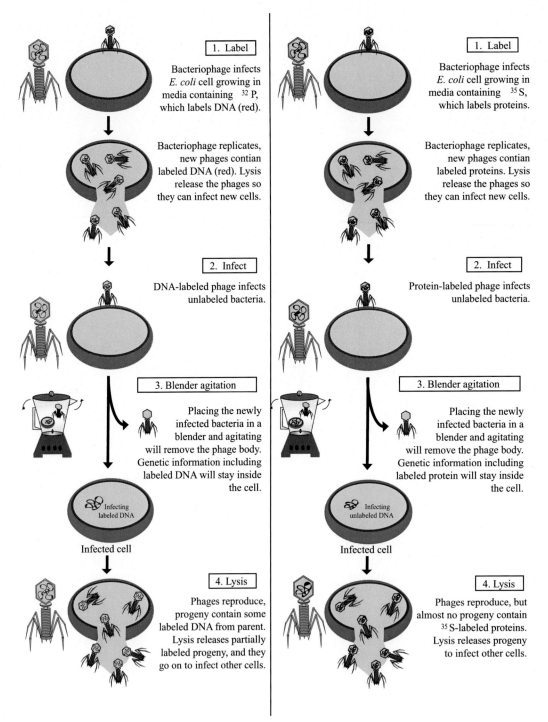

FIG. 1.3 Schematic diagram of Hershey-Chase experiment which demonstrates that DNA is directing reproduction of T2 phage in infected *E. coli* cells.

should reside on the chromosome and be found in the nucleus. Both DNA and protein fit these criteria, but only DNA is enriched inside the nucleus, whereas protein is enriched in the cytoplasm. Furthermore, DNA is also found in both chloroplasts and mitochondria, which are also known for performing genetic functions. As such, DNA is only found where primary genetic functions occur. On the other hand, protein is found everywhere in the cell.

Direct evidence that DNA is the genetic material in eukaryotes comes from recombinant DNA technology. For example, a segment of a DNA fragment corresponding to a specific gene is isolated and ligated to bacterial DNA which can self-replicate inside the bacterial cell. The resulting complex is sent into a bacterial cell, and its genetic expression is examined.

The subsequent production of the eukaryotic protein derived from that specific DNA segment in the bacterial cell demonstrates that DNA is the genetic material in the eukaryotic cells. This so-called gene cloning technique is now widely used in current biomedical research and pharmaceutical production (Fig. 1.4).

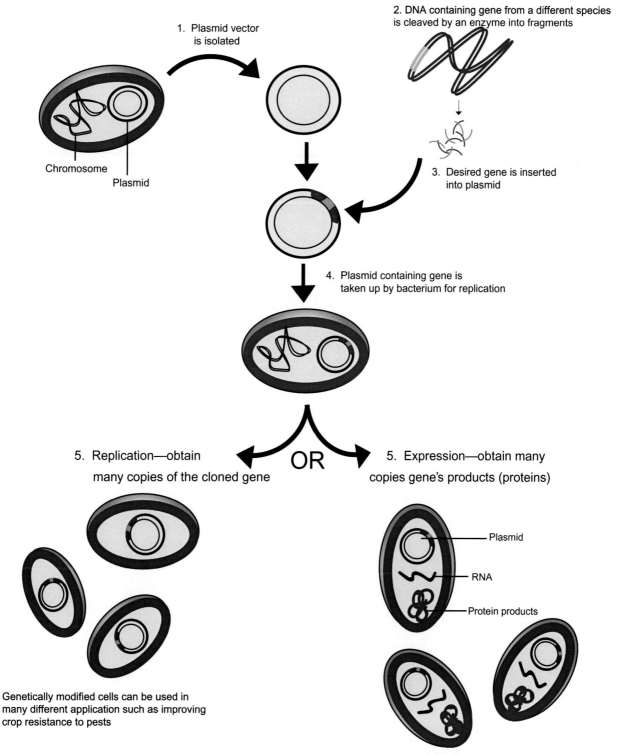

FIG. 1.4   Schematic diagram of a typical gene cloning process and the application of the gene cloning. The production of the specific eukaryotic protein derived from that introduced eukaryotic DNA segment proves that DNA is the genetic material in the eukaryotic cells.

RNA Is the Genetic Material for Viruses

Although DNA is the genetic material for most organisms, it has been demonstrated that the other type of nucleic acid, RNA, can also be genetic material. It was first demonstrated that when purified RNA from tobacco mosaic virus was spread on tobacco leaves, the leaves showed lesions of viral infection. Thus, it was concluded that RNA can be used as genetic material in viruses. Some groups of viruses are known to use DNA as their hereditary material, such as the T2 bacteriophage in Hershey and Chase's experiment. Those that use RNA as genetic material are called retroviruses. Retroviruses use a unique strategy, reverse transcription, to replicate their genetic material by using RNA as the template to synthesize complimentary DNA. This DNA intermediate can be incorporated into the genome of the host cell, and when the host DNA is transcribed, copies of the original retroviral RNA are produced. This type of RNA virus includes human immunodeficiency virus (HIV), which causes AIDS.

THE COMPONENTS OF NUCLEIC ACIDS

DNA and RNA are composed of various combinations of nucleic acids. Nucleic acids are macromolecules that exist as polymers called polynucleotides. A polynucleotide consists of many monomers called nucleotides, and is considered the building block of all nucleic acid molecules. These structural units of nucleic acids consist of three essential components: a nitrogenous base, a phosphate group, and a pentose sugar (a 5-carbon sugar).

Nitrogenous Base

The five-carbon sugar ring and the content of the nitrogenous base between DNA and RNA are slightly different from each other. Four different types of nitrogenous bases are found in DNA: adenine (A), thymine (T), cytosine (C), and guanine (G). In RNA, the thymine is replaced by uracil (U). The chemical structures of A, G, C, T, and U are shown in (Fig. 1.5A). Because of their structural similarity, we usually refer the nine-member double rings adenine and guanine as purines, and six-member single-ring thymine, uracil, and cytosine are pyrimidines.

Pentose Sugar

For the pentose sugar ring, Fig. 1.5B depicts the structure of sugar found in nucleic acids. The difference between DNA and RNA lies in the C-2′-position of the ribose sugar ring. In RNA, the carbon at the C-2 position is attached to a hydroxyl (OH) group. In DNA, the carbon at the C-2 does not contain this hydroxyl group; rather it is replaced by a hydrogen (H) atom. Therefore, the pentose ring in DNA is considered a deoxyribose (it is a deoxygenated five-carbon sugar ring). In the absence of the C-2′ hydroxyl group of DNA, the sugar is more specifically named 2-deoxyribose.

Phosphate Group

If a molecule is composed of a purine or pyrimidine base and a ribose or deoxyribose sugar, this chemical unit is called a nucleoside (Fig. 1.6). The nitrogenous base and the pentose sugar are linked by glycosidic bond between C-1′-position of sugar and nitrogenous base. If the base is a purine, the N-9 atom is covalently bonded to the sugar. If the base is a pyrimidine, the N-1 atom bonds to the sugar. When a phosphate group attaches to a nucleoside through a phosphoester bond, the entire complex becomes a nucleotide. This phosphoester bond is linked between 5′-hydroxyl group of the sugar and a phosphate group. Because it involves a phosphate group and only one sugar, it is a phosphomonoester bond.

The Formation of a Nucleotide

Single units within nucleotides are also called nucleoside monophosphates. The addition of one or two phosphate groups results in nucleoside diphosphates and triphosphates, respectively. The triphosphate form is significant because it serves as the precursor molecule during nucleic acid synthesis within the cell. Fig. 1.7 depicts the structures of adenosine-5′-triphosphate (ATP), adenosine-5′-diphosphate (ADP), and adenosine-5′-monophosphate (AMP). The formation of ADP from AMP requires the addition of one molecule of inorganic phosphate and is accompanied by the release of a molecule of water. Similarly, the formation of ATP from ADP requires the addition of one molecule of inorganic phosphate and is

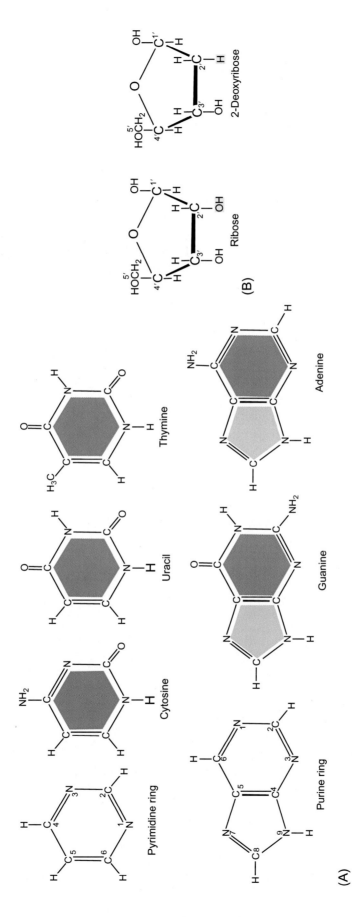

FIG. 1.5 (A) Chemical structure of pyrimidines and purines nitrogenous bases in DNA and RNA. (B) Chemical structure of ribose and 2-deoxyribose that are found in RNA and DNA, respectively.

FIG. 1.6 Formation of a nucleotide by adding phosphate group(s) to a nucleoside.

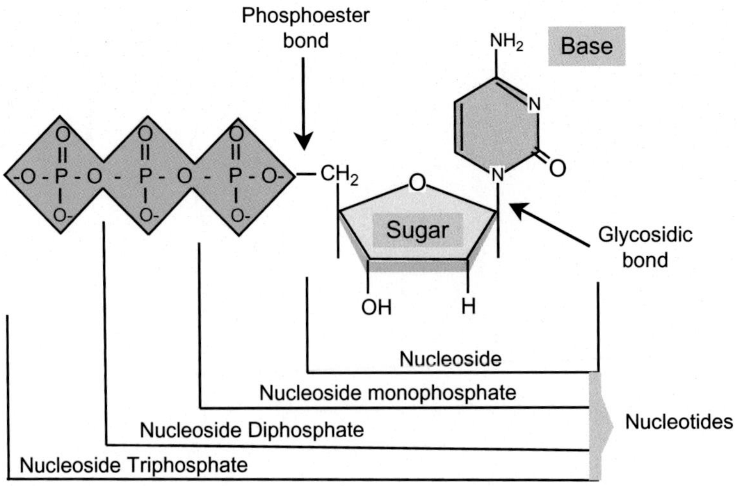

accompanied by the release of a molecule of water. On the other hand, the hydrolysis of ATP to ADP, releasing one molecule of inorganic phosphate (Pi), is accompanied by the release of large amount of energy in the cell. When these chemical conversions are coupled to other reactions, the energy produced is used to drive the reactions, and sustain life. During DNA synthesis, the two phosphate groups, which are β and γ phosphates, are removed from dATP, dGTP, dCTP, and dTTP. Thus, all four nucleotides contain only monophosphates in a polynucleotide chain.

In a polynucleotide chain, nucleotides are joined together through phosphodiester bond to form a long chain of nucleotides. The formation of a phosphodiester bond involves a dehydration reaction (removing a molecule of water) through the linkage between a phosphoric acid and two sugars, which is between 5′carbon of one sugar and 3′ carbon of another sugar

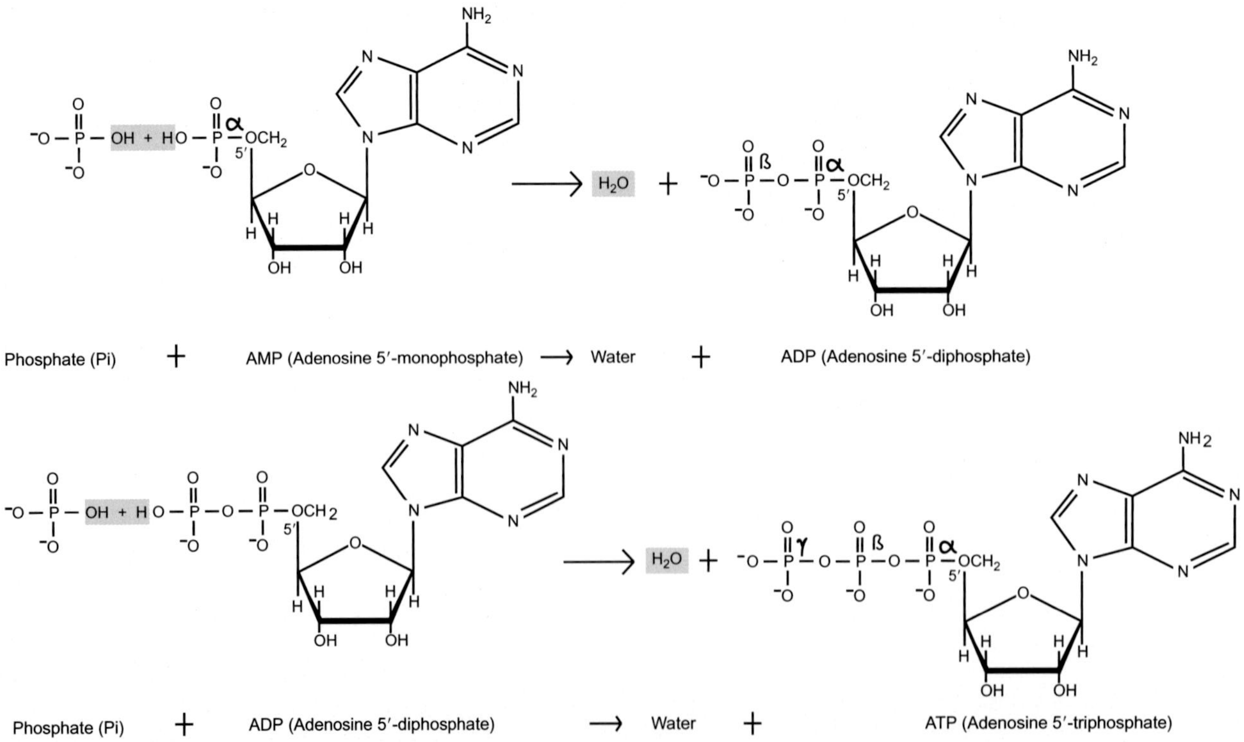

FIG. 1.7 Formation of ADP and ATP by the successive addition of phosphate groups via phosphoric anhydride linkage.

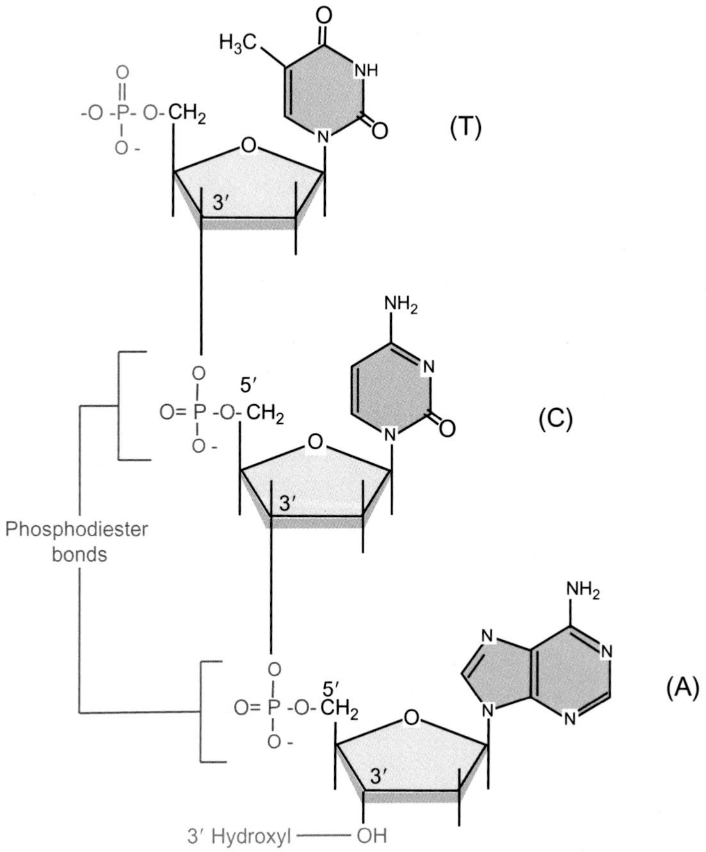

FIG. 1.8 An example of three nucleotides linked together by phosphodiester bonds between the 5'- and 3'-hydroxyl groups of the sugars.

(Fig. 1.8). Such a phosphodiester bond results in a repeating pattern of the sugar-phosphate units called a sugar-phosphate backbone, and this provides for the polynucleotide chain with a 3'→5' phosphodiester linkage direction. Phosphodiester bonds can be found in both DNA and RNA. One end of the polynucleotide has a free 5'-phosphate group, so it is called the 5'-end. The other end of the polynucleotide has a free 3'-hydroxyl group, and it is called 3'-end. It is conventional to write nucleic acid sequences in the 5' to 3' direction, which is from the 5' terminus at the left to the 3' terminus at the right. For example, the polynucleotide in Fig. 1.8 is would be read "TCA."

THE STRUCTURE OF DNA MOLECULES

Chargaff's Rules

In 1953, James Watson and Francis Crick proposed the theory that DNA molecules exist as a double helix, which has since been supported by various studies (Fig. 1.9). The double helix model is mainly based on the X-ray diffraction data collected by Rosalind Franklin and Maurice Wilkins, and DNA composition studies observed by Erwin Chargaff. X-ray diffraction data showed that DNA is a regular helix, and the repeat distance in the helix is 34 angstroms (Å) with a diameter of ~20 Å. The distance between adjacent nucleotide is 3.4 Å. The discovery of double helical model of DNA relied on the critical data from Chargaff's findings, which is also called Chargaff's rules:

1. Two long polynucleotide chains are coiled around a central axis, forming a right-handed double helix.
2. The two DNA strand are antiparallel, that is, their 5'→3' orientation runs in opposite directions.
3. The base of both chains lie perpendicular to the axis, and they are stacked on one another.
4. The nitrogenous bases of opposite chains are paired as the result of the formation of a hydrogen bond in DNA.
5. Each complete turn of helix is 34 Å long.
6. The double helix has a diameter of 20 Å.

FIG. 1.9 Two models of DNA structure. (A) The DNA double helix is presented as a twisted ladder. The two side long bars represent the sugar-phosphate backbones of the two strands and the rungs are base pairs. The curved arrows indicate the 5′ to 3′ orientation of each strand. (B) A space-filling model of DNA structure.

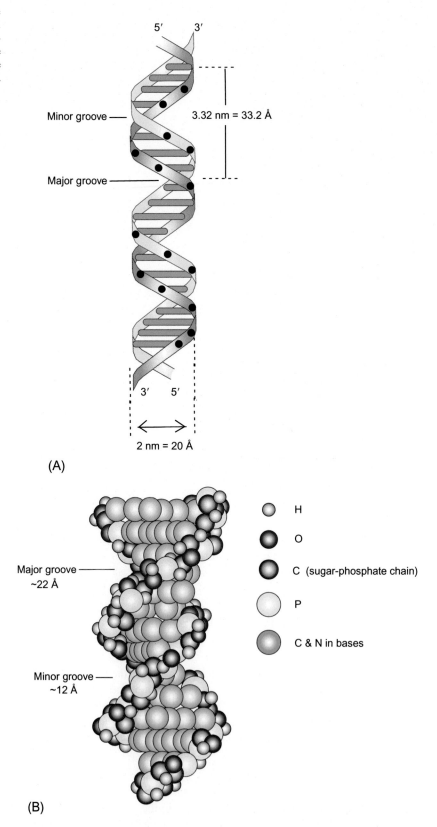

7. The amount of adenine (A) residues is proportional to the amount of thymine (T) residues in DNA. Also, the amount of guanine (G) residues is proportional to the amount of cytosine (C).
8. The sum of the purines equal to the sum of pyrimidine.
9. The percentage of (G+C) is not necessarily equal the percentage of (A+T).

Chargaff's rules indicate a definite pattern of base composition in DNA molecules. In combination of Chargaff's rules with the fact that DNA structure has a 3.4 Å periodicity, Watson and Crick proposed that DNA must be a double helix with its sugar-phosphate backbone on the outside and its nitrogenous bases on the inside. The two polynucleotide chains in the double helix are held by hydrogen bonding between the nitrogenous bases, known as the base pairing. In general, for DNA, G can hydrogen bond specifically only with C, whereas A can bond specifically only with T (Fig. 1.10). The paired bases are said to be complementary. Complementary base pairing occurs because of the geometrical location and interactions between functional groups in the nitrogenous bases so that a hydrogen bond can form.

The Arrangement of Each Subunit in the DNA Structure

The Watson-Crick model places the sugar-phosphate backbones on the outside of the double helix and carries the negative charges on the phosphate group. Because of its negative charges, positive-charged chromosomal proteins or regulatory proteins can easily provide neutralizing force either to determine the higher order DNA structure or to regulate gene expression. The two polynucleotide chains run in opposite directions known as antiparallel. One strand runs in the 5′ to 3′ direction, whereas its complement runs 3′ to 5′ (Fig. 1.11).

The nitrogenous bases are on the inside of the double helix. They are flat and perpendicular to the axis of the helix. Bases are stacked above one another around the axis like a spiral staircase (Fig. 1.12). The curving sides of spiral staircase represent the sugar-phosphate backbones of the two DNA strands; the stairs are the nitrogenous bases. As mentioned earlier,

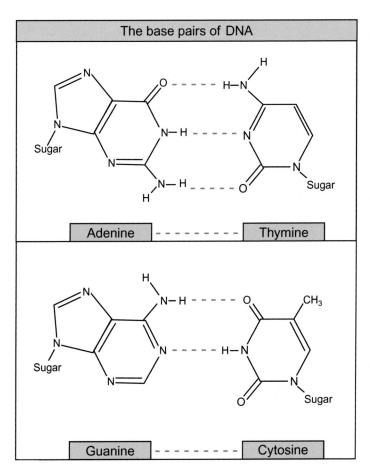

FIG. 1.10 The base pairs of DNA. A guanine-cytosine pair is held by three hydrogen bonds (dashed lines and an adenine-thymine pair is held by two hydrogen bonds).

X-ray diffraction data showed each repeat double helix is 34 Å, and the distance between adjacent nucleotides is 3.4 Å. These suggest that each helix turn has ∼10 base pairs, and each base pair rotates 36 degrees around the axis of the helix relative to the next base pair, so that ∼10 base pairs make a complete turn of 360 degrees. The twisting of the two strands around one another forms a double helix with a minor groove and a major groove. In a minor groove, the distance between the two DNA strands is ∼12 Å, whereas the distance becomes ∼22 Å in a major groove. The double helix in DNA is normally right-handed, which means the turns run clockwise as viewed along the helical axis. It is important to know that ∼10 base pairs per turn is an average structure. If it has more base pairs per turn, it is said to be overwound. On the other hand, if it has fewer base pairs per turn, then it is underwound. The degree of local winding can be affected by the overall conformation of the DNA double helix or by the binding of proteins to specific sites on the DNA.

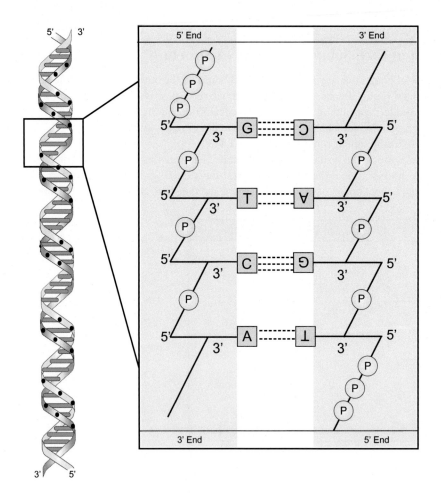

FIG. 1.11 An example of antiparallel nature of the DNA double helix.

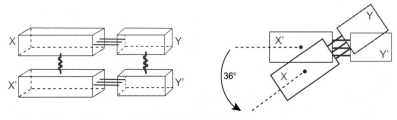

FIG. 1.12 Schematic presentation of the stacked base pairs X/Y and X'/Y' and the twist angle.

Alternative Forms of DNA Exist

Under different conditions of isolation, different conformations of DNA are seen. The Watson-Crick DNA molecule represents the DNA molecule in solution, which is the DNA molecule that exists in a very high relative humidity environment (92%). This is called "B-DNA" (Fig. 1.13). In B-DNA, the double helix is said to be right-handed, because the turns run clockwise as viewed along the helical axis, which means the helix winds upward in the direction in which the fingers of the right hand curl when the thumb is pointing upward (Fig. 1.14). The inside diameter of the sugar-phosphate backbone of the double helix is about 11 Å (1.1 nm). The distance between the points of attachment of the bases to the two strands of the sugar-phosphate backbone is the same for the two base pairs (A-T and G-C), about 11 Å (1.1 nm), which allows for symmetrical stacking and a double helix with a smooth backbone and no overt bulges. Base pairs other than A-T and G-C are possible, but they do not have a correct hydrogen bonding pattern (A-C or G-T pairs) or the right dimensions (purine-purine or pyrimidine-pyrimidine pairs) to allow for a smooth double helix. The outside diameter of the helix is

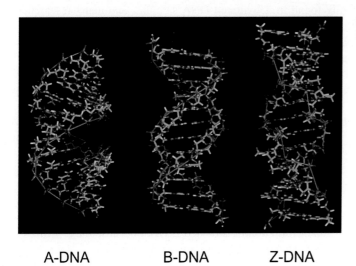

FIG. 1.13 Comparison of the A, B, and Z forms of the DNA double helix.

A-DNA   B-DNA   Z-DNA

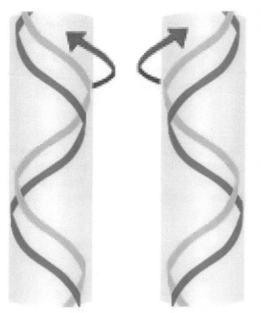

FIG. 1.14 Right- and left-handed helices are related to each other in the same way as right and left hands.

20 Å (2 nm). The length of one complete turn of the helix along its axis is 34 Å (3.4 nm) and contains 10 base pairs. The atoms that make up the two polynucleotide chains of the double helix do not completely fill an imaginary cylinder around the double helix; they leave empty spaces known as grooves. There is a large major groove and a smaller minor groove in the double helix; both can be sites at which drugs or polypeptides bind to DNA. At neutral, physiological pH, each phosphate group of the backbone carries a negative charge. Positively charged ions, such as Na^+ or Mg^{2+}, and polypeptides with positively charged side chains must be associated with DNA in order to neutralize the negative charges. Eukaryotic DNA, for example, is complexed with histones, which are positively charged proteins, in the cell nucleus. B-DNA is the most common form in vivo and in solution in vitro.

Another form of DNA is A-DNA, which is observed when DNA is dehydrated or under high salt conditions. An important shared feature of A-DNA and B-DNA is that both are right-handed helices. A-DNA is both shorter and thicker than B-DNA. Each repeat double helix in A-DNA is 24.6 Å, and each turn has about 11 base pairs (bp). The major group of DNA is deep and narrow, while the minor group is shallow and broad. Although the vast majority form of DNA molecules in the cell are the B-DNA, there are some local variations of DNA structure. For example, A-DNA can be found in both DNA-RNA and RNA-RNA hybrids. A-DNA was originally found in dehydrated DNA samples, and many researchers believed that the A form was an artifact of DNA preparation. DNA:RNA hybrids can adopt an A formation because the 2'-hydroxyl on the ribose (that, as previously discussed, distinguished DNA from RNA) prevents an RNA helix from adopting the B form. RNA:RNA hybrids may also be found in the A form.

A third DNA structure is Z-DNA, which is longer and narrower than B-DNA and is a left-handed helix. A left-handed helix turns counterclockwise away from the viewer when viewed down its axis. Because the backbone formed a zig-zag structure, it is named Z-DNA.

In Z-DNA, the repeat helix is 45.6 Å and each helical turn has 12 bp. The minor groove is very deep and narrow. In contrast, the major groove is shallow to the point of being virtually nonexistent. Z-DNA is formed under conditions of high salt or in the presence of alcohol. Z-DNA is also known to occur in nature when there is a sequence of alternating purine-pyrimidine. Because sequences with cytosine methylated at the number 5 position of the pyrimidine ring can also be found in the Z form, it may play a role in the regulation of gene activation.

As we discussed earlier, RNA differs chemically from DNA in two respects. First, RNA contains C-2' ribose instead of the C-2' deoxyribose in DNA. Second, RNA contains the base uracil instead of thymine. These two small differences make RNA form unique structure in the cell. For example, RNA cannot form double-stranded B DNA because the 2'-hydroxyl group on the ribose sugar hinders formation of the B form RNA. However, RNA adopts an A-form helix when it forms double-stranded regions.

Various Forms of RNA

Unlike DNA, cellular RNA molecules are almost always single-stranded. However, all of them typically contain double-stranded regions formed, when stretched, of nucleotides, with complementary base sequences align in an antiparallel fashion which we will discuss later. Several kinds of RNA play an important role in cellular activities. Three major classes of cellular RNA molecules function during the expression of genetic information: ribosomal RNA (rRNA), transfer RNA (tRNA), and messenger RNA (mRNA). These molecules all originate as complementary copies of one of the two strands of DNA segments during the process of transcription. Ribosomal RNA usually constitutes about 80% of all RNA in *E. coli* cells. They are important structural components of ribosome, which functions as non-specific sites of protein synthesis during translation. Messenger RNA molecules carry genetic information from the DNA of the gene. The mRNAs vary in size, reflecting the range in the sizes of the proteins encoded by the mRNA. Transfer RNA accounts for up to 15% of the RNA in a typical cell. It carries amino acids to the ribosome during translation, aiding in the translation of DNA to mRNA to protein. Although ribosomal RNA and transfer RNA molecules are also synthesized by transcription of DNA sequences, unlike mRNA molecules, these RNAs are not subsequently translated to form proteins, and they remain in RNA form.

A single bacterial mRNA may contain the information for the synthesis of several polypeptide chains within its nucleotide sequence. In contrast, eukaryotic mRNAs encode only one polypeptide, but are more complex in that they are synthesized in the nucleus in the form of much larger precursor molecules called heterogeneous nuclear RNA (hnRNA). hnRNA molecules contain stretches of nucleotide sequence that have no protein-coding capacity, and therefore, will not be translated into protein by a ribosome. These noncoding regions are called intervening sequences or introns because they intervene between coding regions, which are called exons. Intron interrupts the continuity of the information specifying the amino acid sequence of a protein and must be spliced out before the message can be translated.

The other important RNA includes small nuclear RNA (snRNA), small nucleolar RNA (snoRNAs), micro RNA (miRNA), and small interfering RNA (siRNA). The snRNA participates in processing hnRNAs into mRNA and snoRNA primarily functions as RNA chaperones in the processing of ribosomal RNA (rRNA). Both are neither tRNA nor small rRNA molecules, although they are similar in size (100–200 nucleotides) to these species. The snRNA is always found in a stable complex with specific proteins forming small nuclear ribonucleoprotein particles (snRNP). Both miRNA and siRNA are involved in gene regulation. The siRNAs disrupt gene expression by blocking specific protein production hence their name "interfering," even though the mRNA encoding the protein has been synthesized. The miRNAs control developmental timing by base pairing with and preventing the translation of certain mRNAs, thus, blocking synthesis of specific proteins. However, unlike siRNAs, miRNAs (22 nucleotides long) do not cause mRNA degradation. We will discuss the function of mRNA, snRNA, miRNA, and siRNA in much great detail later in Chapter 3, while rRNA, tRNA, and snoRNA will be discussed in Chapter 4.

PHYSICAL PROPERTIES OF NUCLEIC ACIDS

DNA Denaturation

An important feature of double helix DNA is the ability to separate the two strands (a process called denaturation) and to base pair the two strands together, a process called renaturation, without disrupting the covalent bonds that make up the sugar-phosphate backbone. These processes occur at the very rapid rate needed to sustain genetic functions. Because the complementary strands are held by hydrogen bonds, the lack of covalent bonds makes it possible to denature and renature double helix DNA without affecting its properties. The hydrogen bonds can be disrupted by high temperature, low salt concentration, or high pH in vitro. For example, when a DNA solution is heated enough, the hydrogen bonds that hold the two strands together weaken and finally break, a process called DNA denaturation, or DNA melting. On the other hand, when the temperature of a DNA solution is low enough, the hydrogen bonds form again, and finally, all hydrogen bonds are restored: DNA renaturation.

Denaturation of DNA occurs over a narrow temperature range. The midpoint of the temperature range over which the DNA strands are half denatured is called the melting temperature, or T_m. Fig. 1.15 presents a melting curve for DNA from *E. coli*. The amount of denatured DNA is measured by the increase in absorbance at 260 nm. The sudden rise of the curve shows the narrow range of the temperature when two strands hold fast, until the temperature reaches the T_m, and then they rapidly let go. The midpoint of the curve indicates the point at which half of the DNA population is denatured and the other half is still in double helix form. Denaturation and renaturation can occur with the combinations of DNA-DNA, DNA-RNA, and RNA-RNA as the intermolecular or intramolecular interaction.

Secondary Structure of DNA

Through advanced techniques, it has been shown that DNA molecules can have dynamic changes in secondary and tertiary structures, so that they can regulate expression of their linear sequence information. Unusual secondary structures, such as slipped structures, cruciform, and triple-helix DNA, are generally sequence-specific. Slipped structures usually occur at tandem repeats and are usually found upstream of regulatory sequences in vitro. Cruciform structures are paired stem-loop formations. They can be found in vitro for many inverted repeats in plasmids and bacteriophages. In a triplex helix, a third strand of DNA joins the first two to form triplex DNA. Triplex helix DNA occurs at purine-pyrimidine stretches in DNA and is favored by sequences containing a mirror repeat symmetry.

Tertiary Structure of DNA

Many naturally occurring DNA molecules are circular, with no free 5′ or 3′ end. For example, prokaryotic DNA is circular, and this DNA forms a supercoil. To understand how supercoiled and relaxed circular DNA differ, you can take a covalently closed, circular DNA molecule, cleave both strands in one place, and then rotate the DNA end around several times before covalently closing the circle again. You can imagine this by taking a rubber band with two hands and twisting it together. The resulting DNA circle is supercoiled and will wind around itself. This is because rotating the DNA end changes the helical twist of the DNA from its preferred state containing about 10.5 base pairs per turn to one with a fewer number of base pairs per turn. When the DNA is ligated (meaning joined together), the untwisted portion of DNA will tend to spring back to adopt its favored state, with 10.5 base pairs per turn. This, however, causes the plasmid to wrap around itself, and the closed DNA circle is now supercoiled. An initial unwinding of the DNA (which is the clockwise direction and is equivalent

FIG. 1.15 The melting temperature of *E. coli* DNA. The temperature at the midpoint of the curve is approximately 87°C.

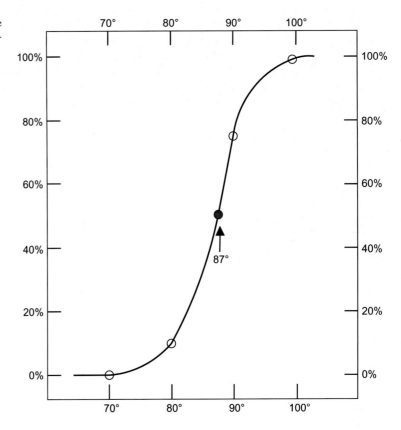

to separating the DNA strands) leads to accumulation of negative supercoils, whereas an initial twisting of the DNA in the counterclockwise direction leads to formation of positive supercoils. Negative and positive supercoils simply differ in direction of rotation.

For closed circular DNA or linear duplex DNA that is topologically constrained (as in the case of prokaryotic circular DNA), the linking number (Lk) is a quantitative descriptor of DNA topology that includes the number of times the helix winds around its central axis and the number of times the helix crosses itself (Fig. 1.16). The linking number can only be altered by breaking and then rejoining a strand of DNA. Lk_0 designates the linking number of DNA when it is relaxed, and ΔLk designates the difference between Lk and Lk_0 under the same experimental conditions. "Twist" (Tw) is the number of helical turns in the duplex DNA. When the helix is overwound, then ΔLk is positive. On the other hand, when the helix is underwound, then the ΔLk is negative. Thus, underwound duplex DNA has fewer than the normal number of turns, whereas overwound DNA has more. DNA supercoiling is analogous to twisting or untwisting a rubber band so that it is torsionally stressed. Negative supercoiling introduces a torsional stress that favors unwinding of the right-handed B-DNA double helix, whereas positive supercoiling overwinds such a helix (Fig. 1.17).

Alternatively, "writhe" (Wr) occurs when the DNA helix buckles into loop-like structures called plectonemic supercoils, or when the DNA wraps around protein complexes, such as nucleosomes. Lk is the sum of Tw and Wr (Lk = Tw + Wr), and thus, changes in the value of ΔLk may partition between changes in Tw and Wr, as illustrated in Fig. 1.16, for a relaxed 210-base pair DNA circle with an average 10.5 base pairs per helical turn. In this case, Tw = 20, Wr = 0, and Lk = Lk_0 = Tw. A ΔLk of −4 could be accommodated at the two extremes by (i) a pure change in Tw, leading to local denaturation of 4 helical turns, or (ii) a pure change in Wr with the formation of 4 plectonemic supercoils.

The balance between DNA twisting and supercoiling has important biological consequences. Negative supercoils in DNA can be removed by unwinding one or more turns of DNA, which is equivalent to separating the strands. In this way, the unfavorable energy of strand separation is partially compensated for. This compensation is derived from relieving superhelical tensions due to negative supercoiling. This is why negatively charged DNA templates are needed for many biological processes that require separation of the DNA strands. For example, virtually all naturally occurring DNA within both prokaryotic and eukaryotic cells is negative supercoiled except during replication, when it becomes positively supercoiled (Fig. 1.18). However, the supercoiling of the nuclear DNA of eukaryotes is more complicated than the supercoiling

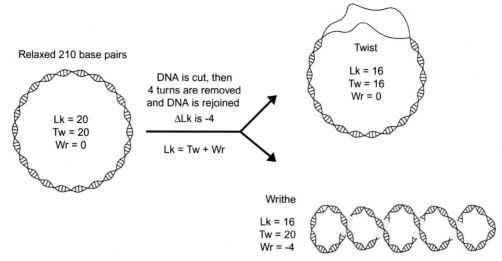

FIG. 1.16 An example of DNA topology. The topology of a double-stranded DNA is described by its linking number (Lk), which is the sum of twist (Tw) and writhe (Wr). (Left) A torsionally relaxed DNA molecule with a length of 210 basepairs contains 20 turns (10.5 bp/turn) or Tw = 20. Hypothetically, if the DNA were cut, then one end was twisted by four turns in the direction opposite to the natural helicity of the DNA, and subsequently resealed, the resulting linking number of the DNA would equal Lk = 20 − 4 = 16. (Right) The upper and lower panels show the topology of the DNA molecule when the removal of these turns is at the expense of twist (Tw = 20 − 4 = 16) and writhe (Wr = 0 − 4 = −4), respectively.

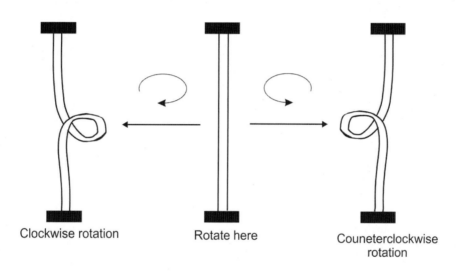

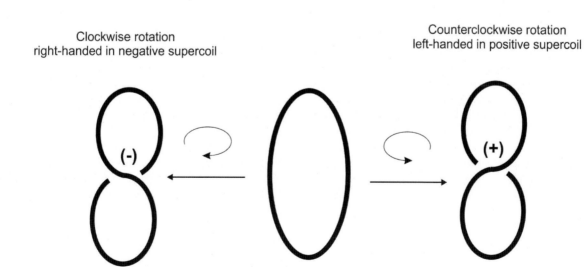

FIG. 1.17 The topology of supercoiled DNA. The DNA double helix can be considered as a two-stranded, right-handed coiled rope. If one end of the rope is rotated counterclockwise, the strads begin to separate and we call this as negative supercoiling. On the other hand, if the rope is twisted clockwise which is a right-handed fashion, the rope becomes overwound and this is positive supercoiling.

of the circular DNA from prokaryotes. Eukaryotic DNA is complexed with a number of proteins, especially with basic proteins, that have abundant positively charged side chains at physiological pH. Electrostatic attraction between the negatively charged phosphate groups on the DNA and the positively charged group on the proteins favors the formation of complexes of this sort. The resulting structure is called chromatin. The principal proteins in chromatin are histones, of which there are five main types, called H1, H2A, H2B, H3, and H4. All these proteins contain large numbers of basic amino acid residues, such as lysine and arginine. In the chromatin structure, the DNA is tightly bound to all types of histone except for H1. Therefore, the topological changes induced by supercoiling must be accommodated by the histone-protein component of chromatin. The chromatin structure will be discussed in Chapter 5.

Secondary and Tertiary Structure of RNA

RNA molecules are typically single-stranded. The course of a single-stranded RNA in three-dimensional space conceivably would have six degrees of freedom per nucleotide, represented by rotation about each of the six single bonds along the sugar-phosphate backbone per nucleotide unit. Compare this situation with DNA, whose separated strands would obviously enjoy the same degrees of freedom. However, the double-stranded nature of DNA imposes great constraint on its conformational possibilities. Compared with dsDNA, an RNA molecule has a much greater number of conformational possibilities.

RNA structures are determined by nucleotide sequence. The intricate three-dimensional architectures they adopt and the conformational dynamics displayed by such architectures endow RNAs with functional abilities, such as ligand binding and even catalysis. RNA secondary structure usually refers to the portion of the RNA molecule that forms short double-helical stretches. Secondary structure of RNA can be predicted accurately by computer analysis. Common secondary structures that form the building blocks of RNA architecture are bulges, stems single-stranded hairpin/internal loops, and junctions (Fig. 1.19). In RNA, helix formation occurs by hydrogen bonding between base pairs and base stacking hydrophobic interactions within one single-stranded chain of nucleotides. The base-paired RNA primarily adopts a right-handed type A double helix with 11 bp per turn. Regular A type RNA helices have a deep narrow major groove that is not for specific interaction with ligands, whereas the 2′-OH groups in the minor groove provide good hydrogen bond potential for interaction with ligands.

In addition to conventional Watson-Crick base pairs, RNA double helices often contain noncanonical (non-Watson-Crick) base pairs. There are more than 20 different types of noncanonical base pairs, involving two or more hydrogen bonds, that have been encountered in RNA structures. The most common are the GU wobble, the sheared GA pair, and the GA imino pair. Because the GU pair only has two hydrogen bonds (compared with three for a GC pair), this requires a sideways shift of one base relative to its position in the regular Watson-Crick geometry. Weaker interactions from the reduction in hydrogen bonding may be countered by the improved base stacking that results from each sideways base displacement. In addition, RNA structures frequently involve unconventional base pairing such as base triples. These typically involve one of the standard base pairs, and a third base that can interact in a variety of unconventional ways. Noncanonical base pairs and base triples are important mediators of RNA self-assembly and of RNA-protein and RNA-ligand interactions. For example, noncanonical base pairs widen the major groove and make it more accessible to ligands.

RNA chains can fold into unique three-dimensional structures that act similarly to globular proteins. For example, the functional transfer RNA (tRNA) can twist into an L-shaped three-dimensional structure in the cell. The tRNA is about 76 nt long, and all of the different tRNAs of a cell fold into the same general shape. The ribosomal RNA (rRNA) is another example of the RNA that can fold into three-dimensional structure. We will discuss both tRNA and rRNA in Chapter 4.

Large RNAs are composed of a number of structural domains that assemble and fold independently. Similar to DNA, RNA folding relies on hydrogen bonding and base stacking. The three dimensional structure is basically maintained by the interaction between distant nucleotides and between 2′-OH groups. However, these long-range interactions are less stable than standard Watson-Crick base pairs and can be easily broken by mild denaturation conditions. Because RNA is negatively charged, it requires charge neutralization to form the tertiary structure. This can be done through the binding of basic proteins or binding of monovalent and/or divalent metal ions. A number of highly

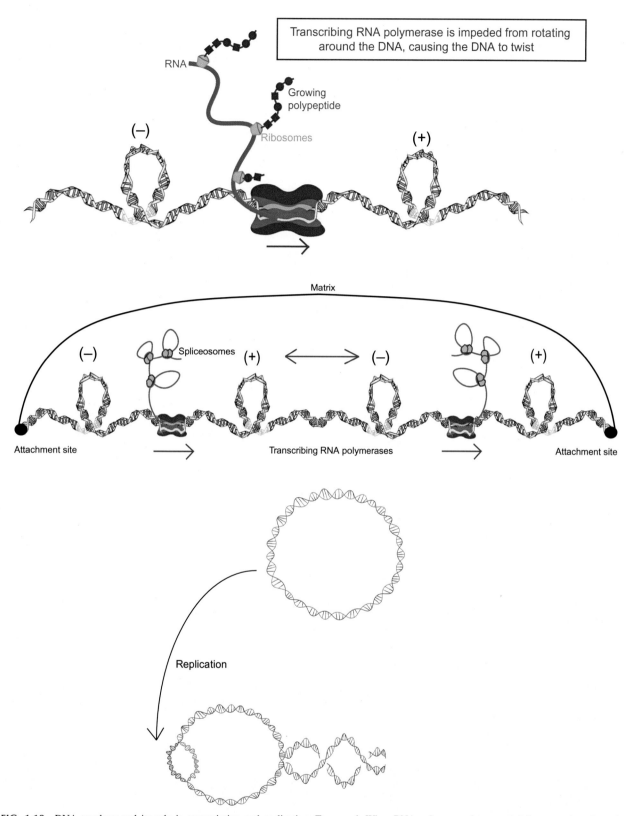

FIG. 1.18 DNA topology and its role in transcription and replication. Top panel, When RNA polymerase is prevented from rotating along the helical axis of the DNA during transcription, positive and negative supercoils accumulate ahead and behind the enzyme, respectively. Multiple factors, including the nascent RNA strand (*blue solid line*), ribosomes on the mRNA (*yellow*), and the growing peptide itself, can impede the rotation of RNA polymerase by increasing its hydrodynamic drag. Middle panel, in eukaryotes, the nascent RNA and its processing factors, such as spliceosomes, increase the rotational drag on RNA polymerase and impede its rotation around DNA's helical axis, leading to supercoiling behind and ahead of the enzyme. When tandem genes are transcribed, RNA polymerase complexes progress in the same direction on duplex DNA. The DNA domain between them contains both negative and positive supercoils that could diffuse towards each other and subsequently annihilate. Bottom panel, when a circular DNA is replicated, two origins move in opposite directions, unwinding the parental DNA. By conservation of linking number, this generates positive supercoils ahead of the forks.

FIG. 1.19 Schematic diagram of the typical structural motifs in a RNA secondary structure.

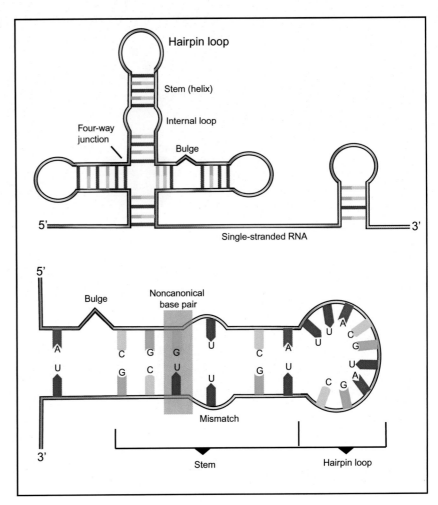

conserved, complex RNA folding motifs have been identified such as pseudoknot, the A-minor motif, tetraloops, ribose zippers, and kink-turns (Fig. 1.20).

DNA TOPOISOMERASE CATALYTIC MECHANISMS

Forms of DNA that have the same sequences yet differ in their linkage number are referred to as topological isomers, or topoisomers. Topoisomers can be visualized by their differing mobilities when separated by gel electrophoresis. Topoisomerases are enzymes that modulate the degree of DNA supercoiling, and they can convert one isomer of DNA to another. Topoisomerases are divided broadly into two families: Type I enzymes transiently cleave and reseal one strand of duplex DNA in the absence of ATP (Fig. 1.21, *top*), and Type II enzymes cleave and religate both DNA strands in the presence of ATP (Fig. 1.21, *bottom*). These two families are divided further into subfamilies, which can be distinguished on the basis of protein architecture (monomer versus oligomer), DNA substrate preference (duplex versus single-strand), reaction outcomes (net loss or gain of supercoils; complete or partial supercoil removal), and requirements for metals and ATP.

Type I Topoisomerases

Type I topoisomerases can be subdivided according to their structure and reaction mechanisms: type IA (TopIA; bacterial and archaeal topoisomerase I, and topoisomerase III), type IB (TopIB; eukaryotic topoisomerase I) and type IC

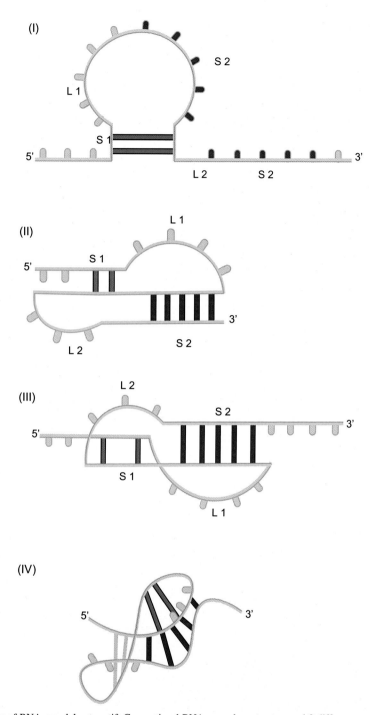

FIG. 1.20 A typical example of RNA pseudoknot motif. Conventional RNA secondary structure and 3 different pseudoknot variations found in the tRNA-like structure of the turnip yellow mosaic virus. S1 and S2 represent double helical stem regions. L1 and L2 are single-stranded loops.

(topoisomerase V). These enzymes are primarily responsible for relaxing positively and/or negatively supercoiled DNA. TopIA enzymes transiently cleave a single strand of supercoiled DNA to form a 5′- phosphotyrosyl intermediate. These enzymes include *E. coli* TopA, which preferentially relaxes negatively supercoiled DNA, and Topoisomerase III, which efficiently unknots and decatenates single-stranded or nicked DNA. TopIA enzymes have a clamp-like structure with a large central cavity in which DNA binds. Cleavage yields a covalent enzyme-DNA intermediate, in which TopA bridges the nick that it created in the DNA. The intact strand is then passed through the nick which results in a change of Lk by one unit per cleavage-religation cycle (Fig. 1.21, *top*).

In contrast, TopIB enzymes form a 3'-phosphotyrosyl intermediate and are structurally unrelated to TopIA. TopIB can switch the DNA back and forth between a nicked and a religated state, with a preference for the religated state over the nicked state. TopIB can relax both negative and positive supercoils. When the DNA is cleaved, torsional energy present in the molecule dissipates by rotation of the DNA about its intact strand (Fig. 1.21, *top*). This mechanism is generally referred to as the "swivel" mechanism. TopIB enzymes engage the DNA duplex as a C-shaped protein clamp (Fig. 1.21, *top*). The tightly closed clamp may hinder DNA swiveling. Thus, DNA strand rotation within the covalent enzyme-DNA complex requires at least some opening of the flexible TopIB protein clamp.

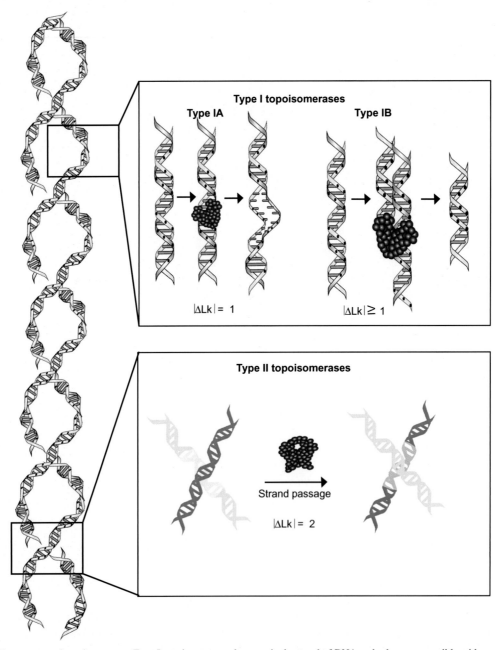

FIG. 1.21   Different types of topoisomerases. Type I topoisomerases cleave a single-strand of DNA and relax a supercoil by either passing the other strand through an enzyme-DNA linked intermediate (type IA enzymes) or by a strand-swivel mechanism (type IB enzymes). Type II topoisomerases cleave duplex DNA and then relax the supercoil by passing a second duplex DNA through the transient enzyme-DNA linked intermediate.

Archaeal topoisomerase V is the sole member of the newly established TopIC family. Although structurally distinct, this enzyme functionally resembles TopIB in terms of its mechanism of action; it forms a 3'-phosphotyrosyl intermediate and relaxes positive and negative supercoils.

Type II Topoisomerases

Type II topoisomerases transiently cleave both strands of a DNA duplex to allow the unidirectional passage of another DNA duplex through the protein-linked DNA gate (Fig. 1.21, *bottom*). Cleavage of the phosphodiester backbone in one segment of duplex DNA (termed the gate or G-segment) by the two active site tyrosines is accomplished by the formation of a covalent 5'-phosphotyrosyl-enzyme adduct on each DNA strand, separated by 4 nucleotides. A second duplex (the transfer or T-segment) is captured by the ATP-bound enzyme and passed along the dimer interface of the enzyme through the double-strand break. The broken DNA strands are then religated. Depending on whether the captured DNA derives from the same duplex cleaved by the enzyme or from a separate DNA molecule, Type II enzymes can catalyze changes in DNA supercoiling, knotting, or catenation. ATP hydrolysis is not required for DNA cleavage or relegation, per se. Rather, Type II enzymes use ATP binding and hydrolysis to drive conformational changes in the dimeric enzyme that are required to change the linkage of the DNA strands or duplexes. DNA gyrase is a bacterial topoisomerase that introduces negative supercoils into DNA. Because it cuts both strands of DNA, it is considered to be a Class II topoisomerase.

Type II topoisomerases are divided into two subfamilies: IIA and IIB. In eukaryotes, IIA enzymes catalyze the relaxation of positively or negatively supercoiled DNA, as well as the decatenation and unknotting of DNA helices. Bacterial IIA topoisomerases, including DNA gyrase and topoisomerase IV (TopIV), can also decatenate and unknot DNA, but they also possess unique activities. Gyrase catalyzes a reduction in Lk, such as the removal of positive supercoils, and it can also introduce negative supercoils into DNA. TopIV preferentially relaxes positive supercoils and is a potent decatenase. These specialized enzymatic activities apparently derive from the unique C-terminal DNA binding domains not found in eukaryotic IIA enzymes. However, the physical basis for the preferential binding of TopIV for positive versus negative supercoils, as well as the role of DNA wrapping in the introduction of negative supercoils by gyrase, is not well understood.

Topoisomerase VI (TopVI) exemplifies the IIB subfamily, which is found in archaea, plants, and algae. It is distinguished from IIA enzymes on the basis of its primary structure and domain architecture. In particular, TopVI lacks the third dimerization interface present in IIA enzymes. Nevertheless, TopVI still bears some structural and mechanistic similarities to TopIIA. TopVI is a homodimer, cleaves two strands of DNA by staggered 5'-phosphotyrosyl linkages, and uses ATP to drive a second DNA duplex through the protein-linked DNA gate to change Lk by steps of two. TopVI can also catalyze DNA decatenation and the relaxation of positive and negative supercoils.

The Cellular Roles of the Various Topoisomerases

All DNA exists in the negative supercoiled state within both prokaryotic and eukaryotic cells. If the DNA is unrestrained, there is equilibrium between tension and unwinding of the helix. If the supercoiled is restrained with proteins, they are stabilized by the energy of interaction between the proteins and the DNA. It has been suggested that DNA supercoiling plays an important role in many genetic processes, such as replication, transcription and recombination. These cellular activities require topoisomerases to remove constraints and stabilize the DNA. All topoisomerases catalyze changes in the linkage of DNA strands or helices by a conserved mechanism of transient DNA strand cleavage and religation, yet the different types of topoisomerases carry out distinct roles inside the cell.

In *E. coli*, for example, the antagonist actions of the type IA topoisomerase TopIA (i.e., removal of negative supercoils to increase Lk) and the Type II enzyme gyrase (i.e., introduction of negative supercoils to decrease Lk) provide a homeostatic mechanism to regulate global DNA supercoiling in chromosomes. The other Type II enzyme in this organism, Topoisomerase IV (TopIV), acts to remove positive supercoils in advance of the replication fork and is a potent decatenase to resolve chromosomal intertwining. In eukaryotes, TopIB and TopII enzymes provide the major DNA relaxation activities during transcription and replication to remove positive and negative supercoils. TopIB uses a mechanism with a protein-linked DNA swivel whereas TopII enzymes act through a strand passage mechanism. The Type II enzymes also act during mitosis to decatenate newly replicated sister chromatids. Topoisomerase III (TopIII), a type IA enzyme, resolves recombination intermediates and acts as a decatenase on nicked DNA during replication. Reverse gyrase is a type I topoisomerase occurring predominantly in archaea, where it has the ability to introduce positive supercoils. Reverse gyrase can thus act in concert with TopIA, a function that may be particularly useful for maintaining genomic stability at the high environmental temperatures at which most archaea thrive.

KINETICS OF RNA FOLDING

Although DNA in cells is normally in a helical, double-stranded form, RNA has more versatile compact structures in cells. This is because single-stranded RNA molecules can fold to form compact three dimensional structures. As a result, they carry out structural, catalytic and regulatory roles in many cellular processes including protein synthesis.

The precise conformation or fold of an RNA molecule is determined by its nucleotide base sequences. To build the macromolecular RNA structure, it is necessary to satisfy stereo-chemical bonding constraints and avoid steric collisions. Currently, the secondary structure of an RNA in its native folded state can be determined by experimental-computational approaches (Fig. 1.22). However, the RNA secondary structure alone imposes significant topological constraints on the allowed range of conformations even in the absence of tertiary interactions, and ignoring electrostatic interactions. For example, stable A-form helices exist in great abundance, and their backbone conformation can be modeled assuming an A-form helix geometry. However, these helical elements are, in turn, tethered together at two positions on either side of the helix perimeter by single strands that are short in comparison to the local diameter of the A-form helix. Therefore, the connectivity constraints are severely restricted by the short length of the two tethers on either side of the helix. The specific nature of these connectivity constraints vary depending on the junction topology, relative lengths of individual strands, and whether bases in junctions adopt a looped-in or looped-out conformation. In addition, steric constraints between RNA

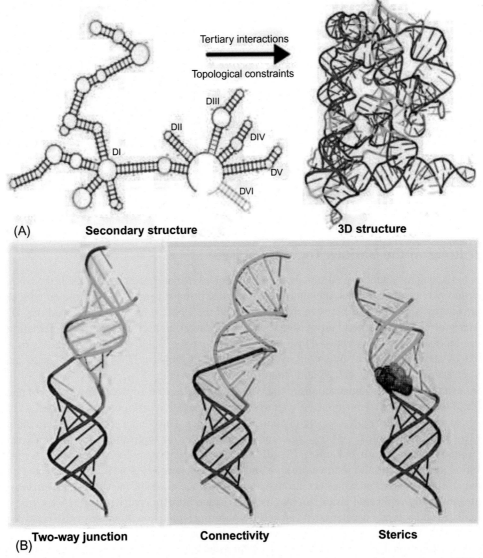

FIG. 1.22   Topological constraints encoded by the RNA secondary structure. (A) Secondary and tertiary structure of group IIC intron determined by experimental-computational approaches. (B) Connectivity and steric constraints in a two-way junction helices. The constraints shown in red restrict the flexibility of helices.

helices and other elements, further restrict the allowed translation and reorientation of helices with respect to one another. These steric collisions can be long-range in nature, occurring between helices that are far apart in sequence. Together, the connectivity and steric constraints define the topological constraints encoded by an RNA secondary structure.

A growing number of computational models are taking advantage of topological constraints in making useful inferences about RNA backbone conformation based primarily on secondary structure information.

The formation of competing helices during RNA co-transcriptional folding is expected to give rise to topological constraints that may, in turn, steer folding pathways. Most regulatory RNAs not only have to fold into specific 3D structures, they also have to retain a degree of flexibility to allow their structure to change adaptively during ribonucleoprotein assembly, recognition, catalysis, and signaling. Topology not only helps define dynamic aspects of RNA global structures, but also which structures are selected and stabilized upon binding to specific ligands. As a result, small changes in the chemical structure of small molecules can lead to stabilization of slightly different RNA structures within the allowed topological space. Conversely, changing other aspects of the small molecule, such as its charge, without affecting size, will not lead to sizeable changes in the bound RNA conformation. Therefore, different RNA conformations from a topologically allowed distribution can be selected by the added topological constraints arising from the interactions with a small molecule ligand.

Chapter 2

Nucleic Acid-Based Cellular Activities

Chapter Outline

DNA REPLICATION

The flow of genetic information in a cell can be demonstrated by the example of a bacterium in which only a single circular chromosome is present, and the genetic information can be used within and also passed between cells (Fig. 2.1). During the transmission of genetic information, all these cellular activities, including DNA replication, recombination and gene expression, are nucleic acid-based activities. For example, each cell needs a complete and faithful set of genetic materials. As a result, a parental cell must duplicate its entire genome before dividing. Subsequently, each of the two new daughter cells that are created then receives one copy of the genome through a process called cell division. Thus, DNA replication is a process to ensure that the DNA is duplicated before cell division so that each offspring cell receives chromosome(s) identical to the parent's. However, in multicellular organisms, cell division does not automatically result in the creation of the new organism. Instead, they increase the size and complexity of the original organism. In each cell cycle, eukaryotic cells must replicate large amounts of genomic DNA distributed on multiple chromosomes. To accomplish this feat in a reasonable period of time, replication initiates throughout S phase (a stage in the cell cycle) at multiple origins along each chromosome. Initiation from these origins must be coordinated so that no region of the genome is left unreplicated, and no region is replicated more than once. DNA replication must also be coordinated with chromosome segregation to ensure that each daughter cell receives a complete and unaltered complement of genetic information. Mistakes in either DNA replication or chromosome segregation can result in loss or duplication of this genetic information, events that can play an important role in the genesis of cancer cells or diseases.

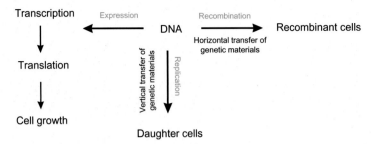

FIG. 2.1 The flow of genetic information. DNA can either be obtained from a parent cell for division, or from another cell of the same generation. It is then expressed within that cell, or transferred to another.

The DNA Replication Hypothesis

Originally, three DNA replication models were proposed—the semiconservative model, the conservative model, and the dispersive model. In the semiconservative model, the two parental strands separate, allowing each separated strand to serve as a template for the synthesis of a complementary strand (Fig. 2.2A). In this replication mechanism, each double-stranded daughter DNA molecule will have a conserved DNA strand that is derived from the parental DNA and a newly synthesized strand. In the conservative model, the two strands of the double helix unwind at the replication site to the extent needed for synthesizing new base sequence. The two original strands remain entwined after replication so that one of the two DNA molecules present after replication contains both original strands and the other DNA molecule is made of the two new

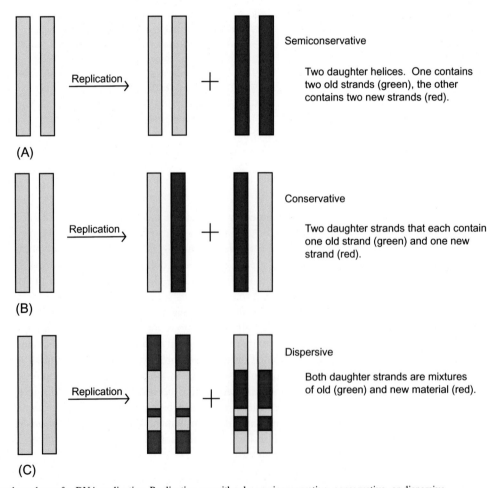

FIG. 2.2 Three hypotheses for DNA replication. Replication can either be semiconservative, conservative, or dispersive.

strands (Fig. 2.2B). The dispersive model shares some of the features of the conservative model but predicts that each strand of the daughter DNA molecules and parental DNA molecules has interspersed sections of both old and new DNA (Fig. 2.2C). Although these three models predict different outcomes in terms of the composition of daughter DNA molecules, they all follow the complementary base-pairing rule for replication.

DNA Replication Follows the Semiconservative Replication Mechanism

In 1958, Matthew Meselson and Franklin Stahl used equilibrium density gradient centrifugation experiments to demonstrate that DNA replication is semiconservative. In their experiments, they cultured *E. coli* cells for many generations in a growth medium containing $^{15}NH_4Cl$ as the sole source of nitrogen (Fig. 2.3). In this way, all nitrogenous bases in DNA were labeled with the heavy isotope ^{15}N. Subsequently, they transferred the bacteria to a new growth medium containing $^{14}NH_4Cl$ as the sole source of nitrogen.

Bacterial cells were removed at various times for DNA extraction, followed by centrifugation for the cesium chloride equilibrium density gradient. If the replication mechanism is conservative, there should be two bands in the gradient after one generation, one band representing $[^{15}N/^{15}N]$ parental duplex, and the other band representing $[^{14}N/^{14}N]$ newly synthesized DNA (Fig. 2.3). If the replication follows a semiconservative mechanism, the two $[^{15}N/^{15}N]$ parental strands will separate to serve as the template for replication, and each will be supplied with a new $[^{14}N]$ strand. One $[^{15}N/^{14}N]$ hybrid band would appear in the gradient.

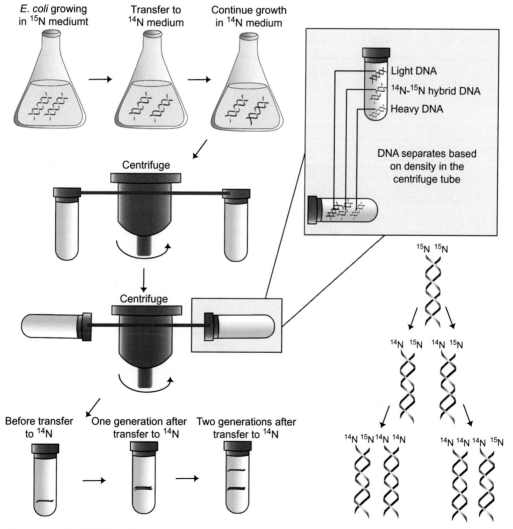

FIG. 2.3 The three hypothesis of DNA replication.

Critical to the success of this experiment, DNA containing ^{15}N can be distinguished from DNA containing ^{14}N. The experimental procedure involves the use of a technique referred to as sedimentation equilibrium centrifugation or buoyant density gradient centrifugation. Samples are forced by centrifugation through a density gradient of cesium chloride. Molecules of DNA will reach equilibrium when their density equals the density of the gradient medium. As such, ^{15}N-DNA will reach the point at a position closer to the bottom of the tube than will ^{14}N-DNA.

Their results showed that there is one DNA band representing [$^{15}N/^{14}N$] in the gradient for the cells incubated in the $^{15}NH_4Cl$-containing medium after one generation of incubation in the $^{14}NH_4Cl$-containing medium. As such, this result rules out the conservative hypothesis which should have two bands.

In two rounds of replication, the dispersive mechanism would give one hybrid product with one-fourth [^{15}N] and three-fourths [^{14}N]. On the other hand, the semiconservative hypothesis predicts that [$^{15}N/^{14}N$] would separate and serve as the template for replication and produce [$^{15}N/^{14}N$] and [$^{14}N/^{14}N$] DNAs. It should give 1:1 ratio for these two DNA molecules. After two generations of incubation in the $^{14}NH_4Cl$-containing medium, Meselson and Stahl observed that there are two bands in the gradient in which one represents [$^{15}N/^{14}N$] and the other one represents [$^{14}N/^{14}N$]. Therefore, this experiment proves that DNA replication follows the semiconservative mechanism.

THE PROCESS OF DNA REPLICATION

The Overview of Three Different Phases of Replication

The process of DNA replication can be divided into three phases: initiation, elongation, and termination. DNA replication starts at specialized sites called origins of replication and moves away from an origin in both directions, creating a structure known as a replication bubble (Fig. 2.4). The DNA double helix is opened at the origins of replication and unwound on both sides of the origin to form two structure called replication forks that unwind the double helix in opposite directions. The replication forks are the sites at which single-stranded DNA is exposed, and at which DNA synthesis occurs.

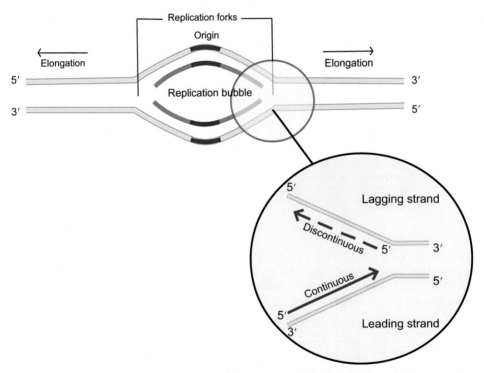

FIG. 2.4   Bidirectional DNA replication. The origin (*red*) indicates the initiation site for DNA replication. Both strands are copied from the origin outward, creating the replication bubble. Replication forks form at either end of the bubble, and continuously unwind the double helix. The leading strand is always the strand synthesized in the 5′ to 3′ direction. The other is considered the lagging strand and is replicated discontinuously as fragments which are later joined to form a continuous strand (see dashed line).

Once the DNA double helix is opened, replication enzymes and proteins are loaded to the single strand, and these will form the templates for the daughter strands that are to be synthesized (Fig. 2.5). After the replication machinery is in place, and the DNA has been opened up, replication enters the elongation phase. During the elongation phase, the replication machinery moves along the parent DNA strands and forms the daughter strands as it proceeds. Termination of DNA

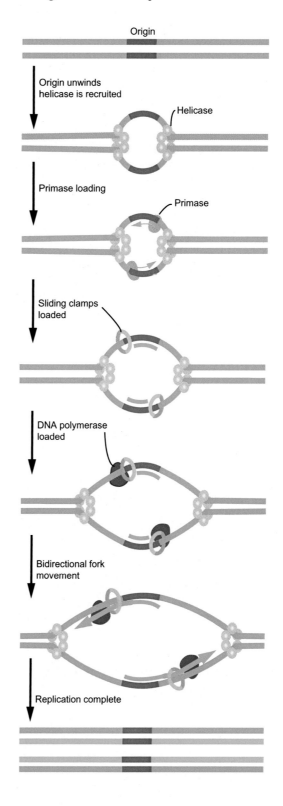

(1) Helicase (orange rings) is recruited to the origin Replication bubble allows both strands to be copied in opposing directions

(2) Helicase recruits primase, which will synthesize RNA primers (green) on both strands. For simplicity, only leading strand replication is shown.

(3) Sliding clamp is recruited and binds

(4) DNA polymerases (blue) are recruited. This interacts with the sliding clamp to elongate 3′ end of primers.

(5) Both replication forks travel in opposing directions, helicase works to unwind the DNA, allowing synthesis of new strands to occur. Replication of the leading strand is continuous, but replication of the lagging strand is discontinuous, and produces short fragments of DNA that are joined later (not shown).

(6) Replication is complete. RNA primers are removed and DNA polymerases fill in nucleotides. DNA ligases (not shown) seal any gaps that remain.

FIG. 2.5 The process of DNA replication.

replication occurs when the two replication forks moving in opposite directions meet, and the replication complexes are disassembled. In bacteria, this occurs at a specific site called *ter*, whereas in eukaryotes, no specific termination sequence is required.

Semidiscontinuous DNA Replication Mechanism

Due to the nature of the DNA synthesis reaction, all synthesis of nucleotide chains occurs in the $5' \rightarrow 3'$ direction. In a growing chain, the incoming nucleotide has a $5'$-triphosphate on its sugar. The last nucleotide added to a growing chain has a $3'$-hydroxyl group on the sugar. Because this $3'$-hydroxyl group is a nucleophile, it attacks the α phosphate group, leading to the elimination of pyrophosphate and the formation of a new phosphodiester bond, which links two nucleotides together and allows DNA strand synthesis to occur (Fig. 2.6). This reaction is catalyzed by DNA polymerase. All known DNA polymerases so far can only add nucleotides in a $5' \rightarrow 3'$ direction. As such, DNA replication can only occur in the $5' \rightarrow 3'$ direction.

During replication, the double helix DNA would unwind to create a fork. Due to the antiparallel nature of the two strands of DNA, if one runs $5' \rightarrow 3'$ direction left to right, the other must run $3' \rightarrow 5'$ direction left to right. Therefore, strand synthesis occurs in opposite directions for the two templates in a double helix. Following the fact that DNA synthesis can only make one direction $5' \rightarrow 3'$, Okazaki and his colleagues proposed a semidiscontinuous replication mechanism. In this model, both strands could not replicate continuously. DNA polymerase could make one strand which is the leading strand continuously in the $5' \rightarrow 3'$ direction at the replication fork on the exposed $3' \rightarrow 5'$ template strand (Fig. 2.7). Its direction of synthesis is the same as the direction in which the replication fork is moving.

The other strand, which is the lagging strand, would have to be made discontinuously in small fragments—Okazaki fragments. These fragments are typically 1000–2000 nucleotides long, as shown in Fig. 2.7. The discontinuity of synthesis of the lagging strand is because its direction of synthesis is opposite to the moving direction of the replication fork. The $5'$ end of each of these Okazaki fragments is closer to the replication fork than the $3'$ end. The small Okazaki fragments of the lagging strand are then linked together by an enzyme called DNA ligase, and together these will constitute the second daughter strand.

Eukaryotic DNA Replication

Eukaryotic DNA replication is more complex than prokaryotic DNA, due to several features of eukaryotic DNA. For example, eukaryotic cells contain much more DNA, and their DNA is complexed with nucleosomes. Furthermore, eukaryotic chromosomes are linear rather than circular. To overcome these problems, replication from a single origin on a typical eukaryotic chromosome would take days to complete. However, replication of entire eukaryotic genomes is usually accomplished in a matter of minutes to hours. To facilitate the rapid synthesis of large quantities of DNA, eukaryotic chromosomes contain multiple replication origins. Eukaryotic replication origins not only act as sites of replication initiation, but also control the timing of DNA replication.

PROTEINS REQUIRED FOR DNA REPLICATION IN *E. COLI*

DNA replication is carried out in all organisms by a multiprotein complex called the replisome. DNA replication begins at the replicon, where the two DNA strands unwind at the replication fork. Therefore, we start from the separation of two DNA strands to discuss the participation of each enzyme and protein at each step in the process.

Helicase and Single-Strand DNA Binding Proteins

During the initiation phase, the double helix parental DNA is opened up in order to give replication enzymes and other proteins access to the single strands that will form the template for the daughter strands that are to be synthesized. In fact, a specific initiator protein can recognize the origin of replication and recruit helicase to the origin of replication. Helicase is the enzyme that harnesses the chemical energy from the ATP hydrolysis to separate the two DNA strands at the replication fork (Fig. 2.8). When the two strands are separated, single-strand DNA binding proteins (SSBs) bind selectively to single-stranded DNA as soon as it forms. The binding of SSBs can stabilize the single-stranded DNA so they will not

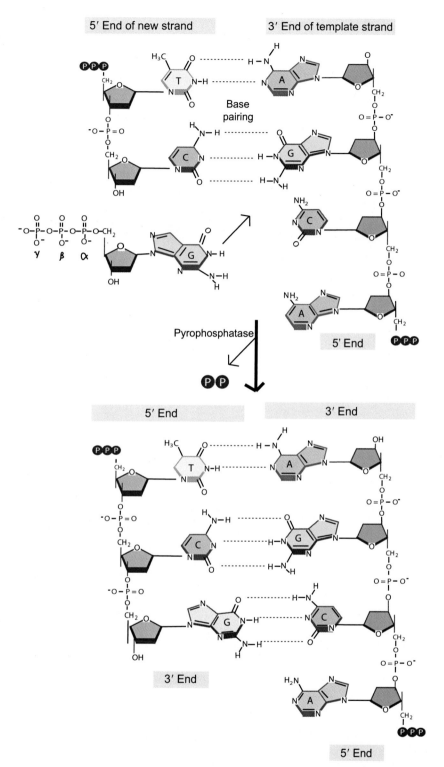

FIG. 2.6   DNA synthesis always occurs 5′ to 3′. The template strand directs the addition of the appropriate dNTP to the growing strand. Synthesis of newly replicated DNA begins once the α-phosphate of the incoming dNTP, which will elongate the 3′ end and release a pyrophosphate, which breaks down pyrophosphate into two phosphates molecules.

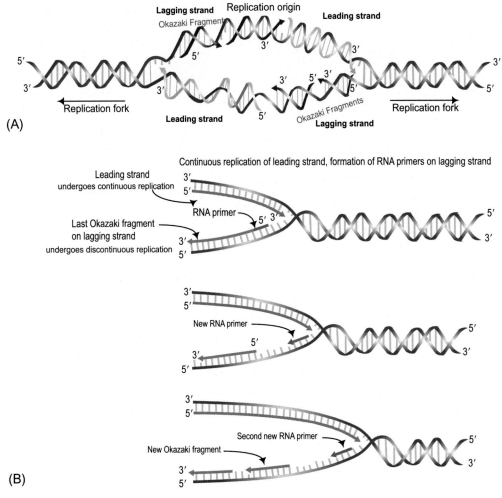

FIG. 2.7 The mechanism of semi-discontinuous DNA replication. DNA replication occurs bidirectionally. (A) Two replication forks form at the origin of replication, and move in opposing directions. (B) The diagram only shows one direction of replication. Replication of the leading strand is continuous, but replication of the lagging strand is discontinuous in that it forms short fragments of DNA that are later sealed together. In both strands, replication always occurs 5′ to 3′.

anneal to reform the double helix and protect the single-stranded DNA from hydrolysis by nucleases. This also allows enzymes to attach to the newly opened single strand and initiate elongation. SSBs are removed as DNA synthesis proceeds.

DNA Gyrase

In prokaryotes, DNA exists in a negatively supercoiled, closed circle form. When two DNA strands are separated during replication, positive supercoils are introduced ahead of the replication fork. This is because the strands of duplex DNA unwind to allow each single strand of DNA to serve as a template for the synthesis of a complementary strand; a replisome that is prevented from rotating will accumulate positive supercoils in front of the replication fork (Fig. 2.9), further twisting the DNA onto itself. If the replication fork continued to move, the torsional strain of the positive supercoils would eventually make further replication impossible, because the DNA would be too compacted. At this moment, DNA gyrase (Type II topoisomerase) fights these positive supercoils by putting negative supercoils ahead of the replication fork. Thus, the torsional strain is released, and replication can continue.

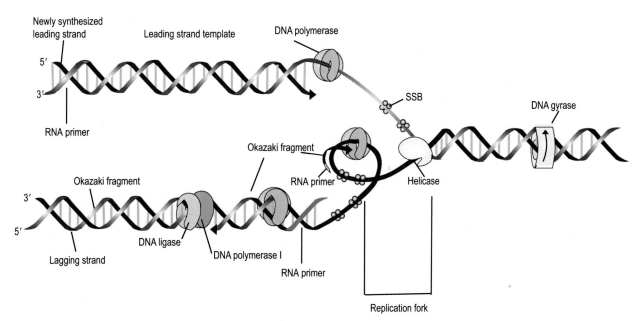

FIG. 2.8 General features of the replication fork. DNA gyrase and helicase work to unwind the DNA so that it can be replicated. The newly unwound, single strands are coated with ssDNA-binding protein, SSB for short. Replicative DNA polymerase is bound to the template strand (either leading or lagging) via the β-subunit. Downstream, DNA polymerase II and DNA ligase remove the RNA primers, replace them with the correct DNA fragments and also ligate (seal together) the Okazaki fragments on the lagging strand.

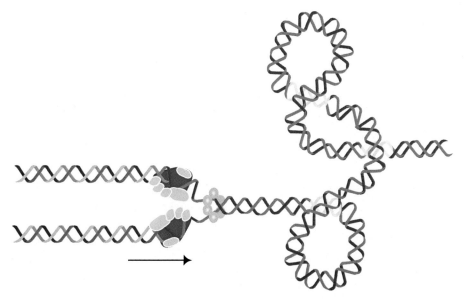

FIG. 2.9 Supercoiling is formed ahead of the replication fork. As the helicase unwinds DNA, the twist ahead of the fork increases in tension. This is similar to twisting a rubber band, then using two fingers to separate the strands. The supercoil is resolved by topoisomerase I or II, allowing the continuous movement of helicase.

Primase

One unusual feature of DNA polymerase is that it cannot synthesize a new DNA strand from the very start of the parent strand. It can only add nucleotides to the 3′ end of a nucleotide fragment that already exists. It is necessary to have a short strand of RNA termed primer for the beginning of DNA replication. The primer must have a free 3′-hydroxyl to which the growing chain can attach. Primase is responsible for copying a short stretch of the DNA template strand to produce the RNA primer sequence (Fig. 2.10). The use of RNA primer for DNA replication is common to all prokaryotes. The leading strand requires one primer. On the other hand, each Okazaki fragment in the lagging strand would need one primer. The primer is

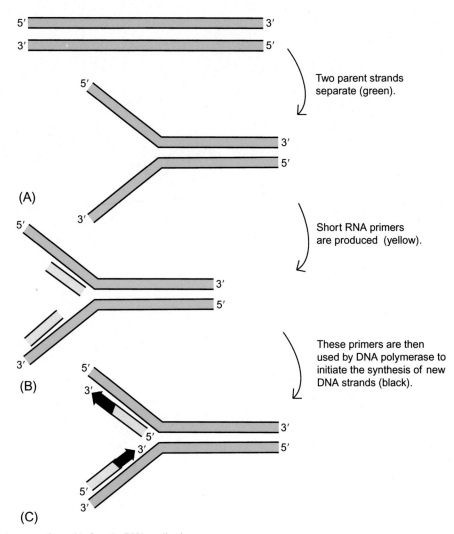

FIG. 2.10 RNA primers are formed before the DNA replication.

hydrogen-bonded to the template so it can provide a stable framework to which the nascent chain starts to grow. In bacteria, a short RNA primer of around 10–30 bases is synthesized by a primase, followed by the replication directed by DNA polymerase III.

The eukaryotic primase complex contains four subunits. Two of them function as a primase, then there is an α catalytic subunit and an accessory subunit. Primases and α catalytic subunit bind in a complex with the DNA polymerase. Normally, the polymerase-α subunit-primase complex synthesizes a stretch of 10–30 nucleotides of RNA. Subsequently, α catalytic subunit continues to synthesize a short stretch of DNA before the DNA polymerase takes over the replication process. This phenomenon is called polymerase switching.

DNA Polymerase

Once the replication is initiated, the synthesis of new DNA strand is directed by one of the polymerases -DNA polymerase III. The first nucleotide is added to the 3′-hydroxyl of the RNA primer, and synthesis proceeds from the 5′ end to the 3′ end on both the leading and the lagging strand. As the replication fork moves, the RNA primer is removed and is replaced by newly formed deoxynucleotides. These reactions are performed by DNA polymerase I's exonuclease activity and polymerization activity. This is a so-called Okazaki maturation process. Just after RNA removal, DNA polymerase I uses its polymerase activity to fill in the gap left by the RNA with new DNA. When it reaches the beginning of the next DNA segment, the DNA polymerase cannot seal the final nick in the DNA backbone that remains. Finally, DNA ligase is the enzyme responsible for sealing the nick between the new strands synthesized by polymerase III and polymerase I.

TABLE 2.1 Properties of Three Different DNA Polymerases of *E. coli*

Property	Pol I	Pol II	Pol III
Mass (kDa)	103	90	830
Structural gene	*pol*A	*pol*B	*pol*C
Polymerization 5′ → 3′	Yes	Yes	Yes
Exonuclease 5′ → 3′	Yes	No	No
Exonuclease 3′ → 5′	Yes	Yes	Yes
Number of subunits	1	More than 4	More than 10

Three DNA polymerases in *E. coli* have been studied intensively (Table 2.1). All three DNA polymerases have the 5′→ 3′ polymerization activity and 3′→ 5′ exonuclease activity. The 5′→ 3′ polymerization activity can add nucleotides to a growing chain during DNA synthesis. On the other hand, the 3′→ 5′ exonuclease activity is part of proofreading function which can remove incorrect nucleotides one at a time in the process of replication and replace them with the correct one. Only DNA polymerase I has 5′→ 3′ exonuclease activity which can remove short stretches of nucleotides during repair. After polymerase III has produced the new polynucleotide chain, DNA polymerase I removes RNA primer through 5′→ 3′ exonuclease activity. It then fills in behind it with its polymerase activity (Fig. 2.11). This is a cut-and-patch process which is also called nick translation. Polymerase I also uses nick translation in the repair process.

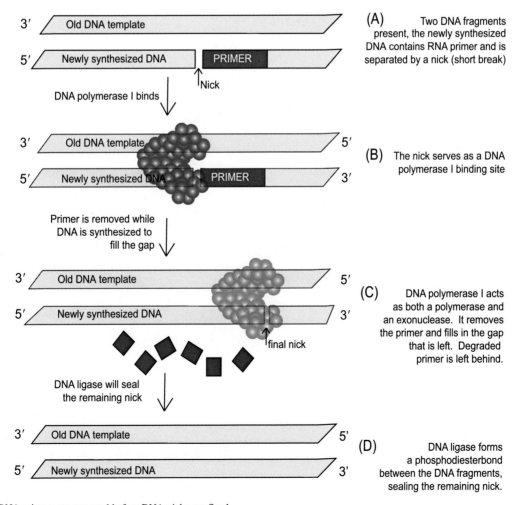

FIG. 2.11 RNA primers are removed before DNA nicks are fixed.

COUPLING DNA REPLICATION

So far, we have discussed that the leading strand synthesis proceeds continuously, and the lagging strand synthesis is discontinuous due to the formation of Okazaki fragments. The movement of polymerases is coordinated to allow the polymerases on both strands to travel together in the direction of fork movement. Therefore, if the leading strand replication stalls, lagging strand replication will also be halted.

The Replisome

The coupling of DNA replication on the leading and lagging strands in bacteria is achieved by physically associating the proteins replicating each strand into one large protein called the replication complex, or replisome. The replisome associates with the DNA, and this complex moves together as the replication fork moves. Protein within the complex can engage and disengage from the DNA itself, but remains associated with the fork.

In *E. coli*, the multisubunit assembly of DNA polymerase, sliding clamp, and clamp loader is called the DNA polymerase III holoenzyme or the replisome (Fig. 2.12). It contains two copies of the multisubunit DNA polymerase III, which contains the α subunit (the DNA polymerase activity), the ε subunit (the 3′-5′ proofreading exonuclease), and the θ subunit (which stimulates the exonuclease) together with sliding clamps and a clamp loader that continually reloads sliding clamps on the lagging strand.

The Assembly of the Replisome

The clamp loader and sliding clamp play an important role in recruiting DNA polymerase to the appropriate location on the DNA template. They localize specifically at the region where DNA synthesis needs to commence (Fig. 2.5, step 3). The

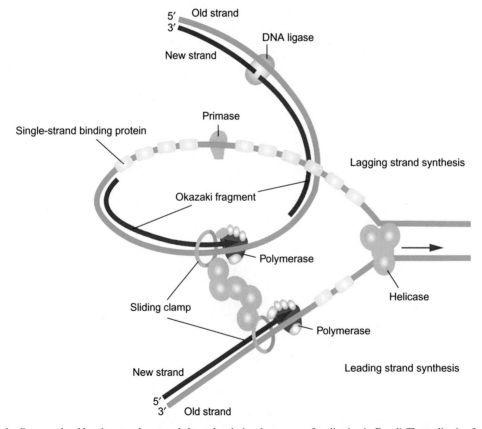

FIG. 2.12   Both leading strand and lagging strand are coupled together during the process of replication in *E. coli*. The replication fork is moving to the right. The leading strand synthesized continuously, and the lagging strand is synthesized discontinuously. Two DNA polymerase III molecules (*blue*) are bound to the sliding clamps (*orange rings*).

sliding clamp is responsible for holding catalytic cores onto their template strands. The clamp loader places the clamp on DNA.

One polymerase replicates the leading strand and one replicates the lagging strand as the fork moves along the double helix. Polymerases are linked together by a protein called tau (τ). Tau protein is associated with the clamp loader and links this polymerases-clamp loader complex to the helicase. Therefore, the replication complex keeps the lagging strand polymerase associated with the fork even as it is being released from the DNA at the end of each Okazaki fragment. Furthermore, through this association of polymerase, clamp loader, and sliding clamp, the loading of a new clamp and a polymerase onto DNA at the beginning of each Okazaki fragment can be efficiently coordinated with the progress of the polymerase on the leading strand.

REPLICATION PATTERNS

Bidirectional Replication

Most eukaryotic and bacterial DNAs replicate bidirectionally (Fig. 2.4). The DNA under the control of the origin of replication is called a replicon. The circular *E. coli* chromosome has a single replicon because it replicates from a single starting point—the origin of replication. DNA replication begins with the creation of a bubble, which is a small region where the parental strands have separated and progeny strands have been synthesized. Two forks arise from this bubble and move in opposite directions. As the bubble expands, the replicating DNA begins to take on the theta (θ) shape until both replication fork meet on the other side of the circle (Fig. 2.13). Eukaryotic chromosomes have many replicons, and the replication in these replicons begins simultaneously. Again, the replication forks move bidirectionally, and they keep moving until they reach the end of the chromosome, which is known as the telomere. The replication at the telomere adopts a special mechanism which we will discuss later.

Rolling Circle Replication

Some circular DNAs are replicated by a rolling circle mechanism instead of the θ mode (Fig. 2.14). Replication by rolling circles is common among bacteriophage. The bacterial phage named ϕX174 has a single stranded circular DNA genome

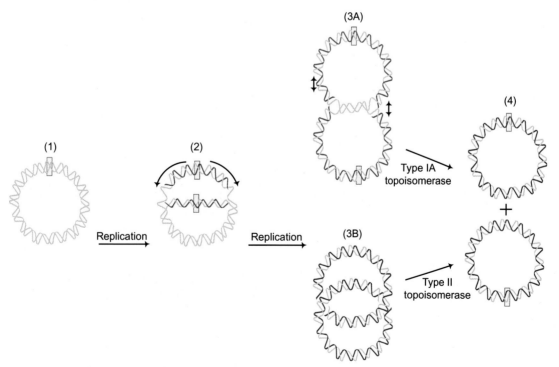

FIG. 2.13 The process of bacterial DNA replication. Two interlinked double stranded daughter molecules are produced by replication of circular DNA. Topoisomerase IA or II will unlink the circular chromosomes, allowing for separation.

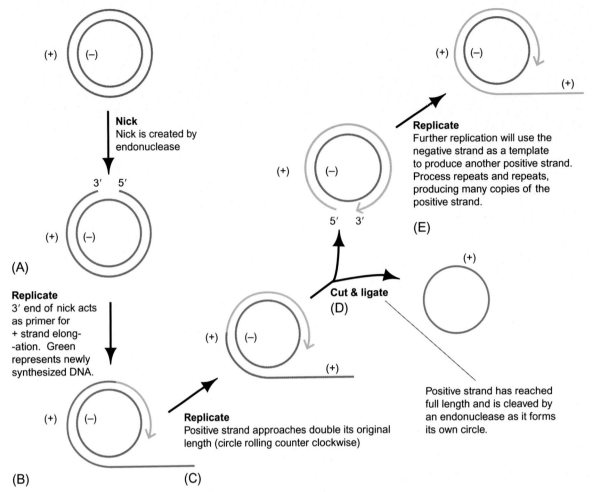

Nick
Nick is created by endonuclease

(A)

Replicate
3′ end of nick acts as primer for + strand elong--ation. Green represents newly synthesized DNA.

(B)

Replicate
Positive strand approaches double its original length (circle rolling counter clockwise)

(C)

Cut & ligate
(D)

Replicate
Further replication will use the negative strand as a template to produce another positive strand. Process repeats and repeats, producing many copies of the positive strand.

(E)

Positive strand has reached full length and is cleaved by an endonuclease as it forms its own circle.

FIG. 2.14 A schematic diagram of rolling circle replication for single strand DNA.

and adopts a simple form of rolling circle replication. Basically, replication of only one strand is used to generate copies of some circular molecules. For example, a double-stranded replicative form gives rise to many copies of a single-stranded progeny DNA. One strand of a double-stranded DNA is nicked and the free 3′-OH end generated by the nick is extended by the DNA polymerase. The chain is elongated around the circular strand template until it reaches the starting point. Subsequently, the newly synthesized strand displaces the original parental strand. Meanwhile, the displaced strand is free as a circle. This displaces the 5′ end. During the replication process, the intermediates that have the double-stranded part of the replicating DNA can be considered to be rolling counterclockwise and trailing out the progeny single-stranded DNA. During the replication phase of viral infection, it may be used as a template to synthesize the complementary strand. The duplex circle may then be used as a rolling circle to generate more progeny. During phage morphogenesis, the displaced strand is packaged into the phage virion.

The rolling circle mechanism can be used to replicate double-stranded DNA. For example, λ phage adopts the θ mode of replication to produce several copies of circular DNA (Fig. 2.15). Subsequently, these circular DNAs serve as the template for rolling circle synthesis of linear λ DNA molecules. The intact DNA strand serves as the template for the leading strand synthesis, whereas the displaced strand serves as the template for the lagging strand synthesis.

Termination of Replication

Termination of DNA replication occurs when the two replication forks moving away from the single origin of replication meet on the opposite side of the circle in the circular chromosome of bacteria (Fig. 2.16). When the forks approach each

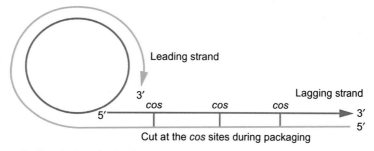

FIG. 2.15 A schematic diagram of rolling circle replication for double strand DNA. The leading strand elongates continuously, and the lagging strand elongates discontinuously, and uses the unrolled leading strand as a template. RNA primers allow DNA polymerase to create Okazaki fragments. Progeny dsDNA will contain many lengths worth of the genome, and will be clipped off and packaged into a phage head.

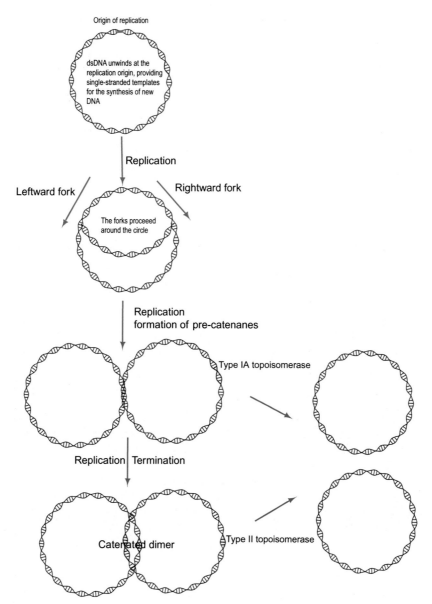

FIG. 2.16 The process of circular DNA replication results in two interlinked or catenated daughter molecules. The circular chromosome can be unlinked or decatenated by topoisomerases.

FIG. 2.17 Meeting of replication forks create interlinked DNA on linear chromosomes. As two replication forks converge, the DNA between them is replicated, and the normally two new daughter strands become intertwined. This is fixed by topoisomerase before their separation.

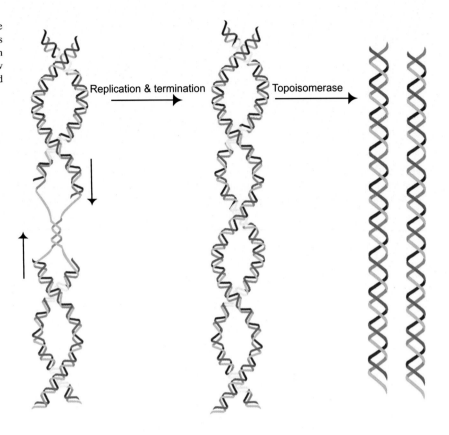

other, the replication complex disassembles and two growing strands join and this results in the interlinking of the two new daughter DNAs. The two circular DNA molecules are separated by topoisomerases. Sequences that are involved with termination are called *ter* sites. A *ter* site contains a short, ~23-bp sequence. The termination sequences are unidirectional and can be recognized by a unidirectional contrahelicase. This helicase known as *Tus* in *E. coli* can recognize the consensus sequences and prevent the replication fork from proceeding.

In eukaryotes, termination is completed at multiple points. This is because the linear chromosomes have multiple origins of replication, each of which initiate a replication bubble. During the replication elongation, the two replication forks form different bubbles will meet in the same way as the two bacterial forks. For the purpose of termination, the resulting interlinked chromosomes will be unlinked by topoisomerase (Fig. 2.17).

Replication at the Telomere

Eukaryotic chromosomes end in distinctive sequences called telomeres that help preserve the integrity and stability of the chromosome. The telomeres consist of simple sequence repeats. The length of the whole tract of telomere repeats varies from about 100 bp to 20,000 bp, depending on the species. Usually, the DNA at the protozoan's chromosome ends consists of the short tandem repeating sequence TTGGGG, which is known as the G-rich strand. On the other hand, its complimentary strand displays the repeated sequence AACCCC, called the C-rich strand. In a similar way, all vertebrates contain the sequence TTAGGG at the end of the G-rich strand. The G-rich strand is always at the 3' end, and this has special significance during telomere replication.

During the replication, the 5' to 3' synthesis on the leading strand template may proceed to the end. However, there is a problem on the lagging strand when the RNA primer is removed. Because the gap cannot be filled in the absence of the 3'—OH group, there will be a gap in the end of replication, which will, therefore, leave a single-stranded region of DNA. With the help of telomerase, a ribonucleoprotein, the 3' end of G-rich strand can be extended by using the RNA as the template to add telomeric sequences to the 3' end of the telomeric DNA (Fig. 2.18). This results in net telomere elongation and allows telomeres to be maintained around an equilibrium length.

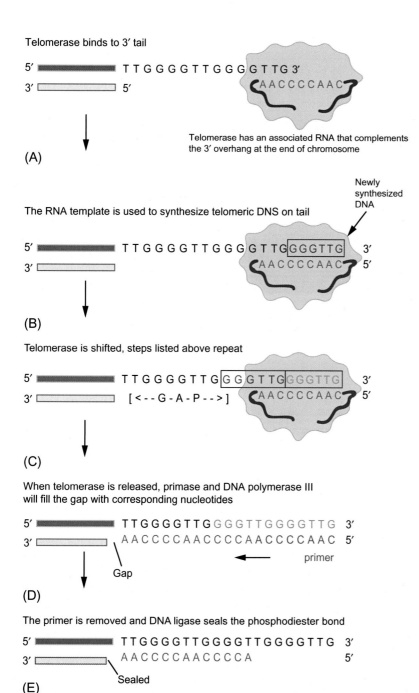

Telomerase binds to 3′ tail

5′ ▬▬▬▬▬▬ T T G G G G T T G G G G T T G 3′
3′ ▭▭▭▭▭ 5′

Telomerase has an associated RNA that complements
the 3′ overhang at the end of chromosome

(A)

The RNA template is used to synthesize telomeric DNS on tail

Newly
synthesized
DNA

5′ ▬▬▬▬▬▬ T T G G G G T T G G G G T T G GGGTTG 3′
AACCCCAAC 5′
3′ ▭▭▭▭▭

(B)

Telomerase is shifted, steps listed above repeat

5′ ▬▬▬▬▬▬ T T G G G G T T G GG GTTG GGGTTG 3′
3′ ▭▭▭▭▭ [<--G-A-P-->] AACCCCAAC 5′

(C)

When telomerase is released, primase and DNA polymerase III
will fill the gap with corresponding nucleotides

5′ ▬▬▬▬▬▬ T T G G G G T T G GGGTTGGGGGTTG 3′
3′ ▭▭▭▭▭ AACCCCAACCCCAACCCCAAC 5′
Gap   primer

(D)

The primer is removed and DNA ligase seals the phosphodiester bond

5′ ▬▬▬▬▬▬ T T G G G G T T G G G G T T G G G G G T T G 3′
3′ ▭▭▭▭▭ AACCCCAACCCCA 5′
Sealed
(E)

FIG. 2.18 Telomerase binds to 3′ GC rich tail. Repeated TTGGGG sequences are synthesized by enzyme telomerase, facilitating the synthesis of DNA on the opposite strand. This also elongates the ends of the chromosomes, and prevents degradation of important genetic material during replication.

Telomerase is an unusual DNA polymerase because it contains both a protein component and an essential RNA component. In the RNA component, it has a short template region that specifies the sequence of the telomere repeats that are added onto the chromosome ends. On the other hand, the protein component synthesizes the telomeric DNA through the RNA template. In human stem cells, telomerase is active, and the telomeres are maintained. In many human tissues, on the other hand, telomerase is not as active as it is in stem cells, so progressive telomere shortening occurs.

Mitochondrial Replication

Mitochondria must be duplicated during the cell cycle and segregated to the daughter cells. The process at each stage of mitochondria duplication (including DNA replication, DNA segregation to duplicated mitochondria, and organelle segregation to daughter cells) is stochastic, meaning that there is no control over which particular copies are replicated. It is governed by a random distribution of each copy. Mitochondria divide by developing a ring around the organelle that constricts to pinch it into two halves. This mechanism is similar to the principles of bacterial division. The combination of replication and segregation mechanisms can result in stochastic assignment of DNA to each of the copies. Thus, the distribution of mitochondrial genomes to daughter mitochondria does not depend on their parental origins.

FIDELITY OF DNA REPLICATION

Proofreading Mechanism

DNA replication takes place once each generation in each cell. As such, it is essential that the fidelity of the replication process be as high as possible to prevent mutations, which are errors in replication and can lead to diseases. For the bacteriophage and *Escherichia coli* replication in the absence of DNA mismatch repair and external environmental stress, the base substitution error rate of the replication machinery in vivo is in the range of once in every 10^7 to 10^8 base pairs. Eukaryotic DNA replication is likely to be at least this accurate. Typically, errors in hydrogen bonding lead to the incorporation of an incorrect nucleotide into a growing DNA chain once in every 10^5 base pair. At this rate, replication would introduce errors into a significant percentage of genes every generation in *E. coli*, whose genome contains over four million base pairs.

Fortunately, the $3' \rightarrow 5'$ exonuclease activity can remove the incorrect nucleotide, and replication resumes when the correct nucleotide is added. Errors in replication occur approximately once in every 10^9–10^{10} base pairs after the combination results of polymerase activity and proofreading activity. As such, the proofreading activity can improve the fidelity of replication significantly.

Polymerase Exonuclease Activity

Existing DNA can also be repaired by polymerase I $5' \rightarrow 3'$ exonuclease activity. For example, if one or more bases have been damaged by an external agent, or if a mismatch was missed by the proofreading activity, polymerase I can use its nick translation process to repair such DNA mistake as it moves along the DNA. As such, polymerases can achieve faithful copying of the template. From structural and biochemical studies polymerase I, it is shown that fidelity arises both from constraints imposed on base pairing at the polymerase active site as well as the editing of mismatched base pairs at a $3'$-exonuclease active site. The polymerase active site and the exonuclease active site are usually spatially separated on the polymerase. The advantage of this arrangement is that when a correct base is added, the polymerase simply moves on to the next template position and proceeds with polymerization. On the other hand, when a mismatch nucleotide is incorporated, the rate of incorporation of the next nucleotide by the polymerase is slowed. At this moment, the exonuclease can correct the mistake before the polymerase resumes its regular polymerization speed.

The mechanism whereby the exonuclease domain exerts its editing function is based on the shuttle mechanism (Fig. 2.19). The exonuclease domain involves a competition with the polymerase active sites for the $3'$ end of the primer template strand and a rapid shuttling of the primer terminus between them. The $3'$-exonuclease active site binds single-stranded DNA, whereas the polymerase active site binds duplex DNA with the ratio of about 1–10 for correctly

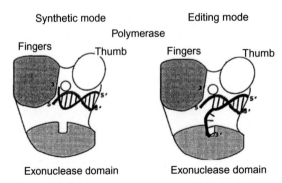

FIG. 2.19 The shuttle mechanism of editing in DNA polymerases. The equilibrium between the $3'$ end of the primer strand being bound as a single strand in the exonuclease active site (right) and bound as duplex at the polymerase active site (left) is shifted toward the editing mode by mismatched base pairs, which destabilize duplex DNA and retard addition of the next nucleotide. The shuttling of the $3'$ end between the two active sites is fast compared with the rate of next nucleotide addition.

Watson-Crick base-paired duplex DNA. Mismatched base pairs destabilize the duplex DNA and thereby enhance the binding of the 3′ single-stranded DNA to the exonuclease active site. Furthermore, polymerization is stalled after incorporation of mismatched base pairs, presumably because of misorientation of the 3′-hydroxyl group of the primer terminus onto which the next nucleotide is to be added. Once again, this stalling of the polymerization reaction serves to enhance the probability of excision by the exonuclease activity.

DNA MUTATION

Point Mutation

Changes can occur in the nucleotide sequence of a DNA molecule if the changes escape the proofreading and repair. Such a genetic change is called a mutation. Mutations are classified by the kind of change in the DNA molecule. Point mutations arise when a base pairs with an inappropriate partner during DNA replication. Transitions, in which one purine (or pyrimidine) is replaced by another, and transversion, in which a purine is substituted for a pyrimidine or vice versa, are the two possible kinds of point mutations. A change of one nucleotide of a triplet within a protein coding region of a gene may result in the creation of a new triplet that codes for a different amino acid in the protein product. If this occurs, the mutation is known as missense mutation (Fig. 2.20). A second possible outcome is that the triplet is changed into a stop codon, resulting in the early termination of the protein synthesis. This is known as the nonsense mutation. The third possibility is that the point mutation changes the nucleotide, but the triplet still codes for the same amino acid; this is known as the silent mutation, because here, there is no effect on the final protein product.

Insertion and Deletion

In addition to point mutation, DNA replication can lead to the introduction of small insertions or deletions. The addition or removal of one or more base pairs leads to insertion or deletion mutations, respectively. As illustrated in Fig. 2.20, the loss or addition of a single nucleotide causes all of the subsequent three-letter codons to be changed. These are called frameshift mutations because the frame of the triplet reading during translation is altered. A frameshift mutation will occur when any number of bases are added or deleted, except multiples of three, which will reestablish the initial reading frame, and just add another reading frame into the sequence, not altering the composition of those reading frames thereafter. It is possible, however, that the frameshift causes early termination of the translation, for example, if a 3-nucleotide sequence stop codon is introduced. Therefore, the results of frameshift mutation can be very severe if they occur early in the coding sequence.

Tautomeric Shifts

The fact that bases can take several forms, known as tautomers, increases the chance of mispairing during DNA replication. The tautomers differ by only a single proton shift in the molecule. The shifts change the bonding structure of the molecule and allow the hydrogen bonding with noncomplementary bases. Therefore, tautomeric shift may lead to permanent base-pair changes and mutations (Fig. 2.21).

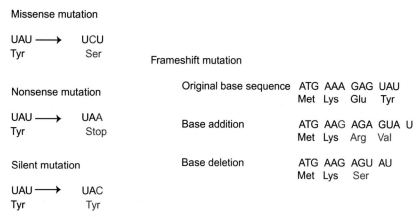

FIG. 2.20 Different types of point mutations in open reading frame.

FIG. 2.21 An example of mutation formation. A mutation occurs during DNA replication when a transiently formed tautomer in the template strand pairs with a noncomplementary base. In the next round of replication, the "mismatched" members of the base pair are separated, and each becomes the template for its normal complementary base and the end result is a point mutation.

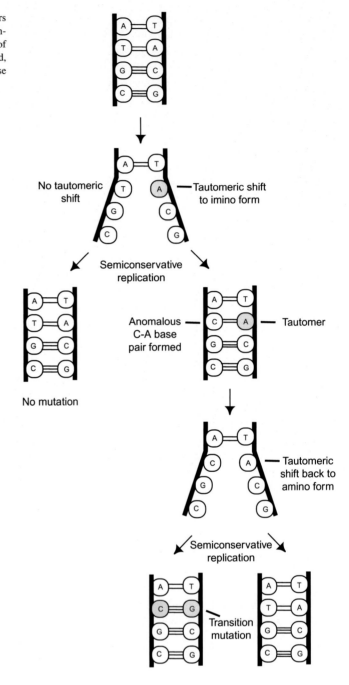

Mismatch Repair

When mismatch has managed to escape the normal exonuclease activities of DNA polymerase I and III, *E. coli* has a mismatch repair system to correct the mistake. The key of the mismatch repair system is to identify the strand that contains methylated adenine in the GATC sequence as the parental strand, so it leaves that strand alone and corrects the nearby mismatch in the unmethylated strand.

The *E. coli* methyl-directed pathway of mismatch repair relies on methylation pattern in the DNA to determine which strand is the newly synthesized one and which one was the parental (template) strand. DNA methylation, which is a process to add methyl groups to certain nucleotides along the new DNA strand, often occurs just after DNA replication. However, a window of opportunity exists between the end of replication and the start of methylation where only the parental strands are methylated. The system assumes that the strand bearing the methylated base is the parental strand and its sequence the

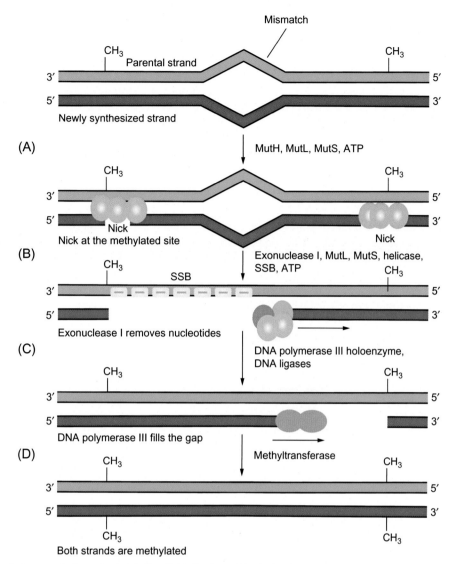

FIG. 2.22 A schematic diagram of mismatch repair in *E. coli*. (A) Products of the depicted genes combined with ATP will recognize mismatched nucleotides (raised band). They use the absence of methyl groups to identify the newly synthesized strand (*purple*), and will introduce a nick on this strand opposite a methyl group on the original (*green*) strand. (B) MutL, MutS, Exonuclease I, helicase, ATP will remove the region of DNA on this (*purple*) strand that houses the nucleotide of the mismatch pair. (C) DNA polymerase III holoenzyme will fill the gap (with help from SSB). Then, DNA ligase is used to remaining gap. (D) Methyltransferase will methylate GATC sequences in the daughter (*purple*) strand. After this action, the original and daughter strands will be indistinguishable, and as a result, mismatch repair will no longer work.

correct one, and the strand without methylation is the new strand. Therefore, this window provides opportunity for mismatch repair system to examine the newly synthesized strands for any replication error.

In this process, MutH, MutS, and MutL enzymes form a loop between the mistake and a methylated site (Fig. 2.22). DNA helicase H helps unwind the DNA. Exonuclease I removes the section of DNA containing the mistake. Single-stranded binding proteins protect the template strand from degradation. Finally, polymerase III can fill in the missing piece.

DNA DAMAGE

In addition to experiencing those spontaneous mutations caused by misreading the genetic code, organisms are frequently exposed to mutagens that cause damage to DNA. DNA damage is a chemical alteration to DNA which can be introduced by many different ways. If the DNA damage is left unrepaired, it can lead to mutation. If a particular kind of DNA damage is likely to lead to a mutation, it is deemed genotoxic. Some common examples of DNA damage are base modifications caused by base analogs and alkylating agents, pyrimidine dimers caused by ultraviolet (UV) radiation, and free radicals caused by ionizing radiation.

Base Analogs

Base analogs are mutagenic chemicals that can be substitute for purines or pyrimidines during nucleic acid biosynthesis. For example, the nucleoside analog bromodeoxyuridine (BrdU) is formed when 5-bromouracil (5-BU) is chemically linked to deoxyribose (Fig. 2.23). 5-BU is a derivative of uracil, and its 5 position of the pyrimidine ring is halogenated. If 5-BU is incorporated into DNA in place of thymine, the tautomeric shift will occur, and the 5-BU will base pair with guanine. After the replication, an A-T to G-C transition occurs. Furthermore, the presence of 5-BU within DNA increases the sensitivity of the molecule to UV light.

Alkylating Agents

Alkylation is the process in which electrophiles attack negatively charged DNA molecules and add carbon-containing groups called alkyl group. Some of the alkylations do not change base-pairing, so they are innocuous such as N7 alkylation of guanine (Fig. 2.24). Others cause DNA replication to stall and thus kill a cell, thus they are cytotoxic. However, such stalled replication sometimes can be resumed without repairing the damage. This is usually error-prone and leads to mutations. Many mutagens and carcinogens (cancer-causing agents) are electrophiles that act by attacking DNA and alkylating it. The common laboratory mutagen ethylmethane sulfonate (EMS) is one of the alkylating agents. It usually alkylates the O6 of the guanine (Fig. 2.25). Because it transfers ethyl (CH_3CH_2) groups to O6 of guanine, it changes the tautomeric form of guanine. It then base-pairs naturally with thymine. This leads to the replacement of a G-C base pair by an A-T base pair. Because it changes the base-pairing properties of a base, it is mutagenic, and thus genotoxic.

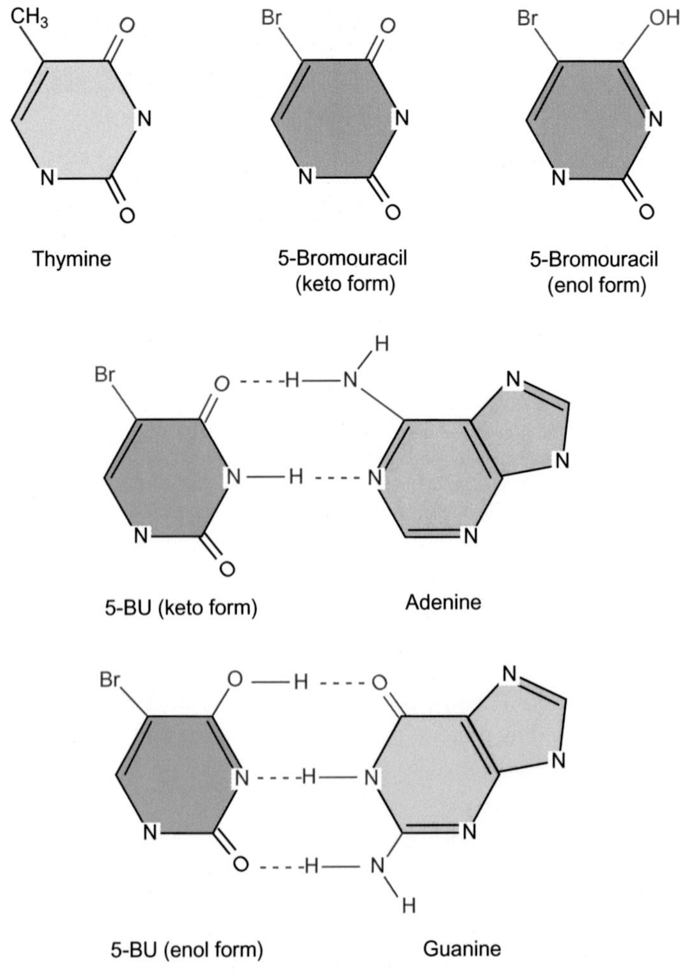

FIG. 2.23 Structural similarity. 5-BU is structurally similar to thymine, therefore it is considered a thymine analog and base-pairs normally with adenine. It is rarely present in its enol form, in which it base pairs with guanine.

Thymine

5-Bromouracil (keto form)

5-Bromouracil (enol form)

5-BU (keto form)   Adenine

5-BU (enol form)   Guanine

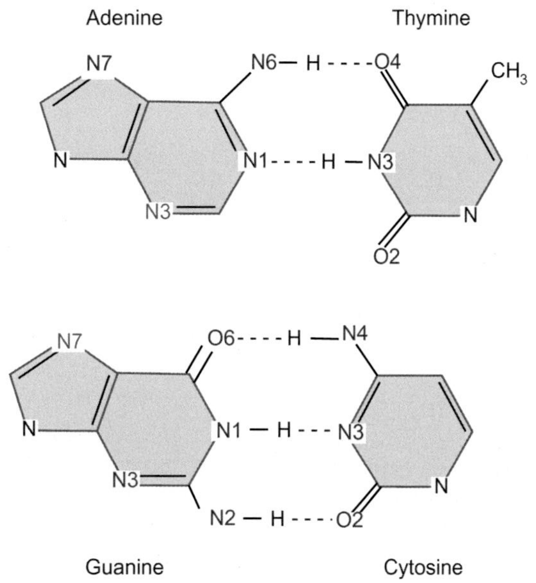

FIG. 2.24 Electron-rich DNA centers are commonly attacked by electrophiles. Phosphodiester bonds are most commonly attacked by electrophiles. In addition, other targets are shown in red, as well as N7 of guanine and N3 of adenine, which are shown in blue.

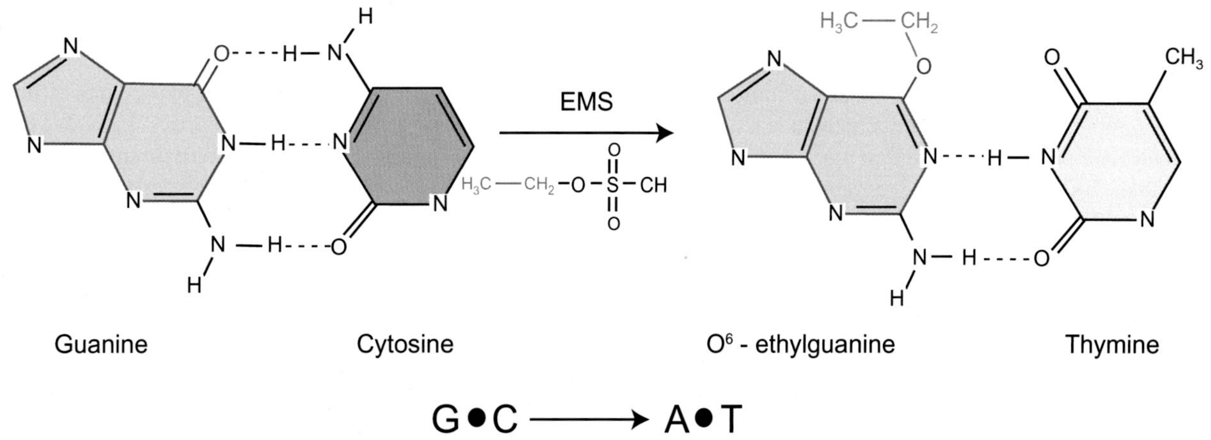

$$G \bullet C \longrightarrow A \bullet T$$

FIG. 2.25 Alkylation of guanine by ethylmethane sulfonate (EMS). A normal G-C pair is shown on the left. EMS will donate an ethyl group (denoted in *blue*) to the free oxygen (*red*) of the guanine. This allows it to base pair with thymine rather than cytosine. A-T will replace this after one more replication.

UV Radiation

UV radiation cross-links adjacent pyrimidines on the same DNA strand, forming two major lesions. The vast majority of the lesions are pyrimidine dimers in which the two carbons on each of the two pyrimidines form a cyclobutyl ring (Fig. 2.26). Only 10%–20% of the lesions are photoproducts, in which the 6-carbon of one pyrimidine is linked to the 4-carbon of an adjacent pyrimidine. The shape of the DNA of both produces is distorted, and thus, both replication and transcription are interfered with.

Ionizing Radiation

Ionizing radiation (including gamma rays and X-rays) can interact with the DNA molecules in a similar manner to UV rays. However, they usually cause damage by ionizing the molecules surrounding the DNA especially water. This forms free radicals which are extremely reactive, especially those containing oxygen. As a result, they can attack the neighboring DNA molecules, this can change a base or it can cause a single- or double-stranded break. One of the common oxidatively damaged DNA bases is 8-oxoguanine (oxoG). DNA polymerase in bacteria and eukaryotes misreads oxoG as thymine and

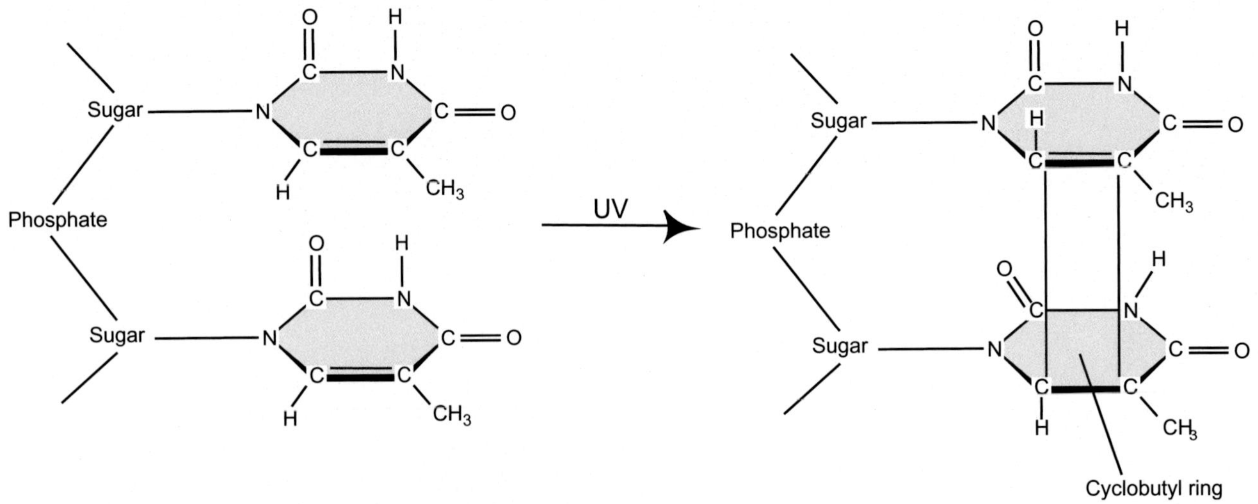

FIG. 2.26 Formation of thymine dimers. Ultra-violet radiation exposure can cause the formation of adjacent thymine dimers of carbon 5 and 6 in pyrimidine rings. This disrupts normal base pairing.

inserts adenine instead of cytosine, resulting in an oxoG-A pair. Because they will lead to mutations if they are not removed before the DNA replication happens again, both bases in this pair are genotoxic.

Byproducts Damage

In addition to the aforementioned DNA damage mutagens, DNA may also suffer damage from the byproducts of normal cellular processes. These byproducts include reactive oxygen species such as superoxides (O_2^-), hydroxyl radicals ($\bullet$OH), and hydrogen peroxide (H_2O_2) that are generated during normal aerobic respiration. These reactive oxidants can produce more than 100 different types of chemical modifications in DNA, including loss of base and single-strand breaks. Therefore, they are considered genotoxic.

DAMAGE REPAIR: SINGLE-STRAND DAMAGE REPAIR

When DNA damage occurs, prokaryotes have a variety of repair mechanisms to reverse the damage. There are two basic ways to do this: (1) directly undo the damage, or (2) remove the damaged section of DNA and fill it in with new, undamaged DNA.

DNA Repair Through Undoing the Damage

In *E. coli*, two methods have been used to undo the DNA damage. UV radiation cross-links adjacent pyrimidines and forms pyrimidine dimers. An enzyme called photoreactivating enzyme, or photolyase, can be activated by absorbing energy from near-UV to blue light to break the bond holding the two pyrimidines together (Fig. 2.27). This restores the pyrimidines to their original independent state. Because it requires the presence of light, it is also called photoreactivation or light repair. The other type of DNA damage caused by alkylation of the O6 of guanine can be revered by O6-methylguanine methyltransferase. This enzyme will accept the alkyl group through its sulfur atom of a cysteine residue. Once the alkyl group is transferred to the enzyme, this enzyme becomes inactivated. As such, it is also called suicide enzyme. Although these two reversal methods are very common in most forms of life, placental mammals, including humans, do not have a photoreactivation pathway.

Excision Repair System

There are two types of excision repair systems, base excision repair and nucleotide excision repair. In base excision repair, a base that has been damaged by oxidation or chemical modification can be recognized by DNA glycosylase, which distorts

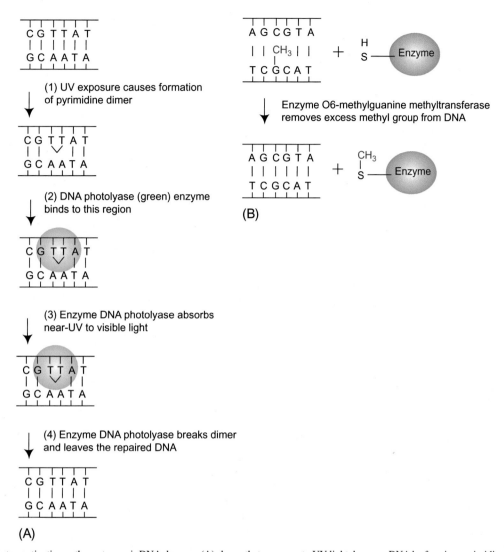

FIG. 2.27 Photoreactivation pathway to repair DNA damage. (A) shows that exposure to UV light damages DNA by forming pyrimidine dimers. These dimers are repaired by enzyme DNA photolyase. (B) depicts the transfer of a methyl group from guanine to the sulfhydryl group of enzyme O6-methylguanine methyltransferase.

the DNA in such a way as to extrude the damaged base out of its association with its base-paired nucleotide, then breaks the glycosidic bond between the damaged base and its sugar (Fig. 2.28). This leaves an AP site, which is an apurinic or apyrimidinic site (without purine or pyrimidine). An AP endonuclease, an exonuclease removes the deoxyribose-P and a number of additional residues, then cuts or nicks the DNA strand on the 5′-side of the AP site. Subsequently, DNA phosphodiesterase removes the sugar and phosphate from the nucleotide. Then, DNA polymerase I performs repair synthesis by degrading DNA in the 5′→ 3′ direction, while filling in with new DNA. Finally, DNA ligase seals the nick in the phosphodiester backbone.

Nucleotide excision repair is common for DNA lesions caused by UV radiation, which distorts DNA structure. In this pathway, the incising enzyme system recognizes the strand with the bulky damage and makes cuts on either side, removing an oligonucleotide containing the damage (Fig. 2.29). In *E. coli*, uvrABC endonuclease is the key incision enzyme system responsible for cutting the damaged DNA and producing an oligonucleotide that is 12–13 bases long. Normally, UvrA and UvrB scan DNA for DNA damage in an ATP-dependent manner. The DNA strands at the site of damage are separated by the helicase activity of UvrB, and UvrB stably associates with the DNA, while UvrA dissociates. UvrC, which has endonuclease activity, is then recruited to the UvrB-DNA complex, and cuts either side of the lesion through the nicks on both 5′

FIG. 2.28 Base excision repair. This multi-step process involves excising a damaged nucleotide and putting a normal corresponding nucleotide in its place.

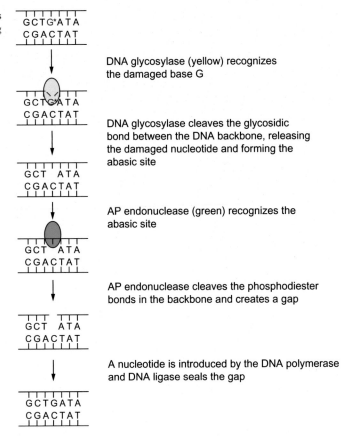

DNA glycosylase (yellow) recognizes the damaged base G

DNA glycosylase cleaves the glycosidic bond between the DNA backbone, releasing the damaged nucleotide and forming the abasic site

AP endonuclease (green) recognizes the abasic site

AP endonuclease cleaves the phosphodiester bonds in the backbone and creates a gap

A nucleotide is introduced by the DNA polymerase and DNA ligase seals the gap

and 3′ side of the lesion. The action of UvrD releases oligonucleotide, UvrC, and UvrB. In eukaryotes, the excision nucleases, excinuclease, removes an oligonucleotide about 24–32 base pair long. In any case, DNA polymerase I fills in the gap left by the excised oligonucleotide, and DNA ligase seals the final nick.

Postreplication Repair and SOS Repair System

Both direct reversal and excision repair systems are true repair processes because they remove defective DNA entirely. However, cells sometimes do not eliminate the defective DNA entirely but simply skirt around it. This phenomenon is referred to as a damage bypass mechanism, and includes recombination repair and error-prone bypass.

The recombination repair is also called post-replication repair because it repairs the pyrimidine dimer that passed the replication. The intact sister chromatid of the damaged DNA duplex guides the process. First, the DNA is replicated (Fig. 2.30). Because the replication machinery stops at the pyrimidine dimer and continues after a pause, the daughter strand has a gap across from the dimer. Next, recombination occurs between the gapped strand and its homolog on the other daughter DNA duplex. This recombination depends on the RecA, which exchanges the homologous DNA strands. In the end, the gap across from the pyrimidine dimer is filled in, but the DNA damage is still there. Furthermore, the new gap that results from recombination can be easily filled in by DNA polymerase and ligase.

When bacteria are subjected to extreme conditions, and a great deal of DNA damage occurs, it will have an SOS response, a repair process that involves error-prone repair. The SOS response is a multifaceted cellular response in which more than 40 genes are induced when DNA is damaged. In this response, the RecA coprotease activity is activated by the UV damage or another mutagenic treatment followed by the cleavage of a transcriptional repressor, LexA repressor. As a result of the inactivity of the repressor, DNA polymerase V (Pol V) is formed and becomes active. Pol V can cause error-prone bypass of the lesions by synthesizing the nucleotides across from the DNA lesions even though the damaged region cannot be read correctly. Pol V can efficiently bypass the three most common types of DNA lesions: pyrimidine dimers, UV lights, and AP sites.

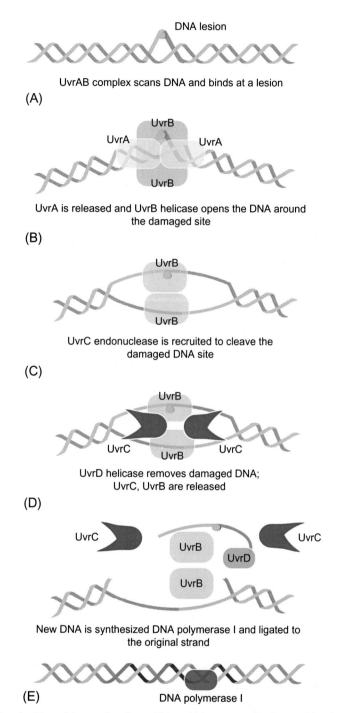

FIG. 2.29 An example of *E. coli* nucleotide excision repair pathway. This multi-step process involves excising damaged DNA and synthesizing new, normal corresponding DNA in its place.

DAMAGE REPAIR: DOUBLE-STRAND BREAK REPAIR

When both strands of DNA are broken, the result is called a double-stranded break (DSB). DSB is the most dangerous form of DNA damage. While exogenous (from sources outside the body) of DSBs are induced by ionizing radiation, endogenous DSBs (from sources within the body) arise as byproducts of normal intracellular metabolism. The spontaneous rate of endogenous DSBs may be as high as 50 breaks per cell per cell cycle. The creation of a DNA DSB represents the principal

FIG. 2.30 Recombination repair. This multi-step process involves the exclusion of damaged DNA (for example, pyrimidine dimers) in DNA replication, resulting in normal daughter strands.

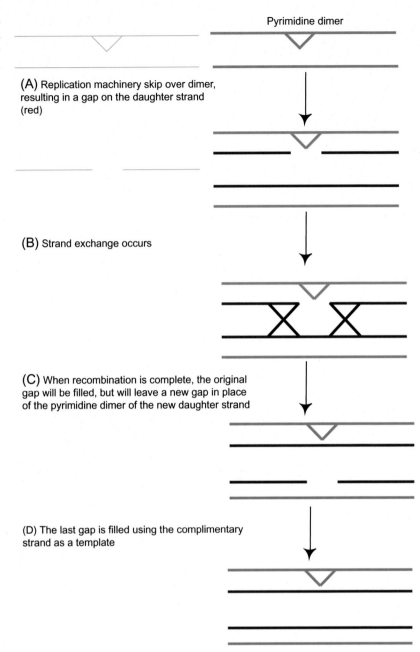

Pyrimidine dimer

(A) Replication machinery skip over dimer, resulting in a gap on the daughter strand (red)

(B) Strand exchange occurs

(C) When recombination is complete, the original gap will be filled, but will leave a new gap in place of the pyrimidine dimer of the new daughter strand

(D) The last gap is filled using the complimentary strand as a template

lesion that can lead to cell death via the generation of lethal chromosomal aberrations or the direct induction of apoptosis (programmed cell death). Alternatively, an inaccurately repaired or unrepaired DSB may result in mutations or genomic rearrangements in a surviving cell, which, in turn, can lead to genomic instability and subsequently result in malignant cell transformation.

Two principal recombinational repair pathways, homologous recombination (HR) and nonhomologous end-joining (NHEJ), which employ entirely separate protein complexes, can protect the cell from the potentially deleterious effects of a DSB. Briefly, DSB repair by HR requires an undamaged template molecule that contains a homologous DNA sequence, typically on the sister chromatid in the S and G2 phases of the cell cycle. In contrast, nonhomologous rejoining of two double-stranded DNA ends, which may occur in all cell-cycle phases, does not require an undamaged partner and does not rely on extensive homologies between the recombining ends. These two mechanisms will come into play under the situation when a cell has not performed real repair of a lesion as mentioned above and continues to replicate its DNA.

Homologous Recombination

The general process of HR repair begins when the enzyme recognizes the double strand break and then digests back the 5′ ends of the broken DNA helix, leaving overhanging 3′ ends (Fig. 2.31). One overhanging end searches for a region of sequence complimentary on the sister chromatid and then invades the homologous DNA duplex, aligning the complementary sequences. Once aligned, DNA synthesis proceeds from the 3′ overhanging ends, using the undamaged DNA strands as template. After DNA repair synthesis, the resulting heteroduplex is resolved, and the two chromatids separate.

Homologous recombination is also called general recombination because the enzymes that mediate the exchange can use essentially any pair of homologous DNA sequences. Recombination occurs by the breakage and reunion of DNA strands, so that physical exchange of DNA parts takes place. Recombination does not occur randomly around a chromosome. There are some areas of a chromosome much more likely to show recombination, and these zones are called hot spots.

In *E. coli*, the RecBCD homologous recombination pathway is well-studied. This recombination process (Fig. 2.32) begins with the induction of a double-strand break. The RecBCD protein binds to a DNA double helix-stranded break and uses its DNA helicase activity to unwind the DNA toward a so-called Chi site which has been found on average every 5 kb in the *E. coli* genome. RecBCD protein can produce a single-stranded tail and which is then coated by RecA protein, which allows the tail to invade a double-stranded DNA duplex and search for a region of homology. This creates a

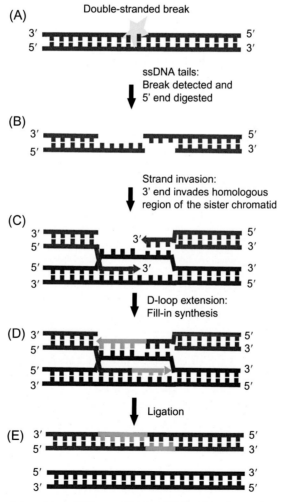

FIG. 2.31 Overview of DSB repair through homologous DNA recombination. (A) Two homologous chromatids; (B) a double-stranded break in one chromatid; (C) a D-loop is formed through homologous recombination; (D) steps for sister chromatid-directed DNA replication to restore the damaged area; (E) the products of noncrossover (bottom) or crossover (top) recombinants.

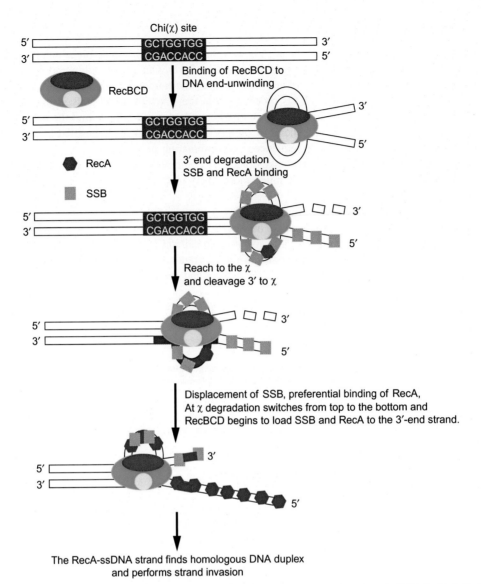

FIG. 2.32 Model of RecBCD-dependent recombination. (A) RecBCD binds to a duplex DNA end unwinds the DNA double helix. The appearance of "Rabbit ears loop" of ssDNA is because the rate of DNA unwinding exceeds the rate of ssDNA release by RecBCD. (B) As the DNA is unwound, SSB binds to single-stranded regions and the RecBCD cleaves the ssDNA. (C) RecBCD continues to cleave the 3′-end of c site, recombination hotspot site. (D) RecBCD directs the binding of RecA to the 3′-terminal strand. (E) A nucleoprotein filament consisting of RecA-coated 3′-strand ssDNA is formed for homologous pairing with s dsDNA and strand invasion.

displacement loop (D-loop). Once the tail finds a homologous region, a nick occurs in the D-looped DNA, possibly with the help of RecBCD. This nick allows RecA and SSB to create a new tail that can pair with the gap in the other DNA. DNA ligase seals both nicks to generate a Holiday junction, which is also called half chiasmas and Chi structures. The branch in the Holiday junction can migrate in either direction simply by breaking old base pairs and forming new ones in branch migration process.

Nonhomologous End Joining (NHEJ)

The process of NHEJ occurs simply by the rejoining of broken DNA ends and does not require another template duplex such as sister chromatid, which is only available in replicating DNA. This repair is an error-prone repair because of a lack of a template. It can occur any time in the cell cycle; therefore, NHEJ is the only readily available repair pathway for cells in G1 phase of the cell cycle. Because most mammalian cells spend the majority of the time in G1, NHEJ is the predominant mode of double-strand break repair in mammalian cells.

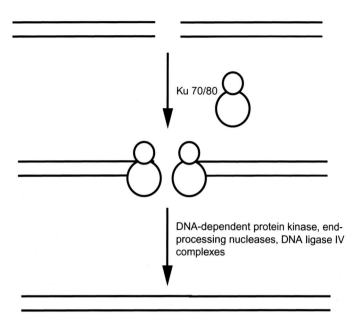

FIG. 2.33 DSB repair through nonhomologous DNA end joining (NHEJ). Ku70/80 binds the ends and recruits a set of proteins that juxtaposes the broken ends. Processing of the ends to generate proper substrates for DNA ligase IV then occurs, followed by DNA-ligased-mediated end joining.

The NHEJ process in bacteria and eukaryotes requires a heterodimeric protein Ku70/80 which binds to the broken ends of the DNA to protect the DNA ends from degradation and to recruit several other proteins to promote the synapsis of the region with homology (Fig. 2.33). Subsequently, Ku's helicase activity is activated through phosphorylation which then can promote the unwinding of the DNA ends followed by the base-pairing of the homology region in the DNA ends, leaving flaps composed of the ends of the other nonpairing strand. The phosphorylation is activated by the binding of a protein kinase, DNA-PKcs, to Ku70/80. Finally, the flaps are removed by nucleases, gaps are filled in, and the DNA strands are ligated together. Ku is conserved from bacteria to mammals. By contrast, DNA-PKcs seems to be found only in vertebrates and may play a part in the relatively high frequency of repair by NHEJ in mammals. DNA-PKcs can also activate cell cycle arrest and programmed cell death (apoptosis) in mammals.

Chapter 3

Gene Expression: Transcription of the Genetic Code

Chapter Outline

GENERAL FEATURES OF RNA SYNTHESIS

Template Strand

Of the two strands of DNA, one of them is the template for RNA synthesis of a particular RNA product. The enzyme that catalyzes the process is DNA-dependent RNA polymerase. RNA polymerase reads the template from 3′ to 5′ (Fig. 3.1). All four ribonucleotides (ATP, GTP, CTP, and UTP) are required. The strand that is used as template is called template strand, because it directs the synthesis of the RNA. It is also called the antisense strand, because its code is the complement of the RNA that is produced. It is sometimes called the (−) strand by convention.

Coding Strand

The other strand is called the coding strand, because its sequence is the same as the RNA sequence that is produced, with the exception of U replacing T. It is also called sense strand, because the RNA sequence is the sequence that we use to determine what amino acids are produced through mRNA. It is also called (+) strand, or nontemplate strand. As the RNA polymerase moves along the template strand in 3′→5′ direction, the RNA chain grows in 5′→3′ direction. The nucleotide at the 5′ end of the chain retains its triphosphate group. Unlike DNA replication, a primer is not needed in RNA synthesis. The RNA synthesized by the RNA polymerase is, therefore, named messenger RNA (mRNA) for its role in carrying a copy of the genetic information to the ribosome.

Diagnostic Molecular Biology. https://doi.org/10.1016/B978-0-12-802823-0.00003-1

FIG. 3.1 The overview of gene expression. When RNA polymerase binds to DNA, it uses one strand as a template to create a corresponding RNA transcript. This transcript has the same code as the non-template strand. However, thymine (T) is replaced by uracil (U). If the transcript is mRNA, it may be transcribed into protein.

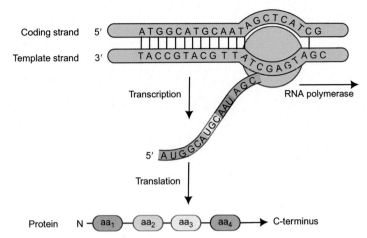

PROMOTERS

Promoters are DNA sequences that provide signal for RNA polymerase, and they are where RNA polymerase binds. By convention, the binding site for polymerase is upstream of the transcription start site, which means to the 5' side of the coding strand and to the 3' side of the template strand. The promoter region is given based on the coding strand, even though the RNA polymerase is actually binding to the template strand.

Normally, the start of transcription, which is the first base to be incorporated into the RNA chain, is said to be at position +1 and is called the transcription start site (TSS). All the nucleotides upstream from this start site are given a negative number. Sites located 3' to TSS are said to be located downstream and are given a positive number.

Promoter Structure—Consensus Sequences

Most bacterial promoters have at least three components: TSS, a -10 box, and a -35 box. The -10 box is also called a Pribnow box. The area from the -35 box to the TSS is called the core promoter, so both -10 box and -35 box are also called core promoter elements. In *E. coli*, both -10 box and -35 box bear a greater or lesser resemblance to two consensus sequences: TATAAT and TTGACA, respectively (Fig. 3.2).

The consensus sequence represents the probabilities of a specific base being found in the given position. In general, as a promoter sequence varies from the consensus sequence, the binding of RNA polymerase becomes weaker, and thus, there is less frequency of transcription. A strong promoter usually produces more transcript than a weak promoter, and the perfect match of the consensus sequence can be found in the strong promoter. The spacing between promoter elements is also important. Any deletions or insertions that move the -10 and -35 boxes unnaturally close together or far apart are detrimental to the transcription efficiency.

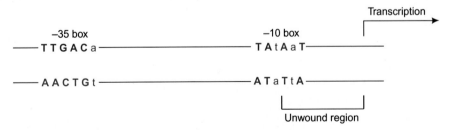

FIG. 3.2 The typical feature of bacterial promoter. The -10 and -30 boxes of a bacterial promoter as well as the unwound region in respect to the origin of transcription in an *E. coli* bacterial promoter is shown. Bases shown in bold capital blue letters are found in these positions in over 50% of all promoters examined.

Promoter Structure—UP Element

Some very strong promoters have an additional element farther upstream called an UP element, which enhances the binding of RNA polymerase (Fig. 3.3). UP elements usually extend from -40 to -60. As the σ subunit of RNA polymerase (please see next section) recognizes the core promoter, the α subunit is responsible for recognizing the UP element. The region from the end of the UP element to the TSS is also known as the extended promoter.

RNA POLYMERASE IN *ESCHERICHIA COLI*

Bacterial RNA polymerase is the smallest and simplest of the cellular RNA polymerases. The molecular weight of the RNA polymerase is about 470,000 Da. It has a multisubunit structure which contains five different types of subunits, including α_2, ω, β, β′, and σ. In other words, two molecules of α and one of all the others are present on the RNA polymerase.

The Content of the Holoenzyme

The content of RNA polymerase holoenzyme is σ subunit and a core enzyme which includes α_2, ω, β, and β′ subunits. As we described in the previous section, the σ subunit is involved in the recognition of specific promoter, the core promoter which is a DNA sequence that signals the start of RNA transcription. At the heart of all polymerase enzymes is a set of subunits called the core enzyme. The core enzyme of RNA polymerase is catalytically active, which is responsible for the polymerization. As such, depriving the holoenzyme of its σ subunit leaves a core enzyme with basic RNA synthesizing capability, but lacking specificity. Adding σ subunit back can restore its specificity. The loosely bound σ subunit is released after transcription begins and about 10 nucleotides have been added to the RNA chain (Fig. 3.4).

The Role of Sigma (σ) in Transcription

Sigma factor can direct the tight binding of RNA polymerase to promoters and place the enzyme in a position to initiate transcription so that it can stimulate initiation of transcription. Because initiation is a rate-limiting step in transcription, the σ factor can also stimulate elongation because stimulation of initiation can provide more initiated chains for core polymerase to elongate.

After the start of the transcription, the σ factor dissociates from the core polymerase at the beginning of elongation or later stage, leaving the core enzyme to carry out elongation (Fig. 3.5). The σ factor that had been released from the original holoenzyme can then associate with a new core enzyme to initiate another round of transcription. Therefore, the σ factor can be reused again and again, and this is known as the σ cycle.

THE PROCESS OF TRANSCRIPTION

Initiation

The process of transcription is usually broken down into three phases: initiation, elongation, and termination. There are usually four steps in the transcription initiation: (1) formation of a closed promoter complex, (2) conversion of the closed promoter complex to an open promoter complex, (3) polymerizing the first few nucleotides (up to 10) while the polymerase remains at the promoter, and (4) promoter clearance, in which the transcript becomes long enough to form a stable hybrid with the template strand (Fig. 3.6). This helps to stabilize the transcription complex, and the polymerase moves away from the promoter and the transcription starts the elongation stage. Below is the detail of these four steps in the initiation stage.

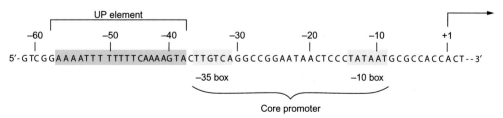

FIG. 3.3 An example of core promoter and UP element. The UP element (*green*) as well as core promoter elements (-10 box and -35 box, *yellow*) are depicted with complete base sequences.

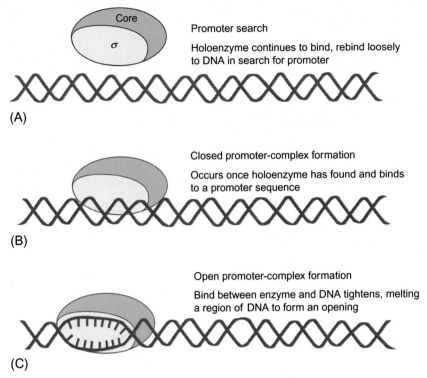

FIG. 3.4   A schematic diagram for RNA polymerase to locate, bind and open a sequence of DNA at the promoter.

FIG. 3.5   Bacterial RNA polymerase holoenzyme can recognize the core promoter. The sigma subunit of RNA polymerase allows it to bind to a specific promoter sequence (*yellow*) in DNA. The alpha-c-terminal domains of the enzyme can then contact the UP element depicted in green.

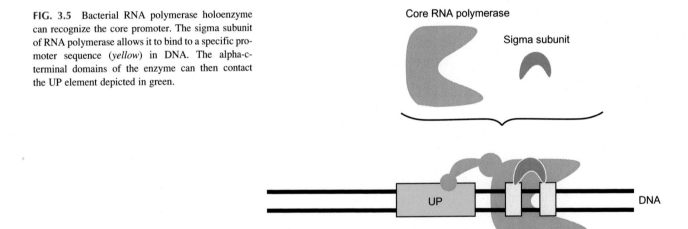

The initiation begins when RNA polymerase binds to the promoter and forms the closed complex. The σ subunit directs the polymerase to the promoter by binding to specific sequences upstream of the start site of transcription. Subsequently, both β′ and σ subunits initiate separation of the two downstream DNA strands, melting about 10–17 bp surrounding the TSS, to form the open complex-transcription bubble which is also known as the transcription bubble (Fig. 3.7).

Next, the polymerase starts building the RNA chain. The first base in RNA is a purine ribonucleoside triphosphate and A tends to occur more often than G (Fig. 3.8). After the first nucleotide is in place, the polymerase joins a second nucleotide to the first, forming the initial phosphodiester bond in the RNA chain. Both β and β′ subunits are involved in phosphodiester bond formation as well as DNA binding. Several nucleotides may be joined before the polymerase leaves the promoter and elongation begins. Within the transcription bubble are the nine most recently added ribonucleotides of the RNA transcript,

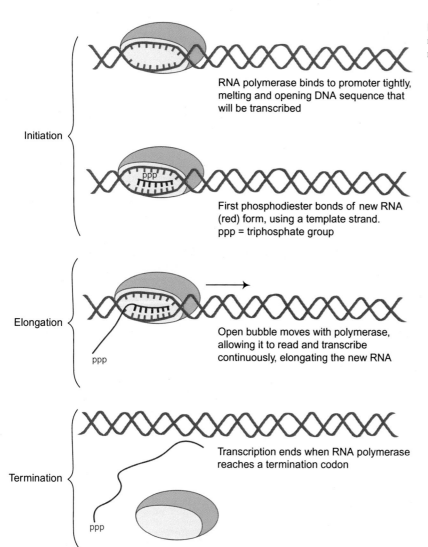

FIG. 3.6 A schematic diagram of RNA polymerase binding to and transcribing DNA to produce new single-stranded RNA.

Initiation

RNA polymerase binds to promoter tightly, melting and opening DNA sequence that will be transcribed

First phosphodiester bonds of new RNA (red) form, using a template strand. ppp = triphosphate group

Elongation

Open bubble moves with polymerase, allowing it to read and transcribe continuously, elongating the new RNA

ppp

Termination

Transcription ends when RNA polymerase reaches a termination codon

ppp

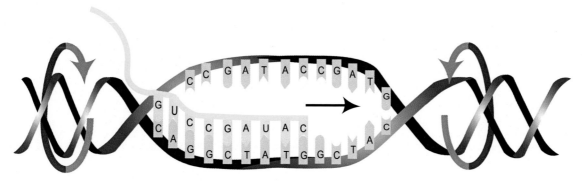

FIG. 3.7 A typical example of transcription bubble which consists of 14 unpaired bases, the first 9 have been used to transcribe a new singled stranded RNA.

FIG. 3.8 Steps of transcription initiation and elongation. (A) Sigma subunit of RNA polymerase recognizes the promoter and allows the enzyme to bind to the DNA. (B) Formation of an RNA polymerase closed-promoter complex. (C) At the promoter, DNA unwinds and open-promoter complex forms. (D) RNA polymerase synthesizes mRNA, almost always begins with a purine. (E) RNA polymerase holoenzyme catalyzes elongation of mRNA by around 4 nucleotides. (F) The sigma subunit is released as the core proceeds for elongation of the RNA transcript.

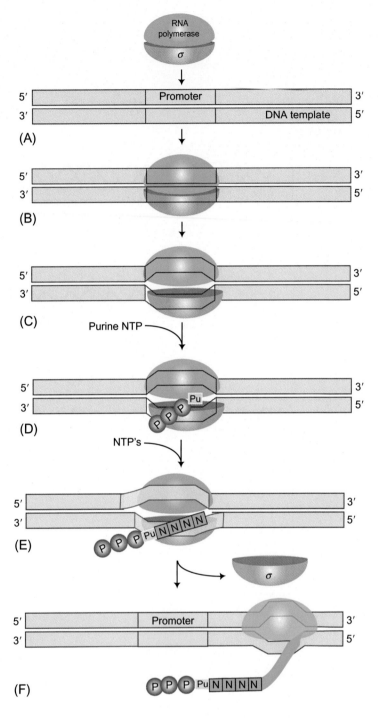

which remain base-paired to the template strand. The association of the polymerase with the RNA-DNA hybrid within the transcription bubble contributes to the stability of the elongation complex.

All cells have a primary sigma factor which directs transcription from the promoters of gene encoding essential proteins that are needed by growing cells. These genes are usually referred to as housekeeping genes. On the other hand, bacteria also have a variety of alternative sigma factors. These alternative sigma factors' levels or activities regulated by the response to specific signals or stress conditions. Therefore, these alternative sigma factors direct transcription of genes that are only required under certain conditions. We will discuss the alternative sigma factor later.

Elongation

During the elongation phase, RNA polymerase directs the sequential binding of ribonucleotides to the growing RNA chain in the $5' \rightarrow 3'$ direction, while the RNA polymerase and transcription bubble move along the template DNA in $3' \rightarrow 5'$ direction. As the RNA polymerase moves along the template DNA, the transcription bubble also moves with it. This melted region exposes the bases of the template DNA one by one so they can pair with the bases of the incoming ribonucleotides.

When about 9-10 nucleotides have been incorporated, the σ subunit dissociates from the holoenzyme and is later recycled to bind to another core enzyme for another initiation process. The core enzyme continues to elongate the RNA, adding one nucleotide after another to the growing RNA chain. The core subunit β lies near the active site of the RNA polymerase where phosphodiester bonds are formed, and involves the bond formation.

As the RNA polymerase travels along the template DNA, the polymerase maintains a short melted region of template DNA. This requires that the DNA unwind ahead of the advancing polymerase and close up again behind it. During this process, positive supercoiling is produced ahead of the transcription bubble, and negative supercoiling is produced behind the transcription bubble (Fig. 3.9). As such, topoisomerases come in to relax the supercoils in front of and behind the advancing transcription bubble. As soon as the transcription machinery passes, the two DNA strands wind around each other again, reforming the double helix.

The process of elongation is far from uniform and steady. Instead, RNA polymerase frequently pauses, or even backtracks, while elongating an RNA chain. Pausing is physiologically important for two reasons: first, it allows translation to keep pace with the RNA polymerase. This is important for attenuation and aborting transcription if translation fails. Attenuation is a mechanism to regulate the expression by causing premature termination of transcription of the operon when the operon's products are abundant. The second important aspect of pausing is that it is the first step in transcription termination. The backtracking of RNA polymerase aids proofreading by extruding the 3'-end of the RNA out of the polymerase, where misincorporated nucleotides can be removed by an inherent nuclease activity of the polymerase, stimulated by auxiliary factors.

Termination

Termination of RNA transcription involves specific sequences downstream of the actual gene for the RNA to be transcribed. There are two types of termination mechanism—intrinsic termination and rho (ρ)—dependent termination. For

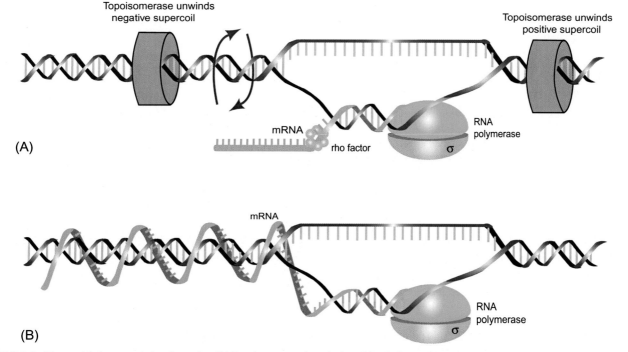

FIG. 3.9 Two models for transcription elongation. (A) Topoisomerase relaxes both positive (before replication bubble) and negative supercoiling (after) of DNA. (B) RNA polymerase follows the template strand around the duplex, so that it can avoid any supercoiling of DNA.

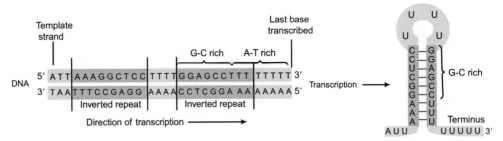

FIG. 3.10   Intrinsic terminator sequence features—inverted repeats and a series of uracils.

the intrinsic termination, regions at the end of genes called terminator or termination sites can signal the termination of transcription. The termination sites are characterized by two inverted repeats spaced by a few other bases (Fig. 3.10). The DNA then encodes a series of uracils. When the RNA is created, the inverted repeats form a hairpin loop. This tends to stall the advancement of RNA polymerase. Meanwhile, the presence of the uracils causes a series of A-U base pairs between the template strand and the RNA. Because A-U base pairs between the template strand and the RNA are weakly hydrogen-bonded compared with G-C pairs, the RNA dissociates from the transcription bubble and ends the transcription.

The ρ-dependent termination involves an inverted repeat, so it also causes a hairpin loop to form but no string of U. In this case, the ρ protein binds to the RNA and chases the polymerase (Fig. 3.11). When the polymerase transcribes the RNA that forms a hairpin loop, it stalls, giving the σ protein opportunity to catch up. When the σ protein reaches the termination site, it facilitates the dissociation of the transcription machinery. The result of the dissociation terminate the transcription.

THE REGULATION OF PROKARYOTIC TRANSCRIPTION

The *E. coli* genome has more than 3000 genes. Some of these are active all the time because their products are in constant demand. But some of them are not transcribed most of the time because their products are rarely needed. The reason that the cell does not leave all its genes on all the time is because gene expression is an expensive process. It takes a lot of energy to produce RNA and protein. Thus, control of gene expression is essential to life. In prokaryotes, transcription is controlled by seven major strategies: alternative σ factors, RNA polymerase switching, antitermination, enhancers, operon, transcription attenuation, and riboswitch. Alternative σ factors, RNA polymerase switching, and antitermination involve the change of transcription machinery while operon, transcription attenuation and riboswitch involve the regulation of groups of functionally related genes. Enhancers are the DNA elements that bind protein factors.

Alternative σ Factors

When a phage infects a bacterium, it usually subverts the host's transcription machinery to its own use. In the process, it establishes a time-dependent, or temporal, program of transcription by producing different σ subunits that direct the RNA polymerase to different genes. SPO1 is a virus that infects the bacteria *Bacillus subtilis*. The virus has a set of gene called early genes, which are transcribed by the host's RNA polymerase, using its regular σ subunit (Fig. 3.12). One of the viral early genes codes for a protein gp28, which is a σ subunit. This σ subunit directs the RNA polymerase to transcribe more of viral genes in the middle phase. Products of two middle genes are gp33 and gp34, which together make up another σ subunit that directs the transcription of the late genes. In the end, viral σ subunits control the transcription machinery for the virus instead of the bacterium.

When bacterial cells experience an increase in temperature, or a variety of other environmental insults, they mount a defense response called the heat shock response to minimize damage. A set of heat shock genes are expressed to protect the cells. In this situation, the heat shock response in *E. coli* is governed by an alternative σ subunit, σ^{32} (σ^{H}) which displaces σ^{70} (σ^{A}) and directs the RNA polymerase to the heat shock gene promoters.

In addition to the alternative σ subunit mechanism, *E. coli* has evolved ways of controlling transcription using anti-σ-subunit. When cells are stressed by such insults as loss of nutrients, high osmolarity, or high temperature, they stop growing and enter the stationary phase. At this point, an Rsd protein is produced. It can bind to σ^{70} subunit and block its binding to the core polymerase. Thus, this anti-σ-subunit can supplement the σ replacement mechanism by inhibiting the activity of σ^{70} subunit activity.

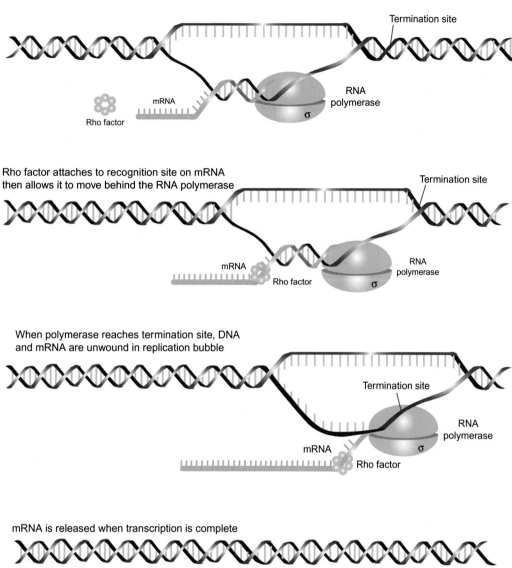

FIG. 3.11 Rho-dependent transcriptional termination.

RNA Polymerase Switching

Phage T7 is a simple *E. coli* phage, and it has three phases of transcription: an early phase called Class I, and two late phases called Classes II and III. Instead of making a new σ subunit, a specific Class I gene encodes a new RNA polymerase with absolute specificity for the Class II and III genes of phage T7 (Fig. 3.13). Thus, the switching mechanism in this phage is quite simple.

Antitermination

Phage λ can replicate in either lytic or lysogenic cycle. In the lytic mode, almost all of the phage genes are transcribed and translated, and the phage DNA is replicated, leading to new progeny. In the lysogenic mode, the λ DNA is incorporated into the host genome. Only *cI*, which encodes λ repressor protein, is expressed and then prevents transcription of all the rest of the phage genes. The immediate early, delayed early, or late transcriptional switching in the lytic cycle of λ phage is

FIG. 3.12 An example of temporal control of transcription in phage SPO1-infected *B. subtilis*. (A) RNA polymerase holoenzyme (*green*) including sigma-subunit (*light blue*) directs early transcription. New phage protein product gp28 (*purple*) is shown, which acts as the new sigma factor. (B) New protein product gp28 directs middle transcription. Two middle phage proteins gp33 and gp34 (*pink* and *green*) are produced. (C) Late transcription occurs depending on the host core polymerase and gp33 and gp34.

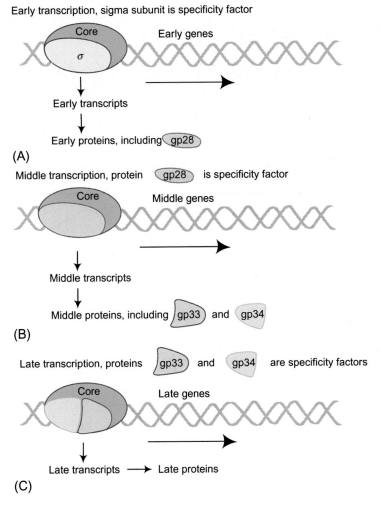

FIG. 3.13 Temporal control of transcription in phage T7-infected *E. coli*. (A) *E. coli* RNA polymerase holoenzyme and sigma subunit control early (class I) transcription (B) T7 RNA polymerase controls late (class II and III) transcription.

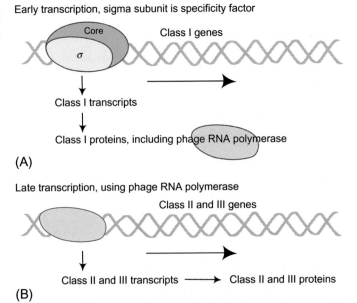

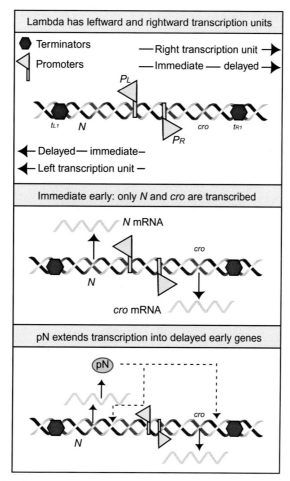

FIG. 3.14 Transcription units in phage lambda genes N and cro are immediate early genes and are separated from delayed early genes by terminators. Protein N (pN) allows RNA polymerase to pass the terminators t_{L1} to the left and t_{R1} to the tight.

controlled by antitermination. One of the two immediate early genes is cro, which codes for a repressor of the cI gene that allows the lytic cycle to continue (Fig. 3.14). The other, N, codes for an antiterminator, N, that overrides the terminators after the N and cro genes. Transcription then continues into the delayed early genes. One of the early delayed genes codes for another antiterminator that permit transcription of the late genes. These antiterminators act to alter the polymerase so it reads through the terminators to continue to the transcription of genes in the next phase.

Operons

In prokaryotes, genes that encode enzymes of certain metabolic pathways are often controlled as a group, with the genes encoding the proteins of the pathway being close together and under the control of a common promoter. Such a group of contiguous, coordinately controlled genes is called an operon (Fig. 3.15). Usually these genes are not transcribed all the time. The expression of these genes can be triggered by the presence of a suitable substance called inducer. This phenomenon is called induction. Because all genes in an operon are transcribed together to produce one message, it is also called polycistronic message, which is simply a message with information from more than one gene. Each cistron in the mRNA has its own ribosome binding site, so each cistron can be translated by separate ribosomes that bind independently of each other.

Operons can be controlled by positive or negative regulation mechanism. Depending on how they respond to the molecules that control their repression, they can be inducible, repressible, or both. There are four general possibilities (Fig. 3.16). In a negatively controlled inducible system, a repressor protein stops transcription when it binds to promoter. It is inducible when it releases the repression under the presence of inducer or co-inducer. If the repressor is mutated in some way that stops its function, the operon is always expressed. Genes that are always expressed are called constituted. For a

FIG. 3.15 An overview of transcription of an operon. In an operon, structural genes are located next to each other, and are transcribed into a single polycistronic mRNA. During translation, several proteins that usually have a similar function are produced from this single mRNA.

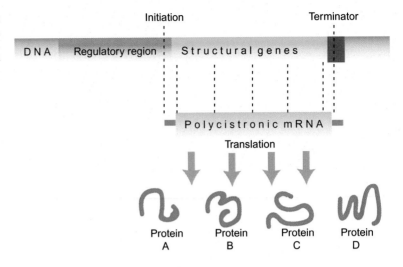

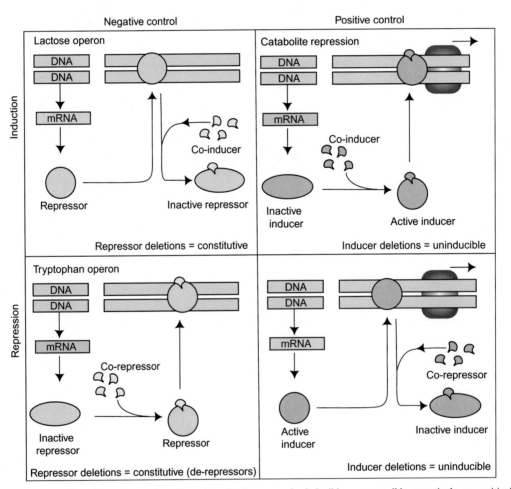

FIG. 3.16 The basic regulation of gene expression. Control mechanisms may be inducible or repressible, negatively or positively controlled. Co-repressors can inactive inducers or active repressors whereas co-inducers can inhibit repressors or stimulate inducers.

positively controlled inducible system, the controlling protein is an inducer that binds to the promoter, stimulating transcription, but it will work only when bound to its co-inducer. If the gene for the inducer is mutated, it becomes uninducible. In a negatively controlled repressible system, a repressor stops transcription, but this repressor functions only in the presence of co-repressor, which can be a protein or small molecule. In a positively controlled repressible system, an inducer protein binds to the promoter, stimulating transcription. However, the inducer is inactivated in the presence of the co-repressor.

Lactose metabolism in *E. coli* is carried out by three structural enzymes. Structural genes encode the gene products that are involved in the biochemical pathway of the operon. The three genes are *lacZ*, which encodes the enzyme β-galactosidase, *lacY*, which encodes the enzyme galactoside permease, and *lacA*, which encodes the enzyme galactoside transacetylase. The *lac* operon is a negatively controlled inducible system. The expression of these structural genes is under the control of a regulatory gene (*lac I*), which encodes the repressor. The repressor can bind to the operator (O). In the absence of lactose, the operator is bound by the repressor, and RNA polymerase cannot bind to the adjacent promoter region, which facilitates the expression of the structural genes. As such, the *lac* operon is repressed as long as no lactose is present, which is a negative regulation (Fig. 3.17).

The *lac* operon becomes inducible when the lactose is the sole carbon source. In the presence of lactose, the inducer-allolactose triggers the expression of *lacZ*. This is because the inducer binds to the repressor, producing an inactive repressor that cannot bind to the operator. RNA polymerase can now bind to the promoter, and transcription and the translation of the structural genes can take place. The product β-galactosidase can break the lactose down into galactose and glucose. On the other hand, galactoside permease can transport the lactose into the cell. Galactoside transacetylase may be responsible for the inactivation of certain antibiotics that may enter the cell through the lactose permease. The *lac* operon is fully active when the lactose is present and the glucose is absent. As we shall see later, the *lac* operon is particularly weak, and RNA polymerase binding is minimal in the presence of glucose and in the presence of lactose. The strong binding of RNA polymerase to the promoter requires another regulatory protein which we will also discuss later.

Enhancers

Enhancers are elements that stimulate transcription and are *cis*-acting DNA elements that are not strictly part of the promoter. They may be found up to several kilobases distant and either upstream or downstream from the promoters they

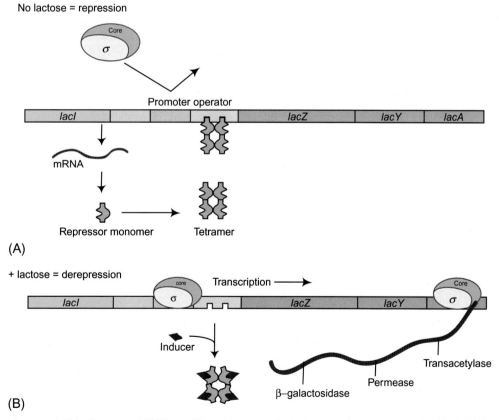

FIG. 3.17 Negative control of the *lac* operon. (A) When no lactose is present, the *lacI* gene produces a repressor that blocks RNA polymerase from transcribing the genes. (B) In the presence of lactose, the inducer binds to the repressor and changes its shape so it can no longer bind to the operator. This allows RNA polymerase to transcribe the *lacZ*, *lacY*, and *lacA* genes.

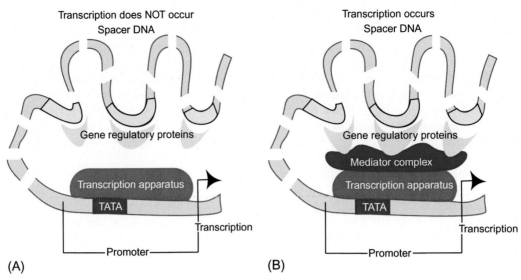

FIG. 3.18 Mediator complex and activator proteins. Because the DNA is folded, activators proteins can reach the transcription apparatus. The mediator complex allows activators and/or repressor proteins to bind to RNA polymerase.

control. The regulation is done by looping DNA around so that the activator proteins bound at the enhancer can make contact with the transcription machinery via a protein known as the mediator complex (Fig. 3.18).

For the example of the *lac* operon, when both glucose and lactose are present, the cell does not make the *lac* proteins. The repression of the synthesis of the *lac* proteins by glucose is called catabolite repression, which involves the two regions in the promoter. One is the binding site for the RNA polymerase and the other is the binding site for another regulatory protein, the catabolite activator protein (CAP). When the glucose is not present, CAP forms a complex with 3′, 5′-cyclic AMP (cAMP) and binds to the CAP site on the promoter (Fig. 3.19). This binding can help RNA polymerase binds to the promoter and the operon is fully active. In the presence of glucose, the level of cAMP is low only very few CAP-cAMP complexes can be formed. It is not enough to activate *lac* operon. The presence of lactose is necessary for the transcription of the operon, but not sufficient. The *lac* operon is active only when the glucose is absent and the lactose is present. As such, the CAP site is an example of an enhancer element, and the CAP-cAMP complex is a transcription factor. The modulation of transcription by CAP is a type of positive regulation.

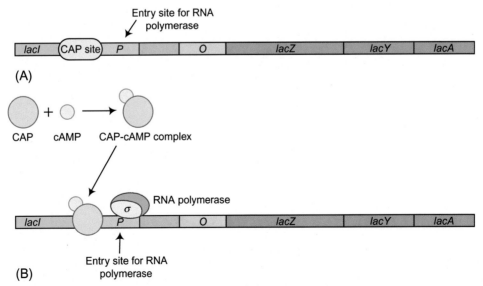

FIG. 3.19 Catabolite repression. (A) The CAP-cAMP complex must bind to CAP site of the promoter before RNA polymerase can bind. (B) In the absence of glucose, cAMP forms a complex with CAP, which binds to CAP site, allowing RNA polymerase to bind to the promoter and transcribe genes.

Transcription Attenuation

The *trp* operon of *E. coli* codes for the enzymes that the bacterium needs to make the amino acid tryptophan. Like the *lac* operon, the *trp* operon is a negative control mechanism. The *lac* operon responds to an inducer that causes the repressor to dissociate from the operator, derepressing the operon. The *trp* operon responds to a repressor protein that binds to two molecules of tryptophan. When the tryptophan is plentiful, this repressor-tryptophan complex binds to the *trp* operator. This binding prevents the binding of RNA polymerase, so the operon is not transcribed (Fig. 3.20). On the other hand, when tryptophan levels are reduced, the repressor will not bind the operator, so the operon is transcribed. This is an example of a system that is repressible and under negative regulation.

In addition to the standard negative regulation, the *trp* operon is regulated by another mechanism of control called transcription attenuation. This mechanism operates by causing premature termination of transcription of the operon when tryptophan is abundant. As shown in Fig. 3.20, there are two loci, the *trp* leader and the *trp* attenuator, in between the operator and the gene *trpE*. Secondary structures formed in the mRNA of the leader sequence are responsible for this premature termination. The formation of such secondary structures comes from the transcription stop signals—an inverted repeat and a string of 8 A-T pairs in the attenuator. When tryptophan is scarce, the operon is translated normally. When it is plentiful, transcription is terminated prematurely after the leader sequences have been transcribed. Thus, attenuation imposes an extra level of control on an operon, over and above the repressor-operator system.

Riboswitches

A riboswitch is a region, usually in the 5′-untranslated region (5′-UTR) of an mRNA, that contains two modules: an aptamer and an expression platform. An aptamer is a region that binds to a ligand. On the other hand, the expression platform can be a terminator, a ribosome-binding site, or another RNA element that affects gene expression upon the change in conformation. For example, the *ribD* operon in *B. subtilis* controls the synthesis and transport of the vitamin riboflavin and one of its products, flavin mononucleotide (FMN). The *rib* operon contains a conserved element in their 5′-UTRs known as the *RFN* element, which is an aptamer. There are two alternative conformations of the *RFN* element, antitermination and termination conformation. FMN binds directly to the *RFN* element and causes it to form termination conformation, blocking the expression. In the absence of FMN, the element forms an antitermination conformation which can complete the transcription. In another example, the *glmS* mRNA of *B. subtilis* contains a riboswitch which is a ribozyme -RNase. It can respond to the product of the enzyme encoded by the mRNA. When this product builds up, it binds to the riboswitch, changing the conformation of the RNA to become an effective RNase. The RNase destroys the mRNA, so less of the enzyme is made. As such, these examples of riboswitches demonstrate that the riboswitch can operate by depressing gene expression.

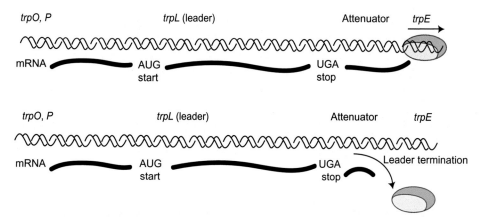

FIG. 3.20 An overview of *trp* operon regulation. When tryptophan is low, RNA polymerase (*blue*) reads through the attenuator and genes are transcribed. Attenuation of the trp operon of *E. coli*. When tryptophan is high, the attenuator causes premature termination of transcription, so the genes that produce more tryptophan are not transcribed.

TRANSCRIPTION IN EUKARYOTES

Eukaryotic RNA Polymerase

The transcription process is more complicated in eukaryotes. Three RNA polymerases with different activities are known to exist, and each one transcribes a different set of genes and recognizes a different set of promoters. RNA polymerase I synthesizes precursors of most ribosomal RNA. RNA polymerase II synthesizes mRNA precursors. RNA polymerase III found in the nucleoplasm and synthesizes the tRNA, precursors of 5S ribosomal RNA and some small RNA molecules involved in mRNA processing and protein transport.

All three eukaryotic RNA polymerases are large complex and contain 10 or more subunits. They are all related to the prokaryotic polymerase and to one another. The subunits that are unique to a particular eukaryotic polymerase are presumably responsible for functions specific to each type of polymerase, including the regulation of transcription. Of the three RNA polymerases, RNA polymerase II is the most extensively studied. In the yeast *Saccharomyces cerevisiae*, the RNA polymerase II has 12 subunits which are called RPB1 through RPB12. The core subunits are RPB1, RPB2, and RPB3, which are required for enzyme activity. They are homologous to the β′-, β-, and α-subunits, respectively, of *E. coli* RNA polymerase.

The RNA polymerase II can couple transcription to processing of the RNA transcript to produce the mature mRNA which is a unique function that is not found in other polymerases. This coupling activity depends on the functional domain—C-terminal domain (CTD). CTD comprises the C-terminus of the RPB1 and consists of repeats of seven amino acids. The CTD has the central role in the transition from initiation to elongation and in processing of the RNA transcript.

Promoter Structure

Three eukaryotic RNA polymerase have different structures and transcribe different classes of genes. It is, therefore, reasonable to predict that these polymerases would recognize different promoters. The promoters recognized by RNA polymerase II (class II promoters) are the most complex and best studied. Class II promoters usually have two parts: the core promoter and the proximal promoter (Fig. 3.21). The core promoter can attract general transcription factors and RNA polymerase II, thereby determining the start site and direction of transcription. It is modular, and it contains both the initiator box and the TATA box lying within about 37 bp of the transcription start site, on either side. The initiator box is a sequence found at the site where the transcription starts. About 25 base pairs upstream from this is the TATA box, an AT rich sequence, which is recognized by the TATA binging proteins (TBP). The proximal promoter elements are usually found upstream of core promoters. These upstream elements are the recognition sites for specific transcription factors. For example, GC box which has consensus sequence of GGGCGG, binds the transcription factor Sp1. Apart from those elements mentioned above, some genes might have a possible downstream regulator, although these are more rare than upstream regulators.

RNA Polymerase I and III-Directed Transcription

Eukaryotic RNA polymerases I and III rely on a distinct set of proteins to initiate transcription. Although both RNA polymerases I and III share several identical core enzyme subunits with RNA polymerase II, they recognize very different promoter sequences and have unique general transcription factors. The promoters recognized by RNA polymerase I are not well conserved in sequence from one specie to another. However, they all have similar general architecture of the promoter as it consists of a core element surrounding the transcription start site, and an upstream promoter element, which is about 100 bp farther upstream. RNA polymerase I, which transcribes rRNA genes, binds to promoter containing a core promoter element and an upstream control element (UCE). The TBP, which is part of a larger complex called SL1, helps RNA polymerase I to recognize the core promoter (Fig. 3.22). The classical RNA polymerase III promoters are Type I and Type II which have promoters that lie wholly within the genes. These Type I and Type II genes include a variety of RNA genes such as tRNA, the 5S RNA subunit of the ribosome, and adenovirus VA RNA genes (Fig. 3.23). The Type III RNA

FIG. 3.21 Four elements of RNA polymerase II promoters.

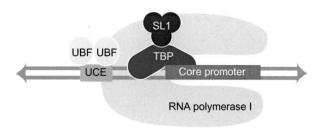

FIG. 3.22 RNA pol I promoters contain a core region as well as an upstream control element (UCE), which interact with transcription factors like UBF and SL1 in humans.

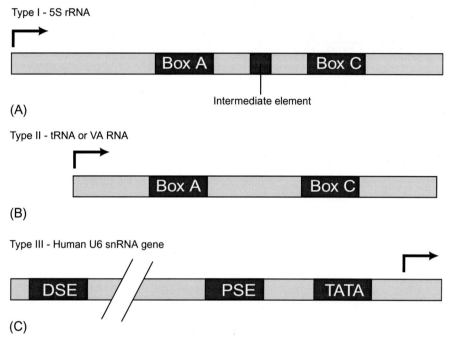

Type I - 5S rRNA

Box A

Intermediate element

Box C

(A)

Type II - tRNA or VA RNA

Box A

Box C

(B)

Type III - Human U6 snRNA gene

DSE

PSE

TATA

(C)

FIG. 3.23 Examples of RNA polymerase III promoters. The red boxes signify promoters of 5S rRNA, tRNA and U6 snRNA genes. DSE represents a distal sequence element, PSE represents a proximal sequence element.

polymerase III promoters is nonclassical and it resembles those of Class II genes including U6 snRNA gene, the 7SL RNA gene, and the 7SK RNA gene.

THE ORDER OF EVENTS IN EUKARYOTIC TRANSCRIPTION INITIATION

The Role of Transcription Factors—General Transcription Factors

Eukaryotic RNA polymerases are not capable of binding by themselves to their respective promoters. They rely on proteins called transcription factors to bind to their promoters. A transcription factor is any protein that regulates transcription but itself is not a subunit of RNA polymerase. There are two types of transcription factors—general transcription factors and gene-specific transcription factors (activators). The general transcription factors combine with RNA polymerase to form a pre-initiation complex that is competent to initiate transcription. Although the general transcription factors can support only a basal level of transcription in the absence of activators, the preinitiation complex is the most important step to start the transcription. Furthermore, less is known about the elongation and termination in eukaryotes than in prokaryotes. Here we present the sequence of events of initiation in RNA polymerase II transcription as an example to understand the eukaryotic transcription.

The preinitiation complex normally contains RNA polymerase II and six general transcription factors – TFIIA, TFIIB, TFIID, TFIIE, TFIIF, and TFIIH. The first step in the formation of the preinitiation complex is the recognition of the TATA box by TFIID (Fig. 3.24). The primary protein of TFIID is the TATA-binding protein (TBP), and it is

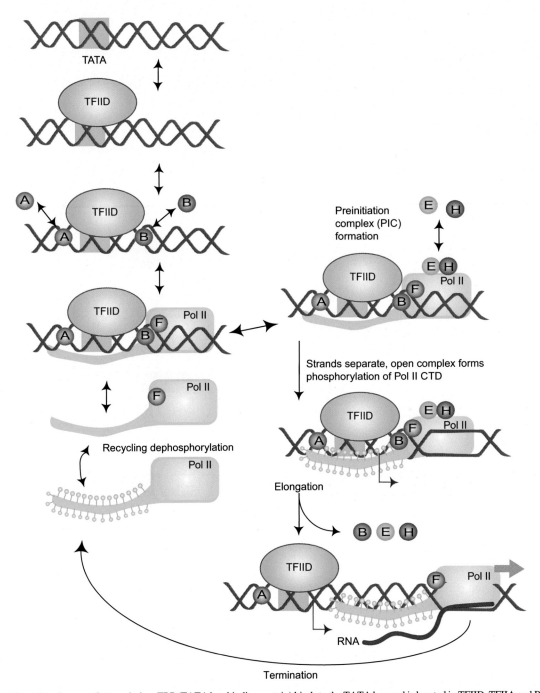

FIG. 3.24   The order of events of transcription. TBP (TATA box binding protein) binds to the TATA box and is located in TFIID. TFIIA and B bind to the TATA box, then TFIIF RNA polymerase II is recruited so that RNA can be produced. TFIIH and TFIIE form a preinitiation complex. As polymerase II ad TFIIF leave the promoter, RNA is produced during elongation. During termination, polymerase II dissociates and phosphorylation is recycled.

associated with many TBP-associated factors (TAFIIs). The TBP binds to TATA box at the promoter. TBP is a saddle-shaped protein that binds to the minor groove of the DNA with its concave face, inducing profound bending and partial unwinding of the DNA double helix. Once TFIID is bound, TFIIA binds followed by TFIIB. TFIIA interacts with the DNA and TFIID, and TFIIB binds TBP and the RNA polymerase II. TFIIB is critical for the assembly of the initiation complex and for the location of the correct transcription start site. Subsequently, TFIIF binds tightly to RNA polymerase II, and this complex binds stably to the promoter. Finally, TFIIE and TFIIH bind to the RNA polymerase II to form the complete preinitiation complex. To start the transcription, the RNA polymerase II is phosphorylated, and the DNA is separated.

The Role of Transcription Factors—Transcription Activators

Activators can either stimulate or inhibit the transcription by RNA polymerase II and they have two functional domains—a DNA-binding domain and an activation domain. Each DNA-binding domain has a DNA-binding motif, which is part of the domain that has a characteristic shape specialized for specific DNA binding. Most DNA-binding motifs fall into three classes: zinc-containing modules, homeodomains, and bZIP and bHLH motifs.

The zinc fingers are one of the zinc-containing modules. It is composed of an antiparallel β-sheet, followed by a α-helix (Fig. 3.25). Two cysteins from the β-sheet and two histidines from the α-helix are coordinated to a zinc ion. This coordination of amino acids to the metal helps form the finger-shaped structure. On the other hand, some zinc-containing DNA binding domain do not have zinc fingers. For example, the GAL4 protein does not have zinc fingers, but its DNA-binding motif has six cycteins that coordinates two zinc ions in a bimetal thiolate cluster.

Homeodomains contain about 60 amino acids and resemble in structure and function the helix-turn-helix DNA binding domains of prokaryotic proteins such as λ phage repressor (Fig. 3.26). Each homeodomain contains three α-helices; the second and third of these form the helix-turn-helix motif while the third serves as the recognition helix.

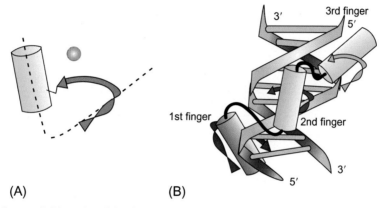

(A) (B)

FIG. 3.25 Examples of zinc fingers. (A) A beta-pleated sheet is represented by the curved, twisted arrow, whereas an alpha-helix is represented by a cylinder and a zinc ion is the gray circle. (B) Multiple fingers are shown simultaneously.

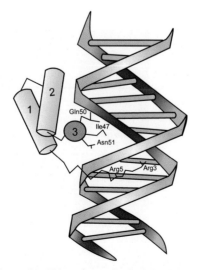

FIG. 3.26 An example of the homeobox and DNA complex. This representation includes three numbered helices (left) and the target double-stranded helix on the right. Number 3, the recognition helix rests on the major groove of the target, and key amino acid side chains interact with the DNA.

The bZIP proteins dimerize through a leucine zipper. Similarly, the bHLH proteins dimerize through a helix-loop-helix motif (Fig. 3.27). This dimerization pattern can promote the basic region to interact with DNA.

Epigenetic Regulation of Eukaryotic Transcription

Eukaryotic DNA is packaged into chromatin, the basic structural repeat unit of which is the nucleosome in which contains 147 bp of DNA wrapped in about 1.75 superhelical turns around a central octamer composed of two each of the four core histones H2A, H2B, H3, and H4 (Fig. 3.28). The histone octamer does not bind to DNA randomly, but rather exhibits preferences for some sequences over others, a phenomenon called nucleosome positioning.

Most eukaryotic promoters and regulatory elements are organized into precise architecture within extensive nucleosome positioning sites. Sequence-specific DNA-binding proteins often cannot recognize their sites when they are tightly wrapped in a nucleosome. To activate transcription, the transcription machinery has to compete with histones for DNA sites that are blocked or obscured by nucleosomes on the promoter. It is also possible that the wrapping of nucleosomal DNA over the surface of the histone octamer may be sterically incompatible with the assembly of a stable transcription preinitiation complex. Therefore, it is clear that the nucleosome is not just a DNA packaging system but also a gene

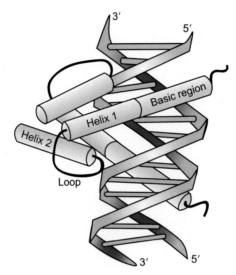

FIG. 3.27   Crystal structure of MyoD and its target DNA. Alpha-helices are represented by cylinders.

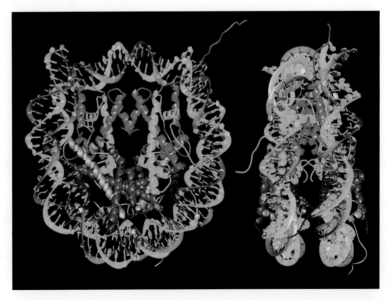

FIG. 3.28   The crystal structure of the nucleosome. DNA double helix is shown in blue and orange, whereas the core histones are shown in purple, pink, green, and lighter blue on the inside.

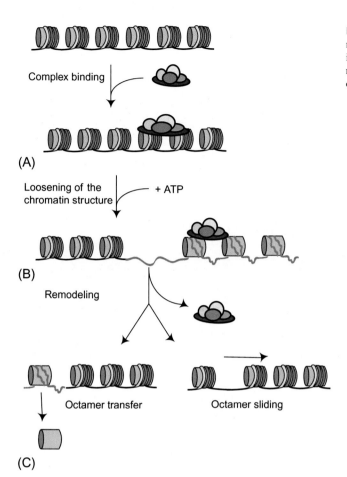

(A)

(B)

(C)

FIG. 3.29 The mechanism of chromatin remodeling activity. ATP is required for the remodeling complex to reconfigure the chromatin. ATP is hydrolyzed in order to loosen the chromatin structure, and then the remodeling complex is released. The remodeling activity can cause octamers to slide, or to be transferred.

expression regulator. The issue about how chromatin structure is disrupted or remodeled in order to facilitate access to sequence-specific transcription factors and the general transcription machinery is known as the epigenetic regulation, and it has been a focus of vigorous research in recent years.

The local changes that involved the disruption or movement of nucleosomes can be mediated by two different mechanisms. One mechanism is mediated by large multi-protein assemblies, called chromatin remodeling complexes. These remodeling complexes use ATP hydrolysis to reconfigures the chromatin structure, increasing the accessibility of DNA within the chromatin template (Fig. 3.29). Covalent modification of histone is another mechanism by which nucleosome structure is modified. The relatively unstructured and highly charged N-terminal tail domains of core histones extrude from the histone octamer, and these tails are important for DNA-histone interactions within and between nucleosomes, as well as for nucleosome-nucleosome interactions (Fig. 3.30). Post-translational acetylation of lysine residues within these N-terminal tails facilitates binding of transcriptional activators to nucleosomal DNA, at least in part through neutralization of the positive charge associated with the lysine ε-amino group by histone acetyltransferases (HATs), and thereby reducing their affinity for negatively charged DNA or other chromatin proteins. This process can be reversed by histone deacetylase (HDAC). Other modifications also play a role in transcription regulation through histone, including phosphorylation of serine residues and methylation of lysine and arginine residues. The combination of variable modification pattern to regulate transcription is called histone code (Fig. 3.31).

TRANSCRIPTION ELONGATION THROUGH THE NUCLEOSOMAL BARRIER

Although chromatin is remodeled upon transcription initiation, the DNA remains packaged in nucleosomes in the coding region of transcribed genes. As RNA pol II moves through the gene-coding region, the enzyme encounters a nucleosome approximately every 200 bp. RNA pol II overcomes this nucleosomal barrier through the existence of two distinct mechanisms for the progression of RNA polymerases through chromatin: (i) nucleosome mobilization or "octamer transfer" (i.e., movement of the octamer on the DNA) and (ii) H2A-H2B dimer depletion.

FIG. 3.30 Covalent modification of histones tails. Modification of histone proteins regulates transcription due to acetylation of lysine residues, methylation of arginines and lysines, and phosphorylation of serines.

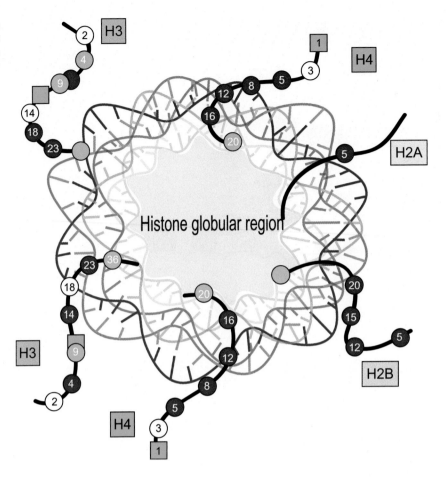

FIG. 3.31 The histone code. The presence of histone remodeling complexes (SWI/SNF) and histone acetyl transferase (HAT) allow transcription to occur because the acetylation of a histone loosens the DNA that is wrapped tightly around it. This way, transcription factors and RNA polymerase have access to the DNA. The top diagram depicts acetylated histones (*purple squiggly lines*), and the DNA around them is "switched on" because it is accessible to be transcribed. Phosphorylated nucleotides are represented by red circles, and unmethylated cytosines are represented by purple circles. The bottom diagram represents non-acetylated chromatin, which is "switched off" because the DNA is wound too tightly to be transcribed.

Transcription will proceed

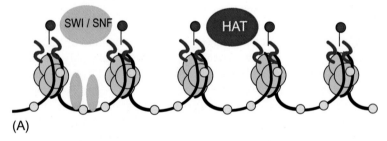

(A)

Transcription will NOT proceed

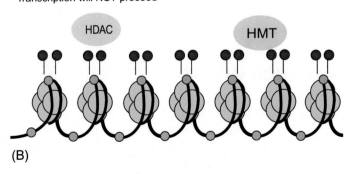

(B)

Nucleosome Mobilization

Polymerases such as RNA polymerase II (RNA pol II) and bacteriophage SP6 RNA polymerase appear to use the nucleosome mobilization or "octamer transfer" mechanism for overcoming the nucleosome barrier. In this mode of action, nucleosomes are translocated without release of the core octamer into solution. In other words, DNA is displaced from the nucleosomes, and the nucleosomes are transferred to a region of DNA already transcribed by RNA pol II. This process may be facilitated by the elongation factor FACT (facilitates chromatin transcription). FACT plays a role in the elongation of transcripts through nucleosome arrays by promoting transcription-dependent nucleosome alterations. FACT enables the displacement of a dimer of H2A-H2B in front of RNA pol II, leaving a histone "hexamer" at the same location as the initial octamer. FACT does not require ATP hydrolysis for its mode of action. After passage of the polymerase, FACT enables the immediate reassembly of the H2A-H2B dimer.

H2A-H2B Dimer Removal

Nucleosomes are disrupted on active RNA pol II transcribed genes by H2A-H2B dimer removal. This mode of overcoming the nucleosome barrier requires a number of auxiliary elongation factors. The first protein factor that has been established to promote RNA pol II elongation in vitro is FACT. Other complexes that facilitate chromatin transcription in vitro are elongator and TFIIS. Even in the presence of FACT, elongator, or TFIIS, the rate of RNA pol II transcription through nucleosomes in vitro is much slower than the rate of transcript elongation in the cell. As such, overcoming the nucleosome barrier requires the coordinated action of additional, unidentified factors.

POSTTRANSCRIPTIONAL MODIFICATION OF mRNA

The genes of prokaryotes are continuous; every base pair in a continuous prokaryotic gene is reflected in the base sequence of mRNA. The genes of eukaryotes are not necessarily continuous. Instead, eukaryotic genes frequently contain intervening sequences that do not appear in the final mRNA for that gene produced. The DNA sequences that are expressed are called exons, and the intervening sequences are called introns. As such, the expression of eukaryotic gene involves not just its transcription, but also the processing of the primary transcript to its final form.

The Cap at the 5′ End of Eukaryotic mRNA

The posttranscriptional modifications include capping of the 5′ end, polyadenylating the 3′ end, and splicing of coding sequence (Fig. 3.32). The cap at the 5′ end of eukaryotic mRNA is a guanylate residue that is methylated at the N-7 position. The presence of 5′ cap can protect the mRNA from exonucleases degradation. The 5′ cap is essential for the efficient elongation and termination of transcription as well as the subsequent processing of the mRNA. It also functions as a binding site

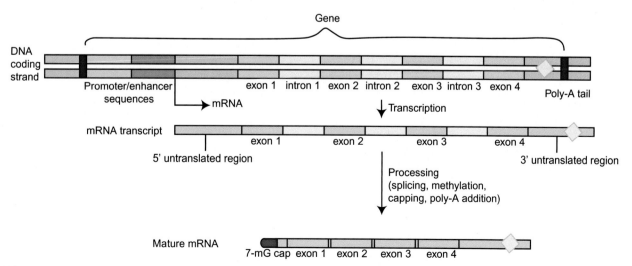

FIG. 3.32   A typical example of an eukaryotic gene structure and the process of posttranscriptional modifications.

for proteins that export the mRNA from the nucleus to the cytoplasm, and for directing the initiation of protein synthesis from the mRNA.

The polyadenylate tail (poly-A) at the 3′ end of a message is added before the mRNA leaves the nucleus. It is thought that the tail protects the mRNA from nucleases and phosphatases, which could degrade it. Some mRNAs have more than one site, called the polyadenylation site, at which the precursor mRNA is cleaved and the poly(A) tail added. These multiple polyadenylation sites contribute to the regulation of protein synthesis and expand the range of protein products made from a single mRNA species.

The removal of introns (known as splicing) takes place in the nucleus, where RNA forms ribonucleoprotein particles through association with a set of nuclear proteins. Splicing requires cleavage at the 5′ end 3′ end of introns and the joining of the two ends. The actual splicing involves the spliceosome. Gene expression can also be controlled at the level of RNA splicing. This is because some proteins can be spliced in different ways to give different isoforms of the proteins to be produced. Regulatory proteins can affect the recognition of splice sites and direct the alternative splicing.

Capping and Polyadenylation Processes

The 5′ capping occur soon after the growing RNA emerges from RNA polymerase II, and it is accomplished in three successive steps (Fig. 3.33). First, an RNA 5′ triphosphatase catalyzes the removal of one phosphate from the triphosphate at the 5′ end of the mRNA. A guanyl transferase then attaches a molecule of guanosine monophosphate (GMP) to this end. Finally, the transferred guanine is methylated by a guanine-7-methyltransferase.

Polyadenylation of the 3′ end of eukaryotic mRNAs begins with an initial cleavage of the mRNA (Fig. 3.34). The cleavage is usually occurred after a CA nucleotide pair that lies somewhere between a conserved AAUAAA hexamer

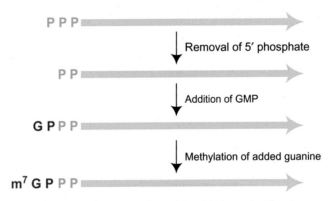

FIG. 3.33 Three steps are involved in the addition of the 5′ cap. At the 5′ end of the mRNA, one phosphate is removed. Next, guanyl transferase attaches a GMP molecule to the end of the mRNA. The newly transferred guanine is methylated by a guanine-7-methyltransferase.

FIG. 3.34 Polyadenylation occurs between a conserved AAUAAA sequence and a downstream U- or GU-rich region. CA nucleotide near the AAUAAA sequence marks the cleavage point for precursor mRNA. Next, up to 200 adenine nucleotides are added by poly(A) polymerase.

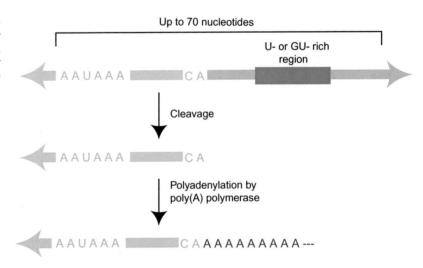

sequence and a U- or GU-rich region further downstream. After cleavage, a tail of approximately 200 adenosines is added by poly(A) polymerase. This process is more complex than 5′ capping because it involves recognition and processing of different polyadenylation sites in different mRNAs.

The 5′ capping and 3′ polyadenylation are intimately coupled to each other and to transcription by RNA polymerase II. For example, capping is required during transcription to allow RNA polymerase II to continue elongation of the mRNA. Polyadenylation is required for the efficient termination of transcription. These mRNA processing events are regulated by C-terminal domain of RPB1. However, RNAs synthesized by RNA polymerase I and III are not capped and polyadenylated, as they don't have domain equivalent to C-terminal domain of RPB1.

RNA Splicing

The mature transcript for many genes is encoded in a discontinuous manner in a series of discrete exons. RNAs includes mRNAs, tRNA, and rRNA; all contain introns that must be removed from precursor RNAs to produce functional molecules. tRNA precursors undergo splicing that is catalyzed by protein factors and rRNA precursors catalyze their own removal and are termed self-splicing. On the other hand, mRNA splicing is carried out by the spliceosome. Spliceosome is composed of several individual small nuclear ribinucleoproteins (snRNP) and many more additional proteins that come and go during the splicing reaction. snRNP removes the intervening sequences from pre-mRNAs, and they contain a large number of proteins and five small nuclear RNAs (snRNAs), including U1, U2, U4, U5, and U6 snRNAs. They are post-transcriptionally modified proteins.

During the splicing, the 2′ OH of an adenosine residue located within the introns itself attacks the previous exon-intron boundary, detaching the intron from the previous exon and producing a branched intron structure (Fig. 3.35). Next, the terminal 3′ OH of the newly released exon attacks the intron-next exon junction, splicing together the two exons and releasing the intron. Before splicing can occur, the spliceosome must identify the splice sites between introns and exons—the site at which exons are separated from their neighboring introns, and at which two exons are subsequently attached. The spliceosome identifies splices sites by recognizing short sequence motifs found in each pre-mRNA. Key sequences are the 5′ and 3′ splice sites, a branch-point nucleotide within the intron and a polypyrimidine tract before

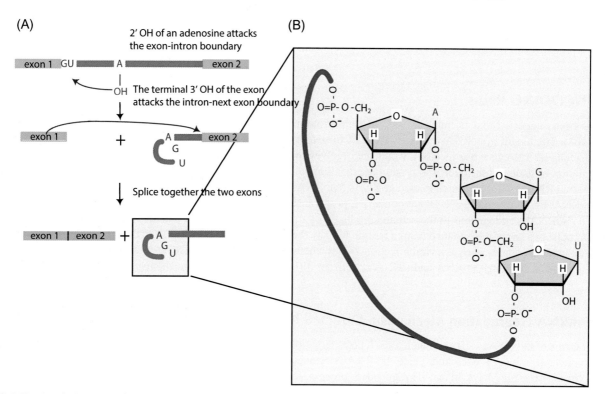

FIG. 3.35   A typical example of splicing pathway. (A) Introns use OH groups of an internal A nucleotides to attack the boundary that joins exon 1 and the intron. Similarly, to group I, the newly open end of exon I will attack the boundary between the intron and exon II, releasing the second exon and joining it with the first. This now forms a lariat, which is shown in (B).

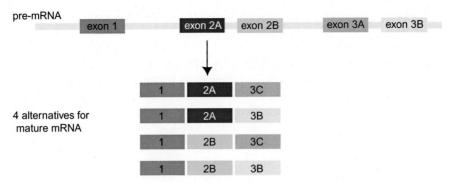

FIG. 3.36 Alternative splicing creates protein diversity. The pre-mRNA shown has a first exon, 2 possible second exons as well as 2 possible third exons. Alternative splicing will generate four different combinations (1+2A+3C, 1+2A+3B, 1+2B+3C and 1+2B+3B).

the 3′ splice site. Some splice sites are recognized by branch-point-binding protein and the polypyrimidine tract. One snRNA, U6, contains highly conserved sequences that are thought to be the functional counterparts of the RNA elements that form the active site of self-splicing group II intron ribozymes. An in vitro-assembled, protein-free complex of U6 with U2, the base-pairing partner in the spliceosomal catalytic core, can catalyze a two-step splicing reaction in the absence of all other spliceosomal factors, suggesting that the two snRNAs may form all or a large share of the spliceosomal active site. On the other hand, several spliceosomal proteins are thought to help in the formation of functionally required RNA-RNA interactions in the catalytic core.

One of the important contributions of RNA splicing is alternative splicing. Many genes contain multiple introns and exons. Exons can be spliced together in a variety of combinations to produce different proteins (Fig. 3.36). Alternative splicing does not change the nucleotide sequences in an RNA but determine whether a given sequence is present or not in the synthesized product. In addition to the potential exclusion or inclusion of a particular exon, the splicing pattern of an mRNA can be disrupted by the utilization of alternative 5′ or 3′ splice sites or even through the utilization of alternative transcriptional start or termination sites. These alternations all depend on a variety of factors such as cell type, the developmental stage or the environmental conditions. Therefore, alternative splicing represents an immense source of potential genetic diversity.

NONCODING RNAs

It has been suggested that as much as 98% of transcriptional output of our genome is comprised of non-coding RNA (ncRNA). The length of ncRNAs can vary from 21 to several thousand nucleotides (nt), and these molecules are divided into (i) long nc RNAs, such as lncRNA, which are involved in epigenetic regulation of protein-coding gene expression and modulation of gene transcription and protein degradation, and (ii) small ncRNAs, such as microRNAs (miRNAs), small interfering RNAs (siRNAs), small nucleolar RNAs (snoRNAs), small nuclear RNAs (snRNAs), and others. Both miRNA and siRNA are involved in gene expression at the post-transcription stage.

The miRNA is endogenous to the cell and produced by transcription of the cell's gene. It is about 22 nucleotides long and is cut from a larger RNA that contains a hairpin loop by the enzyme dicer (Fig. 3.37). On the other hand, the siRNAs are from exogenous sources, such as a viral infection, or a synthetic dsRNA. They can also be cut by the dicer to between 21 and 25 nucleotides. Although miRNA and siRNA arise from different sources, their mechanisms of action are similar.

The mRNA Degradation Mechanism Directed by siRNA

Dicer is one of the RNAase III endonucleases. It takes larger dsRNAs and cut them to their characteristic small size leaving a two nucleotide-long 3′ overhang. Once the RNA is cleaved, it is passed to a protein complex RNA-induced silencing complex (RISC) (Fig. 3.37). In an ATP-dependent process, RISC unwinds the double-stranded RNA and selects the antisense strand. Subsequently, a nuclease protein in the RISC complex known as argonaut guides the complex to the targeted mRNA. In the case of siRNA, the matching of the antisense strand to the mRNA is perfect. As such, the argonaut protein cleaves the mRNA, and the mRNA is silenced. This is a process of RNA interference.

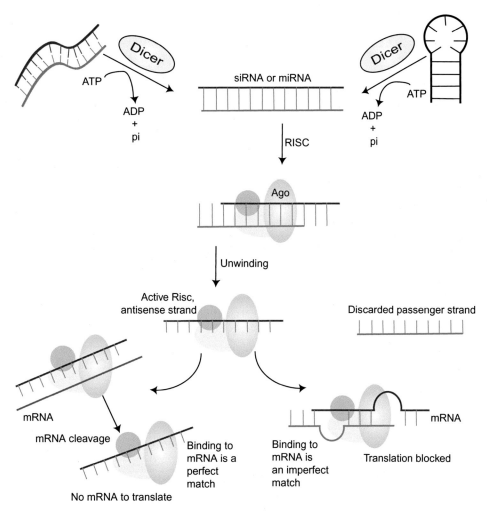

FIG. 3.37   RNA interference. Dicer enzymes cleave exogenous RNA into siRNA, and cleaves pre-miRNA to yield miRNA. These bind to RISC complex including Argonaut proteins (Ago), then detaches and discards the passenger strand. RISC is recruited to mRNA that matches the antisense strand. When the binding is perfect, the slicer enzyme will cleave the RNA so there's nothing left to translate. When binding is imperfect, hairpin loops form (bottom right) and the complex cannot be translated.

The mRNA Degradation Mechanism Directed by miRNA

In the case of miRNA, miRNAs are transcribed by RNA polymerase II from what are called *MIR* genes. The first intermediate transcripts are "hairpin" molecules called pri-miR. Similar to mRNAs, these transcripts undergo capping and polyadenylation at the 5' and 3' ends of the transcript, respectively. Unlike mRNAs, miRNA maturation begins in the nucleus and finishes in the cytoplasm. In animals, the nuclear processing is performed by DROSHA, an RNase type III enzyme with endonucleolytic activity, which, in combination with PASHA, recognizes the hairpin structure of the pri-miR and cleaves it. This generates a pre-miR of approximately 60–70 nt in length. Pre-miRs are transported to the cytoplasm by exportin-5 and the cofactor Ran-GTP. Once in the cytoplasm, pre-miRs are processed by the enzyme dicer. Dicer binds to the pre-miRNA and cleaves it to miRNA. The miRNA then binds to RISC as it did with the siRNA. The miRNA bound to the RISC structure is guided to the complementary sequence of the target mRNA. The process is the same as RNA interference; if the pairing is complete (100% sequence complementarity between the miRNA and mRNA), the mRNA will be degraded. If the miRNA is not a perfect match for the mRNA target, there is no cleavage of the mRNA, but the RISC continues to bind to the mRNA, interfering with the ability of ribosomes to translate mRNA. One unique aspect of miRNA regulation is its complexity. It has been observed that a single miRNA can regulate expression of different mRNAs. Additionally, one mRNA can be regulated by multiple miRNAs.

Both siRNA and miRNA Can Repress Transcription

In addition to repressing mRNA translation and triggering mRNA degradation, siRNA and miRNA can also repress the transcription of specific genes and larger regions of the genome through associating with a different complex—the RNA-induced initiation of transcription silencing complex (RITS). The antisense RNA strand within the RITS targets the RITS complex to specific gene promoters or large regions of chromatin. RITS then recruits chromatin remodeling enzymes to the promoters and these enzymes methylate histones and DNA, resulting in heterochromatin formation and subsequent transcriptional silencing (Fig. 3.38).

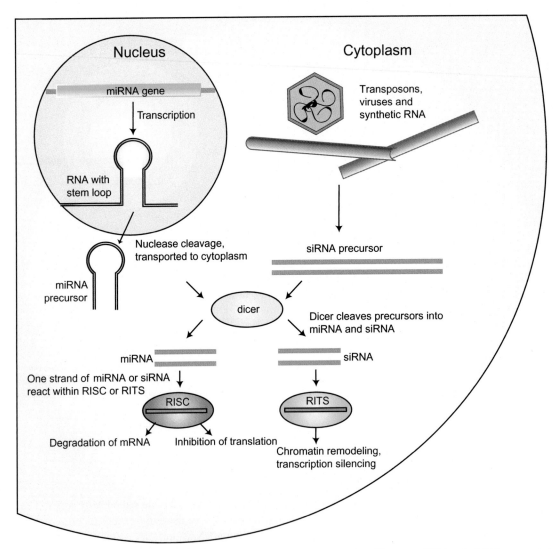

FIG. 3.38 Gene regulation via RNA-induced gene silencing. In the cytoplasm, a dicer complex will process siRNA and miRNA precursors into shorter double-stranded RNA molecules. Now, they can be recognized by RISC and RITS complexes. The RITS complex will modify chromatins and repress transcription by recruiting chromatin remodeling proteins. The RISC complex either marks mRNA for degradation or translation inhibition.

Chapter 4

Gene Expression: Translation of the Genetic Code

Chapter Outline

THE STRUCTURE AND PROPERTIES OF AMINO ACIDS

General Structure of Amino Acids

The general structure of amino acids includes an amino group, a carboxyl group, an α-carbon, a hydrogen, and an R group (Fig. 4.1). The R group, which is the side chain, determines the identity of the particular amino acid. Because the α-carbon is located in the center with the other four different groups bonded to it, the α-carbon is the chiral center in all amino acids except glycine, whose R group is hydrogen. Glycine is not chiral because of this symmetry. In all other commonly occurring amino acids, the α-carbon has four different groups bonded to it, giving rise to two nonsuperimposable mirror-image forms, which are stereoisomers or enantiomers (Fig. 4.2).

Diagnostic Molecular Biology. https://doi.org/10.1016/B978-0-12-802823-0.00004-3

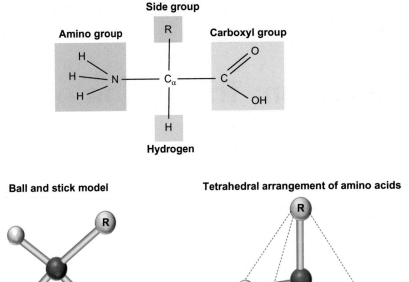

FIG. 4.1 The standard component of amino acids represented as a chemical structure, a ball-and-stick model, and a tetrahedral diagram.

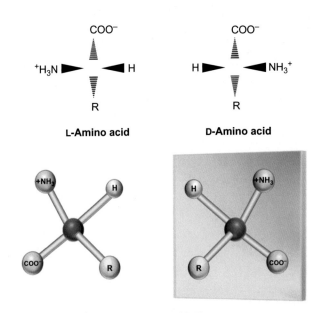

FIG. 4.2 The stereochemistry of amino acids, which can be found in L or D chirality.

Enantiomeric Molecules

Enantiomeric molecules display a special property called optical activity: the ability to rotate the plane of polarization of plane-polarized light. Clockwise rotation of incident light is referred to as dextrorotatory behavior, which displays D-configuration, and counterclockwise rotation is called levorotatory behavior, which displays L-configuration. The two possible stereoisomers of one chiral compound, L- and D-amino acids, represents left and right form amino acids. The position of the amino group on the left or right side of the α-carbon determines the L or D designation, respectively. The amino acids that occur in protein are all of the L form. The D form also occurs in nature, but they are not found in proteins. Instead, they are found mostly in bacterial cell walls.

CLASSIFICATIONS OF AMINO ACIDS

There are several ways to classify the common amino acids. The most useful of these classifications is based on the polarity of the side chains. There are four types of amino acids based on the features of the side chain: (1) nonpolar or hydrophobic amino acids, (2) neutral (uncharged) but polar amino acids, (3) acidic amino acids, which have a net negative charge at pH 7.0, and (4) basic amino acids, which have a net positive charge at pH 7.0.

Nonpolar Amino Acids

The nonpolar amino acids contain a nonpolar side chain. The nonpolar amino acids are critically important for the processes that drive protein chains to fold, that is, to form their natural and/or functional structures. This group consists of glycine, alanine, valine, leucine, isoleucine, proline, phenylalanine, tryptophan, and methionine (Fig. 4.3). Proline has a benzene ring. Because its nitrogen is bonded to two carbon atoms, the amino group of proline is a secondary amine, and proline is often called an imino acid. Both phenylalanine and tryptophan are aromatic. The side chain of methionine contains a sulfur atom in its hydrocarbon group. The side chain of all other amino acids including alanine, valine, leucine, and isoleucine is just the hydrocarbon group.

(A) Nonpolar hydrophobic

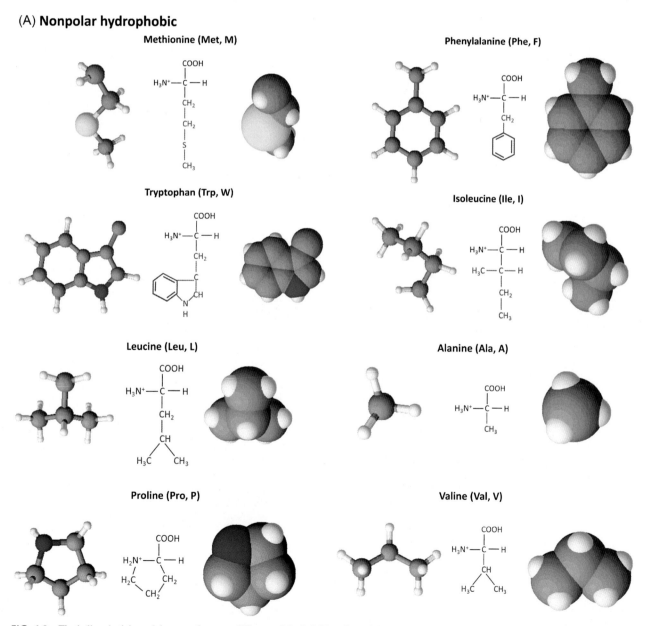

FIG. 4.3   The ball-and-stick model versus the space-filling model of all 20 amino acids categorized by their chemical properties.

(Continued)

Glycine (Gly, G)

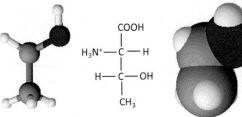

(B) Polar uncharged

Threonine (Thr, T)

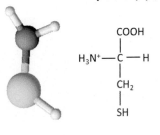 

Tyrosine (Tyr, Y)

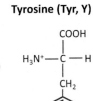 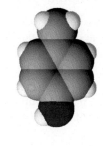

Cysteine (Cys, C)

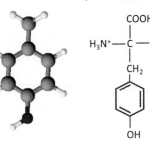

Serine (Ser, S)

Asparagine (Asn, N)

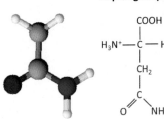 

Glutamine (Gln, Q)

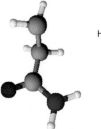 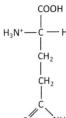 

FIG. 4.3, CONT'D

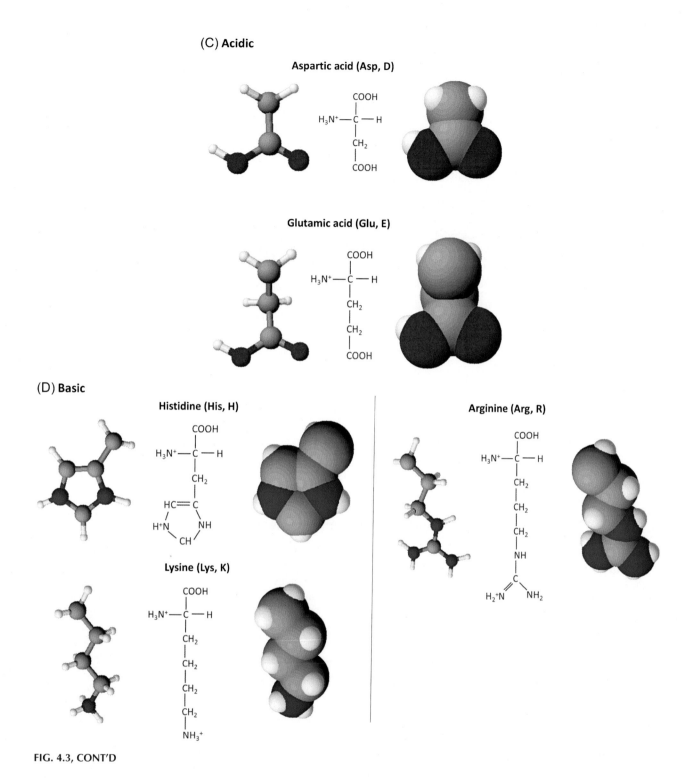

(C) Acidic

Aspartic acid (Asp, D)

Glutamic acid (Glu, E)

(D) Basic

Histidine (His, H)

Arginine (Arg, R)

Lysine (Lys, K)

FIG. 4.3, CONT'D

Neutral (Uncharged) but Polar Amino Acids

The amino acids that have polar side chain are uncharged at neutral pH. The R group of polar uncharged amino acids can form hydrogen bonds with water and play a variety of nucleophilic role in enzyme reactions. These amino acids are usually more soluble in water than the nonpolar amino acids. In serine and threonine, the polar group is a hydroxyl (—OH) bonded to hydrocarbon groups. The hydroxyl group in tyrosine is bonded to an aromatic hydrocarbon group, which is a phenol.

The hydroxyl group in cysteine contains a thiol group (—SH), which can react with other cysteine thiol groups to form disulfide (—S—S—). Such disulfide bond bridges in proteins through an oxidation reaction. Both glutamine and asparagine have amide groups that are derived from carboxyl groups in their side chains.

Acidic Amino Acids That Have a Net Negative Charge at pH 7.0

Both glutamic acid and aspartic acid are acidic amino acids. They have carboxyl groups in their side chains in addition to the one present in all amino acids. A carboxyl group can lose a proton, forming the corresponding carboxylate anion. These forms are appropriately referred to as aspartate and glutamate. Because of the presence of the carboxylate, the side chain of each of these two amino acids is negatively charged at neutral pH. These negatively charged amino acids play several important roles in proteins. Many proteins that bind metal ions for structural or functional purposes possess metal-binding sites containing one or more aspartate and glutamate side chains.

Basic Amino Acids That Have a Net Positive Charge at pH 7.0

Histidine, lysine, and arginine have basic side chains, and the side chain in all three is positively charged at the neutral pH. In lysine, the side-chain amino group is bonded to a hydrocarbon tail. In arginine, the side-chain guanidino group is also bonded to a hydrocarbon tail. The histidine has an imidazole side chain. The side chains of lysine and arginine are fully protonated at pH 7.0, but histidine is only partially protonated at pH 7.0. Therefore, histidine-containing peptides are important biological buffers, and the side chains of both arginine and lysine participate in electrostatic interactions in proteins.

Alternative Classification of Amino Acids

There are alternative ways to classify the 20 common amino acids. For example, these 20 amino acids are usually grouped into hydrophobic, hydrophilic, or amphipathic. Hydrophobic amino acids include alanine, glycine, isoleucine, leucine, phenylalanine, proline, and valine. Hydrophilic amino acids are arginine, asparagine, aspartic acid, cysteine, glutamic acid, glutamine, histidine, serine, and threonine. Finally, lysine, methionine, tryptophan, and tyrosine are amphipathic amino acids.

Amino Acids Can Join Through Peptide Bonds to Form polymers

The crucial feature of amino acids that allows them to polymerize to form peptides and proteins is the existence of their two chemical groups: the amino (—NH$_3^+$) and carboxyl (—COO$^-$) groups. In the linear polymers of amino acids, the carboxyl group of one amino acid is linked to the amino group of another amino acid. The peptide bond is an amide bond formed in a head-to-tail fashion between the —COO$^-$ group of one amino acid and the -NH$_3^+$ group of another amino acid. This reaction eliminates a water molecule and forms a covalent amide linkage. The reaction shown in Fig. 4.4 for the two glycines is a condensation reaction, because a water is lost as the amide bond is formed.

The molecule formed by condensing two amino acids is called a dipeptide. The amino acid with a free -NH$_3^+$ group is known as the amino-terminal or simply the N-terminal amino acid residue, or N-terminus. On the other hand, the amino acid with a free —COO$^-$ group is known as the carboxyl or C-terminal amino acid residue, or C-terminus. Structures of proteins are conventionally written with their N-terminal amino acid on the left. Repetition of this condensation reaction produces polypeptides and proteins.

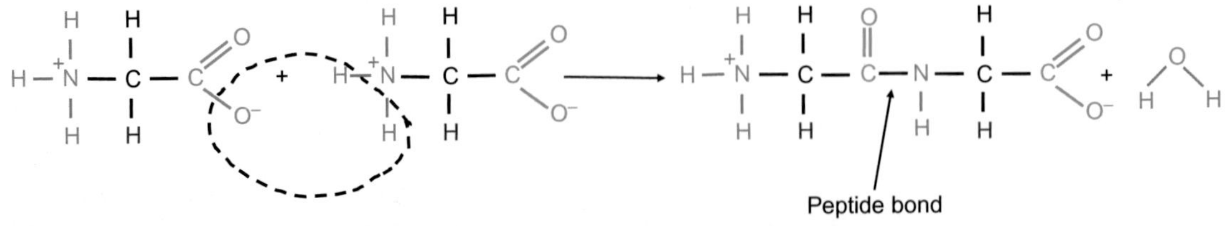

FIG. 4.4 Dehydration synthesis to form a peptide bond between amino acids.

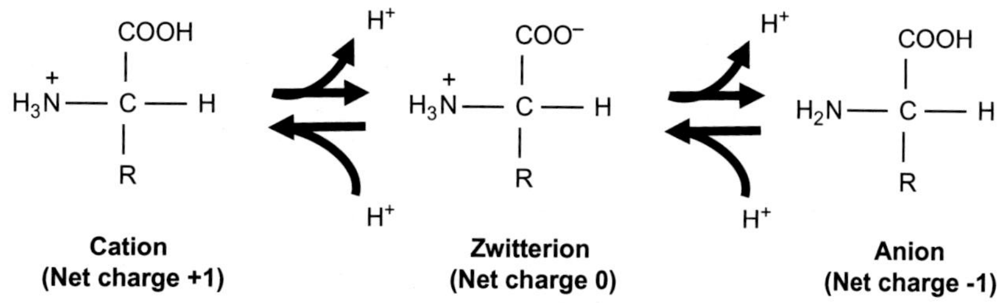

FIG. 4.5 The ionic forms of an amino acid. At the p*I*, a molecule has no net charge.

Zwitterions

In a free amino acid, the carboxyl group and amino group of the general structure are charged at neutral pH: the carboxylate portion negatively and the amino group positively. Amino acids without charged groups on their side chains exist in neutral solution as zwitterions with no net charge. The pH at which a molecule has no net charge is called the isoelectric pH, or isoelectric point, which is termed p*I* (Fig. 4.5). At the p*I*, a molecule will not migrate in an electric field.

PROTEIN STRUCTURE

Biologically active proteins are polymers consisting of amino acids linked by covalent peptide bonds. Many different conformations are possible for a molecule as large as a protein. Of these many structures, only one or a few have biological activity. These are called the native conformation, and this is defined by four levels of structure: primary structure, secondary structure, tertiary structure, and quaternary structure.

Primary Structure

Primary (1°) structure is the order in which the amino acids are covalently linked together through the peptide bond (Fig. 4.6). The amino acid sequence defines the primary structure which is the one-dimensional first step in specifying the three dimensional structure of a protein and thus its properties. Because the polypeptide chains are unbranched, a polypeptide chain has polarity which has the amino terminus, or N-terminus on the left and a free carboxyl group on the right which is called carboxyl terminus, or C-terminus (Fig. 4.6). Therefore, the amino sequence is read from the N-terminal end of the polypeptide chain through to the C-terminal end.

The astounding sequence variation possible within polypeptide chains provides a key insight into the incredible functional diversity of protein molecules in biological systems. One of the most striking examples of the importance of primary structure is found in the hemoglobin associated with sickle-cell anemia. In this disease, the red blood cells cannot bind oxygen efficiently is because the change of one amino acid residue in the sequence of the primary structure. The glutamic acid in sixth position of normal hemoglobin is replaced by a valine residue in the sickle-cell anemia. Such change makes cells to become trapped in small blood vessels, cutting off circulation and thereby causing organ damage.

Secondary Structure

The secondary (2°) structure of proteins is the hydrogen-bonded arrangement of the polypeptide chain backbone. This is the local conformations of the polypeptide. In nearly all proteins, the hydrogen bonds that make up secondary structures involve the amide proton of one peptide group and the carbonyl oxygen of another peptide group (Fig. 4.7). These structures tend to form in cooperative fashion and involve substantial portions of the peptide chain.

Two most important secondary structures are the repeating α-helix and β-pleated sheet. The α-helix is stabilized by hydrogen bonds parallel to the helix axis within the backbone of a single polypeptide chain (Fig. 4.8). Counting from the N-terminal end, the C—O group of each amino acid residue is hydrogen bonded to the N—H group of the amino acid four residues away from it in the covalently bonded sequence. All of the hydrogen bonds point in the same direction along the α-helix axis. Each peptide bond possesses a dipole moment that arises from the polarity of the N—H and C—O groups. Because these N—H and C—O groups are all aligned along the helix axis, the helix itself has a substantial dipole moment, with a partial positive charge at the N-terminus and a partial negative charge at the C-terminus (Fig. 4.9).

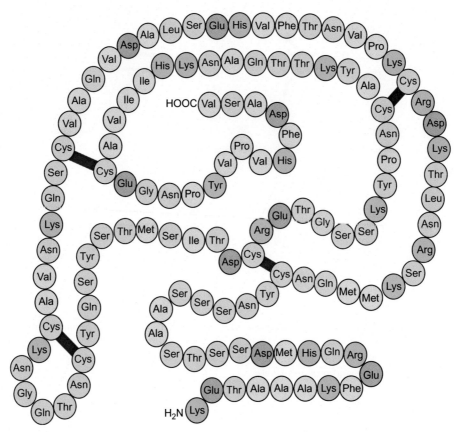

FIG. 4.6 An example of the primary structure: Bovine pancreatic ribonuclease A amino acid sequence of 124 amino acid residues including 4 disulfide bonds that further strengthen the tertiary structure of this polypeptide.

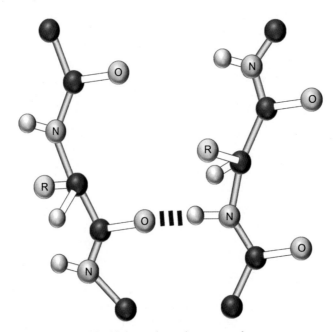

FIG. 4.7 Hydrogen bond formation between amino acids: this interaction makes up secondary structure.

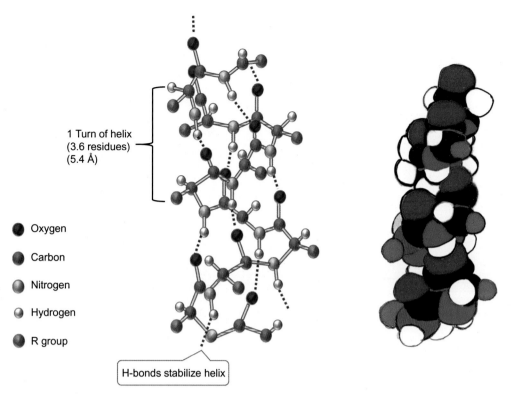

FIG. 4.8 The formation of an alpha helix secondary structure represented in a ball-and-stick model, as well as a space-filling model. The hydrogen bonds are parallel to the helix axis within the backbone of a single polypeptide chain.

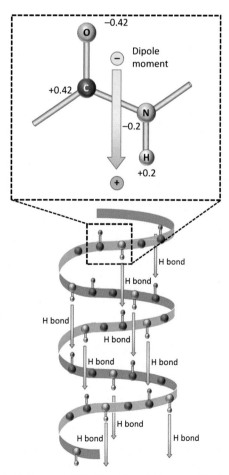

FIG. 4.9 The large net dipole created by the helical alignment of N—H and C=O side groups. The helix itself has a substantial dipole moment, with a partial positive charge at the N-terminus and a partial negative charge at the C-terminus.

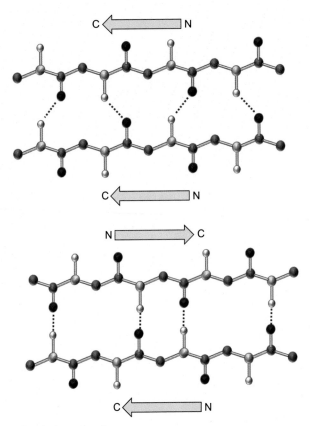

FIG. 4.10   Parallel and antiparallel beta-sheet hydrogen bonds.

The β-pleated sheet also forms because of local, cooperative formation of hydrogen bonds. The peptide backbone in the β-pleated sheet is almost completely extended. Hydrogen bonds can be formed between parts of a single chain that is doubled back on itself or between different chains. If the peptide chains run in the same direction, a parallel pleated sheet is formed (Fig. 4.10). When the peptide chains run in opposite directions, an antiparallel pleated sheet is formed.

Parallel sheets are typically large structures; those composed of fewer than five strands are rare. Antiparallel sheets, however, may consist of as few as two strands. Parallel sheets characteristically distribute hydrophobic side chains on both sides of the sheet, whereas antiparallel sheets are usually arranged with all their hydrophobic residues on one side of the sheet. This requires an alternation of hydrophilic and hydrophobic residues in the primary structure of peptides involved in antiparallel β-sheets because every other side chain projects to the same side of the sheet.

The α-helix, β-sheet, and other secondary structures are combined in many ways in a protein. The combination of α- and β-strand produces various kinds of domains in proteins. Such independently folded domains may have structural-functional motifs that may be involved in DNA binding, as we discussed in the Chapter 3.

Tertiary Structure

The tertiary (3°) structure of a protein is the three-dimensional arrangement of all the atoms in a single polypeptide. The interactions between the side chains play an important role in the folding of proteins, including the peptide backbone hydrogen bonding, hydrogen bonds between polar residues, hydrophobic interactions between nonpolar residues, electrostatic interaction between oppositely charged groups, and disulfide bonds between the side chains of cysteines (Fig. 4.11).

Quaternary Structure

Quaternary (4°) structure is the final level of protein structure, which is the way in which two or more individual polypeptide chains fit together in a complex protein. Each of these polypeptide chains is called a subunit. The number of subunits can range from two to more than a dozen, and subunits may be identical or different. Commonly occurring examples are dimers, trimers, and tetramers, consisting of two, three, and four polypeptide subunits, respectively. The subunits

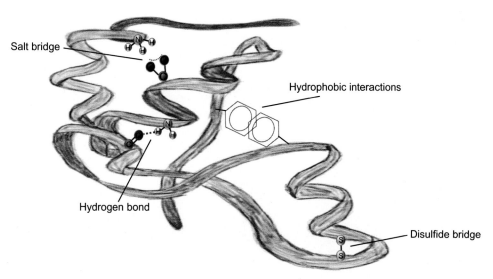

FIG. 4.11 Diagram of the weak interactions that contribute to the tertiary structure of proteins.

interact with one another noncovalently via electrostatic attractions, hydrogen bonds, and hydrophobic interactions. Most intracellular enzymes are oligomeric and may be composed either of a single type of monomer subunit (homomultimers) or of several different kinds of subunits (heteromultimers). The way in which separate folded monomeric protein subunits associate to form the oligomeric protein constitutes the quaternary structure of that protein. Because the quaternary structure is maintained by noncovalent interactions, subtle changes in structure at one site of the protein molecule may cause dramatic changes in properties at a distant site, known as allosteric modulation.

FOLDING INTO THREE-DIMENSIONAL PROTEIN STRUCTURE

All of the information needed to fold the protein into its native tertiary structure is contained within the primary structure of the peptide chain itself. Sometimes chaperones assist in the process of protein folding in the cell. Proteins in dilute solution can be unfolded and refolded without the assistance of such chaperones. Proteins can be grouped into three large classes based on their structure and solubility: fibrous proteins, globular proteins, and membrane proteins.

Fibrous Proteins

Fibrous proteins contain polypeptide chains organized approximately in parallel along a single axis, producing long fibers or large sheets. Such proteins tend to be mechanically strong and resistant to solubilization in water. Fibrous proteins often play a structural role in nature. For example, α-keratin is composed of α-helical segments of polypeptides and is the predominant constituent of claws, fingernails, hair, and horn in mammals. The structure of the α-keratin is dominated by α-helical segments of polypeptide. The structure of the central rod domain of a typical α-keratin is shown in Fig. 4.12. Pairs of right-handed helices wrap around each other to form a left-twisted coiled coil.

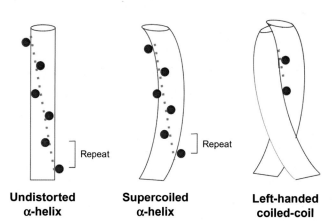

Undistorted α-helix

Supercoiled α-helix

Left-handed coiled-coil

FIG. 4.12 The structure of the central rod domain of a typical α-keratin is shown. Pairs of right-handed helices wrap around each other to form a left-twisted coiled coil. Shown also is the periodicity of hydrophobic groups in helical arrangements.

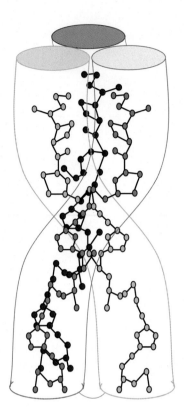

FIG. 4.13 Triple helix structure of a collagen fiber.

Collagen is a rigid, inextensible fibrous protein that is composed of three intertwined polypeptide chains and is a principal constituent of connective tissue in animals, including tendons, cartilage, bones, teeth, skin, and blood vessels. Collagen is the most abundant protein in vertebrates. It is organized in water-insoluble fibers of great strength. A collagen fiber consists of three polypeptide chains wrapped around each other in a ropelike twist, or triple helix (Fig. 4.13). The three individual collagen chains are themselves helices that differ from the α-helix. They are twisted around each other in a superhelical arrangement to form a stiff rod, and the three strands are held together by hydrogen bonds involving the hydroxyproline and hydroxylysine residues.

Globular Proteins

Globular proteins are named for their approximately spherical shapes and are the most abundant proteins in nature. The globular proteins exist in an enormous variety of three-dimensional structures. Nearly all globular proteins contain substantial numbers of α-helices and β-sheets folded into a compact structure that is stabilized by both polar and nonpolar interactions (Fig. 4.14). The globular three-dimensional structure forms spontaneously and is maintained as a result of interactions among the side chains of the amino acids. Most often, the hydrophobic amino acid side chains are buried, closely packed, in the interior of a globular protein, out of contact with water. Hydrophilic amino acid side chains lie on the surface of the globular proteins exposed to the water. Consequently, globular proteins are usually very soluble in aqueous solutions. The diversity of protein structures reflects the remarkable variety of functions performed by the globular proteins: binding, catalysis, regulation, transport, immunity, cellular signaling, and more.

Proteins range in molecular weight from 1000 to more than 1 million daltons (Da), but the folded size of a globular protein is not necessary correlated to its molecular weight. Proteins composed of about 250 amino acids or less often have a simple, compact globular shape. Larger globular proteins are usually made up of two or more recognizable and distinct structures, termed domains or modules. These are compact, folded protein structures that are usually stable by themselves in aqueous solution. Typical domain structures consist of hydrophobic cores with hydrophilic surfaces. Individual domains often possess unique functional behaviors and often perform unique functions within the larger protein in which they are found.

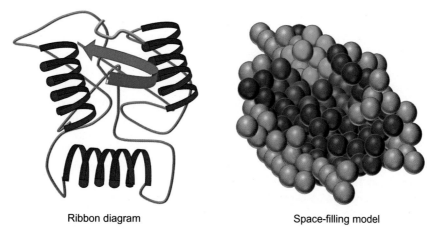

Ribbon diagram   Space-filling model

FIG. 4.14   The three-dimensional structure of a polypeptide as a ribbon diagram and a space-filling model. This typical globular structure contains substantial amounts of alpha-helices and beta-sheets folded into a compact structure.

Membrane Proteins

Membrane proteins are found in association with the various membrane systems of cells. For interaction with the nonpolar phase within membranes, membrane proteins have hydrophobic amino acid side chains oriented outward (Fig. 4.15). For example, proteins in a biological membrane can be associated with the lipid bilayer in either of two ways: as peripheral proteins on the surface of the membrane, or as integral proteins within the lipid bilayer. Peripheral proteins are usually bound to the charged head groups of the lipid bilayer by polar interactions, electrostatic interactions, or both. They can be removed by increasing the ionic strength of the medium. On the other hand, integral proteins are insoluble in aqueous solutions but can be solubilized in detergents.

Membrane proteins have a variety of functions. For example, transport proteins help move substances in and out of the cell, and receptor proteins are important in the transfer of extracellular signals. Some membrane proteins responsible for aerobic oxidation reactions are tightly bound to mitochondrial membrane. Some of these proteins are on the inner surface of the membrane, and some are on the outer surface. Typically, these proteins are unevenly distributed on the inner and outer layers of all cell membranes.

FIG. 4.15   Ribbon model representation of a globular protein versus a membrane protein. Membrane proteins fold so that hydrophobic amino acid side chains are exposed in their membrane-associated regions.

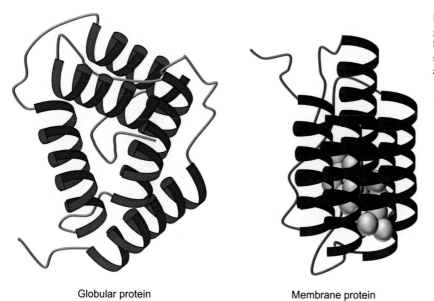

Globular protein   Membrane protein

Protein Denaturation

The primary structure of proteins arises from the amide covalent bonds, whereas the secondary, tertiary, and quaternary structures of proteins are maintained by weak, noncovalent forces as discussed previously. These noncovalent interactions that maintain the three-dimensional structure of a protein can be disrupted easily. Therefore, a variety of environmental conditions, including temperature and pH, can disrupt these weak forces in a process known as denaturation, the loss of protein structure and function (Fig. 4.16). Chemical agents, such as organic solvents, detergents, and particular denaturing solutes, can also denature proteins. For example, sodium dodecyl sulfate (SDS) tends to disrupt hydrophobic interactions and electrostatic interactions. Urea and guanidine hydrochloride form hydrogen bonds with the protein that are stronger than those within the protein itself. As such, they can disrupt hydrophobic interactions as SDS does. The reducing agent β-mercaptoethanol is often used to reduce disulfide bridges to two sulfhydryl groups. Denaturation in all these cases involves disruption of the weak forces that stabilize proteins. On the other hand, covalent bonds that maintain the primary structure are not affected. The process of denaturation and refolding is a dramatic demonstration of the relationship between the primary structure of the protein and the forces that determine the tertiary structure.

THE GENETIC CODE

Ribonucleotide Bases Are Used as Letters in the Genetic Code

The genetic code is written in linear form, using the ribonucleotides that compose mRNA molecules as letters. The ribonucleotide sequence is derived from the complementary nucleotide bases in the DNA template strand. Therefore, the nucleotide sequence is exactly the same as the DNA coding strand. Each genetic code consists of three ribonucleotide letters, thus referred to as a triplet code. As such, a genetic code is a triplet code in which a sequence of three bases is needed to specify one amino acid. The genetic code translates the RNA sequences into the amino acid sequence (Fig. 4.17). Each group of three ribonucleotides, called a codon, specifies one amino acid. These codes are unambiguous, as each triplet specifies only a single amino acid. Thus, one would imagine that a codon would be at least three bases long. With three bases, there are $4^3 = 64$ codons, which is more than enough to encode the 20 amino acids. Therefore, the genetic code is degenerate, which means more than one triplet can encode the same amino acid. Each amino acid can have more than one codon, but no codon can encode more than one amino acid. Furthermore, the genetic code is universal, as the code can be used by all viruses, prokaryotes, archaea, and eukaryotes.

Wobble Rules

All 64 codons have been assigned meaning, with 61 of them coding for amino acids and the remaining 3 serving as the termination signals, also called nonsense codons (Table 4.1). Multiple codons for a single amino acid are not randomly distributed but have one or two bases in common. The bases that are common to several codons are usually the first

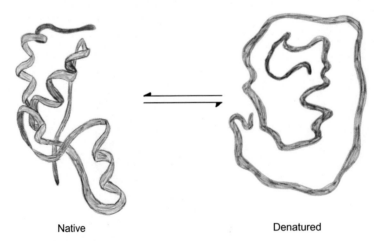

Native   Denatured

FIG. 4.16 The reversible nature of denaturing a protein.

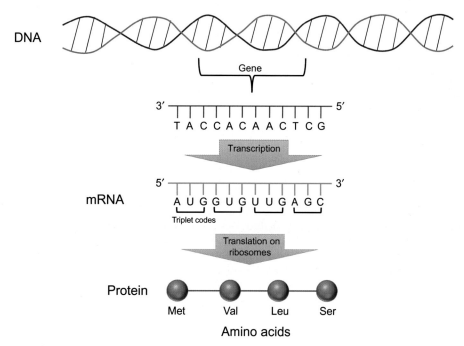

FIG. 4.17 Flowchart demonstrating the central dogma of biology in which DNA is transcribed to mRNA, which is then translated into an amino acid sequence of a protein.

TABLE 4.1 Degenerative Genetic Code

First letter		Second letter				Third letter
		U	**C**	**A**	**G**	
U	U	UUU UUC }Phe UUA UUG }Leu	UCU UCC UCA UCG }Ser	UAU UAC }Tyr **UAA Stop** **UAG Stop**	UGU UGC }Cys **UGA Stop** UGG Trp	U C A G
	C	CUU CUC CUA CUG }Leu	CCU CCC CCA CCG }Pro	CAU CAC }His CAA CAG }Gln	CGU CGC CGA CGG }Arg	U C A G
	A	AUU AUC AUA }Ile **AUG Met**	ACU ACC ACA ACG }Thr	AAU AAC }Asn AAA AAG }Lys	AGU AGC }Ser AGA AGG }Arg	U C A G
	G	GUU GUC GUA GUG }Val	GCU GCC GCA GCG }Ala	GAU GAC }Asp GAA GAG }Glu	GGU GGC GGA GGG }Gly	U C A G

TABLE 4.2 The Wobble Hypothesis Base Pairing at Third Codon

Base on the Anticodon	Base Recognized on the Codon
U	A,G
C	G
A	U
G	U,C
I (inosine; a modified base in tRNA)	U,C,A

and second bases, with more room for variation in the third base, which is called the wobble base. The wobble rules indicate that a first-base anticodon U could recognize either an A or G in the codon third-base position, and a first-base anticodon G might recognize either U or C in the third-base position of the codon (Table 4.2). Because the degenerate codons for a given amino acid differ in the third base, a given tRNA can base-pair with several codons. Thus, fewer different tRNAs are needed.

TRANSLATIONAL MACHINERY: RIBOSOME

Translation of mRNA is the biological polymerization of amino acids into polypeptide chains. This process requires mRNA, tRNA, and a ribosome, which is composed of rRNAs. In association with a ribosome, mRNA presents a triplet codon that calls for a specific amino acid. A specific tRNA molecule contains within its nucleotide sequence three consecutive ribonucleotides complementary to the codon, called the anticodon, which can base-pair with the codon. Another region of this tRNA is covalently bonded to the codon's corresponding amino acid. Hydrogen bonding of tRNA to mRNA holds amino acids in proximity to each other so that a peptide bond can be formed. This process occurs repeatedly as mRNAs run through the ribosome, and amino acids are polymerized into a polypeptide. Before we further discuss this process, we will need to understand the structure of both rRNAs and tRNA.

The Components of Ribosomes

A bacterial ribosome is a complex molecular machine and it consists of two subunits, one large and one small. Both subunits consist of one or more molecules of rRNA and ribosomal proteins. These two distinct subunits perform the disparate tasks of mRNA reading, tRNA recognition, and polypeptide chain elongation. The small subunit (30S in prokaryotes and 40S in eukaryotes) contains the decoding site where the codon-anticodon interaction is deciphered, whereas the large subunit (50S in prokaryotes and 60S in eukaryotes) contains the active site where the peptidyl transfer and hydrolysis reactions occur.

Despite their conserved molecular function, eukaryotic and prokaryotic ribosomal subunits differ significantly in size and complexity. These differences may reflect an additional regulation of translation and is also the foundation of several antibiotics that specifically block the function of the prokaryotic subunits. In general, two-thirds of the mass of the ribosome is made up of rRNA, and the other third is composed of ribosomal proteins. The subunit and rRNA components are most easily isolated and characterized on the basis of their sedimentation behavior in sucrose gradient. The velocity of sedimentation has been standardized in units called Svedbergs (S). Because the mass and shape of a molecule determine its migration, the greater the mass, the greater the sedimentation velocity.

The large subunit in prokaryotes consists of a 23S rRNA molecule, a 5S rRNA molecule, and 34 ribosomal proteins; the small subunit consists of a 16S rRNA component and 21 ribosomal proteins (Fig. 4.18). On the other hand, the large subunit in eukaryotes consists of a 28S rRNA, a 5.8S rRNA, an 5S rRNA, and 46 ribosomal proteins, and the small subunit consists of 18S rRNA and 33 ribosomal proteins. The small subunit mediates the interactions between mRNA and tRNA. The large subunit catalyzes peptide bond formation. These distinct events are integrated at the interface between the two subunits, where gross movement of the interface can shift the position of the tRNA-mRNA complex as amino acids are sequentially added to the growing polypeptide chain. These movements are coordinated by specific bridge elements between the two subunits, which are composed of RNA and proteins.

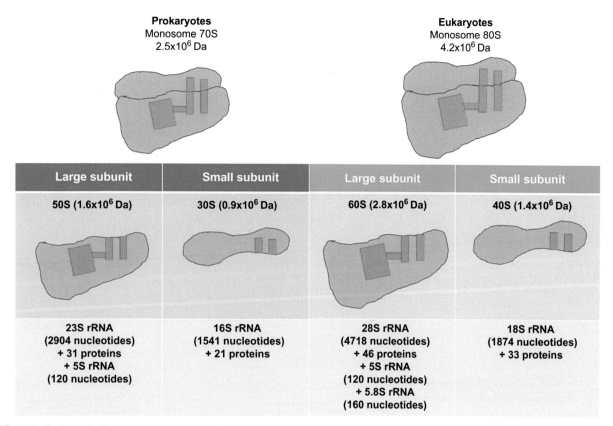

FIG. 4.18   Prokaryotic ribosome composition versus eukaryotic ribosome composition.

Ribosomal Structure

Each of the major rRNAs has a secondary structure with several discrete domains. Four general domains are formed by 16S rRNA, in which just under half of the sequence is base paired. Six general domains are found by 23S rRNA. The three-dimensional structure of both the 30S and 50S ribosomal subunits shows that the general shapes of the ribosomal subunits are determined by the conformation of the rRNA molecules within them. Ribosomal proteins serve a largely structural role in ribosomes, and their primary function is to brace and stabilize the rRNA conformations within the ribosomal subunits.

The organization of the rRNA domains in the two ribosomal subunits is strikingly different. In the small subunit each rRNA domain is largely limited to one area of the subunit, whereas in the large subunit the rRNA elements are intricately interwoven. This organization reflects the functional differences of the subunits. In the small subunit the interdomain flexibility is suited to large-scale movements. On the other hand, the stability provided by the interwoven RNA domains is suited to protection of the active site for protein synthesis by the large subunit. Therefore, the central domain of ribosome binds mRNA and the anticodon stem-loop end of aminoacyl-tRNAs, providing the framework for decoding the genetic information in mRNA by codon-anticodon recognition.

The central domain of the 30S subunit that serves as the decoding center is composed only of 16S rRNA. 16S rRNA molecules fold into characteristic secondary structures as a consequence of intramolecular base-pairing interactions (Fig. 4.19). The 3′ end of the 16S rRNA interacts directly with mRNA at initiation and the specific regions of 16S rRNA interact directly with the anticodon regions of tRNA in both the A site and the P site (see detail later). On the other hand, the 50S subunit binds the aminoacyl-acceptor ends of the tRNAs and is responsible for catalyzing formation of the peptide bond between successive amino acids in the polypeptide chain. This catalytic center is called peptidyl transferase and is located at the bottom of a deep cleft. The 23S rRNA interacts with the CCA terminus of peptidyl-tRNA in both the P site and A site. Therefore, these two distinct subunits perform the disparate tasks of tRNA recognition and polypeptide chain elongation. During translation initiation, the two subunits come together to form the 70S (80S in eukaryotes) ribosome and launch the elongation cycle.

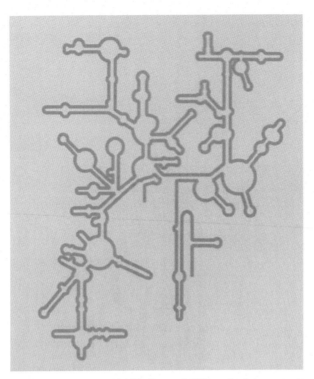

FIG. 4.19   Hydrogen-bonded complimentary base sequences of 16S ribosomal RNA maintaining secondary structure. Aligned regions represent hydrogen-bonded complementary base sequences.

TRANSLATIONAL MACHINERY: tRNA

The Structure of tRNA

To understand how a tRNA carries out its functions, we need to know the structure of the molecule. The tRNA is composed of only 75 to 90 nucleotides, displaying a nearly identical structure in both prokaryotes and eukaryotes. Just as a protein has primary, secondary, and tertiary structure, so does a tRNA. The tRNA is transcribed from DNA as large precursors. The primary structure is the linear sequence of bases in the RNA; the secondary structure is the stem-loops structure derived from the base pairing between different regions of the sequence; and the tertiary structure is the overall three-dimensional shape of the molecule.

The Secondary Structure of tRNA

In general, the tRNA secondary structure adopts a cloverleaf shape. The cloverleaf has four base-paired stems that define the four major regions of the molecule (Fig. 4.20). The first, seen at the top of the diagram, is the acceptor stem. The acceptor arm has both 5′ and 3′ ends of the tRNA, and these two ends are base-paired to each other. The 3′ end, bearing the invariant sequence CCA, protrudes beyond the 5′ end. On the left is the dihydrouracil loop (D loop). This is named for the modified uracil bases in this region. At the bottom is the anticodon loop. This region base-pairs with an mRNA codon and therefore allows decoding of the mRNA. At right is the T loop because of its invariant sequence of TΨC. The Ψ is a modified uridine. The region between the anticodon loop and T loop is called the variable loop because of its variable length. Tertiary structure in tRNA arises from base-pairing interactions between bases in the D loop with bases in the variable and TΨC loops (Fig. 4.21). These base-pairing interactions involve the invariant nucleotides of tRNAs. These interactions fold the D and TΨC arms together and bend the cloverleaf into the stable L-shaped tertiary form. Many of these base-pairing interactions involve base pairs that are not canonical A:T or G:C pairing.

The Active Form of tRNA

For accurate translation to occur, codon recognition must be achieved by aminoacyl-tRNA, which is a charged, activated form of tRNA. The process of tRNA activation, catalyzed by the aminoacyl-tRNA synthetases, is discussed in the next

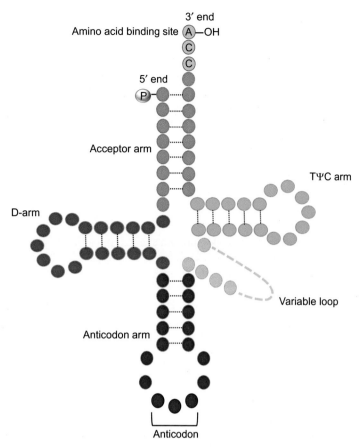

FIG. 4.20 Cloverleaf model of tRNA with hydrogen bonds represented as dotted lines.

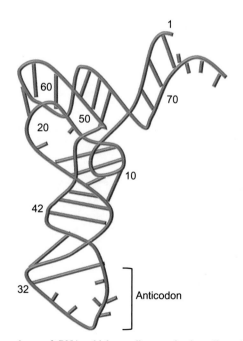

FIG. 4.21 Example of hydrogen bonds between bases of tRNA, which contributes to its three-dimensional structure.

section. The appropriate aminoacyl-tRNA reads the codon through base-pairing via its anticodon loop. Subsequently, the specific amino acid is passively added to the growing peptide chain. Therefore, it is important that synthases be able to discriminate between the various tRNAs and aminoacylate their cognate tRNAs. This is relevant to the so-called second genetic code, which we will also discuss later in this chapter.

PREPARATION FOR THE TRANSLATION

As we discussed previously, tRNA itself does not carry aminoacyl and an amino acid needs to be incorporated into tRNA so it becomes functional. Thus two important events must occur before translation initiation can take place. One is to generate a supply of aminoacyl-tRNA, which is the tRNA activation, and the other is the dissociation of ribosomes into their two subunits.

Activation of tRNA

Before an amino acid can be incorporated into a growing protein chain, it must first be activated. This process is also called tRNA charging, which involves both tRNA and aminoacyl-tRNA synthetases. All tRNAs have the same three bases (CCA) at their 3′-ends, and the terminal adenosine is the target for charging. In the first step, the amino acid forms a covalent bond to an adenine nucleotide, providing an aminoacyl-AMP (Fig. 4.22). The free energy of hydrolysis of ATP provides energy for bond formation, which is trapped in the aminoacyl-AMP. This is why we call this an activated amino acid. In the second part of the reaction, an ester linkage is formed between the amino acid and either the 3′-hydroxyl or the 2′-hydroxyl of the ribose at the 3′-end of the tRNA, forming aminoacyl-tRNA. The aminoacyl-tRNA is now ready to join the ribosome for peptide formation.

Dissociation of Ribosomes

In *Escherichia coli*, the 70S ribosomes consist of one 30S subunit and one 50S subunit. The 30S subunit binds the mRNA and the anticodon ends of the tRNAs. It is the decoding agent of the ribosome that reads the genetic code in the mRNA and allows binding with the appropriate aminoacyl-tRNAs. The 50S subunit binds the ends of the tRNAs that are charged with amino acids and has the peptidyl transferase activity that links amino acids together through peptide bonds. Thus, *E. coli* ribosomes need to dissociate into subunits at the end of each round of translation for a new initiation complex to form.

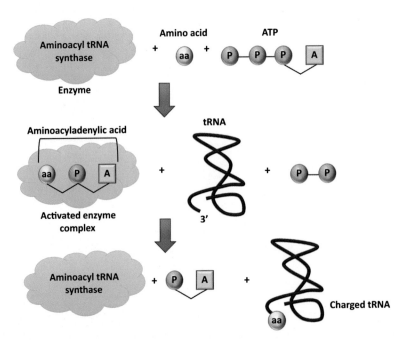

FIG. 4.22 Steps involved in charging of tRNA with an amino acid.

Bacteria have a ribosome release factor that acts in conjunction with an elongation factor to separate the subunits. In addition, an initiation factor, IF3, binds to the small subunit and keeps it from reassociating with the large subunit before the formation of the initiation complex.

AMINOACYL-tRNA SYNTHETASES

Classification of Aminoacyl-tRNA Synthetase

Although all aminoacyl-tRNA synthetases catalyze common enzymatic functions, they are a diverse group of proteins in terms of size and amino acid sequence. They can be divided into class I and class II families based on mutually exclusive sets of sequence motifs, structural domains, and acylation function (Table 4.3). Class I and class II synthetases are functionally differentiated in a number of ways. Class I aminoacyl-tRNA synthetases first add the amino acid to the 2'-OH of the terminal adenylate residue of tRNA before shifting it to the 3'-OH. On the other hand, class II aminoacyl-tRNA synthetases add the amino acid to the 3'-OH directly (Fig. 4.23). These two classes of enzyme appear to be unrelated and indicate a convergent evolution.

Second Genetic Code

Several tRNAs can exist for each amino acid, but a given tRNA does not bond to more than one amino acid. The aminoacyl-tRNA synthetase enzyme requires Mg^{2+} and is highly specific both for the amino acid and for the tRNA. Each synthetase exists for each amino acid, and this synthetase functions for all the different tRNA molecules for that amino acid. Eukaryotic cells and some bacteria have 20 different aminoacyl-tRNA synthetases, one for each amino acid. Each of these enzymes catalyzes ATP-dependent attachment of its specific amino acid to the 3'-end of its cognate tRNA molecules. The synthetase ensures that the right amino acid pairs up with the right tRNA. Once the aminoacyl-tRNA has been synthesized, the amino acid part makes no contribution to accurate translation of the mRNA. This is achieved by the codon recognition through base pairing via anticodon of aminoacyl-tRNA. Therefore, it is important that the right amino acid be matched with its proper tRNA.

A second genetic code is the code by which each aminoacyl-tRNA synthetase matches up its amino acid with tRNAs that can interact with codons specifying its amino acid. The presence of a second genetic code can help an aminoacyl-tRNA synthetase discriminate among the 20 amino acids and the many tRNAs and uniquely pick out its proper substrate. The appropriate tRNAs are those having anticodons that can base-pair with the codons specifying the particular amino acid. The second genetic code is crucial to the fidelity of amino acid transfer. Therefore, the specificity of the enzyme contributes to the accuracy of the translation process.

TABLE 4.3 Aminoacyl-tRNA Synthetase Classes

Class I	Class II
Arg	Ala
Cys	Asn
Gln	Asp
Glu	Gly
Ile	His
Leu	Lys
Met	Phe
Trp	Pro
Tyr	Ser
Val	Thr

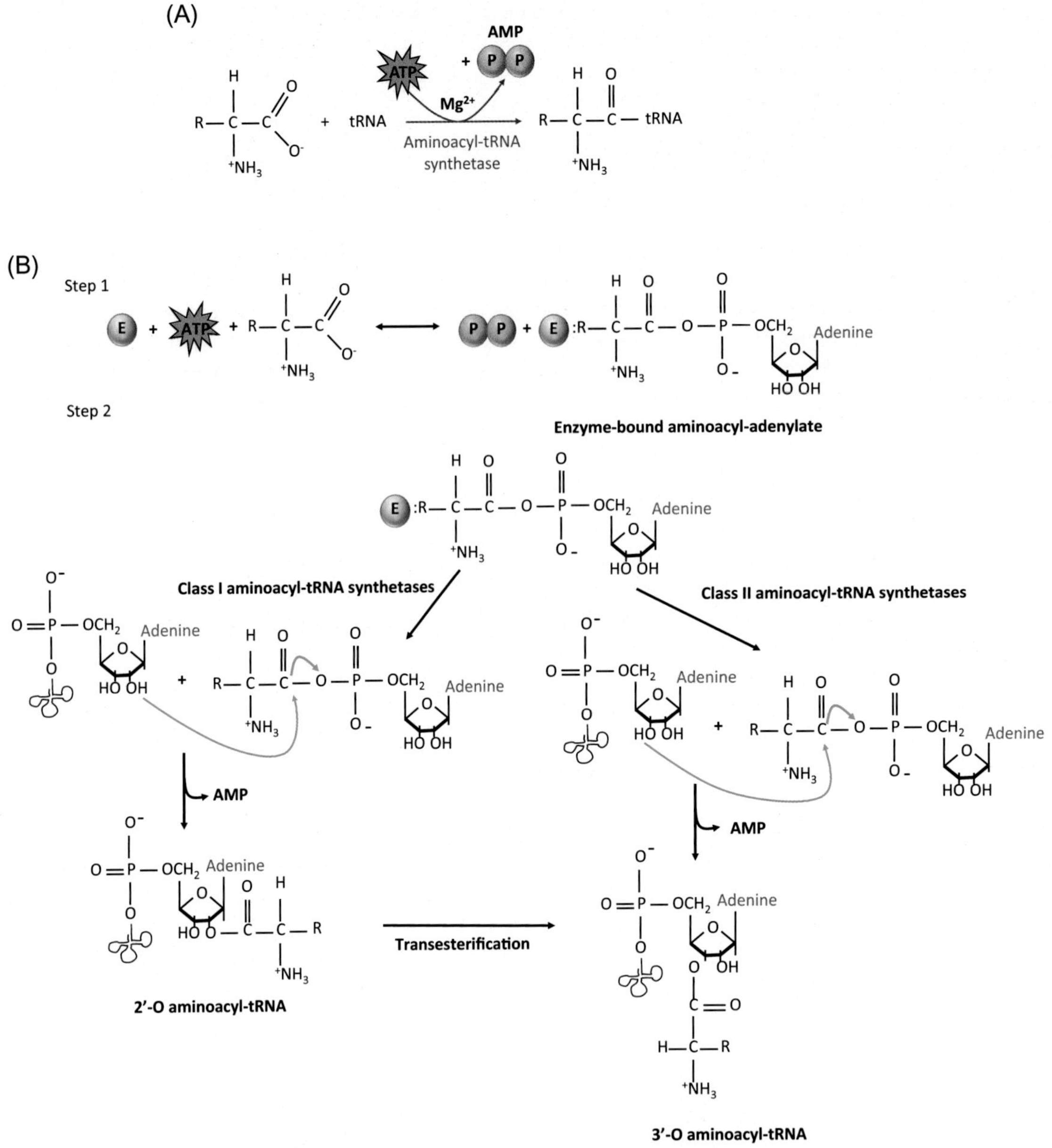

FIG. 4.23 The overall aminoacyl-tRNA synthetase reaction (A) the overall reaction. (B) Aminoacyl-tRNA formation proceeds in two steps: (Step 1) formation of an aminoacyl-adenylate and (Step 2) transfer of the activated amino acid moiety of the mixed anhydride to either the 2′-OH (class I aminoacyl-tRNA synthetases) or 3′-OH (class II aminoacyl-tRNA synthetases) of the ribose on the terminal adenylic acid at the 3′-CCA terminus common to all tRNAs. Those aminoacyl-tRNAs formed at the 2′-aminoacyl esters undergo a transesterification that move the aminoacyl function to the 3′-O of tRNAs. Only the 3′-esters are substrates for protein synthesis.

For most tRNAs, a set of sequence elements is recognized by its specific aminoacyl-tRNA synthase, rather than a single distinctive nucleotide or base pair. These elements include one or more of the following: (1) at least one base in the anticodon; (2) one or more of the three base pairs in the accepter stem; and (3) the unpaired base preceding the CCA end (canonical position 73), referred to as the discriminator base because this base is invariant in the tRNAs for a particular amino acid. The location of nucleotides that contribute to specific recognition by the respective aminoacyl-tRNA synthases for each of the 20 amino acids are shown in Fig. 4.24. These tRNA features are examples of the second genetic code that can help aminoacyl-tRNA synthetases discriminate between the various tRNAs.

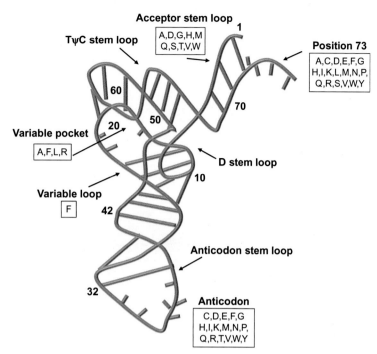

FIG. 4.24   tRNA tertiary structure as represented by a ribbon diagram. Numbers represent nucleotides recognized by aminoacyl-tRNA synthetases, while letters represent the amino acids of the synthetase interacting at that location.

Proofreading Activity of Aminoacyl-tRNA Synthetases

The aminoacyl-tRNA synthetases display an overall error rate of about one in 10,000. This very low frequency of errors is achieved by two mechanisms. First, the aminoacyl-tRNA synthetases make an intricate series of contacts with both their amino acid and tRNA in the enzyme active site, which ensures for the most part that only the correct substrates are selected from the large cellular pool of similar candidates. Second, the enzymes possess a variety of proofreading (editing) activities that serve to hydrolyze the mismatched amino acid either before or after transfer to tRNA. The proofreading mechanism involves water-mediated hydrolysis of the mischarged tRNA. The correct product is not hydrolyzed because of steric exclusion. As such, the error rate is very low.

PROKARYOTIC TRANSLATION

Translation Initiation

Protein synthesis requires the specific binding of mRNA and aminoacyl-tRNAs to the ribosome. The stages of translation can be divided into chain initiation, chain elongation, and chain termination. For the mRNA to be translated correctly, the ribosome must be placed at the correct start location. Base pairing between a pyrimidine-rich sequence at the 3'-end of 16S rRNA and complementary purine-rich tracts at the 5'-end of mRNAs positions the 30S ribosomal subunit in proper alignment with an initiation codon on the mRNA. The purine-rich mRNA sequence is called the Shine-Dalgarno sequence (5'-GGAGGU-3') and is the ribosome binding site (Fig. 4.25). The start of translation requires the formation of an initiation complex, including mRNA, the 30S ribosomal subunit, fmet-tRNA^{fmet}, GTP (guanosine-5'-triphosphate), and three protein initiation factors, named IF-1, IF-2, and IF-3.

In prokaryotes, the initial N-terminal amino acid of all proteins is *N*-formylmethionine (fmet), which is usually removed after the synthesis. There are two different tRNAs for methionine in *E. coli*, one for unmodified methionine (tRNA^{met}) and one for *N*-formylmethionine (tRNA^{fmet}). The tRNA^{fmet} recognizes the start signal (AUG) when it occurs at the beginning of the mRNA sequence that directs the synthesis of the polypeptide. The tRNA^{met} recognizes the AUG when it is found in an internal position in the mRNA sequence. The aminoacyl-tRNAs that they form with methionine are called met-tRNA^{fmet} and met-tRNA^{met}, respectively. The met-tRNA^{fmet} is further formylated to fmet-tRNA^{fmet}.

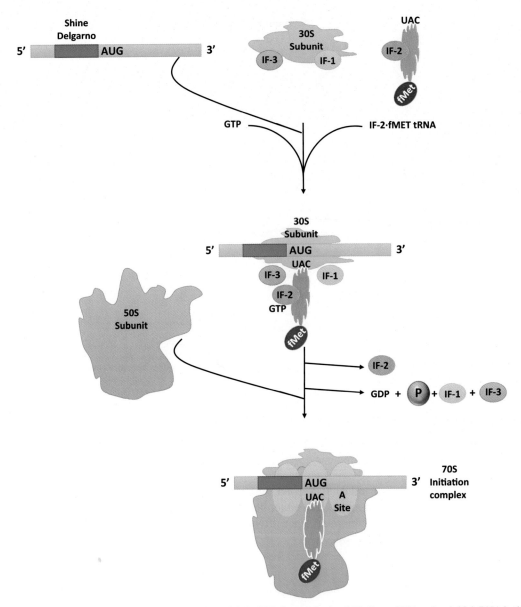

FIG. 4.25 The formation of the initiation complex of translation with the 30S ribosomal subunit binding mRNA and an initial tRNA in the presence of initiation factors, IF-1, IF-2, and IF-3. The 50S ribosomal subunit is then added to form the 70S complex.

The IF3 protein facilitates the binding of mRNA to the 30S ribosomal subunit (Fig. 4.25). It also prevents premature binding of the 50S subunit. IF2 binds GTP and aids in the selection of the initiator tRNA (fmet-tRNA^{fmet}). Although the function of IF-1 is less clear, it appears to bind to IF-3 and IF-2 and to catalyze the separation of the 30S and the 50S ribosomal subunits being recycled for another round of translation. The combination of mRNA, the 30S ribosomal subunit, and fmet-tRNA^{fmet} is the 30S initiation complex. A 50S ribosomal subunit binds to the 30S initiation complex to produce the 70S initiation complex.

Translation Elongation

Chain elongation begins with the addition of the second amino acid specified by the mRNA to the 70S initiation complex (Fig. 4.26). The elongation phase uses the fact that three binding sites for tRNA are present on the 50S subunit of the 70S ribosome. The three tRNA binding sites are called the P (peptidyl) site, the A (aminoacyl) site, and the E (exit) site.

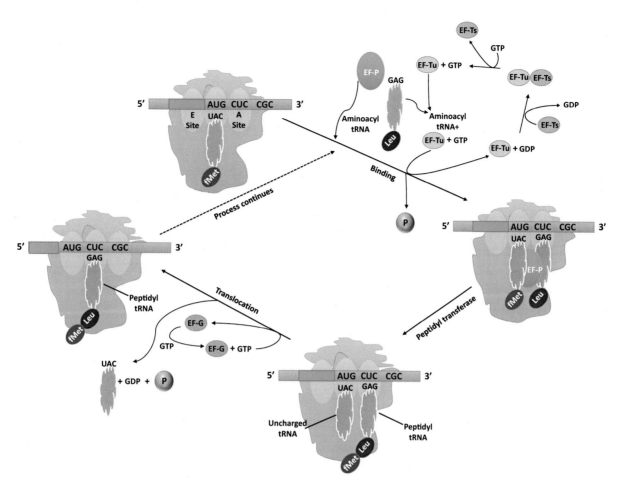

FIG. 4.26 The steps of translational elongation. The initiator tRNA, fMet, is bound to the ribosomal P-site, while a second tRNA carrying the next amino acid arrives at the A-site. Peptidyl transferase then catalyzes the removal of the initiator amino acid to the second amino acid, with the growing chain now located in the A-site. In the translocation step, the ribosome now undergoes a shift so that the uncharged tRNA in the exit, while the growing peptidyl chain is located at the P-site. The A-site is now ready to take in the next charged tRNA.

The P site binds a tRNA that carries a peptide chain. Initially, the P site is occupied by the fmet-tRNA^{fmet} in the 70S initiation complex. Because the A site binds an incoming aminoacyl-tRNA, the second aminoacyl-tRNA binds at the A site. A triplet of tRNA bases forms hydrogen bonds with a triplet of mRNA bases. In addition, GTP and three elongation factors, EF-P, EF-Tu, and EF-Ts, are required. EF-Tu guides the aminoacyl-tRNA into part of the A site and aligns the anticodon with the mRNA codon. When the match is correct, the aminoacyl-tRNA inserts into the A site completely. GTP is hydrolyzed and EF-Tu dissociates. EF-P is bound adjacent to the P site and E site and is thought to help the formation of the first peptide bond. EF-Ts helps the regeneration of EF-Tu-GTP. A peptide bond is formed through a reaction catalyzed by peptidyl transferase. Then the amino acid is transferred to the aminoacyl-tRNA at the A site so it becomes a dipeptidyl-tRNA and a tRNA with no amino acid attached at the P site.

Subsequently, the uncharged tRNA moves from the P site to the E site, and the peptidyl-tRNA moves from the A site to the vacated P site. The E site thus carries an uncharged tRNA that is about to be released from the ribosome. In addition, the mRNA moves with respect to the ribosome. Another elongation factor, EF-G, is required at this point, and once again GTP is hydrolyzed to GDP and P$_\text{i}$. This is the translocation step. Therefore, the chain elongation has three steps: aminoacyl-tRNA binding, peptide bond formation, and translocation.

During the translation elongation, the decoding center is the 30S subunit. This decoding center is where anticodon loops of the A- and P-site tRNAs and the codons of the mRNA are matched up. This is primarily a property of 16S rRNA. The conformational changes of the 16S rRNA triggered by the interaction between 16S rRNA and mRNA are keys to codon-anticodon recognition. On the other hand, peptidyl transfer is the central reaction of protein synthesis, the actual peptide bond–forming step. No energy input is needed, and this reaction is a property of the 23S rRNA in the 50S subunit.

Translation Termination

Chain termination requires a stop signal. The codons UAA, UAG, and UGA are the stop signals. These codons are recognized by released factors RF-1 or RF-2, which bind to UAA and UAG or to UAA and UGA, respectively (Fig. 4.27). On the other hand, RF-3 facilitates the activity of RF-1 and RF-2. Either RF-1 or RF-2 is bound near the A site of the ribosome when one of the termination codons is reached. The released factor not only blocks the binding of a new aminoacyl-tRNA but also affects the activity of the peptidyl transferase so that the bond between the carboxyl end of the peptide and the tRNA is hydrolyzed. Subsequently, the whole complex dissociates, setting

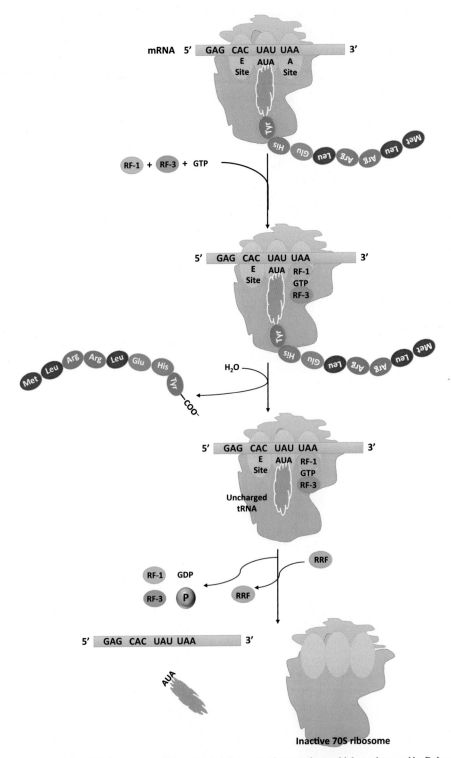

FIG. 4.27   Peptide chain termination in which there are no tRNA molecules that can read stop codons, which are then read by Release Factor (RF) instead of tRNA.

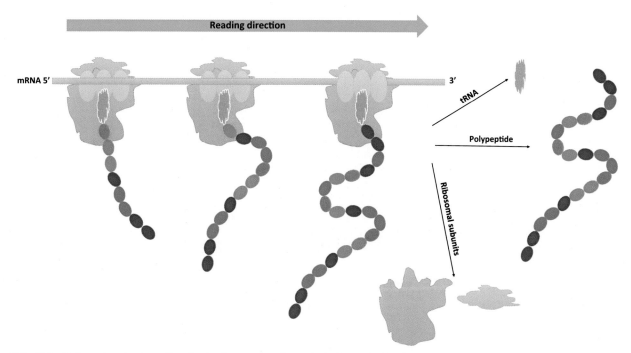

FIG. 4.28   Prokaryotic polysomes allow for simultaneous protein synthesis along a single mRNA molecule.

free the release factors, tRNA, mRNA, and the 30S and 50S ribosomal subunits. All these components can be reused in further protein synthesis.

Coupled Translation

In prokaryotes, translation begins very soon after mRNA transcription. The gene is being simultaneously transcribed and translated, a process called coupled translation. In this situation, several molecules of RNA polymerase are attached to a single gene, giving rise to several molecules of mRNA are transcribed. While each mRNA molecule is still being transcribed, a number of ribosomes attach to it that are in various stages of translating that mRNA. This complex of mRNA with several ribosomes is called a polysome or polyribosome (Fig. 4.28).

EUKARYOTIC TRANSLATION

Initiation

The main features of translation are the same in prokaryotes and eukaryotes except the two major posttranscriptional modifications: 5′ cap and the 3′ poly A tail. The chain initiation of eukaryotic translation is the part that is the most different from prokaryotes. Step 1 in chain initiation involves the assembly of a 43S preinitiation complex that consists of met-tRNA, 40S subunit, GTP, and eukaryotic initiation factors, which include eIF1A, eIF2, and eIF3 (Fig. 4.29). In step 2, the mRNA is recruited. The 5′ cap orients the ribosome to the correct AUG via scanning mechanism. The eIF4A, eIF4B, eIF4E, and eIF4G help the 43S preinitiation complex bind to the mRNA. The 40S subunit, mRNA, and the seven eIFs constitute the 48S preinitiation complex. It moves downstream until it encounters the first AUG in the correct context. In step 3, the 60S ribosome is recruited, forming the 80S initiation complex. GTP is hydrolyzed and the initiation factors are released.

Elongation

Chain elongation in eukaryotes is very similar to that of prokaryotes. The same mechanism of peptidyl transferase and ribosome translocation is seen. However, the structure of the eukaryotic ribosome is different from the prokaryotic ribosome in that there is no E site in eukaryotes, only the A and P site. There are two eukaryotic elongation factors, eEF1 and eEF2. The eEF1 consists of two subunits, eEF1A and eEF1B. The 1A subunit is the counterpart of EF-Tu in prokaryotes, and the 1B subunit is the equivalent of the EF-Ts in prokaryotes. The eEF2 protein is the counterpart of the EF-G in prokaryotes that causes translocation.

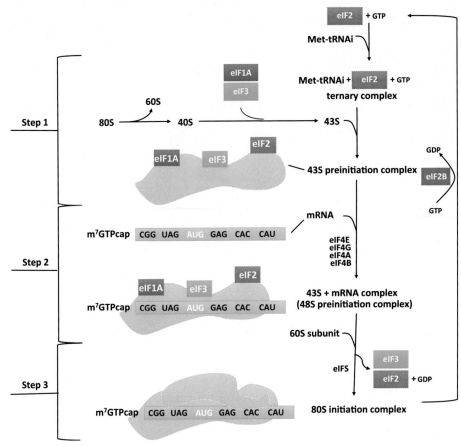

FIG. 4.29 The three stages of eukaryotic initiation of translation.

Termination

As in prokaryotic termination, the ribosome encounters a stop codon: UAA, UAG, or UGA. Unlike prokaryotes, where three release factors are involved, in eukaryotes only one release factor (RF) binds to all three stop codons and catalyzes the hydrolysis of the bond between the C-terminal amino acid and the tRNA. Eukaryotic RF binding to the ribosomal A site is GTP dependent, and RF:GTP binds at this site when it is occupied by a termination codon. Then hydrolysis of the peptidyl-tRNA ester bond, hydrolysis of GTP, release of nascent polypeptide and deacylated tRNA, and ribosome dissociation from mRNA occur in sequence.

POSTTRANSLATIONAL MODIFICATION PROTEINS

Molecular Chaperones

Protein synthesis does not end with the termination of translation. The newly synthesized proteins must fold properly so that they can reach the form in which they have biological activity. Native proteins are folded so that any hydrophobic regions are buried in the interiors of the proteins, away from the aqueous environment in the cell. In principle, the primary structure of the protein conveys enough information to specify its three-dimensional structure. In the cell, the complexity of the process and the number of possible conformations make it less likely that a protein would spontaneously fold into the correct conformation. Instead, proteins require help.

Molecular chaperones can help newly made polypeptides to fold properly so that they will not become misfolded proteins, which may be inactive or deadly toxic to a cell. The ordinary free-standing chaperones can envelop the exposed hydrophobic protein regions in a hydrophobic pocket of their own, preventing inappropriate associations with other exposed hydrophobic regions (Fig. 4.30). Bacteria, archaea, and eukaryotes use this mechanism to avoid misfolding. In *E. coli* cells, they have another special chaperone called trigger factors. It associates with the large ribosomal subunit and catches newly synthesized polypeptide as it emerges from the ribosome's exit tunnel. Thus,

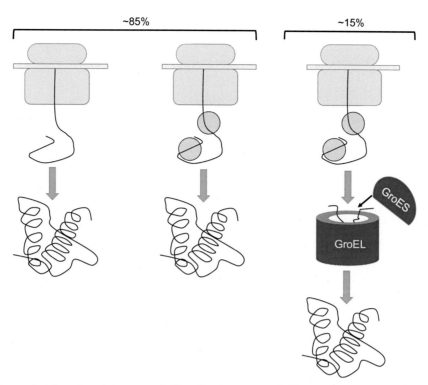

FIG. 4.30   Protein folding can be independent of chaperones, facilitated by chaperones, or facilitated by chaperonin complexes (in this case, both GroEL and GroES).

hydrophobic regions of the nascent polypeptide are protected from inappropriate association until the appropriate partner is available.

Ubiquitinylation: A Process of Protein Degradation

Proteins are in a dynamic state in which they are turned over often. It is believed that a single break in the peptide backbone of a protein is enough to trigger the rapid degradation of the pieces. Protein degradation poses a real hazard to cellular processes. To control this hazard, the degradation pathways are restricted to degradative subcellular organelles, such as lysosomes, or to macromolecular structures called proteasomes. Proteins are directed to lysosomes by specific signal sequences. Once in the lysosome, the destruction is nonspecific. Proteasomes are found in eukaryotic as well as prokaryotic cells. The proteasome is a functionally and structurally sophisticated counterpart to the ribosome.

Regulation of protein levels via degradation is an essential cellular mechanism. In eukaryotes, the most common degradation mechanism in proteasome is by ubiquitinylation. Ubiquitin is a small polypeptide that is highly conserved in eukaryotes. When ubiquitin is linked to a protein, it condemns that protein to destruction in a proteasome. Three proteins in addition to ubiquitin are involved in the ligation process: E1, E2, and E3 (Fig. 4.31). E1 is the ubiquitin-activating enzyme. It becomes attached via a thioester bond to the C-terminal Gly residue of ubiquitin. Ubiquitin is then transferred from E1 to an SH2 group of E2, the ubiquitin-carrier protein. In protein degradation, E2-S-ubiquitin transfers ubiquitin to free amino groups on proteins selected by E3, the ubiquitin-protein ligases. Upon binding a protein substrate, E3 catalyzes the transfer of ubiquitin to this specific protein. Therefore, E3 plays a central role in recognizing and selecting proteins for degradation.

The normal function of the ubiquitin-linked proteasome appears to be quality control. It is estimated that about 20% of cellular proteins are made incorrectly because of mistakes in transcription and translation. These aberrant proteins are potentially damaging to the cell, so they are tagged with ubiquitin and sent to a proteasome for degradation before they can cause any damage.

Sumoylation

Although protein ubiquitination is a signal for protein degradation, it is also used for other nondegradative function in the cells. For example, small ubiquitin-like protein modifiers (SUMOs) are highly conserved proteins and are covalently ligated to target proteins by a three-enzyme conjugating system (Fig. 4.32). SUMO proteins share only limited

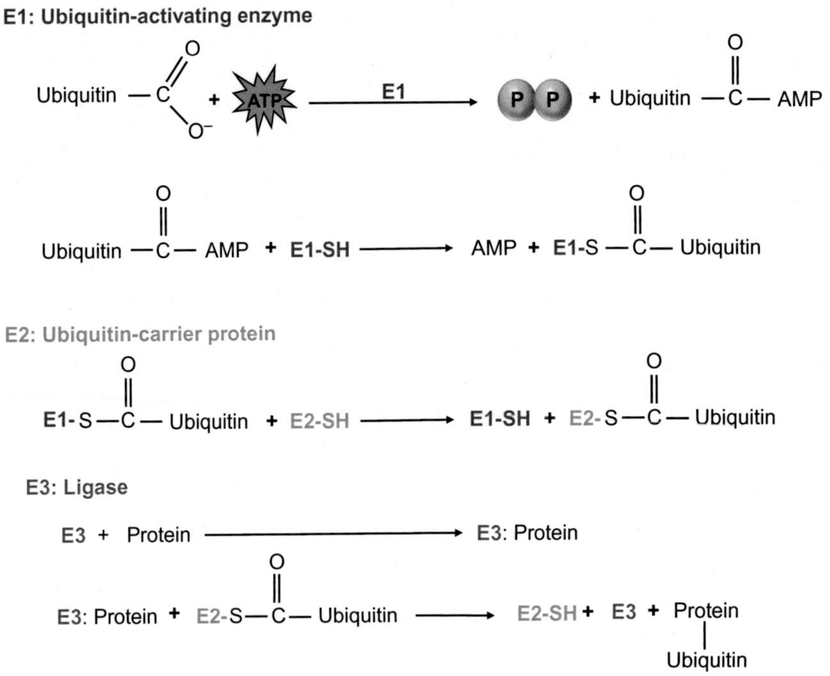

E1: Ubiquitin-activating enzyme

Ubiquitin — C + ATP $\xrightarrow{\text{E1}}$ P P + Ubiquitin — C — AMP

Ubiquitin — C — AMP + E1-SH $\longrightarrow$ AMP + E1-S — C — Ubiquitin

E2: Ubiquitin-carrier protein

E1-S — C — Ubiquitin + E2-SH $\longrightarrow$ E1-SH + E2-S — C — Ubiquitin

E3: Ligase

E3 + Protein $\longrightarrow$ E3: Protein

E3: Protein + E2-S — C — Ubiquitin $\longrightarrow$ E2-SH + E3 + Protein | Ubiquitin

FIG. 4.31 Enzymatic ligation of ubiquitin to proteins via isopeptide bonds formed between the ubiquitin carboxyl-terminus and free amino group on the protein.

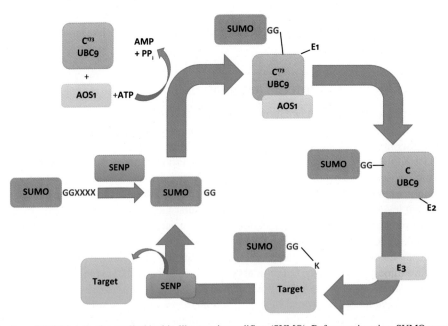

FIG. 4.32 The mechanism of SUMOylation by small-ubiquitin-like protein modifiers (SUMO). Before conjugation, SUMOs need to be proteolytically processed, removing anywhere from 2 to 11 amino acids to reveal C-terminal Gly-Gly motif. SUMOs are then activatived by the E1 enzyme in an ATP-dependent reaction to form a thioester bond between SUMO and E1. SUMO is then transferred to the catalytic Cys residue of the E2 enzyme, Ubc9. Finally, an isopeptide bond is formed between the C-terminal Gly of SUMO and a Lys residue on the substrate protein, through the action of an E3 enzyme. The SUMO-specific protease SENP can deconjugate SUMO from target protein.

homology to ubiquitin, and sumoylation alters the ability of the modified protein to interact with other proteins. Posttranslational attachment of SUMOs is a key regulatory protein modification in eukaryotic cells. These regulatory proteins that are subject to reversible sumoylation may participate in DNA repair, transcription, protein homeostasis, protein trafficking, and signal transduction. Thus, sumoylation represents a protective response.

Chapter 5

The Genome

Chapter Outline

THE EUKARYOTIC CELL CYCLE AND CHROMOSOME DYNAMICS

A functional chromosome not only carries genes for expression but also replicates DNA sequences in the cell cycle. The faithful transmission of genetic information from one generation to the next depends on a cell's ability to make copies of each chromosome and then distribute the complete set of chromosomes to the two daughter cells. Chromosomes therefore have many features to ensure their proper duplication and distribution during the cell-division events.

The Cell Cycle

Before a cell can divide, it must replicate its DNA and segregate the duplicated chromosomes equally to what will become the two daughter cells. This highly coordinated sequence of events that occurs during the division process is called the cell cycle. The cell cycle of a somatic cell has four phases that occurs in sequence: G_1, S, G_2, and M (Fig. 5.1). In the first phase, G_1 (gap phase I), cell growth occurs until cells attain a minimum size that is required to progress to the next phase. During

Diagnostic Molecular Biology. https://doi.org/10.1016/B978-0-12-802823-0.00005-5

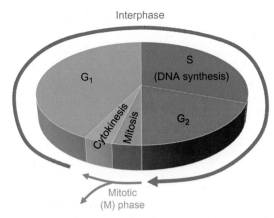

FIG. 5.1 The cell cycle. In a dividing cell, G1, the first part of interphase, is followed by the S phase, in which DNA replication occurs. G2 then completes interphase. In M phase, daughter chromosomes are partitioned, and cytokinesis divides the cytoplasm to produce two identical daughter cells.

the S phase, the DNA is replicated, thereby duplicating all of the chromosomes. The two identical copies of each chromosome are called sister chromatids, and they remain physically associated with one another. Subsequently, the cell prepares for mitosis in G_2 phase. In the M phase, sister chromatids are separated and a complete set of chromosomes is delivered to separate pole of the cell. This process is known as chromosome segregation and is followed by cytokinesis, which completes the division of the mother cell into two daughter cells, each containing the same number of chromosomes.

Chromosome Dynamics During the Cell Cycle

The G_1, S, and G_2 phases are often referred to as interphase, and the M phase is the mitotic phase. During interphase, chromosomes are replicated, and during mitosis they become highly condensed and then are separated and distributed to the two daughter nuclei. The highly condensed chromosomes in a dividing cell are known as mitotic chromosomes. During the portion of the cell cycle when the cell is not dividing, the chromosomes are extended and much of their chromatin exists as long, thin tangled threads in the nucleus (Fig. 5.2).

As the cell enters M phase, the nuclear membrane is disassembled, and sister chromatids condense into compact structures that remain bound together. The centromeres of the condensed sister chromatids bind microtubules, which form the mitotic spindle. The spindle organizes the sister chromatid pair at the center of the cell, which is known as the kinetochore. Subsequently, each sister chromatid moves to the opposite pole on the spindle. The set of chromosomes at each pole become encapsulated by a nuclear membrane, and the cell divides into two daughter cells. Therefore, the chromosome structure is a dynamic structure that can be condensed and extended throughout the cell cycle.

Condensed human mitotic chromosomes have been studied under the microscope for many years. The display of the chromosome set of an individual, lined up from the largest to the smallest, is called a karyotype (Fig. 5.3). Therefore, a karyotype is referred to as the microscope images of metaphase chromosomes when the sister chromatids are maximally condensed but have not yet separated.

THE GENOME IN BACTERIA AND VIRUSES

A cell's genome, the total DNA content of the cell, is divided among one or more chromosomes. Chromosomes organize, store, and transmit genetic information and they must be compacted significantly to fit within cells. For example, a typical bacterial chromosome would be nearly 1 mm long, or approximately 1000 times the length of a bacterium. Not only must cells massively compact their DNA, but they must also organize DNA in a way that is compatible with a myriad DNA-based processes including replication, transcription, repair, and recombination. Each chromosome contains one double-stranded DNA molecule, along with proteins that help to condense and organize DNA within the cell. The DNA molecule can be linear, as in all eukaryotes, or circular, as in most bacteria and archaea.

Bacterial Genome

The first level of DNA packaging in all organisms is the binding of small, basic proteins directly to the DNA along its length. In bacteria, the binding of small basic proteins bends the DNA, which helps pack it into a more compact structure.

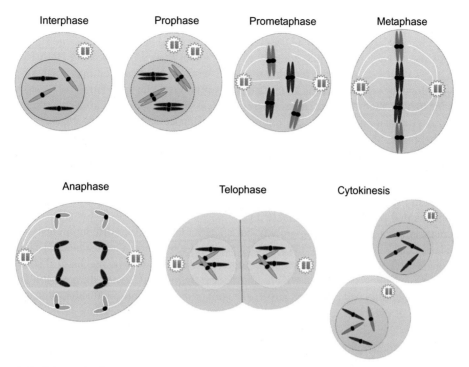

FIG. 5.2 The stages of mitosis in an animal cell, starting with interphase in which chromosomes are extended and uncoiled into chromatin. Prophase then causes chromosomes to coil and condense, while centrioles divide and move apart. In prometaphase, chromosomes are double structures, and centrioles are on opposite poles of the cell while spindle fibers form. Centromeres then align along the metaphase plate during metaphase. The aligned centromeres then split, and individual chromatids migrate to opposite poles of the cell in anaphase. Lastly, daughter chromosomes arrive at the poles in telophase, and the cells separate through cytokinesis.

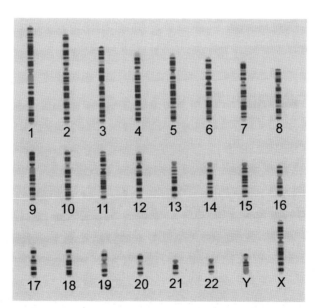

FIG. 5.3 Karyotypes of human chromosomes. This is a representation of a male karyotype visualized with Giemsa dye.

The compact bacterial chromosome is called the nucleoid. The bacterial genomes are usually organized into definite bodies and can be seen as a compact clump that occupies about a third of the volume of the cell.

The bacterial genome has many independent chromosomal domains. Each domain consists of a supercoil loop of DNA. The ends of the domains appear to be randomly distributed instead of located at predetermined sites on the chromosome. Although the majority of bacterial chromosomes are circular, some are linear, including the multiple ~1-Mb chromosomes in *Borrelia* species and the ~8-Mb chromosomes of *Streptomyces* species.

The Organization and Dynamics of Bacterial Chromosome

The pre-replication ("G₁") nucleoid, as defined in *Escherichia coli*, tends to be thinner at the ends than in the middle, that is, to be ellipsoidal, and also to be deformed into a gently curved helical shape. The nucleoid DNA is denser centrally than radially. Underlying this global shape is the fact that the DNA tends to be organized into a pair of parallel bundles that extend longitudinally along the nucleoid length and rotate gently relative to one another to give the curved, helical nucleoid shape. In some circumstances, including late stages of the cell cycle, the nucleoid appears as open ring and/or pairs of well-separated bundles. Longitudinal confinement could potentially influence shape in these stages.

Two types of global nucleoid-wide dynamics make the nucleoid more "fluid." Fluidity can facilitate local movements required for diverse chromosomal processes including, for example, displacement of transcribed regions to move to the nuclear periphery for translation as well as the dynamics of replication, sister chromatids segregation, and organization. Total nucleoid density fluctuates along the length of the nucleoid with a periodicity, probably throughout the cell cycle. These longitudinal density waves might be able to promote internal nucleoid mobility by promoting loss of intersegment tethers or entanglements that would otherwise create a gel.

Cell length increases monotonically during growth, but nucleoid length varies discontinuously, in a cyclic pattern. In each cycle, a period of nucleoid shortening is followed by a period of elongation. Cyclic nucleoid extension and shortening are strikingly consonant with accumulation, release, and dissipation of viscoelastic mechanical stress. This phenomenon implies the existence of mechanical stress cycles, including, but extending well beyond the period of DNA replication. Such cycles could make up a primordial cell cycle that governs the program of chromosomal events and its linkage with the cell division cycle. These global cycles are also implicated specifically in sister segregation.

Nucleoid-Associated Proteins

The organization of bacterial chromosomes is profoundly influenced by DNA-binding proteins, and in particular by a heterogeneous class of abundant proteins called nucleoid-associated proteins (NAPs) (Fig. 5.4). NAPs typically bind relatively nonspecifically across bacterial genomes, wrapping, bending, or bridging DNA. The local action of NAPs ultimately influences global chromosome organization and, in many cases, transcriptional patterns. *E. coli* H-NS is a small protein that can bridge DNA, bringing loci separated on the primary sequence level into close physical proximity. DNA bridging by H-NS likely enables it to constrain negative supercoils, by effectively isolating the intervening, looped region of the chromosome.

HU is another small, abundant (∼30,000 copies/cell) NAP found in many bacteria that coats and wraps chromosomal DNA around itself, grossly similar to histones. HU inserts conserved proline residues into the minor groove of DNA, inducing a sharp bend in the DNA. Two other proteins that can also sharply bend DNA are integration host factor, IHF, and factor for inversion stimulation, FIS. IHF can dramatically alter DNA shape and facilitate the formation of loops, frequently bringing RNA polymerase together with distant regulatory proteins. IHF also impacts a range of other DNA-based processes, including replication initiation and recombination. FIS binds throughout the genome, impacting transcription, replication, and recombination. FIS probably also influences chromosome compaction and organization in significant ways.

Although NAPs are generally small proteins, some large proteins also stably associate with and influence the structure of chromosomes. Most prominent in this category is the widely conserved protein SMC, homologous to eukaryotic condensin. SMC forms an extended, antiparallel coiled coil with a so-called hinge domain at one end and an ATPase domain at

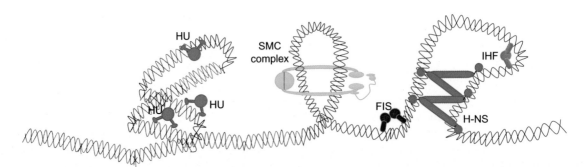

FIG. 5.4 Nucleoid-associated proteins (NAPs) with DNA bridging, wrapping, or bending activities contribute to the organization of the chromosome.

the other. Homodimerization via the hinge domains creates a ringlike structure that may encircle DNA. SMC associates with two regulatory proteins that likely modulate the ATPase activity of SMC, thereby affecting the opening and closing of the homodimeric ring.

Viral Genome

A viral particle is deceptively simple in its superficial appearance. The nucleic acid genome is contained within a capsid, which is a symmetrical or quasisymmetrical structure assembled from one or only a few proteins. In a process to construct a capsid that contains nucleic acid, the protein shell can be assembled around the DNA or RNA by protein-nucleic acid interactions during the process of assembly. A typical example is tobacco mosaic virus (TMV). Assembly starts at a duplex hairpin that lies within the RNA sequence. From the nucleation center, assembly proceeds bidirectionally along the RNA until it reaches the ends. The unit of the capsid is a two-layer disk that is a circular structure. It forms a helix as it interacts with the RNA. The RNA hairpin inserts into the central hole in the disk, and the disk changes conformation into a helical structure that surrounds the RNA (Fig. 5.5). Additional disks are added, with each new disk pulling a new stretch of RNA into its central hole. The RNA becomes coiled in a helical array on the inside of the protein shell.

On the other hand, the spherical capsids of DNA viruses such as phage lambda and T4 are assembled in a different way. In this case, an empty headshell is assembled from a small set of proteins. The duplex genome then is inserted into the head, accompanied by a structural change in the capsid (Fig. 5.6). Inserting DNA into a phage head involves translocation and condensation reactions, which involves the driving of DNA into the head by an ATP-dependent mechanism and providing the scaffold onto which the DNA condenses. The packing of DNA into the head follows a general rule for condensation, and the pattern is not determined by particular sequences.

EUKARYOTIC CHROMOSOME STRUCTURE

Human Chromosome: An Example of a Eukaryotic Genome

Normally, each human cell contains two copies of each chromosome, one inherited from the mother, maternal chromosomes, and one from the father, paternal chromosomes. The maternal and paternal chromosomes of a pair are called homologous chromosomes. The only nonhomologous chromosome pairs are the sex chromosomes in males, where a Y chromosome is inherited from the father and an X chromosome from the mother. Therefore, each human cell contains a total of 46 chromosomes: 22 pairs common to both males and females, plus two sex chromosomes.

The smallest human chromosome contains $\sim 4.6 \times 10^7$ bp of DNA. This is equivalent to 1.4 cm of extended DNA. The DNA must be condensed to fit into the cell, and such condensation of chromosomes during mitosis can be described as the packing ratio, which calculated by taking the length of the DNA divided by the length of the chromosome. At the most

RNA

FIG. 5.5 Helical arrangement for TMV RNA formed from stacking of protein subunits in the virion.

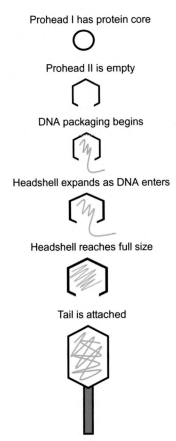

Prohead I has protein core

Prohead II is empty

DNA packaging begins

Headshell expands as DNA enters

Headshell reaches full size

Tail is attached

FIG. 5.6   The process of phage lambda maturation, which begins with a change in the shape of the empty head, which fills with DNA and expands.

condensed moment of mitosis, the chromosome is ~2 μm long. Thus the packing ratio of DNA in the chromosome can be as great as 7000 (Fig. 5.7). This packaging system is regulated by the association of the DNA with proteins. This process includes the packing of DNA into nucleosome followed by the folding of the higher order structure.

Assembly of the Nucleosome Structure

The packing of DNA inside the nucleus is through the association of DNA with histone proteins to form a nucleoprotein complex known as chromatin. Nucleosome is the fundamental repeating unit of chromatin. The formation of nucleosomes convert a DNA molecule into a chromatin thread about one-third of its initial length. This long strings of nucleosomes is the first level of the chromatin structure known as beads on a string. At the center of the nucleosome is a disk-like protein complex known as the histone octamer or core particle. Each nucleosome core particle consists of a complex of eight histone proteins—two molecules each of histones H2A, H2B, H3, and H4—and 147 base pairs (bp) of DNA. The 147 bp of DNA are wrapped in 1.7 left-handed superhelical turns around a histone octamer, whose surface is positively charged. Each nucleosome particle is connected by linker DNA (20–80 bp) to form repetitive motifs of ~200 bp (Fig. 5.8). This nucleosome fiber is also known as the 10-nm fiber.

Because core histones are essential and are required to interact with many chromosomal proteins, they represent some of the most highly conserved proteins in eukaryotes. All four of the core histones have a common protein fold, known as the histone fold. Each core histone protein has an N-terminal tail that extends away from the nucleosome core between the coils of DNA. These tails do not adopt a defined structure and they interact with adjacent nucleosome to condense the chromatin structure further. The N-terminal tails carry lysine and arginine, which help to interact with the negative charged DNA (Fig. 5.9). This electrostatic interaction helps to stabilize the overall histone-DNA assembly. In higher eukaryotic cells, linker histones H1 or H5 bind to the nucleosome core particle to form the chromatosome and promote the organization of nucleosomes into the 30-nm filament. This packaging of the DNA into chromatin, which occurs at a series of different

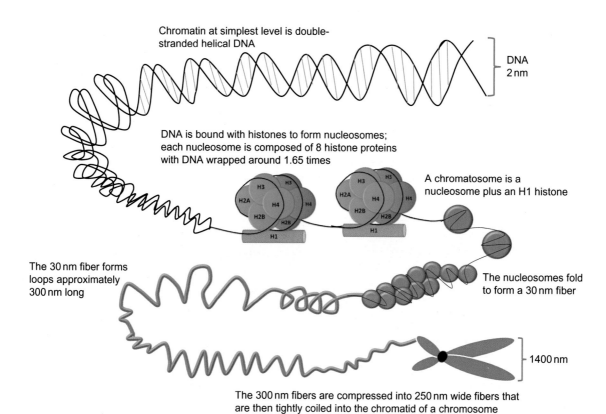

The 30 nm fiber forms loops approximately 300 nm long

FIG. 5.7 Packaging of eukaryotic DNA into highly condensed chromatin fibers.

levels, provides the compaction and organization of the DNA required to accommodate processes such as transcription, recombination, replication, and mitosis.

Histone Variants

During replication of DNA in the S-phase of cell cycle, and in order to maintain the proper chromatin organization, there is a high demand for histone synthesis and deposition onto the newly synthesized DNA. This requires a quick transcription and translation of histone genes and, in metazoan animals, results in the encoding of RNAs that are not polyadenylated and do not contain introns, presumably to reduce the posttranscriptional processing. The histones encoded by these genes are known as canonical histones.

In contrast to canonical histones, the term histone variants is globally used to describe those histones that are expressed throughout the cell cycle in smaller quantities. The histone variants often replace the canonical histones during chromatin metabolism, and hence are also referred to as replacement variants. Since their discovery, it has suggested that histone variants provide an alternative mechanism for introducing small variations into the eukaryotic epigenome (Fig. 5.10). This alternative mechanism is now known to govern fundamental aspects of chromatin structural organization, nucleosomal dynamics, and transcription. For those variants, only one H4 isoform (that is, canonical H4) has so far been identified, and H3 variants are generally less diverse than those arising from the H2A and H2B families. This is probably due to the role of H3 and H4 in the formation of core histone tetramers, which ultimately dictate nucleosomal assembly. H2A variants are generally the most diverse family of the core histone proteins and are accompanied by substantial variation in their amino- and carboxy-terminal tail regions. The linker histone H1 is the most diverse of all histones and is represented by at least 11 isoforms in mice and 10 isoforms in humans, including 7 somatic variants and 3 germline-specific variants. Since histone variants confer unique properties on chromatin structure by promoting differential interactions with various associated complex proteins as well as through alterations in the activity of chromatin binders, mutations in specific histone variants and their associated chaperone machinery can contribute to human disease. This reflects an essential function for regulation of histone variants during crucial periods of cellular development (Tables 5.1 and 5.2).

FIG. 5.8 DNA packaging and the various packing capacities involved to fit eukaryotic DNA into the nucleus.

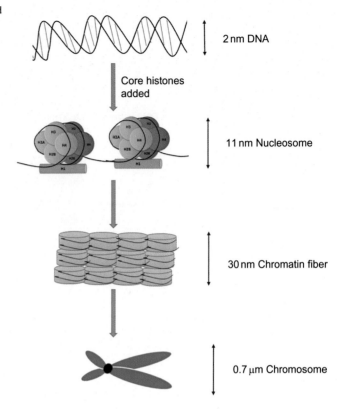

FIG. 5.9 Significant histone tail modifications.

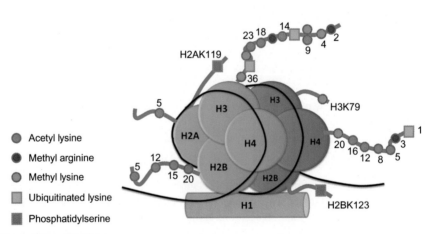

HIGHER ORDER CHROMATIN STRUCTURE

30-nm Chromatin Fiber

In living cells, the nucleosomes are packed on top of one another, generating a 30-nm fiber where the DNA is further condensed. The formation of 30-nm fiber results from both the histone tail interaction between neighboring nucleosomes and the presence of linker histone H1. Two common classical models of 30-nm chromatin fibers are solenoids and two-start helix. In solenoid model, consecutive nucleosomes are located adjacent to one another in the fiber and folded into a simple "one-start helix" (Fig. 5.11A). The second model assumes that nucleosomes are arranged in a zigzag manner, where a nucleosome in the fiber is bound to the second neighbor (Fig. 5.11B).

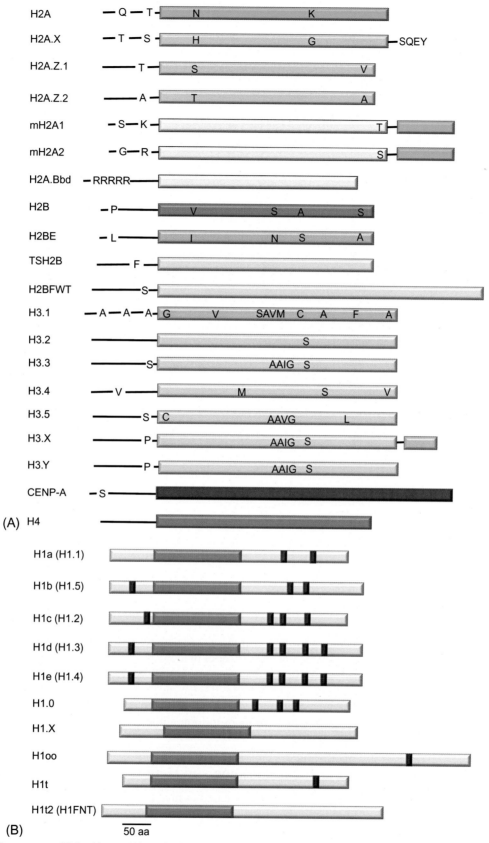

FIG. 5.10 (A) Human core and linker histone variants. Variants of the core histones H2A *(green)*, H2B *(blue)*, H3 *(orange)* and H4 *(pink)* are shown. Unstructured amino- terminal tails are shown as *black lines*. Specific amino acid residues are depicted at key differences among variants of a common histone protein family (for example, H2A.X, H2BE, and H3.3). Different shades of color are used to indicate protein sequences that are highly divergent between canonical histones (which are listed first) and their variants (e.g., H2A.Bbd, histone H2B type W-T (H2BFWT), and histone H3-like centromeric protein A (CENP-A)). (B) Human linker histone variants are shown. Because of the high sequence divergence between variants, specific amino acid differences are not shown. Unstructured amino- and carboxy-terminal domains are shown in *light gray*. Globular domains are shown in *dark gray*. Canonical serine/threonine PXK phosphorylation sites that are targeted by cyclin-dependent kinases are indicated in *magenta*. Alternative names of variants are given in parentheses: *aa*, amino acid; *mH2A1*, macroH2A1.

TABLE 5.1 Core Histone Variants in Vertebrate Development

Histone	No. of Gene Copies	Cell-cycle Expression	Location	Function	Knockout Phenotype
H2A	15	RD	Throughout the genome	Core Histone	ND
H2A.X	1	RI	Throughout the genome	DNA repair	Sperm defect in meiosis
H2A.Z	2	RI	Throughout the genome	Gene activation and silencing	Embryonic infertility

ND, Not determined.

TABLE 5.2 Core Histone Variants in Human Disease

Histone	No. of Gene Copies	Cell Cycle Expression	Mutation & Expression Patterns	Tumorigenic Consequences
H2A.X	1	RI	Reduced Expression	Increased cancer progression in p53 KO mice
H2A.Z	2	RI	Over-Expression (oncogene)	Numerous cancers
MacroH2A	2	Possibly RI	Reduced Expression (tumor suppressor)	Melanoma and numerous cancers
H3.1	10	RD	K27M in H3	Adult and pediatric glomas, including GBMs

GBM, Glioblastoma multiforme; *RD*, replication dependent; *RI*, replication independent.

Loops and Coils

The 30-nm fiber is further folded and compacted into interphase chromatin or into mitotic chromosomes. Further compaction of 30-nm fiber involves the folding of the fibers into a series of loops and coils. Several models have been proposed to describe the structure of higher order chromatin. The "hierarchical helical folding model" suggests that a 30-nm chromatin fiber is folded progressively into larger fibers, including ∼100-nm and then ∼200-nm fibers, to form large interphase chromatin fibers (chromonema fibers) or mitotic chromosomes (Fig. 5.11C). In contrast, the "radial loop model" assumes that a 30-nm chromatin fiber folds into radially oriented loops to form mitotic chromosomes.

Variation in Chromatin Structure: Euchromosome and Heterochromosome

Although chromosomes are in an extended, relatively decondensed form at interphase, most regions of the nucleus demonstrate a ∼1000-fold compaction of fiber, and some regions demonstrate ∼10,000-fold compaction. The relatively decondensed ∼1000-fold compaction region is called euchromatin (Fig. 5.12). Euchromatin can be further packed into ∼10,000-fold compaction in mitotic chromosomes and become heterochromatin. On the other hand, heterochromatin usually maintains ∼10,000-fold compaction in both interphase and mitosis. Heterochromatin contains very few genes, and it is typically found at centromeres.

Interphase chromosomes contain regions of both heterochromatin and euchromatin. Most genes present in euchromatin are being actively expressed. This is because the euchromatin is relatively decondensed, this open structure can allow ready

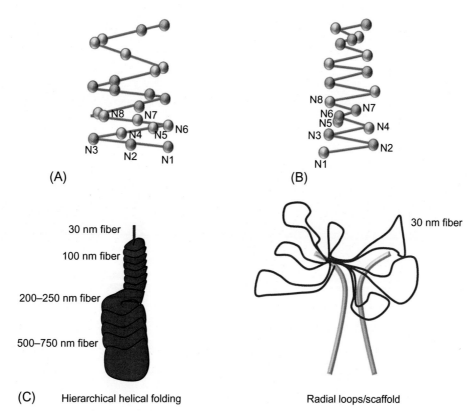

(A)

(B)

(C) Hierarchical helical folding

Radial loops/scaffold

FIG. 5.11 Two classical models of 30-nm chromatin fibers and higher order chromatin structures, including (A) the one-start helix vs. (B) the two-start helix, as well as (C) hierarchal folding versus radial scaffolding.

access to DNA by the proteins required for transcription. By contrast, heterochromatin appears to be more condensed, and therefore genes are typically not transcriptionally active.

The same fibers run continuously between euchromatin and heterochromatin, which implies that these states represent different degrees of condensation of the genetic material. Meanwhile, euchromatin regions exist in different states of condensation during interphase and during mitosis. The expression of those euchromatic genes that become packaged into heterochromatin are turned off because of more condensed compaction. The euchromatin that can be converted to heterochromatin is known as facultative heterochromatin. The common form of heterochromatin is called constitutive heterochromatin. Therefore, the genetic material is organized in a manner that permits alternative states to be maintained side by side in chromatin, and allows cyclical changes to occur in the packaging of euchromatin between interphase and mitosis.

Mitotic Chromosome and Its Structural Organization

The general form of mitotic eukaryotic chromosome displays a form of organization in which loops of DNA of ~60 kb are anchored in a central proteinaceous scaffold. In interphase nuclei, this underlying proteinaceous structure changes its organization to occupy the entire nucleus. The DNA sites attached to proteinaceous structures in interphase nuclei are called MARs (matrix attachment regions). Sometimes the DNA sites can also be called SARs (scaffold attachment regions). This is because that the same sequences appear to attach to the protein substructure in both metaphase and interphase cells. Chromatin often appears to be attached to a matrix, and this attachment may be necessary for transcription or replication.

UNIQUE ANIMAL CHROMOSOMES

Lampbrush chromosomes of amphibians and polytene chromosomes of insects have unusually extended chromosome structures. The packing ratio is usually <100-fold. Both lampbrush and polytene chromosomes demonstrate that the genetic material is dispersed from its usual more tightly packed state for genes to be transcribed.

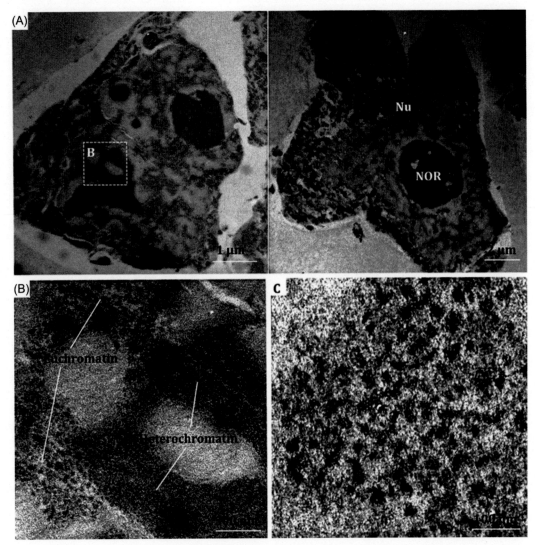

FIG. 5.12   Electron micrographs of isolated HeLa nuclei with high-pressure freezing and freeze substitution. (A) Low-magnification image of sectioned nuclei. Nu represents nucleus and NOR is nucleolus. (B) Enlarged view of the region highlighted in A. Presumptive euchromatin and heterochromatin are indicated. (C) Chromatin-like fibers with longitudinal-section and cross-section orientations are indicated by *red boxes* and *red circles*, respectively.

Lampbrush Chromosome

The lampbrush chromosomes have the lateral loops that extrude from the chromomeres at certain positions. Chromomeres are the beads that appear at the meiosis stage in which the chromosomes resemble a series of beads on a string. The loop is an extruded segment of DNA that is being actively transcribed. Because of their distinctive chromomere-loop organization and intense transcriptional activity of lateral loops, the lampbrush chromosomes have served as a powerful system for exploring the general principles of chromosome organization and function.

Oocytes of all urodele amphibians contain spectacular diplotene prophase chromosomes in which DNA is decondensed into several thousand lateral loop pairs. These lateral loop pairs are displayed along the chromosome axis. Lateral loops, which are responsible for the typical aspect of lampbrush chromosomes, represent regions of intense RNA synthesis, and the nascent RNA transcripts associate with proteins to form a RNA matrix (Fig. 5.13A and B).

Polytene Chromosomes

Polytene chromosomes are usually found at the interphase nuclei of some tissue of the larvae of flies. Polytene chromosomes are considered to be very useful for the analysis of many facets of eukaryotic interphase chromosome organization and the genome as a whole. They develop from the chromosomes of diploid nuclei by successive duplication of each chromosomal element (chromatid). Cells with polytene chromosomes differ in many ways from mitotically dividing cells.

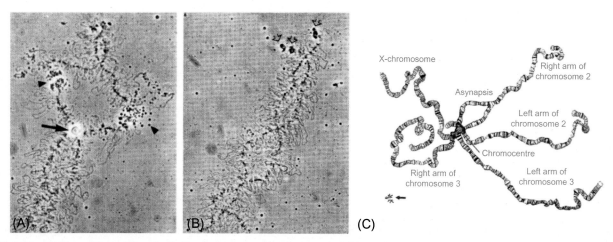

FIG. 5.13 (A) *Pleurodeles waltl* lampbrush chromosomes as seen by phase contrast. Bivalent XI characterized by two kinds of landmarks: a sphere at the midpoint *(arrow)* and two homologous globular loops *(arrowheads)*. (B) Bivalent X characterized by dense matrix loops in a subterminal position on the two homologous chromosomes. Bar, 20 μm. (C) The polytene chromosome set from *Drosophila melanogaster* made under light microscopy. The chromosomes have been spread by squashing them on a microscopic slide. Each parental chromosome is tightly paired with its homologue, and there are regions where two homologous chromosomes are separated (asynapsis). All the chromosomes are linked together by the pericentromeric region to create a single chromocenter. In the left lower corner, mitotic chromosomes (indicated by *arrows*) from ovarian tissue are shown at the same magnification. The polytene chromosomes demonstrate their giant sizes and transversal banding.

Firstly, the formation of polytene chromosomes is associated with the elimination of the final steps of mitosis after each doubling of DNA. This results in the cell cycle consisting of just two phases, synthetic (S) and intersynthetic (G). This peculiar "polytenization" cell cycle is established during mid-embryogenesis in *Drosophila melanogaster*. Secondly, at the end of each S-phase, DNA strands do not segregate; instead they remain paired to each other, forming polytene ("multi-stranded") chromosomes. As a result, giant chromosomes is formed, and these giant chromosomes are 70–110 times longer than typical metaphase chromosomes (Fig. 5.13C).

Variation in the extent of coiling of the DNA and its associated proteins along the linear axis of each chromatid leads to variation in chromatin concentration and compaction. Each member of the polytene set consists of a visible series of bands that contain most of the mass of DNA. As the polytene chromosomes, the centromeres of all chromosomes normally aggregate to form a chromocenter that consists largely of heterochromatin. The remaining ~75% of the genome is organized into alternating bands and interbands in the polytene chromosome. The DNA and protein content of interbands is much lower than that found in bands. Thus, the alternation of compacted and decompacted regions of chromosomes— bands and interbands—appears in polytene chromosomes as black and white transverse stripes (Fig. 5.13C). The band/ interband pattern of polytene chromosomes is constant. The interband regions that are the transcriptionally active sites on the polytene chromosomes can form an expanded state known as puffs, which are the sites where RNA is being synthesized.

SPECIALIZED DNA SEQUENCES IN EUKARYOTIC CHROMOSOMES

Origin of Replication

There are usually three different types of specialized DNA sequence in eukaryotic chromosomes, which can bind specific proteins that guide the machinery that replicates and segregates chromosomes. One type of nucleotide sequences act as a DNA replication origin, which is the location where DNA duplication begins (Fig. 5.14). Once the replication is initiated, DNA replication is bidirectional, proceeding away from the origin of replication in both directions. In bacteria, DNA replication initiates at a single site. In eukaryotes, DNA replication initiates at multiple sites along the chromosome (Fig. 5.14).

Origins of replication are typically found in regions that do not contain genes. In *E. coli* and other bacteria, a distinctive DNA sequence specifies the single origin of replication. The exact sequence differs in different bacteria. In budding yeast, replication origins are defined by a relatively short, specific DNA sequence. Although the well-defined DNA sequences that serve as replication origins in budding yeast have been defined, DNA sequences that specify replication origins in other eukaryotes have been more difficult to define. In higher eukaryotes, specific histone modifications, rather than a unique DNA sequence, play an important role in determining where the replication begins.

FIG. 5.14 Replication origins (represented by *green*) initiate bidirectional replication. A circular bacterial chromosome has a single origin that proceeds bidirectionally, whereas the linear eukaryotic chromosome has many bidirectional independently initiating origins.

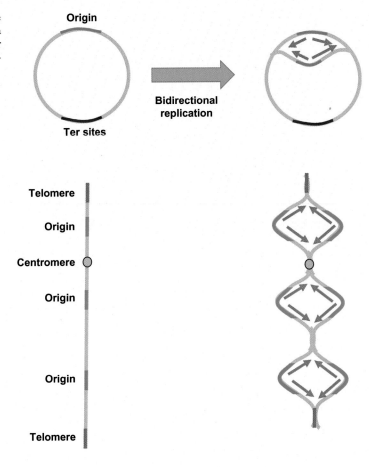

Centromere

The second specialized DNA sequences is centromere, which allows one copy of the duplicated and condensed chromosomes known as sister chromatids to be pulled into each daughter cell when a cell divides (Fig. 5.15). In most species, each chromosome has a single centromere, which is packaged into heterochromatin and whose structure is stably inherited from one generation to the next. The centromere is also rich in satellite DNA sequences, which are hundreds of kilobases of

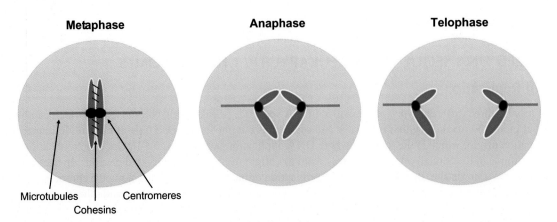

FIG. 5.15 Chromosomes are pulled to the poles via microtubules that attach at centromeres. Sister chromatids are held together until anaphase by cohesins. The centromere is shown as metacentric (in the *middle* of the chromosome), as opposed to the acrocentric (close to *end*) and telocentric (at the *end*) options.

repetitive DNA sequence. Although the sequences of satellite repeats are all very similar, each chromosome contains its own specific variants by which it can be identified.

During mitosis, the centromeric region on each sister chromatid migrates along microtubules to the opposite pole. The microtubules comprise a cellular filamentous system, which is reorganized at mitosis so that they connect the chromosomes to the poles of the cell. The kinetochore provides a microtubule attachment point on the chromosome. Subsequently, the sister chromatids separate at their centromeres, and then they are released completely from one another during anaphase. During anaphase/telophase, the centromere is pulled toward the pole and the attached chromosome appears to be dragged along behind it. Therefore, the chromosome provides a device for attaching a large number of genes to the apparatus for division.

Centromeres of all organisms have a defined region in which nucleosomes contain a histone H3 variant known as CENP-A. The region of the centromere that contains CENP-A is the place where kinetochore proteins assemble. The satellite DNA region has a high probability of binding CENP-A to form a centromere-specific chromatin structure. In the absence of CENP-A, cells fail to recruit the kinetochore components to the kinetochore. Therefore, the presence of CENP-A is crucial to centromere function.

Telomere

The third type of specialized DNA sequences is telomere, which is at the end of a chromosome. Telomeres define the end of eukaryotic chromosomes and are essential for their stability. Telomere sequences have been characterized from a wide range of eukaryotes. The same type of sequence is found in plants and humans, so the structure of the telomere follows a nearly universal principle.

Telomere consists of simple tandem repeated nucleotide sequences that have one strand that is G-rich and the other strand is C-rich. All telomere sequences can be written in the general form $C_n(A/T)_m$, where $n > 1$ and m is 1 to 4 (Fig. 5.16). One unusual property of telomere sequence is the extension of the 14 to 16 bases of G-T-rich single strand. The repeated sequences enable the end of chromosome to be efficiently replicated and the telomere to be extended. If it is not extended, it becomes shorter with each replication cycle. Shortening the telomere leads to the loss of binding sites for proteins. The inhibition of protein binding will leave the chromosome end unprotected. The unprotected end can be recognized by the cell as a DNA break. Subsequently, the DNA damage pathway is activated, causing arrest of cell cycle or cell death. If the telomere length reaches zero, it becomes difficult for the cells to divide successfully.

COMPLEXITY AND GENE NUMBER

Gene Number

The development of next-generation sequencing technologies has led to the sequencing of many genomes. A complete DNA sequence of the genome reveals both the genes an organism possesses and the genes it lacks. One most important piece of information provided by a genome sequence is the number of genes. Normally, the minimum gene number required for any type of organism increases with its complexity (Table 5.3). For example, *Nasuia deltocephalinicola*, a symbiotic bacterium, has the smallest known genome of any organism, with only 137 genes. Normally, the genomes of free-living

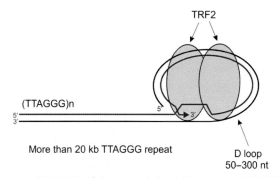

FIG. 5.16 A loop forms at the end of chromosomal DNA. The 3′ single-stranded end of the telomere (TTAGGG)n displaces the homologous repeats from the duplex DNA to form a t-loop. The reaction is catalyzed by TRF2 (Telomeric Repeat-binding Factor 2).

TABLE 5.3 Increase in Minimum Gene Number Required as Organisms Increase in Complexity

Organism	Minimum Gene Number
Intracellular parasitic bacterium	500
Free-living bacterium	1500
Unicellular eukaryote	5000
Multicellular eukaryote	13,000
Higher plants	25,000
Mammals	25,000

bacteria have from ~800 to 7500 genes. *Methanothermus fervidus*, with a genome coding for 1311 proteins and 50 RNA genes, stands as the free-living archaeon with the smallest sequenced genome. A cell with a nucleus requires at least 5000 genes, and a multicellular organism requires at least 10,000 genes.

Genome Size

The sequences of the genomes of prokaryotes show that virtually all the DNA encodes RNA or polypeptide. Thus, the genome size is proportional to the number of genes. The typical gene size averages just under 1000 bp in length (Table 5.4). All the prokaryotes with genome size below 1.5 Mb are symbiotic, and this genome size would suggest the minimum number of functions required for a cellular organism. The smallest symbiotic bacterium is *Nasuia deltocephalinicola*, with a 112-kbp genome and 137 protein coding genes. The smallest free-living nonsymbiotic bacterium is *Candidatus Actinomarina minuta*, with a 0.7-Mb genome and ~800 genes. The gram-negative bacterium *Haemophilus influenzae* has 1.83 Mb and 1743 genes. Archaeans have biological properties that are intermediate between those of other prokaryotes and those of eukaryotes, but their genome sizes and gene numbers fall in the same range as those of bacteria, which is around 1.5 to 3 Mb, corresponding to 1500 to 2700 genes. Through the genome sequencing on those pathogenic bacteria, researchers are able to identify pathogenicity islands. These islands are DNA segments, ~10 to 200 kb, which are

TABLE 5.4 Genome Sizes and Gene Numbers for Various Species

Species	Genome (Mb)	Genes
Nasuia deltocephalinicola	0.11	137
Mycoplasma genitalia	0.58	470
Candidatus Actinomarina minuta	0.7	800
Rickettsia prowazekil	1.11	834
Haemophilus influenza	1.83	1743
Bacillus subtilis	4.22	4420
Escherichia coli	4.64	4498
Saccharomyces cerevisiae	13.5	6034
Saccharomyces pombe	12.5	4929
Oryza sativa	466	30,000
Drosophila melanogaster	165	13,601
Caenorhabditis elegans	97	18,424
Homo sapiens	3,300	25,000

present in the genomes of pathogenic species but are absent from the genomes of nonpathogenic variants of the same or related species. Information on pathogenicity islands might provide important insight into the nature of pathogenicity.

The Genome Size Is Not Necessarily Related to the Gene Number in Eukaryotes

The relationship between genome size and gene number is weaker in eukaryotic genomes than in prokaryotic genomes. The genomes of unicellular eukaryotes fall in the same size range as the largest bacterial genomes. Multicellular eukaryotes have more genes, but the number does not correlate well with genome size. For example, the yeast *Saccharomyces cerevisiae* has a 13.5-Mb genome and ~6400 genes. Its size of genome and gene number is relatively in proportional. On the other hand, the genome of *Caenorhabditis elegans* is 97 Mb with only ~18,000 genes. Furthermore, the genome of *Drosophila melanogaster* is larger than that of *C. elegans*, but it has fewer genes. As such, the genome size is not necessary related to the gene number in eukaryotes.

THE FEATURES OF GENOMIC SEQUENCES

Genomes of higher eukaryotes contain more DNA than expected when estimates are based on the length and number of coding genes in the genomes. The amount of DNA in the unreplicated genome, or the haploid genome, of a species is known as the C-value or constant value. The lack of correlation between size and complexity of eukaryotic genomes, largely due to the presence of noncoding repetitive DNA, is termed the C-value paradox. These repetitive DNAs do not seem to carry critical information. In general, there are two types of genomic sequences—nonrepetitive DNA and repetitive DNA, and the repetitive DNA can be further divided into moderately repetitive sequences and highly repetitive sequences. The genomes of prokaryotes usually contain nonrepetitive DNA almost exclusively. For example, the genome of *E. coli* consists of nonrepetitive DNA only. For unicellular eukaryotes, most of the DNA is nonrepetitive and only ~20% of genomic sequences are repetitive DNA. In plants and amphibians, the moderately and highly repetitive DNA may account for up to 80% of the genome. For example, the genome of *Nicotiana tabacum* consists of 33% of nonrepetitive DNA, 65% of moderately repetitive DNA and 7% of highly repetitive DNA.

Nonrepetitive DNA

Nonrepetitive DNA consists of sequences that are unique, and there is only one copy in a haploid genome. The proportion of the genome occupied by nonrepetitive DNA varies widely among taxonomic groups. The length of the nonrepetitive DNA tends to increase with overall genome size. Meanwhile, the increase in genome size for higher eukaryotes, such as plants and mammals, usually reflects the increase in the amount and proportion of repetitive DNA. Because the nonrepetitive DNA usually corresponds to the protein coding genes, the increase in genome size for eukaryotes would not necessary reflect the increase of the gene number in proportion with prokaryotes and unicellular eukaryotes. As a result, the variation in genome size results from the differences in the amount of repetitive DNA, and therefore the relationship between genome size and gene number is weaker in eukaryotic genomes than in prokaryotic genomes.

Moderately Repetitive Sequences: Interspersed Elements—Repetitive Transposed Sequences

The repetitive DNA consists of sequences that are present in more than one copy in each genome. It is often divided into two general types: moderately repetitive sequences and highly repetitive sequences. Although the repetitive DNA is present in almost all multicellular eukaryotes, its overall amount is extremely variable. Tandemly repeated sequences are especially liable to undergo misalignments during chromosome pairing. Thus the sizes of tandem clusters tend to be highly polymorphic, with wide variation between individuals. This variation can be used to characterize an individual genome in DNA fingerprinting.

The moderately repetitive DNA consists of relatively short sequences that are repeated typically 10 to 1000 times in the genome. This type of sequence is dispersed throughout the genome and is responsible for the high degree of secondary structure formation in pre-mRNA. Although some moderately repetitive DNA does include some duplicated genes, such as ribosomal RNA, most prominent in this category are either noncoding tandemly repeated sequences or noncoding interspersed sequences. No function has been found for these sequences.

The noncoding interspersed transposons are short sequences of DNA that have the ability to move to new locations in the genome and to make additional copies of themselves. These DNAs are transposons that sometimes are called selfish

DNA or junk DNA. This is because they propagate themselves within a genome without contributing to the development and functioning of the organism.

Short interspersed elements also known as SINEs are <500 base pairs long and may be present 500,000 times or more in a human genome. The best characterized human SINE is a set of closely related sequences called the *Alu* family. Members of this family are 200 to 300 base pairs long and are dispersed uniformly throughout the genome, both between and within genes. The importance of *Alu* sequences is their potential for transposition within the genome, which is related to chromosome rearrangements during evolution. SINEs constitute about 13% of the human genome (Fig. 5.17).

The group of long interspersed elements (LINEs) represents another category of repetitive transposable DNA sequences. LINEs are usually about 6 kb in length and may be present 850,000 times in the human genome. The transposition mechanism of LINEs is similar to that used by retrovirus. This is because their DNA sequences are first transcribed into RNA molecules. The RNAs then serve as the templates for synthesis of the DNA complements. Subsequently, the new copies integrate into the chromosome at new sites. Therefore, LINEs are also referred to as retrotransposons. LINEs constitute about 21% of the human genome.

Moderately Repetitive Sequences: Tandem Repeated DNA—VNTRs and STRs

According to the length of the repeated unit and array size, tandem repeated DNA sequences can be classified into two groups: (1) microsatellites with 2- to 5-bp repeats and an array size on the order of 10–100 units, and (2) minisatellites with 10- to 100-bp (usually around 15-bp) repeats and an array size of 0.5–30 kb.

An example is DNA described as variable number tandem repeats (VNTRs). These repeating DNA sequences may be 15 to 100 bp long and are found within and between genes. Many such clusters are dispersed throughout the genome. These DNA sequences are often referred to as minisatellites. The number of tandem copies of each specific sequence at each location varies from one individual to the next. This variation in size of these regions between individual humans is the basis for the forensic technique known as DNA fingerprinting.

Another group of tandemly repeated sequences consists of di-, tri-, tetra-, and pentanucleotides. These DNA sequences are known as microsatellites or short tandem repeats (STRs). Similarly, the STRs are dispersed throughout the genome and vary among individuals in the number of repeats present at any site.

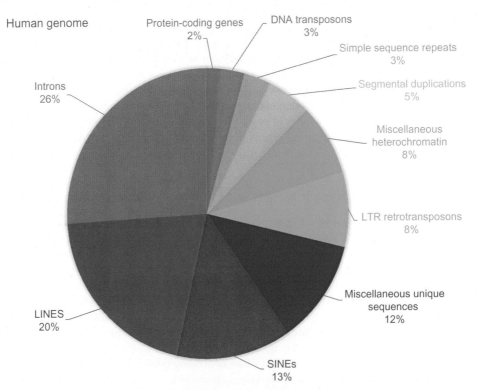

FIG. 5.17 Less than 1.5% of the genome represents protein-coding sequences. The majority of the genome, as represented in this pie chart, consists of DNA encoding intronic sequences, transposons, and long and short interspersed nuclear elements (LINEs and SINEs).

FIG. 5.18 General distribution of a plant chromosome's repetitive sequences. *Gray* represents centromeric tandem repeats; *blue*, telomeric; *pink*, sub-telomeric tandem repeats; *purple*, intercalary repeats; *brown*, dispersed repeats; *white*, genes and low-copy sequences.

Highly Repetitive Sequences

Satellite DNA (satDNA) is the highly repetitive DNA consisting of short sequences repeated a large number of times. It carries a variable AT-rich repeat unit that often forms arrays up to 100 Mb. The monomer length of satDNA sequences ranges from 150 to 400 bp in the majority of plants and animals. satDNA sequences are located at heterochromatic regions, which are found mostly in centromeric and subtelomeric regions in the chromosomes but also at intercalary positions. Fig. 5.18 shows a diagrammatic representation of different types of repetitive sequences on a plant chromosome. Satellite repeats in eukaryotes are likely involved in sequence-specific interactions and subsequently in epigenetic processes. Nonetheless, repetitive satDNA also has a sequence-independent role in the formation and maintenance of heterochromatin. Transcripts from tandem arrays or satellites are processed by RNA-dependent RNA polymerase and Dicer to produce siRNAs. satDNA-derived siRNAs are probably involved in posttranscriptional gene regulation through the action of the RNA-induced silencing complex (RISC).

Eukaryotic Genome Carries DNA Sequences That Do Not Encode Functional Genes

Although various forms of repetitive DNA are present in eukaryotes, the vast majority of the eukaryotic genome consists of single-copy sequences that do not code for proteins. Noncoding RNAs have been found to play an important role in neuronal functions. A large number of functional RNAs from noncoding regions have been reported to play vital roles in a wide variety of organisms. MicroRNAs appear to turn over rapidly, but can be strongly influenced by positive selection. Pseudogenes are conventionally thought of as dead genes that play no functional roles, but may evolve functions in regulating expression of related genes. It is believed that pseudogenes may regulate their parental genes, similar to long noncoding RNAs or microRNAs (miRNAs). The pseudogenes are DNA sequences representing evolutionary vestiges of duplicated copies of genes that have undergone significant mutational alteration.

ORGANELLE GENOMES

Mitochondria and chloroplasts have genomes that show nonmendelian inheritance. This is because the offspring of a mating cannot display mendelian segregation for parental characters. The mitochondrial genes are derived exclusively from the mother. Chloroplasts are generally also maternally inherited, although some plants show parental or biparental inheritance of chloroplasts. Mitochondrial genes and chloroplast genes are extranuclear genes, as they reside outside the nucleus. They are transcribed and translated in the same organelle compartment in which they reside. Organelle genomes can code for some or all of the tRNAs and rRNAs, but code for only some of the polypeptides needed to perpetuate the organelle. The other polypeptides are encoded by nuclear genes, expressed through cytoplasmic protein synthetic apparatus, and imported into the organelle.

Most organelle genomes take the form of a single circular molecule of DNA of unique sequence. There are a few exceptions in unicellular eukaryotes for which mitochondria DNA is a linear molecule. Chloroplast genomes are usually large: ~140 kb in higher plants and >200 kb in unicellular eukaryotes. On the other hand, mitochondrial genomes vary widely in total size. For example, the animal cell has a small mitochondrial genome, about 16.5 kb. Usually there are several copies of the genome in the individual organelle. There are multiple organelles per cell. Therefore, there are many organelle genomes per cell, and thus they can be considered as a repetitive sequence.

GENOME EVOLUTION

Introduction of Mutation

For genome evolution, it is important to consider the processes leading to variation in populations and species. The first step in the process is the introduction of a mutation into the genome due to a replication error or DNA damage. If a mutation occurs in a protein-coding region, it can be characterized by its effect on the polypeptide product of the gene. A substitution mutation that does not change the amino acid sequence of the polypeptide product is a synonymous mutation, which is a specific type of silent mutation. If the mutation does alter the amino acid sequence of the polypeptide product, creating a missense codon or a nonsense codon, then it is a nonsynonymous mutation.

Synonymous mutations accumulate more rapidly than nonsynonymous substitutions. The rate of divergence at nonsynonymous sites usually can be used to establish a molecular clock. The molecular clock can be distilled to the concept of accumulation of substitutions, through time yielding a stable rate from which we can estimate lineage divergence. The difference between two genes is expressed as their divergence, the percentage of positions at which the nucleotides are different. Thus, it can be used to calculate the time of divergence between any two members of the family.

Fixation of Mutation

The second step in the process is the fixation of this mutation over successive generations, thereby making the molecular change a feature of the entire phylogenetic unit such as population, species, or lineage. If a mutation is selectively neutral or near-neutral, then its fate is predictable by probability. The random changes in the frequency of a mutational variant in a population are called genetic drift. Because genetic drift is a random process, a variant may be either lost from the population or fixed, replacing all other variants. Such fixed mutations are usually in the form of nucleotide substitutions.

Evolutionary Rate

Since the molecular variation we typically use for comparing genes or species comes from fixed mutations, the evolutionary rate we estimate is a combination of the mutation rate and the rate of fixation. Under the neutral model, the substitution rate is equal to the mutation rate; however, there are factors that disconnect these two values. When we compare the rate between species or genes, a portion of the difference comes from changes in the mutation rate and a portion from changes in the rate of fixation. The overall influence that a life history trait has on sequence evolution rate is then largely a result of the magnitude and directions of its effects on mutation and fixation rates. Because genetic drift will overcome selection to a greater extent in smaller populations, slightly deleterious mutations are more likely to become fixed in species with small effective population sizes.

MECHANISMS OF GENOME EVOLUTION

Mutation toward a new gene structure is the first step of new gene evolution. Many distinct molecular processes contribute to the formation of new genes, The number of genes present in the genome varies by several thousand between animal species. Deducing the precise number of protein-coding genes in a genome is extremely hard, even with a "complete" genome sequence. This is because it is difficult to recognize short protein-coding genes, to distinguish functional genes from nonfunctional pseudogenes, and to assemble chromosomal regions containing repeats and duplications. Nonetheless, even within a group of relatively closely related species, such as placental mammals, the number of predicted genes varies by several thousand. For example, the US National Center for Biotechnology Information (NCBI) currently lists total protein-coding gene numbers to be: human 20,160; mouse 27,374; dog 20,422; cattle 23,134 (www.ncbi.nlm.nih.gov/gene/; March 2019) (Fig. 5.19). Moving to animals outside the mammals reveals even more differences between species; for example, *Ciona intestinalis* 13,930; *Drosophila melanogaster* 13,931; *Caenorhabditis elegans* 20,186.

What is the basis for these numerical differences among taxa? The differences reflect additions of genes and losses of genes. Change in total gene number represents a net balance between gain and loss, and hence the true rate of gene gain in evolution must be greater than suggested by the raw numbers alone. Although many mechanisms contribute to the genome evolution, in general almost 80% of genes are formed by DNA-based duplication, 5% to 10% by de novo duplication, and ~10% by retroposition. Although these mechanisms may generate the initial gene structures, many new structures (in a large variety of taxa) undergo radical structural renovation to change exon-intron structure, or even recruit new or existing coding sequences into the new locus.

Duplication

Gene duplication is thought to contribute most to the generation of new genes (Fig. 5.20). A single or a few new gene structures can be formed at one time by DNA-based duplication (the copying and pasting of DNA sequence from one genomic region to another) or retroposition. DNA-based duplications are often tandem. On the other hand, retroposed genes most often move to a new genomic environment, where they must acquire new regulatory elements or risk becoming

# of protein coding genes

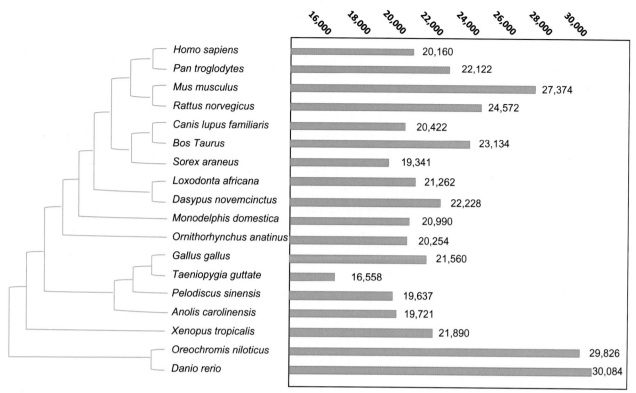

FIG. 5.19   Comparison of protein-coding genes of different species.

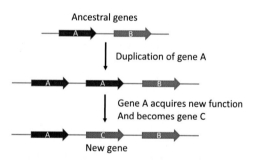

FIG. 5.20   Mechanism of new gene acquisition via duplication-divergence, in which a duplicate of an ancestral gene *(red)* acquires a new function and becomes a new gene *(purple).*

processed pseudogenes. Depending on the organism and the duplication mechanism, the size of the region of DNA that is duplicated can vary from just a few bases up to whole chromosomes (aneuploidy) or genomes (polyploidy).

An important gene duplication mechanism is whole genome duplication (WGD), which has occurred multiple times in eukaryote evolution, particularly in plants. Hundreds to thousands of duplicate genes are formed by a WGD event. Although the vast majority of duplicates are quickly lost, large fractions of duplicated loci can also be retained.

Duplications are often formed by unequal crossing-over mechanisms or rolling-circle amplification (Fig. 5.21). These mechanisms generate tandem arrays that generally are intrinsically unstable because of the presence of long, directly repeated regions with perfect homology that will allow homologous recombination and segregational loss of the amplified region down to one copy. However, if the duplication mechanism generates duplicated copies that are inversely oriented (e.g., by duplication-inversion), or where the copies are located at widely separated sites (e.g., by retrotransposition), the amplified state can be stabilized.

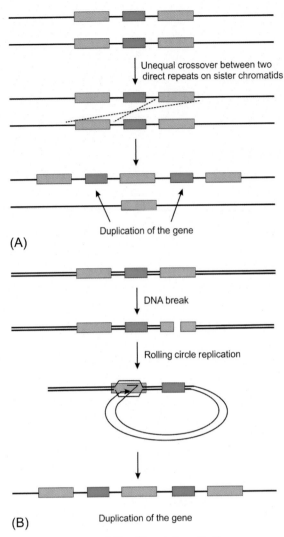

FIG. 5.21 Amplification through (A) unequal crossing-over and (B) rolling circle replication.

De Novo Origination

New gene structures may arise from previously noncoding DNA (Fig. 5.22). De novo origination of protein-coding genes occurs when genes emerge from a nonfunctional DNA sequence that was previously not a gene. Although it would seem highly improbable that functional proteins could emerge from noncoding DNA and translation of any random ORF devoid of genes is expected to produce insignificant polypeptides rather than proteins with specific functions, the advent of large-scale sequencing and comparative genomics has provided increasing evidence that new genes have evolved and continuously are originating from noncoding sequences.

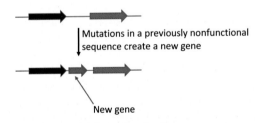

FIG. 5.22 De novo origination in which mutations in a previously nonfunctional sequence create a new gene.

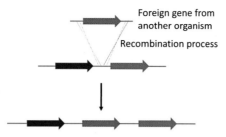

FIG. 5.23 Horizontal gene transfer in which a foreign gene is transferred from another organism and integrated into the genome by recombination.

Horizontal Gene Transfer

Similar to de novo gene origination, horizontal gene transfer (HGT), the exchange of genes between genomes from distantly related taxa, can immediately add new genes and functions to a genome (Fig. 5.23). HGT is a major mechanism for the addition of new genes to prokaryotic genomes, but has also been reported in a number of eukaryotic organisms including plants, insects, and fungi.

Gene Recombination

New gene structures can be generated by modifying existing exons or domains to produce new chimeric genes (Fig. 5.24A and B). Chimeric proteins formed by gene recombination have been found in many organisms, including yeast, *Drosophila*, *Caenorhabditis elegans*, mammals, and plants, and are estimated to have contributed ~19% of new exons in eukaryotes. In addition, retroposed sequences may jump into or near existing genes and recruit existing exons, or be recruited into an existing coding sequence. Conversely, new gene structures may be formed by splitting existing genes. For example, gene duplication is an intermediate stage in an evolutionary process leading to gene fission (Fig. 5.24C). In addition, divergence in alternative splicing patterns between duplicate genes can generate distinct transcripts that produce noncoding RNAs or polypeptides with slightly or entirely different functions and rapidly alter duplicate gene structures and functions.

New Gene Regulatory Systems

New genes must acquire a specific transcription regulatory system to ensure certain temporal and spatial expression patterns. For example, the male-specific expression of *Dntf-2r*, a retroposed gene in the *D. melanogaster-D. simulans* clade, does not contain the parental promoter, but has acquired a new *β2-tubulin*-like promoter by recruiting a novel 5′ regulatory sequence. This regulatory sequence drives testis-specific expression of *β2-tubulin*, and appears to still do so for *Dntf-2r*. In addition, the new retrogene *Xcbp1* recruited existing neuron promoters present at its site of integration. This coopted mode of promoter recruitment is also observed in human retrogenes and may be a general mode for retrogene promoter gain.

Transposable Elements

Transposable elements can contribute to functional divergence between duplicate genes in several ways, all similar to those described earlier. For example, transposable elements can mediate gene recombination by carrying coding sequences from one part of the genome to another, and can even themselves be incorporated into existing coding sequences.

Molecular Evolution of Repetitive Sequences

Tandem repetitive sequences are considered to be generated de novo by the combinatorial action of molecular mechanisms such as mutations, unequal crossing over, gene conversion, slippage replication, and/or rolling circle replication, which create and maintain homogeneity of satDNA sequences within species. Unequal crossing over is considered as the primary evolutionary force acting on satellite sequences. Arrays of satDNA are maintained by unequal exchange and intrastrand exchange, and unequal crossing over accounts for the alterations in copy numbers of satellite monomers. Individual repetitive units do not evolve independently; instead, the arrays evolve in concert. However, unequal crossing over, by itself, cannot create large tandem arrays of satDNA. Gene amplification and subsequent duplications also play a significant role in satDNA evolution. A few repeats may excise from a tandem array and circularize to provide a template for rolling-circle replication, and after amplification into a linear array, the repeats may be inserted into a new location in the genome.

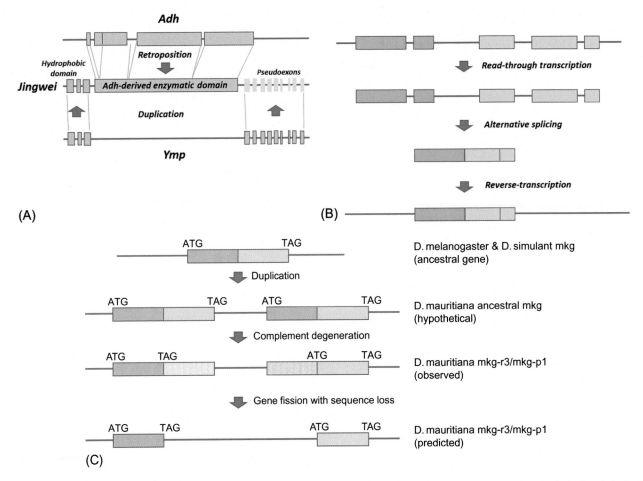

FIG. 5.24 Representative new genes exhibiting various new gene origination mechanisms. (A) Jingwei, a new gene found only in *D. teissieri* and *D. yakuba*, was generated by a combination of retroposition, DNA-based duplication, and gene recombination, which formed a chimeric gene consisting of Adh-derived enzymatic domain and a hydrophobic domain from Ymp. (B) PIPSL in humans is a consequence of gene fusion between two adjacent ancestral genes by read-through transcription and subsequent co-retroposition. (C) Gene fission split the ancestral gene monkey king into two distinct genes in *D. mauritiana*, revealing an intermediate process of gene fission aided by gene duplication and complementary degeneration.

Evolution Rate of Repetitive DNA Sequences

Repetitive elements are considered as fast-evolving components of eukaryotic genomes. The high evolution rates of repetitive sequences can be used to differentiate related species. Sequence homogeneity and evolution of repetitive sequences are correlated with their copy number. Repetitive sequences with a low copy number are homogeneous and evolve slowly. On the other hand, repetitive sequences with a high copy number are more heterogeneous and evolve quickly. Sequence divergence in satDNA proceeds in a gradual manner due to the accumulation of nucleotide substitutions. Evolutionary persistence of large tandem arrays is affected not only by the balance between the rate of amplification and the rate of unequal exchange, but also by a wide range of mechanisms for recombination, replication, and gene amplification. Because satDNA sequences are not transcribable, natural selection does not play a role on satDNA evolution.

EVOLUTION OF THE PROKARYOTIC GENOME

Free-living prokaryotes and endosymbiotic prokaryotes have evolved diverse strategies for living with reduced gene sets. In the case of free-living prokaryotes, natural selection directly favored genome reduction, whereas in the case of endosymbiotic prokaryotes, neutral processes played a more prominent role.

Genome Reduction: The Streamlining Hypothesis

Genome reduction by a process known as streamlining, in which smaller genomes are favored directly by selection as a way to cellular economization, is the most common explanation for genome reduction in free-living bacteria. In the streamlining hypothesis, natural selection directly favors genome reduction and low G+C content in free-living prokaryotes living in low-nutrient environments. The free-living prokaryotes also have small intergenic regions and small cell size. The process of genome reduction in host-associated bacteria is mainly determined by the intracellular environment in which they live. Specifically, genes unnecessary for living in intracellular conditions are not maintained by selection and are lost in the course of evolution.

Genome Reduction: The Muller Ratchet

Genome reduction based on the process known as the Muller ratchet states that in populations undergoing constant bottlenecks and no recombination, genome reduction occurs through the accumulation of slightly deleterious mutations. Under these conditions selection fails to retain genes, which then, by the constant accumulation of mutations, become inactive and are eventually deleted from the genome. As a result, several of the typical characteristics of these genomes, such as their large A+T content or their small genomes, reflect known mutational biases (i.e., G:C to A:T mutations and deletions over insertions) rather than adaptations evolved by selection.

The Impact of Genome Reduction on Host-Associated Bacteria

How host-associated bacteria perform their symbiotic function and all the processes necessary to maintain themselves with a reduced gene set? There are at least three nonmutually exclusive possibilities. First, modifications of some genes coded in the reduced genome could allow the endosymbiont to cope with the loss of otherwise essential genes; second, the presence of complementary genes in the genomes of cosymbionts (if any) may compensate for gene losses in the endosymbiont; and third, genes coded in the genome of the host compensate for gene losses in the genome of the endosymbiont. From an evolutionary point of view, this last group of genes could be of host origin, or originally from the endosymbiont and transferred to the host, or horizontally transferred from unrelated organisms not participating in the symbiosis to the host genome or its endosymbionts.

Chapter 6

Extraction and Purification of Nucleic Acids and Proteins

Chapter Outline

THE IMPORTANCE OF DNA EXTRACTION

In recent years, advances in molecular diagnostic methods have provided fast and accurate detection and identification of clinical samples and microorganisms. After samples are collected, DNA extraction is the first step in laboratory procedures required to perform any chosen molecular applications. Therefore, it is essential to choose the appropriate DNA extraction method to obtain high-yield and high-quality genomic DNA from any selected population of samples. Technical requirements, time efficiency, and cost effectiveness of the chosen method are among the issues to consider when choosing a particular technique, especially when dealing with a large number of samples. The quality and quantity of recovered DNA affect the success of molecular diagnostic applications. In this chapter, we discuss currently available techniques that have been used in the field to prepare good-quality materials for further analysis.

Diagnostic Molecular Biology. https://doi.org/10.1016/B978-0-12-802823-0.00006-7

DNA EXTRACTION FROM WHOLE BLOOD SAMPLE

Genomic DNA obtained from patient whole blood samples is a key element for molecular diagnosis. Isolation of genomic DNA is the first step in running molecular diagnostic assays. Thus, it is essential to obtain highly purified genomic DNA from sample populations using appropriate DNA isolation techniques.

Ficoll-Directed Density Gradient Through Centrifugation

Whole blood can be routinely collected for plasma and peripheral blood mononuclear cells isolation using centrifugation and a density gradient medium such as Ficoll-Hypaque. Fresh blood is collected in the presence of anticoagulants such as EDTA or citrate. After centrifugation, whole blood is separated into four layers, with plasma at the top followed by white blood cells containing peripheral blood mononuclear cells, a Ficoll-Hypaque medium layer, and a bottom layer containing erythrocytes and granulocytes (Fig. 6.1). This is because red blood cells (RBCs) and granulocytes have a higher density than Ficoll and will sediment at the bottom of the Ficoll-Hypaque layer.

Genomic DNA can be obtained from the plasma level through further DNA purification, as described later in this chapter. The typical yield is around 100–150 ng per ml of plasma. For the white blood cells, platelets are separated from the mononuclear cells by centrifugation through a fetal bovine serum cushion gradient that allows penetration of mononuclear cells but not platelets. After centrifugation, the supernatant containing the platelets is discarded and peripheral blood mononuclear cells can be collected for further DNA purification.

After the removal of plasma and white blood cells, the remaining Ficoll, which is now the top layer, can be removed from the residual erythrocytes/granulocytes, a rich source of genomic DNA for molecular genetic studies. However, the residual erythrocytes/granulocytes are often discarded. In fact, the granulocyte purification steps can include addition of ammonium chloride to the erythrocytes/granulocytes followed by lysis of the erythrocytes, washing, and resuspension of the granulocyte pellet. Normally, purification of the granulocytes should be done immediately after blood separation. In this way, DNA can be extracted from the granulocytes.

Quick Extraction Through Proteinase K and Phenol

Two other quick genomic DNA extraction methods that are not involved the Ficoll density gradient technique are proteinase K and phenol methods. For the proteinase K protocol, whole blood is mixed with Tris, EDTA, sodium dodecyl sulfate (SDS), $MgCl_2$, and proteinase K in the presence of high salt for overnight digestion at 37°C. After digestion is complete, samples can be further purified through a standard purification protocol as described later in the section of DNA purification. For the phenol method, whole blood is mixed with Tris-HCl (pH 8.0)-saturated phenol and water followed by shaking for 4 h at room temperature. After centrifugation, the aqueous phase is collected for further standard purification as described later in the section on DNA purification.

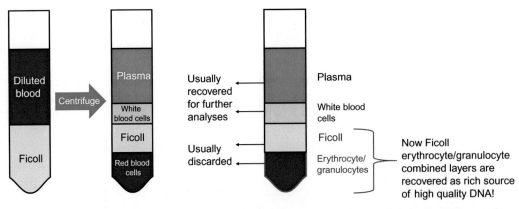

FIG. 6.1 Blood separation via Ficoll gradient centrifugation and diagram representation of granulocyte isolation for DNA extraction.

DNA EXTRACTION FROM DRY BLOOD SPOTS

Dry Blood Spot Preparation

Generally, genomic DNA extracted from whole blood has been widely used for diagnosis, but dry blood spots (DBS) on filter paper have emerged as a convenient way to collect and transfer specimens for genomic DNA extraction in rural areas where effective cold chains are lacking and transport is difficult. There are many advantages to using DBS for genomic DNA preparation. For example, blood collected in tubes should ideally be stored at 4°C and processed as soon as possible to obtain maximum DNA yield and quality. In contrast, the collected blood spots can be dried at room temperature and shipped without refrigeration. Furthermore, DBS can be collected from fingersticks or, for infants and young children, from heelsticks. This blood collection procedure causes minimal risks to the participant and may therefore be of particular interest for diagnosis involving young children and infants, where regular venous punctures may be challenging.

Another advantage is that DBS can be kept in a long-term storage facility (below −20°C) for years prior to DNA extraction without significant loss in quality. For example, DNA extracted from blood spots stored for up to 25 years has successfully been used in whole-genome amplification for various genetic investigations, including direct sequencing and genome-wide association studies.

The simple collection procedure and the stability of DNA in blood spots allows the collection to be handled by adults without requiring the participants to visit a medical facility. This may be of particular use in large epidemiological studies where participants are asked to give biosamples at multiple time points and/or when study participants are geographically widespread.

Release the DNA From the Dry Blood Spot

The DBS can be punched out from filter paper and is treated with 0.5% saponin or phosphate-buffered saline (PBS) overnight at 4°C. Subsequently, the punched filter paper can be subjected to either the Chelex-100 or the InstaGene Matrix method (Table 6.1). For Chelex-100, the punch is placed in 5% Chelex-100 and subject to 100°C for 8 min followed

TABLE 6.1 Chelex-100 and InstaGene Matrix DNA Extraction Methods From Dried Blood on Filter Paper

Chelex-100			InstaGene Matrix	
Soaking in Saponin	**Soaking in PBS**	**No Soaking**	**Soaking in PBS**	**No Soaking**
1. Place punches in 1 mL 0.5% saponin 2. Incubate at 4°C overnight 3. Remove saponin and add 1 mL PBS 4. Incubate at 4°C for 30 min 5. Remove PBS and place punches in 100 µL 5% Chelex-100 6. Incubate at 100°C for 8 min 7. Centrifuge at 10,600 g for 2 min 8. Carefully remove and store supernatant at −20°C if the extract is not used promptly	1. Place punches in 100 µL PBS 2. Incubate at 4°C overnight 3. Centrifuge at 18,000 g for 2 min 4. Remove PBS and place punches in 100 µL PBS 5. Centrifuge at 18,000 g for 2 min 6. Remove PBS and place punches in 100 µL 5% Chelex-100 7. Incubate at 100°C for 8 min 8. Centrifuge at 10,600 g for 2 min 9. Carefully remove and store supernatant at −20°C if the extract is not used promptly	1. Place punches in 180 µL 5% Chelex-100 already heated to 100°C 2. Vortex 30 s 3. Incubate at 99°C for 10 min 4. Centrifuge at 12,000 g for 1.5 min 5. Remove the supernatant and transfer to a clean tube 6. Centrifuge the supernatant at 12,000 g for 1.5 min 7. Carefully remove and store supernatant at −20°C if the extract is not used promptly	1. Place punches in 100 µL PBS 2. Incubate at 4°C overnight 3. Centrifuge at 18,000 g for 2 min 4. Remove and discard supernatant and add 100 µL PBS 5. Centrifuge at 18,000 g for 2 min 6. Remove and discard supernatant 7. Add 200 µL InstaGene Matrix 8. Incubate at 56°C for 30 min. Vortex carefully after 15 min and after completed incubation 9. Boil samples at 100°C for 8 min 10. Centrifuge at 15,000 g for 2 min 11. Carefully remove and store supernatant at −20°C if the extract is not used promptly	1. Place punches in a clean tube 2. Add 200 µL InstaGene Matrix 3. Incubate at 56°C for 30 min. Vortex carefully after 15 min and after completed incubation 4. Boil samples at 100°C for 8 min 5. Centrifuge at 15,000 g for 2 min 6. Carefully remove and store supernatant at −20°C if the extract is not used promptly

by centrifugation to collect the supernatant into which the DNA has been released. On the other hand, the punch can be incubated with InstaGene Matrix at 56°C for 30 min followed by centrifugation to collect the supernatant. The sensitivity and DNA quality of these two DNA extraction methods are good. Therefore, simple and low-cost methods such as Chelex-100 and InstaGene Matrix can be sensitive and useful in extracting DNA from DBS. A DBS with 11- to 15-mm diameter can typically yield 0.8 to 1.6 μg genomic DNA. For the application of next-generation sequencing (NGS) technology in genome-wide association studies (GWAS) or methylome-wide association studies (MWAS), as little as ~1 μg of starting material can be used without sacrificing the quality of the sequencing data and without biasing the methylation profile.

NONINVASIVE HUMAN DNA ISOLATION

Although blood samples are being used as a reliable source of genomic DNA, such sampling is invasive in nature. Furthermore, it often needs alternative sources when only a self-administered collection protocol is logistically or economically feasible, or as a backup source of DNA in studies that collect blood samples. The ease of sampling and noninvasive nature of the collection method avoids constraints associated with sampling from traditional sources.

Genomic DNA Purification From Hair

For the simple purification of genomic DNA from a hair with the root, a simple alkaline lysis method can be used. A hair with root is incubated at 95°C for 10 min in NaOH buffer, and the supernatant is subjected to DNA purification after centrifugation.

Alternatively, a smooth chemical digestion method using dithiothreitol (DTT) can be employed. DTT is a strong reducing agent with relatively high salt content and also an anionic detergent. A hair sample can be incubated 2 h at 56°C with buffer that contains Tris-HCl, EDTA, NaCl, SDS, DTT, and proteinase K, followed by gentle mixing and incubation at 60°C for 2 h or until the hair has dissolved completely. The genomic DNA can then be extracted from the solution.

Genomic DNA Purified From Saliva

Exfoliated buccal epithelial cells and other cells found in saliva are also a very promising alternative source of DNA because they can be obtained using self-administered, noninvasive, and relatively inexpensive techniques. Buccal swabs and mouthwash protocols are the most commonly used protocols for buccal cell collection. Different types of buccal swabs, such as cotton swabs or cytobrushes, can provide similar DNA yields and PCR success rates. The buccal swab samples are first suspended in lysis buffer that includes Tris, EDTA, SDS, and proteinase K. The sample is incubated 1–3 h at 56°C until the tissue is totally dissolved. The DNA is then extracted from the solution. For the mouthwash method, samples from saline rinses need to be processed or frozen immediately after collection. Similarly, cotton swabs stored for 4 days at 37°C also can be contaminated with DNA of predominantly bacterial origin. An alcohol-containing mouthwash can be used to reduce bacterial growth before isolation.

Genomic DNA Extraction From Urine Sample

The urine specimen is inverted or swirled in a specimen cup to create a homogenous suspension of cells followed by the centrifugation. The supernatant is removed, and a dry pellet containing cells is chilled at −20°C for 15 min followed by the addition of lysis buffer that includes Tris, EDTA, SDS, and proteinase K. The sample is incubated 2 h at 56°C. The DNA is then extracted from the solution.

Comparison of Different Extraction Methods

Comparison of the extraction procedures shows that DNA isolated from buccal cells, urine, and hair can be successfully used to perform PCR-based assays. Although DNA from all the samples was suitable for PCR, the blood and hair samples provided a good-quality DNA for restriction analysis of the PCR product compared with the buccal swab and urine samples.

DNA PREPARATION FROM MICROORGANISMS

Bacteria and fungi have cell walls that must be broken to release the nucleic acid. Usually, both chemical methods, where cell lysis is brought about by exposure to chemical agents that affect the integrity of the cell barriers, and physical methods, in which the cells are disrupted by mechanical forces, are used to break the cell walls. Chemical methods involve the use of enzymes and chemical reagents, and physical methods include freezing in liquid nitrogen, grinding with a pestle, enzyme digestion, glass bead milling, and microwaving.

Chemical Methods

For chemical method, a combination of enzymes and chemical reagents are commonly used. Detergent cell lysis is a milder and easier alternative to physical disruption of cell membranes, although it is often used in conjunction with homogenization and mechanical grinding. Detergents break the lipid barrier surrounding cells by solubilizing proteins and disrupting lipid-lipid, protein-protein, and protein-lipid interactions. Detergents self-associate and bind to hydrophobic surfaces. They are composed of a polar hydrophilic head group and a nonpolar hydrophobic tail and are categorized by the nature of the head group as ionic (cationic or anionic), nonionic, or zwitterionic. Their behavior depends on the properties of the head group and tail.

In general, nonionic and zwitterionic detergents are milder and less denaturing than ionic detergents and are used to solubilize membrane proteins where it is critical to maintain protein function and/or retain native protein-protein interactions for enzyme assays or immunoassays. CHAPS, a zwitterionic detergent, and the Triton-X series of nonionic detergents are commonly used for these purposes. In contrast, ionic detergents such as SDS are strong solubilizing agents and tend to denature proteins, thereby destroying protein activity and function.

Animal cells, bacteria, and yeast all have different requirements for optimal lysis because of the presence or absence of a cell wall. Because of the dense and complex nature of animal tissues, they require both detergent and mechanical lysis. For bacteria, lysozyme is very efficient for breaking down the cell walls of Gram-positive bacteria because their cell walls have a high portion of peptidoglycan. Although Gram-negative bacteria are less susceptible to lysozyme, the combination of lysozyme and EDTA can effectively break the cell walls and membranes. On the other hand, the mechanical disruption method can be used to disrupt Gram-positive bacterial cells and spores efficiently.

Fungal cells are difficult to disrupt because the cell walls may form capsules or resistant spores. Lyticase, also known as zymolyase, can be used to digest the cell walls of fungal cells.

Cells can also be broken by alkaline chemicals, detergents, and xanthogenates. Fungal genomic DNA can also be extracted with the CTAB (cetyltrimethylammonium bromide) method. Usually, CTAB mixes with Tris, EDTA, high-salt buffer, and the conidial suspension followed by incubation at 65°C for 1 h. This can break fungal cell walls very efficiently.

Chelex-100 Method

Chelex-100, a chelating resin that has a high affinity for polyvalent metal ions, can also be used to isolate fungi DNA quickly and efficiently. It is frequently used to release DNA from cells by a boiling treatment, at the same time protecting the DNA from the boiling effects with resin beads. Typically, conidial suspension from samples is extracted by adding 5% Chelex-100 resin. The mixture is incubated at 90°C for 30 min followed by centrifugation. Samples are then incubated at 90°C again for 15 min. Subsequently, the supernatant is collected after centrifugation.

The Chelex extraction method is quick to obtain DNA from spores. Chelex-100 efficiently makes extraction of the DNA from spores available for direct use in molecular analyses. Also, the quantity and quality of extracted DNA are adequate for PCR analysis. Comparatively, the quality of DNA samples isolated using Chelex method can obtain good-quality DNA from spores and provides an alternative to the CTAB-based extraction method (Fig. 6.2). The yield is about 28 ng and 17.9 ng per μL from 200 μL conidial suspension which contains $1–5 \times 10^5$ spores/mL for Chelex extraction and CTAB extraction, respectively. Thus, the Chelex method is usually recommended for fungal genomic DNA extraction considering its simplicity and cost effectiveness, especially in the routine processing of large amounts of samples.

NUCLEIC ACID PURIFICATION

The purification of DNA from crude cell extract involves the removal of proteins, carbohydrates, lipids, and cell debris. Various techniques have been developed, including organic, inorganic, and spin column methods.

FIG. 6.2 An example of DNA quality evaluated by PCR amplification. Genomic DNAs derived from two DNA extraction methods are compared. Lanes 1 to 6: DNA samples extracted with Chelex-100. Lanes 7 to 12: The same DNA samples extracted with CTAB method. MW, 100 bp plus DNA ladder (Fermentas Life Sciences), and Lane13 is negative control (no DNA).

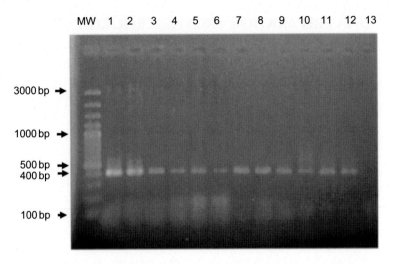

Organic Purification

For the organic method, phenol and chloroform/isoamyl alcohol (25:24:1) are mixed with an equal volume of samples by vortexing. Although phenol, a flammable, corrosive, and toxic carbolic acid, can denature proteins rapidly, it does not completely inhibit ribonuclease (RNase) activity. A mixture of phenol:chloroform:isoamyl alcohol (25:24:1) is normally used to inhibit RNase activity during the purification process. Furthermore, both proteinase K and RNase can be added to the sample at this step to remove lipids and degrade RNA, respectively.

After centrifugation, a biphasic emulsion forms. The organic hydrophobic layer of the emulsion is settled on the bottom and the aqueous hydrophilic layer on top. The upper aqueous hydrophilic layer contains DNA, and the bottom organic layer contains the precipitated proteins (Fig. 6.3). The upper aqueous layer is then mixed with an equal volume of chloroform, which can remove the residual phenol from the aqueous layer. The upper aqueous DNA-containing layer is now pure and can be further concentrated. A high concentration of salt, such as 0.3 M sodium acetate in the presence of 2.5 volumes of 100% ethanol or 1 volume of isopropanol at −20°C or below can precipitate DNA. DNA precipitate is collected by centrifugation, and excess salt is rinsed with 70% ethanol and centrifuged to discard the ethanol supernatant. The DNA pellet is then dissolved with Tris-EDTA solution or sterile distilled water for long-term storage.

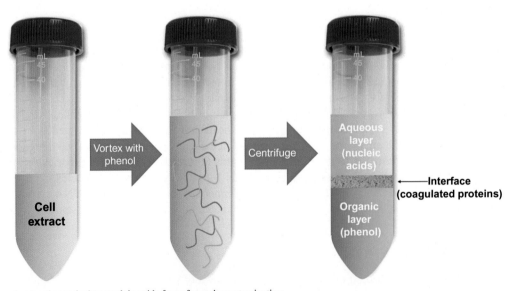

FIG. 6.3 Phenol extraction to isolate nucleic acids free of protein contamination.

With some cell extracts, the protein content is so great that a single phenol extraction is not sufficient to completely purify the nucleic acids. This problem can be solved by carrying out several phenol extractions one after the other, but this is not desirable, as repeated mixing and centrifugation may break the DNA molecules into small fragments. The other alternative is to treat the cell extract with a protease such as proteinase K before phenol extraction. The proteinase can break the polypeptides down into smaller peptides, which can be easily removed by phenol.

Inorganic Purification

The inorganic method involves the incubation of nuclei with only proteinase K at 65°C. It has been shown that proteinase K is more active on denatured protein and that after prolonged incubation at 65°C it can be auto-inactivated. As a result, following incubation for more than 2 h with Tris, EDTA, and proteinase K in a low-salt buffer, the extracted DNA can be used directly for diagnostic analysis without any additional purification. The yield is greater than 90% of theoretical with an average size greater than 300 kb.

Salting out is another simple inorganic DNA isolation method. In this procedure, saturated NaCl is used to precipitate protein. Next, the DNA is purified from the supernatant by addition of 1–2 volumes of −20°C chilled absolute (100%) ethanol. In these approaches, pure DNA is obtained and nontoxic substances are used during sample processing. This technique is fast and inexpensive for use in laboratory settings.

SOLID-PHASE NUCLEIC ACID EXTRACTION

Solid-phase nucleic acid purification can be found in most of the commercial extraction kits on the market. It allows quick and efficient purification compared to conventional methods. Many of the problems that are associated with liquid extraction, such as incomplete phase separation, can be prevented. A solid-phase system will absorb nucleic acid in the extraction process depending on the pH and salt content of the buffer. The absorption process is based on the following principles: hydrogen-bonding interaction with a hydrophilic matrix under chaotropic conditions, ionic exchange under aqueous conditions through an anion exchanger, and affinity and size exclusion mechanisms.

Solid-phase purification is normally performed by a spin column through centrifugation. This method can purify nucleic acid rapidly compared to conventional methods. Silica matrices, glass particles, diatomaceous earth, and anion-exchange carriers are commonly used solid supports in solid-phase extraction.

The major key steps involved in solid-phase extraction are cell lysis, nucleic acid adsorption, washing, and elution. The initial step in a solid-phase extraction process is to condition the column for sample adsorption. Column conditioning can be done by using a buffer at a particular pH or salt concentration to convert the surface or functional groups on the solid into a particular chemical form. Next, the cell extract is applied to the column. The desired nucleic acid will adsorb to the column with the aid of high pH and salt concentration of the binding solution. Other compounds, such as protein, may have strong specific bond with the column surface as well. These contaminants can be removed in the washing step by using washing buffer containing a competitive agent or high salt concentration. For the elution step, TE buffer or water is introduced to release the desired nucleic acid from the column, so that it can be collected in a purified state. Normally, rapid centrifugation, vacuum filtration, or column separation is required during the washing and elution steps of purification process (Fig. 6.4).

Diatomaceous Earth Extraction

Diatomaceous earth, which is also known as kieselguhr or diatomite, has silica content as high as 94%. It has been used for filtration and in chromatography and it is useful for the purification of plasmid and other DNA by immobilizing DNA onto its particles in the presence of a chaotropic agent. The resulting diatomaceous earth-bound DNA can then be washed with an alcohol-containing buffer. The alcohol-containing buffer is then discarded and DNA is eluted out in a low-salt buffer or in distilled water.

Affinity Extraction

Magnetic separation is a simple and efficient method that can also be applied in the purification of nucleic acid. Often, magnetic carriers with immobilized affinity ligands or prepared from biopolymer showing affinity to the target nucleic acid are used for the isolation process.

Magnetic particulate materials such as beads are preferable as supports in the isolation process because of their larger binding capacity. The nucleic acid binding process may be assisted by the nucleic acid "wrapping around" the support.

FIG. 6.4   An example of spin column procedure to isolate pure plasmid DNA with centrifuge or vacuum filter techniques.

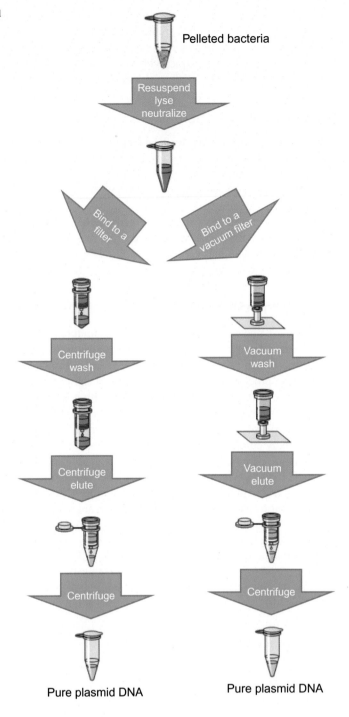

A magnet can be applied to the side of the vessel that contains the sample mixture, aggregating the particles near the wall of the vessel and allowing the remainder of the sample to be poured away. Particles with magnetic or paramagnetic properties are encapsulated in a polymer such as magnetizable cellulose. In the presence of certain concentrations of salt and poly-alkylene glycol, magnetizable cellulose can bind to nucleic acids. Small nucleic acids require higher salt concentrations for strong binding to the magnetizable cellulose particles. Therefore, salt concentration can be selectively manipulated to release nucleic acid bound to magnetizable cellulose on the basis of size. The magnetizable cellulose bound with nucleic acid is washed with suitable wash buffer before being contacted with a suitable elution buffer to separate out the desired nucleic acid with cellulose.

Anion-Exchange Extraction

Anion-exchange resin uses the anion-exchange principle. It is based on the interaction between positively charged diethyl-laminoethyl cellulose (DEAE) groups on the resin's surface and negatively charged phosphates of the DNA backbone. The anion-exchange resin consists of defined silica beads with a large pore size, a hydrophilic surface coating, and a high charge density. The large surface area of the resin allows dense coupling of the DEAE groups. The resin works over a wide range of pH conditions (pH 6–9) and/or salt concentration (0.1–1.6 M) which can optimize the separation of DNA from RNA and other impurities. Therefore, salt concentration and pH conditions of the buffers can play a major role to determine whether nucleic acid is bound to or eluted from the column.

DNA PREPARATION FROM FORMALIN-FIXED, PARAFFIN-EMBEDDED TISSUES

Fresh or fresh frozen tissue samples are the best material for isolation DNA of high quality and quantity. However, storage of frozen tissue samples is expensive and time-consuming. For many retrospective studies, formalin-fixed and paraffin-embedded (FFPE) material is, therefore, the only available tissue for DNA analysis. It is common to archive biological samples from patients as FFPE blocks. Tissues are first immersed for 24–48 h in formalin. Following fixation, tissue blocks are dehydrated using a dimethylbenzene (i.e., xylene) and ethanol series and embedded in paraffin wax. These tissue blocks are therefore referred to as FFPE samples. Since the procedure of tissue fixation and embedding has a profound effect on DNA, the challenge to obtain DNA of satisfactory quality and quantity depends strongly on the type of fixative used for preservation (Table 6.2), fixation times, type of tissue, and the method used for extraction. As a general rule, long fixation times—in particular if longer than 12–24 h highly acidic fixatives, longer periods of time in storage, and storage in warm environments all contribute to DNA decay. Another major problem is DNA protein cross-linking caused by formalin fixation that must be broken up during the extraction process. Because of all of these factors, DNA isolated from FFPE may be impure and degraded, which can cause problems in some molecular applications.

Formaldehyde as a 10% neutral buffered formalin is the most widely used universal fixative because it preserves a wide range of tissues and tissue components. Formaldehyde, in its basic form, is a gas. The liquid is actually a mixture of form-aldehyde gas and water. The most common concentration used is a 37% solution. That solution contains 37 g of

TABLE 6.2 Influence of Tissue Fixatives on Nucleic Acid Quality

Fixative	Relative Quality of Nucleic Acid	Average Fragment Size Range (kb)
10% buffered neutral formalin	Good	2.0–5.0
Acetone (acetone-methylbenzoate-xylene technique)	Good	2.0–5.0
HOPE (Hepes-Glutamic acid buffer mediated Organic solvent Protection Effect)	Good	2.0–5.0
Zamboni's (2% Paraformaldehyde; 15% Picric acid)	Not as good	0.2–2.0
Clarke's (75% Ethanol; 25% Glacial acetic acid)	Not as good	0.8–1.0
Paraformaldehyde (4% Paraformaldehyde in PBS)	Not as good	0.2–5.0
Metharcan (60% Methanol; 30% Chloroform; 10% Glacial acetic acid)	Not as good	0.7–1.5
Formalin-alcohol-acetic acid (85% Ethanol; 4% Formaldehyde; 5% Glacial Acetic acid)	Not as good	1.0–4.0
B-5 (4% Formaldehyde; 6% Mercuric chloride; 1.25% Sodium acetate)	Less desirable	<0.1
Carnoy's (60% ethanol; 30% Chloroform; 10% Glacial acetic acid)	Less desirable	0.7–1.5
Zenker's (5% Mercuric chloride; 2.5% Potassium dichromate; 1% Sodium Sulfate; 5% Glacial acetic acid)	Less desirable	0.7–1.5
Bouin's (75% Picric acid; 10% Formaldehyde; 5% Glacial acetic acid)	Less desirable	<0.1

formaldehyde gas to 100 mL of solution. Formaldehyde solution can polymerize. To prevent polymerization of formaldehyde solution, about 10% to 15% methyl alcohol is added. It is the addition of methyl alcohol that causes the substance to be called formalin as opposed to formaldehyde. Samples treated with formalin results in cross-linking of biomolecules that not only complicates isolation of nucleic acid but also introduces polymerase "blocks" during PCR, and this is directly proportional to the time spent in formalin. It is currently not impossible to reverse the effects of DNA fragmentation, and the key step of successful DNA extraction is the one at which there is reversal of the DNA-protein cross-links and neutralization of the excess formalin in the sample.

The Mechanism of Fixation by Formaldehyde

Formaldehyde reacts reversibly with water to form methylene glycol (Fig. 6.5). The routinely used fixative is 10% formalin, which is 3.7% formaldehyde in water with 1% methanol. It is an inexpensive, commonly available fixative that does not cause excessive tissue shrinkage or distortion of cellular structure. The undiluted commercial formaldehyde solutions contain 10% methanol as a preservative to prevent the spontaneous condensation reaction. Owing to this, the commercial formalin is a two-phase fixative, with an initial alcohol fixation phase, followed by a cross-linking phase mediated by aldehyde. The alcohol initially causes dehydration in the process, hardening the tissues and membrane. However, after long period of storage, formalin can oxidize to form formic acid. At the first stage, tissues are rapidly penetrated by the methylene glycol, but it has no role in fixation. The actual covalent chemical fixation depends on the fraction of the formaldehyde forming bonds with the tissue components, and this formaldehyde comes from dissociation of methylene glycol. Since the equilibrium of this reaction strongly favors methylene glycol, formaldehyde penetrates tissues rapidly as methylene glycol but fixes slowly as carbonyl formaldehyde.

Formaldehyde, in aqueous solution, becomes hydrated to form a glycol (hydrated formaldehyde) called methylene glycol. Methylene glycol hydrate molecules react with one another, combining to form polymers (Fig. 6.6). This can be given by the equation:

$$CH_2O + H_2O \rightleftarrows CH_2(OH)_2 \text{ (methylene glycol)}.$$
$$nCH_2(OH)_2 \rightleftarrows H_2O + HO(CH_2O)_nH \text{ (polyoxy methylene glycol)}.$$

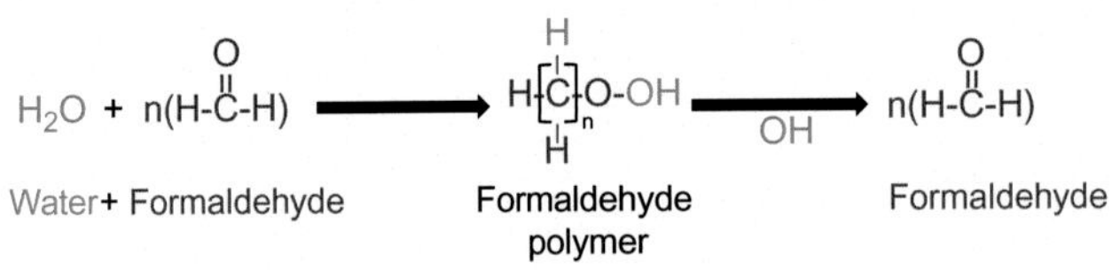

FIG. 6.5 Reaction of formaldehyde with water to form methylene glycol.

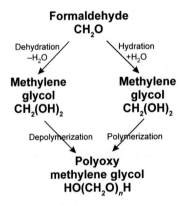

FIG. 6.6 Aqueous formaldehyde cycle.

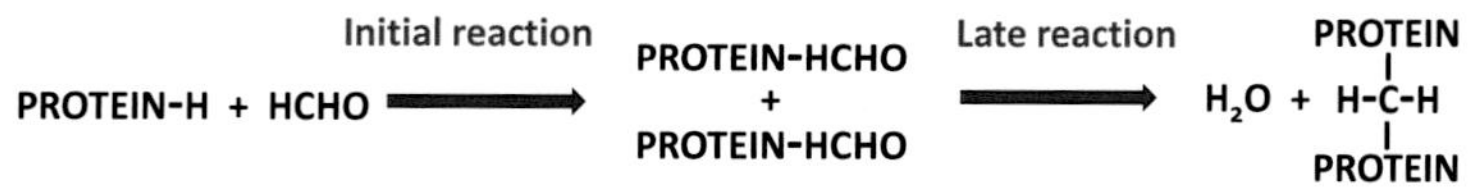

FIG. 6.7 Initial and late reactions in the protein interaction with formaldehyde.

With long standing, methylene glycol polymerizes to form polyoxymethylene glycol. In a neutral to alkaline buffered system such as tissue, it depolymerizes to methylene glycol, which dehydrates into carbonyl formaldehyde (one that contain a C=O group, the dehydrated form). Both the hydrated and nonhydrated forms of formaldehyde fix the tissue.

When tissues are immersed in formalin, they are rapidly penetrated by methylene glycol and the minor quantity of formaldehyde. Actual covalent chemical reaction of the fixative solution with tissue depends on the formaldehyde present being consumed after forming bonds with the tissue components. Consecutively, more formaldehyde forming from the dissociation of methylene glycol leads to a shift of the equation, so more formaldehyde is formed.

Formaldehyde, being a reactive electrophilic species, reacts readily with various functional groups of biological macromolecules in a cross-linking fashion (Fig. 6.7). The most frequent type of cross-link formed by formaldehyde in collagen is between the nitrogen atom at the end of the side chain of lysine and the nitrogen atom of a peptide linkage, and the number of such cross-links increases with time.

During the formaldehyde fixation, penetration means the ability of the solution to diffuse into the tissue, whereas fixation is the ability of the formaldehyde to complete the initial cross linking. For routine histochemical processes, the tissue-block thickness is in the range of 20 mm so that penetration occurs by 24 h at 25°C or 18 h at 37°C.

The completion of fixation depends on the buffering capacity, depth of penetration, temperature, concentration, and time interval. At the tissue level, penetration is dependent on the temperature, pH, volume of solution, and concentration. Alkaline pH and/or presence of hydroxyl ions plays an important role in conversion of polyoxymethylene glycol. In a mild alkaline system it depolymerizes to methylene glycol, which dehydrates to maintain equilibrium with active carbonyl formaldehyde. The presence of formic acid in a long-standing solution will appear to retard this reaction. A sufficient volume of solution will ensure that tissue penetration occurs from all surfaces, ensuring deeper penetration. With more time given, the depth of penetration will increase. Increase in surface area of the tissue is expected to increase the rate of penetration. The presence of crypts, vessels, and clefts appears to increase the rate of penetration. With an increase in pressure, the penetration of the solution is also rapidly increased. Concentrated formaldehyde consistently penetrates more rapidly than less concentrated.

Deparaffinization

Routinely used methods for extracting DNA from FFPE consist of many steps, including deparaffinization in xylene, washing in a descendent series of ethanols, protein digestion, and DNA purification. Typically, the FFPE materials are deparaffinized by incubating the samples with xylene followed by centrifugation. The pellets are washed with ethanol. After dewaxing, the samples are subject to lysis and thorough homogenization followed by incubation at 65°C for 1 h and 95°C for 15 min. Subsequently, proteinase K is added for 48 h digestion at 56°C. This proteinase K digestion is important to increase the DNA yield. After centrifugation, the supernatant can be further purified through phenol/chloroform, spin column, or inorganic methods as described previously.

Another method uses glycine in an alkaline environment. Glycine reacts with formaldehyde to form either N-methyleneglycine or dimethylglycine and shifts the equilibrium toward formaldehyde, thereby depleting the store of methylene glycol, the source of formaldehyde, in the tissue. Alkalinity of the buffer further aids in reversing the cross-linking of DNA and proteins. Subsequently the method involves heat treatment and proteinase K followed by purification of the nucleic acids. These extraction methods can yield DNA amounts of 1–9 μg/mg of tissue and allow stable amplification of PCR products. These materials are then suitable for further molecular diagnosis.

RNA PREPARATION

Handling RNA

There are several types of naturally occurring RNA, including rRNA, mRNA, and tRNA. RNA is an unstable molecule and has a very short half-life once extracted from the cell or tissues. Special care and precautions are required for RNA isolation to prevent degradation. The reason why RNA is especially unstable is because of the presence of RNases, which are enzymes present in blood and all tissues, as well as in most bacteria and fungi in the environment. Strong denaturants have always been used in intact RNA isolation to inhibit endogenous RNases. Therefore, RNA extraction relies on good laboratory technique and RNase-free methods.

RNases are heat stable and refold following heat denaturation. RNases are present throughout both prokaryotes and eukaryotes and even appear in some viruses. Because these enzymes are found in microorganisms, many of which are dispersed throughout the air, they can easily contaminate laboratory samples. In addition, RNases are secreted from the skin and are found in various fluids produced by the human body including tears, saliva, mucus, and sweat. RNases rapidly degrade RNA. RNases are also resilient and resist a vast number of chemical insults. For example, RNase A remains active in conditions such as pH ranges between 2 and 10, solutions of up to 8 M urea, extreme temperatures (15–80°C), and boiling for 30 min. If a clinical sample containing RNA becomes compromised by RNases, subsequent analyses using that sample may produce unreliable results. Therefore, laboratory precautions must always be taken in preparing samples to inhibit degradation of RNA. For example, gloves should be worn at all times and changed frequently to avoid introduction of RNases. Bottles containing tubes and solution should be closed when they are not in use to avoid contamination with dust.

In parallel, endogenous RNases naturally occurring in the tissues have to be inactivated by inhibitors or by chemical treatment as quickly as possible after sample collection. The most common reagent for such a purpose is guanidinium thiocyanate, a strong chaotropic denaturant. Other inactivation reagents for RNase, including guanidine isothiocyanate, 2-mercaptoethanol, or dithiothreitol, are normally added to the extraction buffer so that the RNase molecules can be inactivated immediately after they are released from cells. Inactivation of RNase can also be performed prior to cell lysis: for example, the pretreatment of all equipment and tools with DEPC (diethyl pyrocarbonate). RNA degradation can also be minimized by reducing the sample size (tissues or cells).

Effects of Storage Conditions on RNA Quality

Measuring the level of specific RNA species in the bloodstream has proven to be a powerful tool for monitoring disease occurrence and stratifying patient risk for a variety of oncological, neurological, immunological, and cardiovascular disorders. However, whole blood is a complex mixture of various cell types in which the relative distribution of white blood cells may differ substantially between normal and pathological samples. Furthermore, a variety of physiological changes and disease states can alter gene expression in peripheral blood cells. For optimal results, current methods for extracting RNA from whole blood require sample processing within a few hours of collection to minimize degradation of the RNA.

Normally, mRNAs from blood cells have different stabilities. For example, mRNAs of regulatory genes have shorter half-lives than mRNAs of housekeeping genes. To ensure that the isolated RNA contains a representative distribution of mRNAs, blood samples should not be stored for long periods before isolating RNA. However, it may often be impractical in a clinical setting. More often, the collected blood samples are not readily processed at the time when they are collected. The samples may need to be transported to a laboratory facility for extraction. During the transportation, physiological and environmental factors can influence the quality of blood samples. Therefore, RNAs quality may be affected and can either induce or repress gene expression, or lead to degradation if the blood is not handled properly upon sampling.

Reliable quantification of mRNA levels for epidemiological, diagnostic, prognostic, and therapeutic purposes requires optimal RNA quality and quantity and optimized procedures for blood collection and preservation. In order to prevent both in vitro RNA degradation and gene expression changes that can induce under- or overestimation of gene expression analysis, dedicated blood collection tubes containing RNA stabilizer can be used. For example, the PAXgene Blood RNA System contains a stabilizing additive in an evacuated blood collection tube called the PAXgene Blood RNA Tube. The additive in the PAXgene tube reduces RNA degradation, and the RNA in whole blood can be stable at room temperature for 5 days, following storage for up to 12 months at −20°C and −80°C, and also after repeated freeze-thaw cycles. Furthermore, the RNA purified from the blood samples that are kept at the room temperature for 5 days can be optimized to include deoxyribonuclease (DNase) treatment and provide high-quality RNA for gene expression studies (Fig. 6.8). As such, preservation in PAXgene tubes restricts ex vivo gene expression, allowing meaningful RNA assays and yielding transcript concentrations that are much closer to in vivo responses than can be obtained by other methods

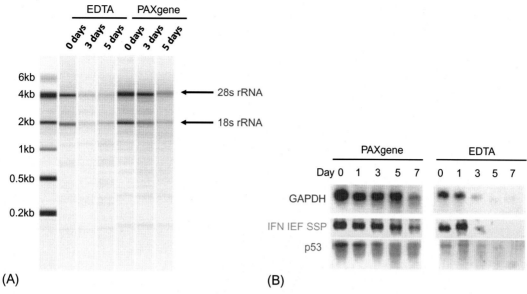

FIG. 6.8 (A) Gel electrophoresis evaluation of total RNA integrity for whole blood samples isolated and stored via EDTA prep versus PAXgene prep at 22°C. (B) Northern blot analysis of GAPDH, IFN IEF SSP, and p53 mRNA from whole blood samples collected and stored in PAXgene or EDTA at 22°C.

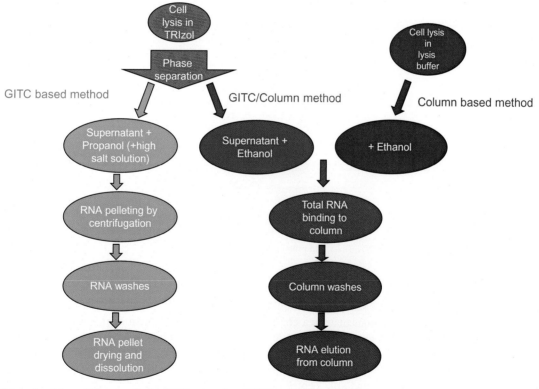

FIG. 6.9 Schematic of three different methods of RNA extraction: GITC-based method, GITC-column-based method, and column-based method.

EXTRACTION OF RNA

The general steps of RNA extraction are similar to DNA extraction. These steps include cell lysis, which disrupts the cellular structure to create a lysate, inactivation of cellular nucleases such as RNase, separation of RNA from cell debris, and purification of RNA. Three commonly used methods can effectively isolate RNA from samples: the guanidinium thiocyanate-based method, column-based method, and the combined guanidinium thiocyanate-column based method (Fig. 6.9). These three methods are compared in Table 6.3.

TABLE 6.3 RNA Extraction Method Comparison

	GITC-Based Method (e.g., TRIzol, TRI-Reagent)	TRIspin (Combined GITC-Column Based Method) (e.g., TRIzol + RNeasy)	Column Based Method (e.g., RNeasy)
Pros	• Possible higher yield for a small number of cells	• Quick and easy	• Quick and easy
Cons	• Longer time • Demands more techniques	• The RNA yield may be low for a smaller number of cells • Higher cost	• The RNA yield may be low for a smaller number of cells • May not be effective for some tissues

GITC, guanidinium isothiocyanate.

FIG. 6.10 An example of layer separation of RNA, DNA, and proteins/lipids in chloroform.

Guanidinium Thiocyanate Phenol-Chloroform Extraction

In the widely used guanidinium thiocyanate phenol-chloroform method, the cell extract is extracted with phenol/chloroform at low pH. Guanidinium thiocyanate is a chaotropic agent used in protein degradation. The principle of this single-step technique is that RNA is separated from DNA after extraction with acidic solution consisting guanidinium thiocyanate and phenol. The guanidinium thiocyanate–phenol solution, which is commercially available as TRIzol, TriFast, or TRI Reagent, disrupts the cells, denatures the proteins, and deactivates the nucleases, thereby stabilizing the DNA, RNA, and protein. Chloroform is then added, and after centrifugation the solution separates into an upper aqueous phase containing the RNA and a lower organic phase containing the DNA and protein under acidic conditions (Fig. 6.10). Recovery of total RNA is then done by precipitation with isopropanol. Another simple method is to use hot acid phenol to obtain RNA from microorganisms. The cell lysates are mixed with acid phenol and incubated at 65°C for 1 h. Subsequently, the aqueous phase is subject to phenol-chloroform extraction.

Spin Column Method

Alternatives to the guanidinium thiocyanate extraction methods are the commercially available kits that use spin column technology. These are much faster, avoid the use of the toxic phenol and chloroform reagents. Although several manufacturers produce kits for simultaneous DNA, RNA, and protein extractions, these kits are all similar in basic concept. After homogenization in a (usually guanidine-based) denaturing buffer, the sample is passed through a spin column that binds the DNA. Alcohol or acetone is then added to the flow-through (containing the RNA and protein), after which it is centrifuged through a second spin column that binds the RNA, leaving just the protein in solution. The RNA is washed on-column and then eluted in water or Tris-EDTA buffer.

Isolation of Poly(A)$^+$ RNA

Poly(A)$^+$ RNA is the template for protein translation, and most of the eukaryotic mRNAs carry tracts of it at their 3′ termini. It makes up 1% to 2% of total RNA and can be separated by affinity chromatography on oligo(dT)-cellulose (Fig. 6.11).

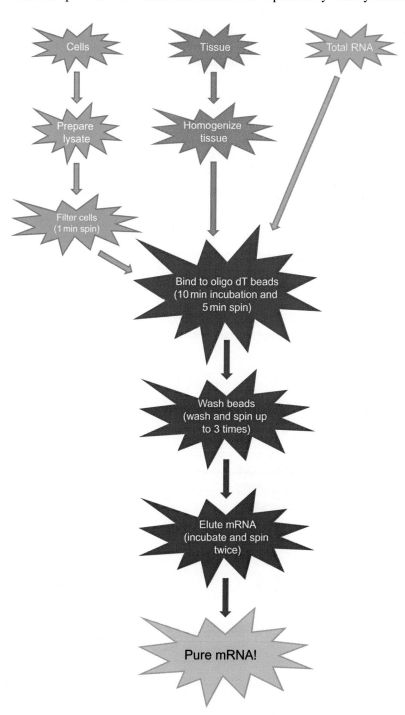

FIG. 6.11 A schematic procedure of mRNA purification by an oligo dT affinity chromatography.

Because mRNA molecules extracted from bacterial cells mostly lack poly-A tails, the methods developed so far for bacterial mRNA preparation have focused on removing non-mRNAs rather than selecting mRNAs. Poly(A) tails form stable RNA-DNA hybrids with short chains of oligo(dT) that attach to various support matrices. High salt must be added to the chromatography buffer to stabilize the nucleic acid duplexes, as only a few dT-dA base pairs are formed. A low-salt buffer is used after nonpolyadenylated RNAs have been washed from the matrix. This buffer helps to destabilize the double stranded structures and elute the poly(A)$^+$ RNAs from the resin.

Two methods are commonly used in the selection of poly(A)$^+$ RNA—column chromatography on oligo(dT) columns and batch chromatography. Column chromatography is normally used for the purification of large quantities (>25 μg) of poly(A)$^+$ RNA isolated from mammalian cells. Batch chromatography is the preferred method when working with small amounts (<50 μg) of total mammalian RNA. It can be used when many RNA samples are to be processed. Batch chromatography is carried out with a fine grade of oligo(dT) cellulose at optimal temperatures for binding and elution.

Magnetic oligo(dT) bead is an alternative for the purification of poly(A)$^+$ RNA from total RNA sample. The poly(A)$^+$ RNA can be extracted by introducing magnetic beads coated with oligo(dT) (Fig. 6.12). RNA with a poly(A) tail attaches to the oligo(dT). The beads will then be drawn to the bottom of a tube, removing mRNA directly from total RNA. The magnetic beads are specially treated to minimize the nonspecific binding of other nucleic acids and ensure the purity of mRNA.

Removal of rRNA

Removal of rRNA from total RNA is essential for mRNA enrichment, which is required by some analyses, especially in mRNA sequencing by next-generation sequencing techniques. To eliminate rRNA, several methods have been developed. These methods include subtractive hybridization with rRNA-specific probes, reverse transcription with rRNA-specific primers followed by RNase H digestion to degrade rRNA in rRNA:cDNA hybrids, preferential polyadenylation of mRNA, recovery of mRNA from gel electrophoresis, and digestion with exonuclease that preferentially acts on RNA molecules with a 5′-monophosphate end (including mature rRNA, tRNA, and fragmented mRNA). For RNA-sequencing transcriptomics (also known as RNAseq) using high-throughput sequencers, subtractive hybridization and exonuclease digestion are popular methods for the removal of rRNA. Subtractive hybridization uses probes that correspond to either conserved sites on 23S or 16S bacterial rRNAs or rRNA operon. Target rRNA could be removed after hybridization with the probes or operons, which are conjugated to magnetic beads for recovery. In order to increase the probe-to-sample specificity, sample specific probes should be applied. Sample rRNA is amplified and appended with a T7 promoter, which then allows transcription of the probes. For the exonuclease digestion, Terminator 5′-phosphate-dependent exonuclease is a processive 5′→3′ exonuclease that digests RNA that has a 5′ monophosphate. The enzyme does not digest RNA that

FIG. 6.12 The use of Dynal bead and poly-T Ologo primer for the purification of poly(A)+ RNA from total RNA sample.

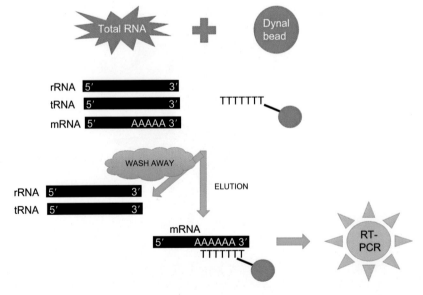

has a 5′-triphosphate, a 5′ cap (present on eukaryotic mRNAs), or a 5′-hydroxyl group. Bacterial rRNAs that are transcribed as a single transcript and then processed to yield rRNAs with 5′ monophosphates are substrates for the enzyme. Thus, this method can be used to isolate prokaryotic mRNA substantially free of 16S and 23S rRNA.

Terminator Exonuclease will also digest single-stranded DNA (ssDNA) and double-stranded DNA (dsDNA) that has a 5′-phosphate group. It does not digest ssDNA or dsDNA having a 5′-triphosphate or a 5′-hydroxyl group. In some instances, these methods have been used in combination to improve rRNA removal. There is no consensus, however, on the best approach.

PROTEIN PREPARATION

Proteins are essential cellular machinery. They can perform and enable tasks within biological systems. Each step of cellular generation, from replication of genetic material to cell senescence and death, relies on the correct function of several distinct proteins. The precision of cellular machinery can be disrupted, and the disruption can lead to disease. As a result, proteomic analysis has become an important technique in molecular diagnostics.

The field of proteomics is the large-scale study of proteins. The rapid development of high-throughput proteomic techniques has provided a valuable approach for new biomarker discovery in the postgenomic era. The proteome is the total set of proteins encoded by a genome. In other words, it is the total protein complement of an organism. Many proteins are processed and modified by other proteins after translation, and the final protein complement depends on complex interactions between these proteins. Therefore, large-scale protein analysis becomes essential for discovering and evaluating biomarkers for personalized medicine. Currently, there are increasing uses of proteomics in the molecular diagnostic field. The field of proteomics encompasses many techniques, such as immunoassays and two-dimensional differential gel electrophoresis (2-D DIGE). Another group of methodologies that are growing in popularity for protein discovery and analyses are mass spectrometry-based approaches.

In order to obtain protein from a biological specimen, the sample must first be harvested from the organism, culture, or patient. Samples can be obtained by several methods. Traditional dissection from biopsies, blood draws, and additional methods can deliver adequate protein for analysis. Samples can be immediately frozen at −80°C after removal or processed fresh. The proteins in the sample must be made readily accessible via lysis and extraction from the cells in the sample. The first step of protein preparation is thus to disrupt the cells. The effectiveness of a particular cell disruption method determines the accessibility of all intracellular proteins to extraction. Different biological materials require individual cell disruption strategies. In general, these methods can be divided into two main categories: gentle and harsher methods. Following is the detail for each method.

Gentle Cell Disruption

Gentle cell disruption protocols are generally employed when the sample of interest consists of cells that lyse easily, such as red blood cells and tissue culture cells. Osmotic lysis is one of the most popular methods. During this process, cells swell and burst when they are suspended in hypotonic solution, releasing all cellular contents. In freeze-thaw lysis, cells can be lysed by subjecting them to one or more cycles of quick freezing using liquid nitrogen and subsequent thawing. For detergent lysis, cellular membranes can be solubilized just by suspending the cells in many detergent-containing solutions. Cells can also be lysed in isosmotic solutions by enzymes that specifically digest the cell wall. These enzymes include cellulase and pectinase for plant cells, lyticase for yeast cells, and lysozyme for bacterial cells.

Mechanical Rupture

Biological material with tough cell walls and many tissue types requires more rigorous methods, which are mainly based on mechanical rupture. In sonication, a cell suspension, cooled on ice to avoid heating, can be disrupted by shear forces using short bursts of ultrasonic waves. In a French press, cells are lysed by shear forces by forcing a cell suspension through a small orifice at high pressure. On the other hand, cells of solid tissues and microorganisms can be broken with a mortar and pestle. Usually, the mortar is filled with liquid nitrogen and the tissue or cells are grounded to a fine powder. For soft, solid tissues, mechanical homogenization can be employed by using handheld devices, blenders, or other motorized devices. Glass bead homogenization can also be employed for cell disruption.

CELL LYSIS CONDITIONS

Enzymatic Degradation During Cell Lysis

During cell lysis, the compartmentalization of a cell will be partly or fully destroyed, depending on the lysis method applied. As a consequence, hydrolases (e.g., phosphatases, glycosidases, and proteases) will be present in a homogeneous protein solution and possibly alter the protein composition of the lysed cells. Therefore, enzymatic degradation must be avoided by placing the freshly disrupted sample in solutions with strong denaturing agents such as 7–9 M urea, 2 M thiourea, or 2% SDS. In this environment, enzymatic activity is often negligible. In addition, cell disruption is often performed at low temperatures to diminish enzymatic activity. Furthermore, because enzymatic activity is pH dependent, unwanted hydrolysis can often be inhibited by lysing the protein samples at a pH above 9 using either sodium carbonate or tris(hydroxymethyl)aminomethane as a buffering agent in the lysis solution.

In contrast to relatively labile phosphatases and glycosidases, some proteases are fairly resistant to denaturation, pH, and temperature change. In these cases it is advisable to consider the addition of protease inhibitors. Because individual protease inhibitors are specific to particular classes of proteases, broad-range protease inhibitor cocktails are normally used. Examples of protease inhibitors are phenylmethylsulfonyl fluoride (PMSF), aminoethyl benzylsulfonyl fluoride, tosyl lysine chloromethyl ketone, tosyl phenyl chloromethyl ketone, EDTA, benzamidine, and peptide protease inhibitors (e.g., leupeptin, pepstatin, aprotinin, bestatin). Phosphorylation is one of the most important posttranslational modifications that can determine protein function. In functional proteomics studies, phosphatase inhibitors such as okadaic acid, calyculin A, and vanadate are necessary to block the phosphatases involved in phosphorylation pathways.

Lysis Buffer

It is very common that lysis methods involve a chemical lysis and extraction agent, along with some mechanical stimulus that physically breaks apart the cell, allowing the chemical agent to solubilize the available protein. A list of common detergents, including critical micellar concentration (CMC), can be found in Table 6.4. The choice of lysis buffer depends on the protein target of extraction, sample size, and experience in preparation methods.

Lysis buffers can differ in CMC, as seen in Table 6.4. The CMC is the concentration at which the detergent forms micelles spontaneously. Therefore, CMC can affect detergents' efficacy and removal in different environments. Above the CMC value, the detergent forms micelles, and detergent added will move directly into micelles. Higher CMC values

TABLE 6.4 A List of Commonly used Cell Lysis Detergents

Detergent Name	Type	Molecular Weight	CMC (mM)	Micelle Molecular Weight	Suggested Removal
Triton X-100	Nonionic	647	0.24	90,000	TCA-acetone
NP-40	Nonionic	617	0.29	90,000	Acetone
Tween 20	Nonionic	1228	0.06		Acetone
Tween 80	Nonionic	1310	0.01	76,000	Acetone
Octyl glucoside	Nonionic	292	23–24	8000	Ethyl acetate
Octyl thioglucoside	Nonionic	308	9		Ethyl acetate
Big CHAP	Nonionic	878	3–4	8781	Filtration
Deoxycholate	Anionic	415	2–6	2000	Acetone, TCA
Sodium dodecyl sulfate	Anionic	288	6–8	17,887	Filtration/FASP
CHAPS	Zwitterionic	615	8–10	6149	Filtration
CHAPSO	Zwitterionic	631	8–10	7000	Filtration

CMC, critical micellar concentration; *FASP*, filter-aided sample prep; *TCA*, trichloroacetic acid.

are associated with weaker hydrophobic binding to monomers. Thus, higher CMC detergents tend to be more easily removed by buffer exchange and dialysis. Solutions with lower CMC values form micelles more easily and generally require less detergent to effectively solubilize protein. Another factor that can affect a lysis buffer is the micelle molecular weight (MMW). Lower-weight micelles are more easily removed than higher-weight micelles. Making use of CMC and MMW, one can more easily determine the best course for cell lysis. Choosing a lysis buffer depends greatly on these detergent factors. On the other hand, most of the detergents listed are incompatible with downstream mass spectrometry analysis, as such, they must be removed.

Protein Quality Under Different Lysis Conditions

Lysis, extraction, and denaturation of protein can occur in the same step with certain procedures, such as with SDS while boiling and agitating the sample. Macroscale proteomic lysis techniques can efficiently extract protein from large amounts of sample. On the other hand, microscale proteomics focuses on the separation, preparation, and analysis of protein samples under 100 μg, which are very sensitive to losses in the proteomic workflow. The issue for preparing a microscale sample is the removal of detergents and contaminants present in the lysate, which interfere with later steps. Catastrophic losses can occur from indiscriminate application of lysis techniques without proper planning and care. With each transfer and processing step, protein can be lost. Reproducibility of sample preparation is a key for reducing and troubleshooting sample loss in microscale proteomics. The concentration of protein plays a large role in sample loss. More concentrated samples tend to have less catastrophic losses in precipitation, desalting, and resuspension, due to the concentration-dependent adsorption maxima of many proteins. Therefore, microscale proteomic techniques need to focus on the efficient lysis of sample and removal of contaminants while retaining maximum sample. For example, the French press has a large surface area where proteins can adsorb and be lost. This mode of lysis is not ideal with microgram quantities of sample. Instead, sonication in microcentrifuge tubes can be a better choice, so as to efficiently lyse cells, maximize sample concentration of protein, and thus minimize loss to surfaces.

PREPARATION OF CLEAN PROTEIN SAMPLES

Contaminant Removal by Precipitation

Removal of contaminants and detergents is necessary after cell lysis. Some detergents will interfere with enzymatic digestion, and most will interfere with reverse-phase separations and mass spectrometry, sometimes damaging instruments and irreversibly ruining columns. Removal of unwanted cellular material, such as lipids and genomic DNA, can prevent chromatographic interference and it provides a much cleaner protein samples for downstream data analysis.

One common approach for contaminant removal is precipitation. The uses of precipitation vary based on three main factors: (1) the detergent or contaminants for removal; (2) native or denatured state of proteins; and (3) the postprocessing analysis.

In a bottom–up mass spectrometry-based proteomics, which investigates the peptides from digested proteins, detergent removal is a must, and the denaturation of the protein has little to no bearing. Some common detergents can be removed using acetone precipitation. Others can be removed using trichloroacetic acid (TCA) precipitation. A chloroform-methanol mixture or ethyl acetate can remove contaminants with high yield. Most interferants can be eliminated using molecular-weight cutoff filters, which is one of the molecular separation techniques, to capture the protein in the sample. Techniques for molecular separation of proteins are discussed in Chapter 8.

There are various other precipitation techniques that have been developed (Table 6.5). In general, TCA, chloroform-methanol, ethyl acetate, and acetone precipitation have similar efficiency for a wide variety of samples. Chloroform-methanol works better for membrane proteins. Ideally, precipitation would be avoided in microscale proteomics, because of the possibility of sample loss. However, several factors that govern precipitation efficiency can be altered to produce successful precipitations, even in the microscale.

Acetone Precipitation

Acetone precipitation sequesters mostly water-soluble proteins. This precipitation method is a concentration-based procedure. Therefore, higher concentration of protein in the original solution results in higher recovery. Acetone precipitation is simple to perform. The standard procedure for acetone precipitation involves the addition of cold (0 to −20°C) acetone to aqueous sample mixtures to a composition of 80%. The protein yield relies on three major factors including the initial

TABLE 6.5 Techniques to Remove Contaminants and Detergent From Proteomic Samples

Technique	Description
"Salting out"	Saturation of salt to precipitate proteins. This may be in the form of an ammonium sulfate precipitation or sodium sulfate
Ultrafiltration	Centrifugation at high speed using molecular weight cutoff filter to remove contaminants
Polyethyleneimine (PEI)	The use of a cationic polymer to precipitate nucleic acids in 1 M NaCl, where proteins remain in supernatant (PEI must be removed before further analyses)
Isoelectric Point (PI)	The pH is adjusted with mineral acid to the PI of most proteins (pH 4–6) to precipitate neutral proteins
Thermal	Cell extracts are denatured with high temperatures causing them to precipitate. Stability is enhanced
Nonionic polymer Polyethylene glycol (PEG)	Concentration of PEG unique to the protein mixture is added. Due to the excluded volume principle and centrifugation, precipitated proteins are pelleted (PEG must be removed before further analyses)

FIG. 6.13   Proteins yield vary at different salt concentration in 80% acetone.

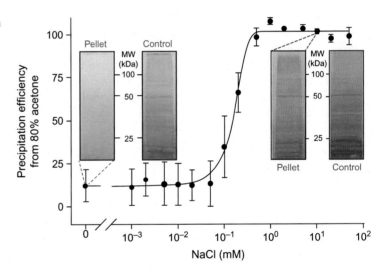

concentration of protein obtained during cell lysis, the percentage of acetone used, and the ionic strength of the pre-precipitation solution. The ionic strength correlated with protein charge and dielectric conditions in the protein sample. Ideal conditions for near-quantitative to quantitative yield of protein were established to include addition of 80% acetone to nearly any protein mixture containing 1 to 100 mM of NaCl or similar salt. The salt concentration necessary for complex mixtures can vary widely, requiring some optimization for particular cell lysates (Fig. 6.13). If acetone precipitation is to be used for microscale proteomics, it is ideal to use Nonidet P-40 (NP-40) as the lysis reagent. The use of NP-40 can provide good protein yields during extraction. Even nanoscale analyses can be performed using acetone precipitation.

TCA Precipitation

Quantitative protein precipitation using TCA is specific to deoxycholate and certain other detergents in solution. Although TCA can be used alone for precipitation, the pellet is not usually readily dissolved, causing sample losses. In a typical TCA-deoxycholate (DOC) precipitation, the protein solution is mixed with a dilute sodium deoxycholate buffer. The protein is intercalated by deoxycholate and then precipitated from solution with TCA. Acidification of the deoxycholate can cause a change in solubility, causing aggregation and precipitation. The deoxycholate can then be preferentially removed using acetone, leaving near-pure protein behind. TCA-DOC precipitation is remarkably efficient, even in dilute samples, with recovery values from 90% to 100% of total protein.

Methanol-Chloroform Precipitation

Methanol-chloroform precipitation is an efficient method of precipitation and concentration for dilute samples, especially those containing membrane proteins. The ratio of methanol:chloroform:water is 4:1:3, with an additional three volumes of methanol added to pellet the protein. Proteins that are only slightly soluble in methanol-chloroform accumulate on the water-organic interface and can then be pelleted by the additional methanol and centrifugation. The efficiency of this procedure approaches 100% for a variety of protein concentrations and detergent solutions.

Problems Caused by the Precipitation

Protein precipitate is usually compacted into a pellet by centrifugation steps and the pellet is often dried to some extent. Exposure of the whole pellet to the dissolution buffer is necessary. Part of the precipitate can easily be left behind if the solution and container are not thoroughly examined. Therefore, dissolving precipitated samples can be a challenge.

Strategies have been developed for fully suspending the pellet in any lysis buffer of choice. These strategies include vigorous vortexing, sonication, shaking, and even two-step, on-pellet trypsin digestion. In microscale, it is best to avoid vigorous agitation, because adsorption and loss occurs with increased exposure to surfaces. On the other hand, sonication avoids much of the splash-up that vortexing or shaking involves, and can efficiently exposes the pellet to buffer.

PROTEIN EXTRACTION FROM ARCHIVAL FORMALIN-FIXED PARAFFIN-EMBEDDED TISSUES

Protein recovery can be difficult for some fresh or frozen tissues, and when used for proteomic analyses, the results generally cannot immediately be related to the clinical course of diseases. In contrast, FFPE tissue samples can be the alternative for diagnostic pathology archives (Fig. 6.14). Use of FFPE tissues for proteomics was traditionally thought to be too problematic, given the presence of formalin-induced protein cross-links and modifications, which create problems for separating, visualizing and characterizing individual proteins. Furthermore, the extraction of proteins from archival FFPE tissues for proteomic analysis has been hampered by the deleterious effects of formaldehyde-induced protein adducts and cross-links that are formed during tissue fixation and subsequent histological processing. However, recent techniques development has proved that FFPE samples represent an excellent resource. There is also the advantage of linking results with relevant clinical data, including long-term outcomes or response to treatment. Importantly, the use of FFPE tissue samples would allow sufficient numbers of cases of some of the rarer diseases or clinical subgroups to be assembled for analysis. Furthermore, proteomics data sets from FFPE tissue are in good agreement with those obtained with fresh frozen tissue samples.

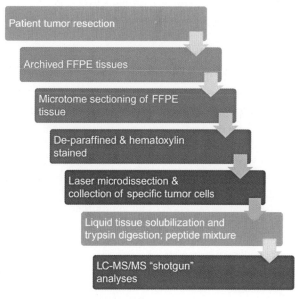

FIG. 6.14 Clinical FFPE tissue proteomic screening process flowchart.

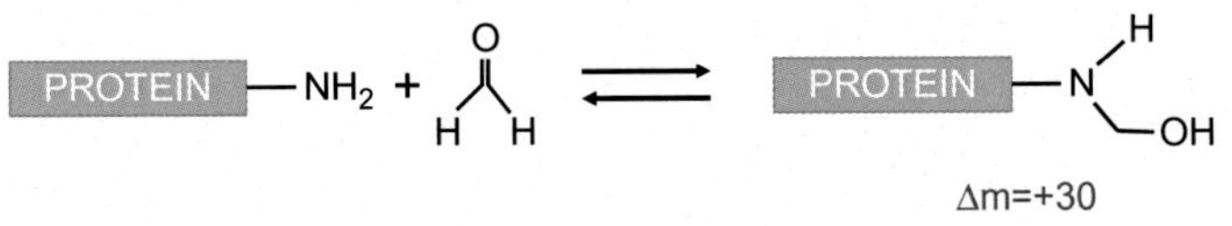

FIG. 6.15 Formaldehyde cross-linking reaction in which a hydroxymethyl-methylol adduct with a + 30 Da mass shift is produced.

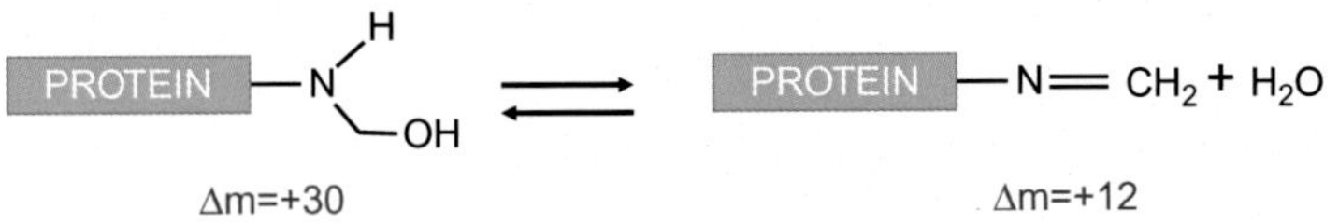

FIG. 6.16 Formaldehyde cross-linking reaction in which water elimination yields a Schiff base and a +12 mass shift.

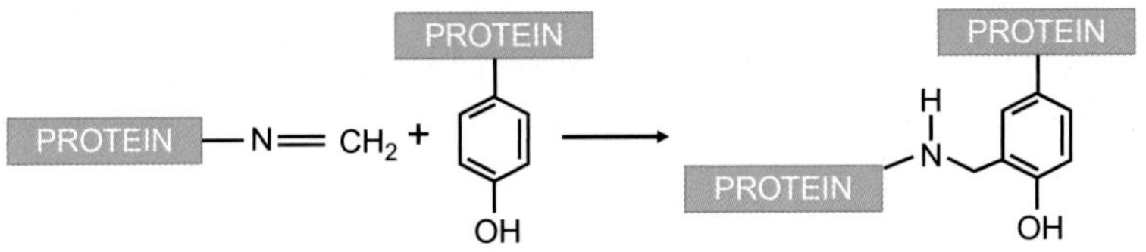

FIG. 6.17 Formaldehyde cross-linking reaction in which a nucleophilic attack yields a methylene bridge.

Problems Caused by the Formalin in Tissue Extraction

Formalin is the most common fixative used in diagnostic laboratories, and formalin is composed of formaldehyde and methanol. Formaldehyde reacts with the amino groups of basic amino acids in proteins to form reactive methylol adducts, which creates a +30-Da mass shift (Fig. 6.15). For example, formaldehyde can react with lysine, cysteine, arginine, tryptophan, histidine, and the N-terminal amine to form methylol adducts. The methylol adduct can subsequently undergo a dehydration reaction to produce a Schiff base, which is seen most frequently in lysine and tryptophan residues. This creates a total mass shift of +12 Da (Fig. 6.16) Furthermore, the protein N-terminal amine can be converted to a stable 4-imidazolidione adduct, and a Mannich reaction can occur between adducted tyrosine and arginine residues in close spatial proximity. A condensation reaction then occurs, resulting in the formation of methylene bridges between these basic amino acids and several other amino acids (Fig. 6.17). The methylene bridges create inter- and intramolecular protein cross-links, and the intrapeptide cross-links can form between lysine and histidine as well as lysine and arginine. These cross-links stabilize the tissue morphology and remain that way for many years. These formaldehyde-induced modifications reduce protein extraction efficiency and may also lead to misidentification of proteins during proteomic analysis. Thus, the most important issue in retrieving protein from FFPE is to reverse formaldehyde-induced protein adducts and cross-links so that the sample can be used in high-throughput proteomics.

Possible Solutions for PPFE Sample Extraction: Heat-Induced Antigen Retrieval (AR) for Tissue Extraction

The majority of the proteomic studies on FFPE tissues employ tissue extraction methods that are derived from heat-induced antigen retrieval (HIAR) methods which are originally developed for immunohistochemistry. The AR-based method has been combined with recovery buffers containing Tris-HCl, detergents such as SDS, and reducing agents such as DTT. Typically, the rehydrated FFPE sections are resuspended in 50 mM Tris-HCl at pH 4, with 2% (w/v) SDS. The samples are homogenized, followed by sonication. Subsequently, the homogenized FFPE samples are then incubated at 100°C for 30 min followed by 80°C for 2 h (Fig. 6.18). HIAR is thought to promote high-temperature cleavage of methylene bridges and protein denaturation which is stabilized by selection of an appropriate buffer pH.

Proteins extracted from FFPE tissue using HIAR protocols are suitable for proteomic analysis. For example, bottom-up mass spectrometry analysis can identify 40% to 90% overlap of proteins compared to fresh tissue samples (Table 6.6). However, the HIAR method may not be ideal for top-down proteomic analysis. This is because it always shows protein fragmentation, incomplete recovery, and incomplete cross-link reversal by multiple top-down gel-based studies, such as one-dimensional polyacrylamide gel electrophoresis and two-dimensional gel electrophoresis. Although heat appeared to be the most important factor for achieving a qualitative and quantitative extraction of proteins from FFPE tissues for

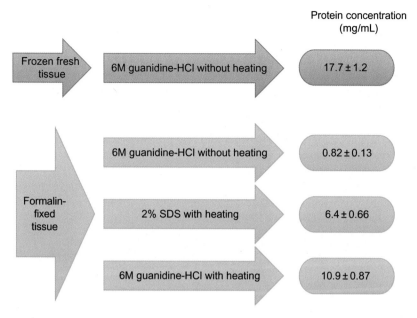

Protein concentration (mg/mL)

FIG. 6.18 Summary of protein extraction protocols from formalin fixed samples.

TABLE 6.6 Frozen and Formalin-Fixed Tissue Identification of Peptides and Proteins

Method	Number of Identified Proteins	Proteins Identified With 2 Unique Peptides Minimum	Number of Unique Peptides
Frozen tissue: 6 M guanidine-HCl without heating	976	480	3207
Formalin tissue: 6 M guanidine-HCl without heating	130	57	352
Formalin tissue: 2% SDS with heating	820	395	2540
Formalin tissue: direct digestion of tissue homogenate	331	106	589
Formalin tissue: 6 M guanidine-HCl with heating	827	470	3005
Formalin tissue: CNBr treatment of pellet	526	202	1129

CNBr, cyanogen bromide; *SDS*, sodium dodecyl sulfate.

bottom-up analysis, for maximum protein retrieval, methodologies that address both of these issues, such as the use of elevated pressure and heat, should improve the proteomic analysis of FFPE tissue. For example, the addition of high hydrostatic pressure (40,000 psi) to augment heat treatment (100°C for 30 min, followed by 80°C for 2 h) dramatically improved protein extraction efficiency from multiprotein FFPE tissue.

The problems observed in the top-down proteomic analysis always lead to gel smearing and background. If a protein has a large number of basic residue, this could lead to less efficient extraction. This decreased efficiency causes increased modifications and cross-links. Therefore, it is required to develop optimum HIAR methods (i.e., relatively short fixation time) and control preanalytical factors that affect protein extraction efficiency of FFPE tissue proteomics. Preanalytical factors can include tissue size, fixative composition, and duration, as well as the age of the FFPE block. For example, formalin fixation may promote selective depletion of basic proteins and also membrane proteins as a result of HIAR-induced aggregation.

To date there is neither a universal protocol for protein extraction from FFPE tissues, nor any basic principle of standardization that is acceptable for practice. It is essential to develop a systematic standardized method for FFPE tissue analysis in top-down proteomics.

Standardization is an important issue, particularly with respect to the measurement of single protein concentrations within total FFPE protein extracts or the relative quantitation of proteins from multiple FFPE tissues for biomarker discovery. Because there is currently no method that can measure the amount of protein in FFPE tissues prior to sample extraction, it is critical that tissue samples be completely solubilized for accurate measurement of either total protein or a single protein within the extract. Complete solubilization of FFPE tissue samples is the best way to achieve the goal of standardization for tissue proteomics. Therefore, approaches to make FFPE samples soluble and the methodologies to control the variance can help to standardize protein extraction from FFPE tissue.

Chapter 7

Detection and Analysis of Nucleic Acids

Chapter Outline

DETERMINING DNA PURITY AND YIELD

Reliable methods to determine the precise concentration of DNA are necessary for numerous molecular applications, ranging from traditional molecular biological manipulations, such as restriction digest analysis, Southern blotting, and polymerase chain reaction (PCR), to diagnostic techniques, including quantification of genetically modified organism content of samples, detection of DNA contamination in drug preparations produced from recombinant organisms, and medical diagnosis of viruses and cancer.

Commonly used methods of DNA concentration measurements are the evaluation of the intensity of a band on an agarose gel, fluorescence measurements using various DNA-binding dyes, measurements of ultraviolet (UV) absorbance at 260 nm, and the real-time PCR method. Every method has its own advantages and disadvantages, and which is right for a particular user depends on choice and convenience.

UV Spectrometry

The standard nucleic acid quantitation method is UV spectrophotometry. In this method, the nucleic acid sample is placed into a quartz cuvette, which is then placed inside the UV spectrophotometer (Fig. 7.1A). UV light is passed through the sample at a specified path length, and the absorbance of the sample at specific wavelengths is measured. Absorbance at 260 nm (A_{260}) is to measure nucleic acid, and A_{280} is to measure contaminating protein in the sample (Fig. 7.1B). The principle of the UV absorbance method is that nucleic acids (DNA or RNA) contain conjugated double bonds in their purine and pyrimidine rings that have a specific absorption peak at 260 nm. The maximum absorbance of nucleic acids occurs at a wavelength of 260 nm (Fig. 7.2). The intensity of this absorbance is proportional to the concentration of nucleic acid. Because of these physical characteristics, the nucleic acid concentration of a sample can be determined.

Based on the absorbance readings, the concentration of the sample is determined, and A_{260}/A_{280} ratios are calculated to indicate sample purity. It is generally accepted that DNA of relative purity will yield an A_{260}/A_{280} ratio of ≥ 1.8 on a scale with a maximum of 2.0. A ratio below 1.8 indicates protein contamination in the DNA samples and a need for further purification.

Diagnostic Molecular Biology. https://doi.org/10.1016/B978-0-12-802823-0.00007-9

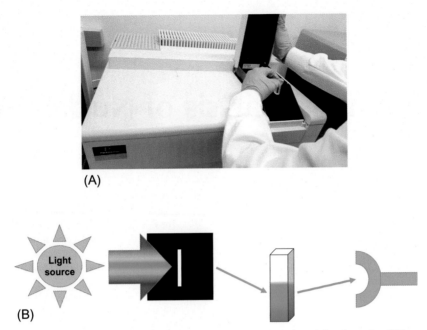

FIG. 7.1 (A) A typical example of a UV spectrophotometer. (B) The principle of the detection of absorbance in a UV spectrophotometer.

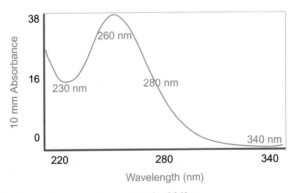

FIG. 7.2 The maximum absorbance of nucleic acids occurs at a wavelength of 260 nm.

The A_{260}/A_{230} ratio is another key measure of relative purity. The absorption wavelength A_{230} represents a wavelength range that can identify the presence of several chemical contaminants. For instance, residual chaotropic salts, which can inhibit PCR, particularly multiplex PCR, are known to absorb in the 230-nm range and below. The A_{260}/A_{230} ratio should generally be greater than 1.4 on a scale of 2.0 in order to maximize multiplex PCR methods. It is a critical measurement to assess when high-purity samples are required.

Using the absorption as a conversion factor from optical density to concentration, one optical density unit at 260 nm is equivalent to 50 mg/L of double-stranded DNA and 40 mg/L of RNA (Table 7.1). Furthermore, the A_{260} reading in conjunction with the specific equations can be used to determine the concentration of DNA present in the solution more accurately (see Table 7.1).

Traditional UV spectrophotometry is a common and simple DNA quantitation method. This method does not require use of a large amount of purified sample. If clean disposable plastic cuvettes are used, nucleic acid extract can be used to measure the concentration and then reused in downstream applications. Further benefits are that the UV method does not require additional reagents or incubation time. Finally, spectrophotometers are widely available.

A drawback to spectrophotometry is that minimum sample volumes of at least 50–75 µL are a common requirement to obtain an accurate instrument reading. This can be problematic if the sample is of low initial concentration and diluting the sample is impractical, or if a certain extraction method requires the elution of nucleic acids into low volumes. Another

TABLE 7.1 Formulas for Estimating Nucleic Acid Concentrations

Nucleic Acid	Concentration Unit	Formula
Single-stranded DNA	pmol/μL	$A_{260}/(10 \times size_{kb})$
	μg/mL	$A_{260}/0.027$
Double-stranded DNA	pmol/μL	$A_{260}/(13.2 \times size_{kb})$
	μg/mL	$A_{260}/0.020$
Single-stranded RNA	μg/mL	$A_{260}/0.025$
Oligonucleotide	pmol/μL	$A_{260} \times ((100)/(1.5*number_{adenines}+0.71*number_{cytosines}$ $+1.2*number_{guanines}+0.84*number_{thymines}))$

disadvantage of this method is that it does not distinguish signals between DNA and RNA, or between double-stranded and single-stranded DNA.

Another problem for the UV method is that the common biological contaminants. For example, if measuring DNA, biological molecules such as proteins, RNA, and chaotropic salts from extraction procedures, can falsely elevate nucleic acid concentration estimations. Buffer salts such as Tris, EDTA, and guanidine isothiocyanate absorb strongly at 230 nm and bleed into the 260-nm absorbance range, which can falsely elevate A_{260}/A_{280} and A_{260}/A_{230} purity ratios for samples (Fig. 7.3). In addition, free nucleotides present in a sample also have the ability to influence the UV quantitation method. Changes in sample pH also alter UV readings. Small shifts in solution pH, resulting simply from the use of different water sources, can cause significant variability of RNA A_{260}/A_{280} ratios. For example, adjusting the pH of water used in UV analysis from 5.4 to 7.5 or 8.5 resulted in an increase in the A_{260}/A_{280} ratio from 1.5 to 2.0. Therefore, it is important to adopt the standard approach and buffer when UV quantitation method is employed.

NanoDrop Spectrometry

An extension and improvement of the UV spectrophotometer is the NanoDrop instrument (Fig. 7.4). The NanoDrop technology is similar in principle to a conventional spectrophotometer but has many additional capabilities. This technique functions by combining fiber optic technology and natural surface tension properties to capture and retain minute amounts of sample independent of traditional containment apparatus such as cuvettes or capillaries. The use of sample surface tension is the enabling technology of the instrument, which requires only 1 μL to 2 μL of sample. Reducing the volume of sample required for spectroscopic analysis facilitates the inclusion of additional quality control steps throughout many molecular workflows, increasing efficiency and ultimately leading to greater confidence in downstream results. The sample is retained on an optical fiber, which assesses the UV absorbance of the sample across an absorbance range of 220 nm to 750 nm. The instrument is accompanied by special software to enable analysis of signal from small quantities of sample.

Although the NanoDrop quantitation method is still susceptible to many of the pitfalls of the traditional UV spectrophotometer method, the NanoDrop instrument is powerful because, in addition to calculating the concentration of the

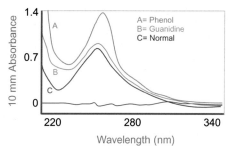

FIG. 7.3 The effects of contaminants on the spectra of nucleic acid sample profiles.

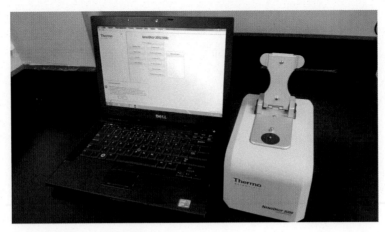

FIG. 7.4 A typical example of a Nanodrop spectrophotometer.

sample and its A_{260}/A_{280} ratio, it displays the entire absorbance spectrum of the sample in graphical form (Fig. 7.2). This allows contaminants to be more readily detected and potentially identified based on their absorbance wavelengths. This is because these contaminants could have been unnoticed if readings were performed with a traditional UV spectrophotometer. Fig. 7.3 depicts two NanoDrop curves, one generated prior to sample cleanup and salt removal (curve B), and the second generated after removal of excess guanidine isothiocyanate salts (curve C). The removal of contaminants, which can limit the amplification of DNA in a PCR reaction, particularly in multiplex PCR, can significantly improve the downstream molecular analysis.

An additional benefit of NanoDrop technology is its capability to determine a wide range of sample concentrations without requiring serial dilutions. The minimum volume required for an accurate instrument reading is also far less than for a conventional spectrophotometer. When using this method, it is of utmost importance to adequately mix the nucleic acid sample before removing an aliquot for measurement to ensure that the aliquot is representative of the entire sample.

Fluorometric Methods

Fluorometric methods are another nucleic acid quantitation method and are widely regarded as among the most sensitive methods available. Because DNA is naturally colorless, the principle of these methods is that the dyes used can intercalate and bind nucleic acid grooves nonspecifically, or selectively bind certain types of nucleic material. After the dyes binding to nucleic acids, dyes are excited at one wavelength of light and emit another wavelength of light. Therefore, different concentrations will show different intensity. Differences in spectral characteristics of nucleic acid-bound fluorophores allow sample concentrations to be determined.

Some commonly used dyes are ethidium bromide, Hoechst 33258, and PicoGreen. Ethidium bromide, a standard fluorometric dye, binds base pairs of DNA by intercalation. Because it has a high degree of intrinsic fluorescence, the sensitivity of the assay is limited. In one of the molecular separation methods, DNA samples can be analyzed with a series of known-concentration DNA samples on an agarose gel stained with ethidium bromide. By comparing the relative dye intensity, one can determine the DNA concentration roughly (Fig. 7.5).

Hoechst 33258 binds dsDNA by intercalating the minor DNA groove. The pitfall for using Hoechst 33258 is that it requires high salt concentrations to detect dsDNA when RNA is present. On the other hand, it requires low salt concentration to detect dsDNA when single-stranded DNA is present. Many fluorometric dyes used to quantitate DNA also bind RNA but, unfortunately, lack the fluorescence intensity and assay linearity needed for a sensitive RNA detection method.

PicoGreen can selectively bind dsDNA, and it exhibits low intrinsic fluorescence when unbound. Therefore, PicoGreen is more sensitive than other dye methods and is able to detect nucleic acid concentrations at 25 ng/mL. Although quantitating dsDNA with PicoGreen is powerful, a limitation is that the PicoGreen dye is sensitive to chaotropic salts and to organic solvents, which tend to decrease signal intensity and to increase signal intensity, respectively.

The drawbacks of fluorometric dye methods include the need for access to relatively costly equipment capable of fluorometric readings, expensive proprietary reagent kits required to perform quantitation assays, lengthy assay setup and dye incubation time, and sample volume consumption. Furthermore, the accuracy of DNA quantification is also significantly affected by DNA fragmentation. Basically, the amount of the DNA that is measured in samples decreases as the level of

1 ng   2 ng   5 ng   10 ng 20 ng   50 ng   100 ng 200 ng 500 ng 1 μg

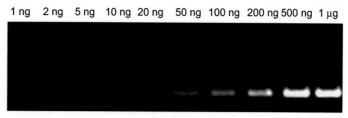

FIG. 7.5   An example of DNA concentration estimation in gel electrophoresis.

fragmentation increases. Meanwhile, contaminants such as sodium chloride and SDS weaken the PicoGreen signal, whereas those of organic reagents such as phenol, ethanol, and chloroform strengthen the PicoGreen signal. On the other hand, the benefits of fluorometric methods over UV spectroscopy include the low sample volume requirement and high degree of sensitivity. Because very small sample concentrations can be adequately detected and accurately quantitated, the maximum amount of sample can be saved for downstream applications.

Real-Time PCR Method

The real-time PCR (qPCR) technique is a powerful nucleic acid quantitation method. This is a good choice for qualitative as well as quantitative analysis of DNA because of its high sensitivity and specificity for typical molecular applications. The basic principle for qPCR is similar to traditional PCR because primers are used to anneal to denatured target DNA for amplification during thermal cycling. The difference is that in qPCR the fluorometric probes are used to bind target DNA during the annealing phase of PCR (Fig. 7.6). The fluorometric probes are then displaced and cleaved, which allows the emission of fluorescent dye upon target extension. The dye can ultimately be detected in order to determine the success of the amplification reaction, and the initial quantity of nucleic acid present in the sample is depicted in graphical form.

The main advantage of the qPCR method is the ability to assess the amount of the target DNA. The concentration of the unknown nucleic acid sample is determined by comparing a standard curve of known concentrations to the amplification plot of the sample. Other benefits include the detection of PCR inhibitors within the reaction mixture and the specificity inherent in the assay by use of fluorometric probes. Drawbacks to this method include the expensive proprietary reagents, primers, and probes needed to perform assays. Furthermore, the specialized instrumentation necessary to amplify nucleic

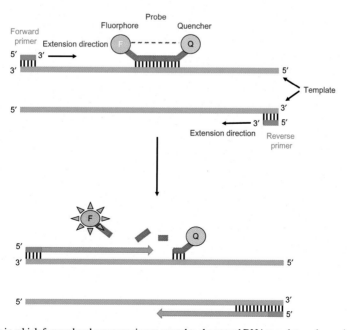

FIG. 7.6   Real-time PCR reaction in which forward and reverse primers anneal to denatured DNA templates, along with a reporter probe composed of a fluorophore and quencher. DNA polymerase nuclease activity then cleaves the probe freeing the fluorophore from the quencher and thus allowing for fluorescent detection of DNA amplification.

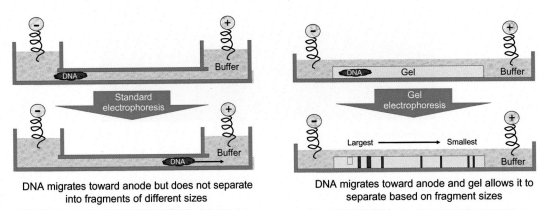

FIG. 7.7 The migration of nucleic acids in an electric with or without the presence of gel matrix.

acid is expensive. A lengthy assay time is required. Measurement accuracy may depend on the qPCR assay design, when shorter amplicons should be less affected because of the DNA fragmentation. Furthermore, the sample volume is expended in order to determine its concentration.

MOLECULAR SEPARATION OF DNA

Gel electrophoresis is routinely used for detection and size analysis of proteins and nucleic acids. Separation by this method uses an electrical current to propel charged biomolecules through a porous gel matrix at a rate that is a function of the charge, size, and shape of the molecules. Proteins are frequently analyzed using vertically oriented gels made of polyacrylamide, whereas DNA and RNA are most commonly analyzed using horizontal agarose slab gels. Here, we will discuss the analysis of DNA/RNA and the molecular separation of protein is discussed in Chapter 8.

Basic Principles of Electrophoresis

DNA and RNA are negatively charged because of their phosphor-sugar backbone. As a result, they migrate toward the positive pole in the electric field (Fig. 7.7). The migration rate of a molecule depends on two factors, its shape and its charge-to-mass ratio. Unfortunately, most DNA molecules are the same shape, and all have very similar charge-to-mass ratios. Fragments of different sizes cannot therefore be separated by standard electrophoresis. In general, DNA molecules larger than ~400 bp migrate with a mobility that is independent of size, because the charge/unit mass is the same for all DNA molecules. The size of the DNA molecule does, however, become a factor if the electrophoresis is performed in a gel, or separation matrix. Separation matrices include agar, agarose, polyacrylamide, and composite agarose-acrylamide gels. The matrix provides a tortuous path through which the DNA molecules migrate in the electric field. The penetration of the DNA molecules through pores of different sizes provides a mechanism for separation by molecular mass. DNA electrophoretic mobility is determined by the interplay of three factors: the relative size of the DNA molecules with respect to the pore size of the matrix, the effect of the electric field on the matrix, and specific interactions of the matrix with the DNA molecules during electrophoresis. Although the relative importance of each of these factors depends on the matrix, the conformation of the DNA, and the conditions, the size of DNA fragment can affect its migration speed at different matrix's pore size dramatically. For example, large DNA molecules can migrate faster at large matrix pore size (low concentration) than at small matrix pore size (high concentration). In the same gel matrix, large DNA molecules migrate slower than the small DNA molecules. As a result, the DNA molecules' migration speed and travel distance are inverse relative to their size, and they can be separated by gel matrix based on their size (Fig. 7.8).

AGAROSE GEL ELECTROPHORESIS

Agarose is a purified linear galactan hydrocolloid isolated from agar or agar-bearing marine algae. It is a linear alternating copolymer of D-galactose and 3,6-anhydro-L-galactose, infrequently substituted with charged groups such as carboxylate, pyruvate, and/or sulfate residues. Agarose is dissolved by heating the fibrous powder in an aqueous solution. The agarose chains have a random coil conformation in solution at high temperatures. Upon cooling, the agarose chains form helical fiber bundles, known as junction zones, held together by noncovalent hydrogen bonds. Subsequently, gelation occurs when

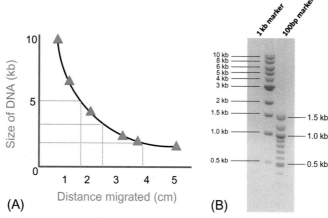

FIG. 7.8 (A) DNA molecules' travel distance is inverse relative to their size. (B) Gel electrophoresis analysis of DNA fragments by size.

the fiber bundles become linked together by the formation of additional hydrogen bonds. Furthermore, strand partner exchange occurs by hydrogen bond rearrangement. The gel point is the temperature at which an aqueous agarose solution forms a gel as it cools. Agarose solutions exhibit hysteresis in the liquid-to-gel transition—that is, their gel point is not the same as their melting temperature. Therefore, it is common that a hot agarose solution is poured into a shallow box known as the electrophoresis gel apparatus with a removable comb that points downward into the agarose. Once the agarose is polymerized, the comb can be removed and leave rectangular holes, also known as slots or wells, in the gel. DNA can be placed in the wells for electrophoresis.

The Gel Matrix

The agarose gel matrix is negatively charged. This is because anionic groups in an agarose gel are affixed to the matrix and cannot move. However, the dissociable cations such as covalently attached sulfate, carboxylate, and/or pyruvate groups, can migrate toward the cathode in the electrophoresis. These negatively charged residues are surrounded by positively charged counterions from the buffer. When an electric field is applied, some of the cations in the diffuse double layer near the gel fibers migrate toward the cathode, carrying along buffer and solvent molecules. The net result is a flow of the solvent toward the cathode, called the electroosmotic flow. DNA molecules are negatively charged and migrate in the opposite direction, toward the anode.

Orientation of the Agarose Gel

The orientation of the agarose gel fibers and fiber bundles affects electrophoresis in unidirectional electric fields. It is useful that an agarose gel is preelectrophoresed in a direction perpendicular to the eventual direction of electrophoresis. This can make linear DNA molecules travel in lanes skewed toward the side of the gel, as though the preelectrophoresis had created pores or channels in that direction. The lanes gradually straighten out and become aligned in the parallel direction as electrophoresis is continued. It is possible that gel fibers and fiber bundles gradually become oriented in the new field direction. Lanes skewed toward the side of the gel are not observed if a gel is preelectrophoresed in the direction in which electrophoresis will be carried out, or if a gel oriented in the perpendicular direction is allowed to stand for 24 h before use. Under this circumstance, the oriented agarose gel fibers and fiber bundles become randomized upon standing. It is likely that rearrangement of the hydrogen bonds in the junction zones occurs and DNA mobility in such randomized gels is identical to that observed in gels that have not been subjected to preelectrophoresis.

Size of DNA Molecules and the Mobility

The mobility observed for DNA molecules in agarose gels can be influenced by the relative size of the DNA compared with the average pore size of the gel matrix. If the end-to-end length of the DNA is smaller than the average pore diameter, the mobility decreases with the decreasing fractional volume of spaces within the gel that are available to the migrating DNA molecules. In low-voltage electric fields, the mobility of the largest DNAs decreases approximately linearly with increasing molecular mass. At higher electric fields, or for DNA molecules larger than ~ 12 kbp, the mobility is nearly independent of molecular mass.

Conformation of DNA Molecules and the Mobility

The topological forms of circular DNA include the supercoiled (SC; interwound) form and the relaxed open circle (OC) form. DNA isolated from bacterial hosts are in the SC forms. SC DNA can be relaxed to form the OC form either by the action of enzymes (topoisomerases) or by making a nick in one of the DNA strands (either enzymatic or chemical actions). A consequence of the physical structure of circular DNA is that it can be trapped on open-ended gel fibers during electrophoresis. Electrophoretic trapping is a balance between the electrophoretic force on a circle (pulling it against the trap) and diffusion (allowing a circle to escape a trap). Large circles have a greater electrophoretic force on them and are trapped at lower critical electric field strengths compared to smaller circles. The electrophoretic trapping of open circles in agarose gels is dependent on the circle size, electric field strength, and buffer ionic strength.

SC DNA molecules have more compact conformations than linear DNAs containing the same number of base pairs, and migrate faster than linear DNA. If the electric field is relatively low, topological isomers (topoisomers) containing different numbers of superhelical turns can be resolved into discrete bands (Fig. 7.9). Positively and negatively supercoiled topoisomers can be separated by two-dimensional electrophoresis if an intercalating agent, such as chloroquine or ethidium bromide, is added to the running buffer in one of the orthogonal directions. Large open circle (relaxed) DNAs have negligible mobilities in high-voltage electric fields, presumably because they are impaled by dangling fibers in the matrix.

The trapping of open circle DNAs in agarose gels can be eliminated by reversing the direction of the electric field or using pulsed unidirectional fields. It is possible that changing the direction of the field allows the circles to migrate out of the trap, or because pulsed fields change the direction of orientation of the agarose gel fibers. However, the trapping of open circle DNAs increases with decreasing agarose concentration, suggesting that the density of traps increases at low agarose concentrations.

Gel Buffer

Electrophoresis of DNA in slab gels is primarily performed using one of two standard buffers. A 1X TAE Buffer solution contains 40 mM Tris-acetate, and 1 mM EDTA at pH 8.3. A 1X TBE buffer contains 890 mM Tris base, 890 mM boric acid, 20 mM EDTA at pH 8.3. These solutions contain a weak acid, acetic or boric acid, that can exist in neutral and anionic forms (e.g., COOH and COO$^-$ species) and a weak base, Tris, that exists in either neutral or cationic forms (Tris-NH$_2$ and Tris $-$ NH$_3$$^+$). These ions carry the electrical current, buffer the pH, and maintain a low-conductivity medium. EDTA

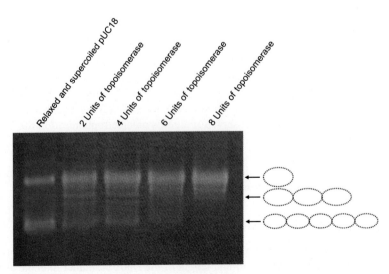

FIG. 7.9   The relaxation of pUC18 plasmid supercoiled DNA via topoisomerase I.

is not absolutely essential, but is added as a preventative because it chelates Mg^{2+} ions and therefore inactivates potential DNA nucleases that may be present.

For separation of DNA molecules ranging in size from several hundred to several thousand base pairs, both TAE and TBE provide good resolution of DNA fragments, with slightly improved separation of smaller fragments in TBE and of larger sizes in TAE (Fig. 7.10). A general practice is that TAE is recommended for resolution of RNA and DNA fragments larger than 1500 bp, for genomic DNA and for large supercoiled DNA. Despite their similar performance in most applications, some buffer-specific effects have been observed, especially in borate-containing buffers. Over the past several years, other conductive media have also been developed for electrophoresis of nucleic acids. Examples include solutions containing L-histidine, sodium borate, sodium threonine, lithium borate, and mixtures of pK_a-matched organic acids and bases.

DNA electrophoresis is usually performed at constant voltage. During an experiment, the current (measured in mA) increases and warming of the running buffer occurs. The voltage is kept low, ~10 V/cm, where cm refers to the length of the gel, to minimize heating effects. For a typical small- or medium-sized horizontal gel rig (10–15 cm length), power supplies are set to between 100 V and 150 V, and gels are run for 50–90 min. Although increasing the voltage above aforementioned voltage range using standard 1 × TAE or TBE buffers increases the speed of movement of the DNAs, this can lead to asymmetric heating of the gel and solution. Such heating is undesirable because it promotes broadening, slanting, and compression of the bands into each other, as well as faster movement of samples in the center than on the outside of the gel and other lane anomalies. Gels prepared with TBE generate less current than TAE gels.

Nucleic acids less than approximately 75 nucleotides in length are poorly resolved on agarose gels, even when high agarose concentrations (e.g., 3%–4%) are employed. This situation is caused primarily by the rapid diffusion of the small molecules in the weak agarose matrix, which leads to strong broadening as the DNAs migrate down the gel. Although polyacrylamide is usually used to resolve small DNAs, performing electrophoresis at high voltage in the absence of EDTA and 0.5 × TB with high agarose concentration can resolve small DNA and RNA effectively (Fig. 7.11). This is important because preparation of agarose gels is faster and easier and agarose gel apparatuses are generally simpler than those used for polyacrylamide gels. In addition, most molecular biology laboratories have agarose gel rigs and power supplies that can achieve high voltages, and therefore experiments can be performed using existing instrumentation.

DNA Size Tracking Dye and Density Agents

Before the nucleic acid sample is loaded onto the gel, tracking dye and a density agent must be added to the sample. The density agents, including Ficoll, sucrose, or glycerol, can increase the density of the sample as compared with the electrophoresis buffer. When the sample solution is dispensed into the wells of the gel below the surface of the buffer, it sinks into the well instead of diffusing in the buffer. The tracking dyes, also known as loading dyes, are used to monitor the progress of the electrophoresis run. The dyes migrate at specific speeds in a given gel concentration and usually run ahead of the smallest fragments of DNA. The dyes are not associated with the sample DNA and thus they do not affect the separation. The movement of the tracking dye can be monitored carefully during electrophoresis, and when the dye approaches the end

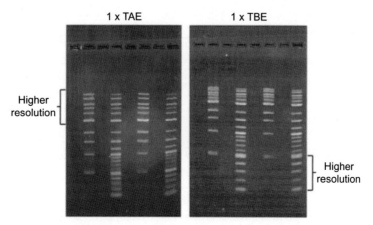

FIG. 7.10 Higher resolution observed in TBE buffer rather than TAE buffer.

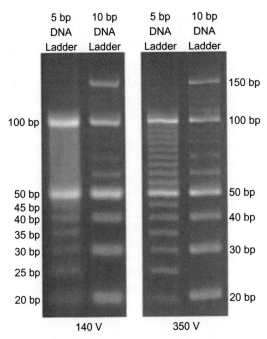

FIG. 7.11 Performing gel electrophoresis at high voltage in the absence of EDTA and 0.5X TB with high agarose concentration can resolve small DNA and RNA effectively than at low voltage.

of the gel, or the desired distance, electrophoresis can be terminated. Both bromophenol blue and xylene cyanol are tracking dyes that are commonly used in many applications for both agarose and polyacrylamide gels. Gel loading dye is typically made at $6\times$ concentration (0.25% bromophenol blue, 0.25% xylene cyanol, 30% glycerol).

Visualization of DNA

The easiest way to see the results of a gel electrophoresis experiment is to stain the gel with a compound that makes the DNA visible. Ethidium bromide (EtBr) is routinely used to stain DNA in agarose and polyacrylamide gels. Ethidium bromide binds to DNA molecules by intercalating between adjacent base pairs. Bands representing the positions of the different size of DNA fragment are clearly visible under UV irradiation after EtBr staining, so long as sufficient DNA is present. Unfortunately, the procedure is very hazardous because ethidium bromide is a powerful mutagen. EtBr staining also has limited sensitivity: if a band contains less than about 10 ng of DNA, it might not be visible after staining (Fig. 7.5). For the safety reason, nonmutagenic dyes that stain DNA green, red, or blue are now used in many laboratories. Most of these dyes can be used either as a poststain after electrophoresis, or alternatively, because they are nonhazardous, they can be included in the buffer solution in which the agarose or polyacrylamide is dissolved when the gel is prepared. Some of these dyes require UV irradiation in order to make the bands visible, but others are visualized by illumination at other wavelengths, for example, under blue light, removing a second hazard, as ultraviolet radiation can cause severe burns. The most sensitive dyes are able to detect bands that contain less than 1 ng DNA.

POLYACRYLAMIDE GELS

Polyacrylamide gels, although more troublesome to cast than agarose gels, are widely used for DNA separations because of their excellent resolving power and high load capacity. Depending on the composition of the gel, very small DNA oligomers or linear double-stranded DNAs ranging up to ~5 kbp in size can be separated. One of the major differences between DNA separations in agarose and polyacrylamide gels is that the mobility observed in polyacrylamide gels depends on DNA sequence, whereas the mobility observed in agarose gels does not. Certain DNA molecules migrate anomalously slowly in polyacrylamide gels, that is, slower than expected from their known molecular mass.

The Chemistry of Polyacrylamide Gel

Polyacrylamide gels are chemically cross-linked gels formed by the reaction of acrylamide with a bifunctional cross-linking agent such as N,N'-methylenebisacrylamide (Bis). The composition of the gel is given by %T, the total (w/v) concentration of acrylamide plus cross-linker, and %C, the (w/w) percentage of cross-linker included in %T. The free radical reaction between acrylamide and Bis is usually carried out using ammonium persulfate as the initiator and N,N,N', N'-tetramethylethylenediamine (TEMED) as the catalyst (Fig. 7.12).

Polyacrylamide gels are not uniform in structure because Bis polymerizes with itself more rapidly than with acrylamide. For this reason, the gels contain highly cross-linked, Bis-rich nodules linked together by more lightly cross-linked, relatively acrylamide-rich fibers. Polyacrylamide gels contain no charged residues, so that electroosmosis is not observed; however, pre-electrophoresis is required to remove polar impurities that can bind to DNA.

The Principle of Polyacrylamide Gel

Polyacrylamide gels, which are chemically cross-linked, are distorted somewhat by the electric field. DNA mobilities observed in polyacrylamide gels are essentially independent of the electric field strength used for electrophoresis. Because of the structural heterogeneity of the polyacrylamide gel matrix, the effective pore size determined by electrophoretic methods depends on the size of the analyte. If proteins are used as the analytes, the apparent pore size corresponds to the pores in the Bis-rich nodules. Larger analytes, such as DNA, do not "see" the small pores in the nodules and instead are retarded by migration through the acrylamide-rich fibers.

Mobility of DNA Molecules in Acrylamide Gel

The mobility of dsDNA molecules is retarded in polyacrylamide gels by a molecular mass-dependent mechanism. The retardation mechanism is most likely the transient interaction of the migrating DNA molecules with the polyacrylamide gel fibers during electrophoresis. This is because DNA molecules recovered from polyacrylamide gels after electrophoresis are complexed with polyacrylamide gel fibers and the DNA can be separated from the gel fibers by diethylaminoethyl (DEAE)-cellulose column chromatography.

Although the polyacrylamide gel matrix is normally uncharged, charged residues can be deliberately added during the polymerization process. For example, copolymerizing acrylamide with acrylic acid creates a negatively charged matrix,

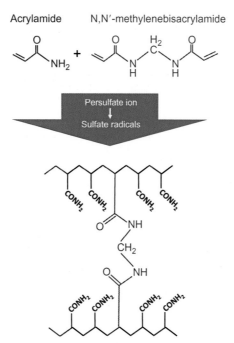

FIG. 7.12 The free radical reaction between acrylamide and bis-acrylamide is usually carried out in the presence of ammonium persulfate and N,N,N', N'-tetramethylethylenediamine (TEMED).

and copolymerizing with allyltriethylammonium bromide (ATAB) creates a positively charged matrix. DNA mobility is significantly reduced in gels containing 0.04%–0.13% acrylic acid (w/v) because of the increased electroosmotic flow in the negatively charged matrix. The mobility is also significantly reduced in gels containing 0.2–0.4 µM ATAB, because the DNA molecules are retarded by interacting with the positively charged gel fibers.

Positively charged polyacrylamide gels can also be made by adding millimolar concentrations of basic Immobilines during polymerization. If the gel contains a concentration gradient of basic Immobilines, large DNA fragments will be arrested in the portion of the gel containing a relatively small concentration of immobilized charges. On the other hand, the smaller fragments will migrate further into the gel and be immobilized there by the higher density of charged residues. Immobiline gradient gels are able to separate ssDNA oligomers with only 1–3 nucleotides difference, as well as 35-base oligomers with different sequences. Such gels may be useful for analyzing single-nucleotide polymorphisms (SNPs) in genomic DNA.

Small single-stranded oligomers containing the same number of nucleotides exhibit highly variable mobilities in native and denaturing polyacrylamide gels. The order of migration of mononucleotides and small homopolymers of the same length varies with oligomer size and depends on the identity of the nucleotide. Because the mobility differences reflect base composition rather than sequence, the individual DNA bases appear to interact differently with the polyacrylamide gel matrix during electrophoresis. Because of this effect, different mobilities are observed for the complementary strands of oligomers containing alternating AT and GC bases, and the mobility observed for DNA fragments in sequencing gels depends on the identity of the terminal nucleotide.

Special Application of Polyacrylamide Gel Electrophoresis: Denaturing Gradient Gel Electrophoresis (DGGE)

The principle of DGGE is that a linearly increasing gradient of denaturants is added to the polyacrylamide gels so that DNA fragments of the same length but with different base-pair sequences can be separated. Separation in DGGE is based on the electrophoretic mobility of a partially melted DNA molecule in polyacrylamide gels, which is decreased compared with that of the completely helical form of the molecule. The melting of fragments proceeds in discrete so-called melting domains, which are stretches of base pairs with an identical melting temperature. Once the melting domain with the lowest melting temperature reaches its melting temperature at a particular position in the DGGE gel, a transition of helical to partially melted molecules occurs, and migration of the molecule will practically halt (Fig. 7.13). Since sequence variation

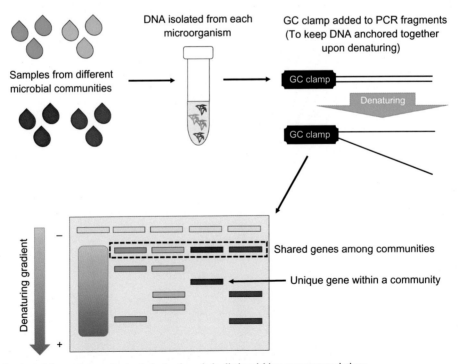

FIG. 7.13 The application of denaturing gradient gel electrophoresis in distinguishing sequence variations.

within the melting domains causes their melting temperatures to differ, sequence variants of particular fragments will therefore stop migrating at different positions in the denaturing gradient. Normally, the denaturing environment is created by a combination of uniform temperature, typically between 50°C and 65°C, and a linear denaturing gradient which consists of 7 M urea and 40% formamide. The denaturing gradient may be formed perpendicular or parallel to the direction of electrophoresis. A perpendicular gradient gel, in which the gradient is perpendicular to the electric field, typically uses a broad denaturing gradient range, such as 0%–100% or 20%–70% (Fig. 7.14). In parallel DGGE, the denaturing gradient is parallel to the electric field, and the range of denaturant is narrowed to allow better separation of fragments.

DGGE has been successfully applied to identifying sequence variations in a number of genes from several different organisms, and it can also be used for direct analysis of genomic DNA from organisms with genomes of millions of base pairs. The DGGE profiling method can also be useful for diagnosing the presence and relative abundance of microorganisms, such as bacteria, yeasts, and fungi, in samples obtained from patients suffering from combined infections. It is therefore expected that this approach will contribute to molecular diagnosis.

PULSED-FIELD GEL ELECTROPHORESIS FOR LARGE DNA MOLECULES

The electrophoretic mobility of DNA molecules larger than $\sim$20 kbp is nearly independent of molecular mass in unidirectional electric fields because of stretching and orientation effects. Pulsed electric fields must be applied to the gel to separate very large DNA molecules. The pulsed fields may be applied intermittently in one direction, in opposite directions, or at obtuse angles. In this way, DNA molecules that are stretched and oriented in the original electric field relax toward their random coil conformations when the field is removed or changed in amplitude and/or direction. The rate at which relaxation occurs depends on DNA size. Consequently, large DNA molecules are trapped in the gel matrix and scarcely move. In the end, separation can occur by molecular mass. The optimal separation window is determined by the ratio of pulse length to DNA reorientation time. However, trapping effects can be observed when the pulse time is close to the reorientation time, causing the DNA molecules to migrate in an order that is not consistent with their molecular mass.

The Basic Principle of Pulsed Field

The mobility observed for DNA molecules in pulse field agarose gels are highly dependent on the electric field applied to the gel. This is because the electric field disrupts the hydrogen bonds in the junction zones constantly, allowing the gel fibers and fiber bundles to orient in the electric field. The oriented gel fibers and fiber bundles are very large, and the gel fibers and fiber bundles orient in the perpendicular direction when the electric field is reversed in polarity. The resulting "flip-flop" orientation and reorientation of agarose fibers and fiber bundles in reversing electric fields provides a mechanism for creating transient large pores in the gel matrix, and this allow very large DNA molecules to migrate through the gel during gel electrophoresis. Whereas standard DNA gel electrophoresis commonly resolves fragments up to $\sim$50 kb in

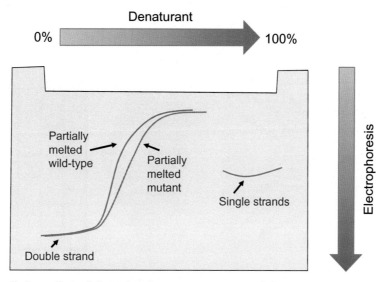

FIG. 7.14 Denaturing perpendicular gradient gel electrophoresis to separate sequence variations.

size, PFGE can separate DNA molecules up to 10 Mb. The mechanism driving these separations demonstrates the fact that very large DNA molecules unravel and penetrate through a gel matrix, and such electrophoretic trajectories are perturbed in a size-dependent manner by carefully oriented electrical pulses. As a result, this produces a DNA fingerprint with a specific pattern. Fig. 7.15 (https://www.cdc.gov/pulsenet/pathogens/pfge.html) is an example of a PFGE to identify unknown

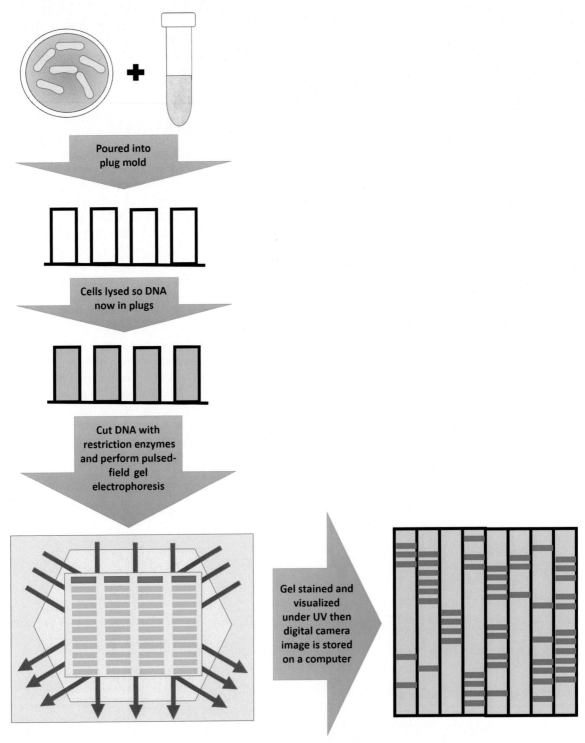

FIG. 7.15 The process of pulsed-filed gel electrophoresis to separate large DNA molecules.

microorganisms' genomic DNA where each lane represents a DNA fingerprint or pattern. PFGE can generate a fingerprint by constantly changing the direction of the electric field. PFGE has enabled the rapid genomic analysis of microbes and mammalian cells, and motivated development of large-insert cloning systems such as bacterial and yeast artificial chromosomes.

Types of PFGE

PFGE is very useful in the field to detect, identify, and classify genome of micro- and macroorganisms accurately and sensitively. Therefore, PFGE is a useful way to discover the causing agents concerning unilateral diseases and tracking the effects of these agents through a single source. The use of electric current with speed and alternate rotations in the magnetic field leads to the development of various types of PFGE. Methods such as FIGE, CHEF, OFAG, RGE, and PACE are all eligible for isolating and typing of the DNA molecule with the large pieces (Fig. 7.16). The selection of these tools is dependent on cost considerations and the purpose of the study.

Contour-clamped homogeneous electric field (CHEF): 24 passive electrodes are arranged hexagonally to create distortion on the edge of the chamber (Fig. 7.16). Since all the electrodes are connected to the power supply through a series of identical ring resistors, they can regulate the voltage on all hexagonally arranged electrodes in a unit electric field. In this system, the size, location, coordination, stability, and continuity of the electric field are precisely controlled. The precisely

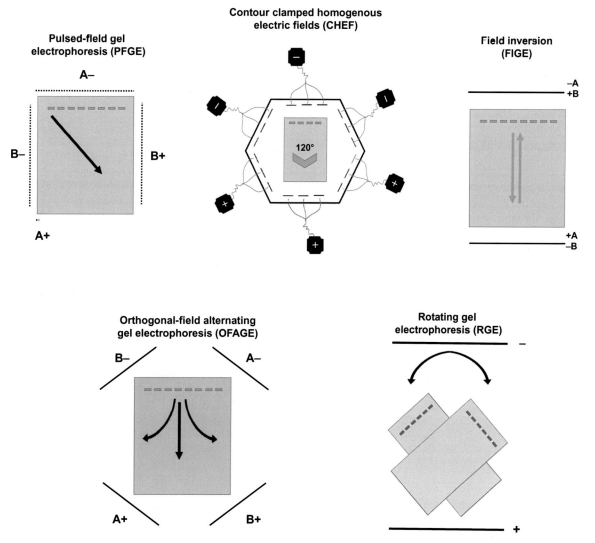

FIG. 7.16   The commonly seen different types of pulsed field gel electrophoresis (PFGE) methods.

controlled electric field can allow the separation of DNA fragments with different sizes. DNA molecules as large as 2 Mb can be well separated with a CHEF system alternating between two orientations 120 degrees apart. DNA smaller than 50 kb can be separated without distortion.

Field-inversion gel electrophoresis (FIGE): Two fields are arranged in separated straight angle (180 degree) so that the direction of the electric field can be periodically reversed (Fig. 7.16). Shorter time pulses make the molecules move forward and longer time pulses make the molecules move reversely. The alternate time pulses create inverse of the electric field periodically. As such, FIGE can overcome the problem caused by the comigration of nucleic acids and protein-detergent complexes. This comigration is common when both nucleic acids and protein-detergent complexes are larger than a threshold size. FIGE provides good resolution, over 800 kb. Asymmetric field-inversion gel electrophoresis (AFIGE) is one of FIGE modifications and this method can be used for the detection of DNA double-strand breaks (DSBs). The rate of DNA breakage can be measured under an inverted and asymmetric electric field, quantitatively.

Orthogonal-field alternation gel electrophoresis (OFAGE): This is a vertical electrophoresis system. In this system, two orthogonal electric fields are alternately supplied to the agarose gel so that nonlinear and dissimilar electric fields are created for DNA molecules migration (Fig. 7.16). DNA molecules between 1000 kb and 2000 kb can be separated by this method.

Rotating gel electrophoresis (RGE): In this method, the agarose gel is rotated between two angles periodically, and the power supply is turned off after switching the angle during electrophoresis (Fig. 7.16). The direct DNA separation is done because only one set of electrodes is used. Because of the monotonous electric field, the intensity of the voltage can be added so that DNA can be isolated in a shorter time. Since it is easy to change the angle of rotation, this method is convenient and suitable for the separation of DNA with 50 kb to 6000 kb.

Programmable autonomously controlled electrodes (PACE): In this system, 24 electrodes are arranged in a closed contour. By independent adjustment of voltage on electrodes, the PACE electrophoresis system controls all the parameters of the electric field. For example, the PACE system can produce an unlimited number of controlled homogeneous electric fields, voltage gradients, and direction and duration of flow. Therefore, the PACE system is flexible, and it is preferable to the other alternating electrophoresis methods (such as FIGE and OFAG). DNA fragments from 100 bp to more than 6 Mb are separable.

Application of PFGE

PFGE is popular in studies for molecular typing and identification of pathogens in the prevalence of certain diseases and are considered the gold standard method to identification of some bacteria. In fact, PFGE methods are able to detect 90% of the bacterial genome and are powerful tools for monitoring genetic changes in bacteria at a global level. For example, any changes in antibiotic-resistant genes in important clinical and hospital infectious bacteria, including *Staphylococcus aureus*, *Pseudomonas aeruginosa*, *Acinetobacter baumannii*, *Mycobacterium tuberculosis*, etc., can be detected at the regional and global level. As such, appropriate solutions can be provided to deal with the phenomenon of antibiotic resistance. Regardless of the high cost of materials and equipment, and the time-consuming nature of this method, PFGE is still presented as a practical and applicable typing method.

In addition to genotyping, PFGE can be used in identification of bacteria isolated from environmental or clinical samples, identification of antibiotic-resistant strains, such as methicillin-resistant *S. aureus* (MRSA) strains, as well as the classification (taxonomy) of bacteria. Furthermore, PFGE can be used in epidemiological studies to identify infectious agents and their origin. The use of this method in the detection of DNA viruses that have large genomes, such as herpes simplex virus (HSV), is also possible. PFGE has often been used for viral DNA fingerprinting of viruses isolated from the environment. PFGE also is a useful tool to determine the relationship among the different strains of a single species. Furthermore, it is an efficient method to estimate the size and chromosomal mechanism, to describe and explain the genome of eukaryotes and prokaryotes, to help with the physical mapping of genes, to allow analysis of large DNA fragments, and to detect primary origin of genes in fungi, protozoa, bacteria, and even mammals.

CAPILLARY ELECTROPHORESIS

Capillary electrophoresis (CE) of DNA is usually carried out in buffers containing entangled hydrophilic polymers that act as the sieving medium. Capillary electrophoresis offers significant advantages over slab gel DNA separations, because of higher resolution, greater speed, online detection, and the minimal use of samples and buffers. However, capillary electrophoresis also has several disadvantages, such as the fragility of the capillaries and capillary coatings and the necessity of running one sample at a time in each capillary to obtain accurate mobility. Multicapillary systems can be used to increase

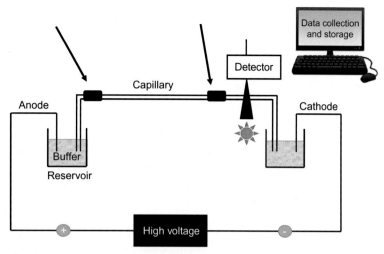

FIG. 7.17 Capillary zone electrophoresis: samples are separated into analytes that flow in directions based on their inherent electrophoretic mobility and are then analyzed when they travel past a detector.

throughput, but they are relatively expensive to purchase and operate. The application of CE in proteomic analysis is discussed in Chapter 8.

Setup of the Capillary Electrophoresis

In general, capillary electrophoresis (CE) experiments are carried out in fused-silica capillaries with a high surface area to volume ratio. An external polymeric coating is usually added to the capillary to produce a surprisingly flexible narrow-bore capillary that otherwise would be extremely fragile. A generalized configuration for CE instrumentation is shown in Fig. 7.17. With capillary zone electrophoresis (CZE), sample is introduced into a buffer-filled capillary either electrokinetically (with low voltage) or hydrodynamically (with pressure or suction). Both ends of the capillary and electrodes are then placed into a buffer solution that also contains the electrodes, and a high voltage is applied to the system. The applied voltage causes the analytes to migrate through the capillary and pass a detector window, where information is collected and stored by an appropriate data acquisition system.

Mobility of the Analytes

Two electrically driven phenomena contribute to the mobility of the analyte: electrophoretic mobility of the analyte itself and electroosmotic flow of the bulk solution. Therefore, when a voltage is applied to the system, individual analytes migrate with an electrophoretic direction and velocity that is generally determined by their charge and mass. With certain analytes under specific conditions, other variables such as shape and hydrophobicity also have an effect.

The net movement of buffer toward the cathode is concurrent with analyte mobility, which is due to electrophoresis. The net movement is due to the electroosmotic force, which is the result of the negative charge imparted to the inner surface of the capillary by the bare silica. Cations along the capillary wall form an organized double layer that, under an applied field, is propelled along the wall in the direction of the cathode. The electroosmotic force is usually adequate to force the movement of all molecules toward the cathode and pass the detector. This includes those with a net negative charge, which are electrophoretically drawn toward the anode when a polarity is going from (anode) through (detector) to (cathode). Subsequently, the analyte order in any given electropherogram has the fastest migrating species passing the detector window first and the slowest passing last and peaks from left to right represent the fastest to slowest migration. The electroosmotic force can be altered through the addition of buffer additives, the use of a coated capillary, or low-pH buffers, Buffer concentration can significantly affect optimum resolution and reproducibility of the results. The increase of buffer concentration can increase the buffering capacity of the system. Therefore, it is less likely to change pH when extra acid or base is added to the system. Proper control of sample buffer concentration and running buffer concentration can result in stacking which can produce sharper peaks and greater sensitivity. For example, reducing the concentration of the sample buffer relative to the run buffer can reach this goal. Fig. 7.18 illustrate the process that analytes reach the stacking through

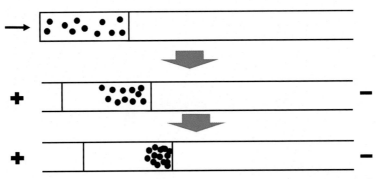

FIG. 7.18 Sample stacking where analyte molecules are injected into a plug and move rapidly under electroosmotic flow causing them to concentrate before migrating out of the plug.

the difference of the buffer concentration. When the buffer concentration is uniform from one end of the capillary to the other end, the voltage drop is linear. However, inserting a plug of dilute sample is like inserting a resistor into the wire. Under this condition, the voltage gradient drop now becomes steeper in the region of the plug than it is elsewhere. Because velocity is proportional to the voltage gradient, the velocity of the analytes will be faster in this region. Analytes will move rapidly until they reach the boundary between the dilute sample plug and the more concentrated running buffer. Therefore, analytes stack in the boundary area (Fig. 7.18).

Although stacking is caused by a difference in conductivity between the running buffer and the sample, one might assume that if the sample is dissolved in pure water and this would create the highest peak efficiencies. However, this is not usually the case. This is because the resistance across a plug of nearly pure water can be very high and generate heat. As a result, it causes de-stacking of the analytes. In fact, the best stacking result is keeping samples at about 1/10 the concentration of the running buffer. Another way to maximize stacking is to increase the concentration of the buffer. This works well for samples with high salt concentration. However, it is important to have an adequate cooling so that the heat generated inside the capillary can be eliminated.

Separation of DNA Molecules

In CE, DNA molecules separations are usually carried out at the neutral pH. Under this system, the capillary walls are negatively charged which can attract positively charged counterions from the buffer, and this attraction creates an electric double layer next to the wall. When an electric field is applied, the cations in the double layer migrate toward the cathode. Therefore, the applied voltage causes the samples to migrate through the capillary and pass a detector window. Meanwhile, information is collected and stored by an appropriate data acquisition system. The resulting electroosmotic flow (EOF) affects the observed mobility, and it can transport the sieving matrix out of the capillary, and, therefore, degrade the separation.

Several different modes of CE have been developed and they are listed in Table 7.2. This classification is based on the basis of separation and the type of analyte for which they are useful. The ability of CE to provide high-efficiency separation by several of these modes offers an attractive alternative for the analysis of low-molecular weight analytes. For example, one of the primary difficulties with the analysis of certain analytes by methods such as high-performance liquid chromatography (HPLC) is the inconveniently long retention times and this may lead to difficulties in quantification because of peak broadening. Among these different modes, capillary zone electrophoresis (CZE) is the most universal of the techniques and it overcomes the aforementioned problem caused by HPLC analysis. As such, it is very useful for the separation of both low-molecular and macromolecular analytes. CZE separation occurs in low-ionic-strength buffer and is primarily based on the differences in analyte mass-to-charge. In CZE, the interaction of analytes with a resin is avoided, and this allows for rapid analysis times, and thus CZE could be called "antichromatography." In addition to its rapidity, CZE can resolve compounds that vary only slightly in structure.

The analysis of dsDNA fragments, derived from PCR products and DNA restriction digests, can be used to track inheritance patterns in a family, diagnose numerous genetic diseases, map genes to specific chromosomes, and clarify forensic applications. CE can provide faster and more efficient separation of dsDNA with easy quantitation. However, dsDNA fragments of various sizes (e.g., 5 and 50 bp) all have the same charge-to-mass ratio. Obviously, separation of dsDNA based on

TABLE 7.2 Mode-Dependent Capillary Electrophoresis Separation Techniques

Mode	Separation Principle	Analytes
Capillary zone electrophoresis	Charge-to-mass ratio	Small ions Small molecules Peptides & proteins Limited DNA
Capillary isoelectric focusing	Isoelectric point	Peptides & proteins
Capillary isotachophoresis	Mobility with buffer	Small molecules Peptides & proteins
Micellar electrokinetic capillary chromatography	Charged species: charge-to-mass ratio and micelles according to hydrophobicity	Small molecules Peptides DNA
	Neutral species: detergent micelle organization based on hydrophobicity	
Capillary gel electrophoresis: nondenaturing	Charge-to-mass ratio sieving	Peptides & proteins DNA
Capillary gel electrophoresis: denaturing (SDS, urea)	Mass sieving	Peptides & proteins DNA

SDS, sodium dodecyl sulfate.

mass-to-charge ratio is not effective, and separation needs to be based on size, which requires a high-resolution separation sieving matrix including low or zero cross-linked polyacrylamide, polyethylene glycol, and methylcellulose derivatives. These sieving matrix can be used in CE so that it can perform high-resolution separations within a narrow DNA size range. Therefore, CE is an effective tool for genomic analysis, carrier detection, prenatal diagnosis of X-linked recessive disorders, forensic applications, post-PCR DNA analysis, analysis of larger DNA fragments, and chromosomal analysis.

Chapter 8

Quantification and Analysis of Proteins

Chapter Outline

PROTEIN QUANTITATION

Various platforms and methods are available to quantitate proteins; however, spectrophotometric assays of protein in solution do not require either enzymatic/chemical digestion or separation of the mixture prior to analysis. There are three major types of spectrophotometric assays: UV absorbance methods, dye-binding assays using colorimetric and fluorescent-based detection. Although these spectrophotometric assays can be run at high throughput, they require an appropriate protein standard or constituent amino acid sequence information to make an estimate of concentration. The choice of method used to determine the concentration of a protein or peptide in solution depends on many factors. For practical procedures, the flowchart in Fig. 8.1 describes the process of selecting the most appropriate assay.

Diagnostic Molecular Biology. https://doi.org/10.1016/B978-0-12-802823-0.00008-0

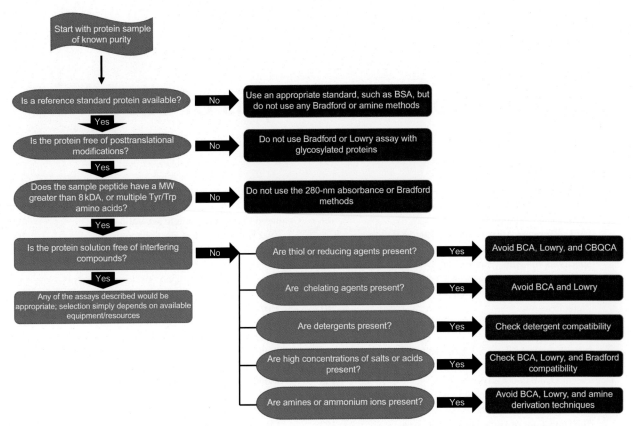

FIG. 8.1 Flow chart for selecting proper techniques for protein quantification upon purification. The chart assumes that the sample for analysis is relatively pure, that is the analyte for quantitation is the major component, for example fractions from affinity chromatography, or extraction from inclusion bodies. The "reference standard protein" refers to a standard that is the same protein that is being quantitated in the same, or similar matrix that is "matched."

Some criteria that need to be considered when selecting an assay include:

- Sample volume: The amount of material available to analyze. Typically, fluorescence-based assays provide the best sensitivity and dynamic range (Fig. 8.2). Microplate assays use lower assay volumes but show improved sensitivity, typically up to 10-fold when compared with cuvette-based assays.
- Sample recovery: If the sample is limited, a nondestructive method such as UV spectroscopy may be more appropriate.
- Throughput: If multiple samples are to be analyzed, a microplate-compatible rapid one-step assay should be considered.
- Robustness: The absorbance-based dye-binding assays appear to display enhanced repeatability and robustness when compared to fluorescent assays.
- Chemical modification: Covalent modification, for example glycosylation or PEGylation, can interfere with specific assays.
- Protein aggregation: The solubility of a protein in solution, often a problem for membrane proteins, or proteins prone to aggregation can alter the expected response for many assays.

UV Spectroscopy Method: Detecting Proteins With Absorbance at 280 nm

At 280 nm, amino acid residues with aromatic rings in a protein absorb light. The amino acids tryptophan (λ_{max} 279.8 nm) and tyrosine (λ_{max} 274.6 nm) have extinction coefficients, ε, of 5.6 and 1.42 $M^{-1}cm^{-1}$, respectively, and thus contribute most to the absorbance at this wavelength, while phenylalanine (λ_{max} 279 nm, ε 0.197 $M^{-1}cm^{-1}$) makes a minor contribution (Table 8.1). The method follows the Beer-Lambert law in which absorbance is proportional to concentration and path length. It should be noted that the method depends on the relative number of aromatic amino acids, primarily tryptophan and tyrosine residues, present in a protein. The method can be used to detect protein in the 20–3000 μg range.

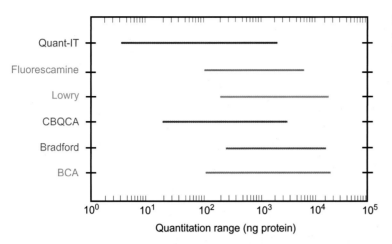

TABLE 8.1 Wavelength Maxima and Molar Absorptivity (ε) of Select Amino Acids

Amino Acid	Wavelength Maxima (nm)	$\varepsilon \times 10^{-3}$ ($M^{-1}cm^{-1}$)
Cysteine	250	0.3
Histidine	211	5.9
Phenylalanine	188	60.0
	206	9.3
	257	0.2
Tryptophan	219	47.0
	279	5.6
Tyrosine	193	48.0
	222	8.0
	275	1.4

The method is particularly useful for the detection of proteins eluting from chromatography columns, as there is no loss of protein. The possibility of nucleic acid contamination should be considered, and compensation can be made.

Advantages of using absorbance at 280 nm are that this method is fast, easily automated, and reasonably sensitive. Furthermore, this method does not destroy protein. However, many buffers and other reagents can interfere with A_{280} spectrophotometric measurements. The concentration limits for those reagents are listed in Table 8.2. The A_{280} measurement is also interfered with by the absorbance from DNA and RNA. DNA and RNA have an absorbance maximum at 260 nm but still absorb at 280 nm and have tenfold higher absorbance values at 280 nm compared to the equivalent concentration of protein. Therefore, if the sample is contaminated with nucleic acids, the absorbance at 280 nm might not adequately reflect the real protein concentration. To resolve the protein concentration in such samples, measure the absorbance at 260 nm and 280 nm and calculate the protein concentration as follows:

$$\text{Protein concentration (mg/mL)} = 1.55 \times A_{280} - 0.76 \times A_{260}.$$

This estimation of protein concentration is valid up to 20% (w/v) nucleic acid or an A_{280}/A_{260} ratio less than 0.6. The UV absorbance at 280 nm is also only usable for concentration determination if the sequence of the protein of interest contains a known amount of tryptophans and tyrosines. If this is not the case, an alternative is to use Fourier transform infrared

TABLE 8.2 Protein Quantification Limits of Interfering Reagents

Reagent	A_{205}	A_{280}
Ammonium sulfate	9% (w/v)	>50% (w/v)
Brij 35	1% (v/v)	1% (v/v)
DTT (dithiothreitol)	0.1 mM	3 mM
EDTA (ethylenediaminetetraacetic acid)	0.2 mM	30 mM
Glycerol	5% (v/v)	40% (v/v)
KCl	50 mM	100 mM
2-ME (2-mercaptoethanol)	<10 mM	10 mM
NaCl	0.6 M	>1 M
NaOH	25 mM	>1 M
Phosphate buffer	50 mM	1 M
SDS	0.10% (w/v)	0.10% (w/v)
Sucrose	0.5 M	2 M
Tris buffer	40 mM	0.5 M
Triton X-100	<0.01% (v/v)	0.02% (v/v)
TCA (trichloroacetic acid)	<1% (w/v)	10% (w/v)
Urea	<0.1 M	>1 M

spectroscopy (FTIR). After subtracting the contribution of water between 1700 nm and 2300 nm, the analysis of the amide band I and II of the IR absorbance spectrum can be used to calculate protein concentration by determining the concentration of amine bonds. The only limitations of FTIR are the minimal and maximal concentrations that can be used (0.2–5 mg/mL) and the incompatibility of several amine-containing buffers (HEPES ≥ 25 mM, Tris ≥ 50 mM) or additives (EDTA ≥ 10 mM).

Another alternative is amino acid analysis (AAA), which is a valuable technique for both identification and quantification of protein. AAA involves hydrolyzing the peptide bonds to free individual amino acids, which are then separated, detected, and quantified, using purified amino acids as standards. Nonetheless, UV-visible spectroscopy remains the most widely spread, cost- and time-efficient technique for total protein concentration determination.

To take full advantage of AAA measurement even in the absence of tyrosine and tryptophan residues, the test solution can be to use FTIR-based protein quantification and AAA measurements at first, and then generate concentration calibration curves for the protein of interest in correlation with UV absorbance (at 280 nm or another wavelength). These calibration curves can then be used to determine the concentration of subsequent samples directly by UV absorbance spectroscopy.

UV Spectroscopy Measurement: Detecting Proteins With Absorbance at 205 nm

Determination of protein concentration by measurement of absorbance at 205 nm (A_{205}) is based on absorbance by the peptide bond. This assay can be used to quantitate protein solutions with concentrations of 1–100 µg/mL protein.

The quantitation of proteins by peptide bond absorption at 205 nm is more universally applicable than A_{280}. Furthermore, the absorptivity for a given protein at 205 nm is several-fold greater than that at 280 nm. Thus, lower concentrations of protein can be quantitated with the A_{205} method. Most proteins have extinction coefficients at A_{205} for a 1 mg/mL solution of between 30 and 35. However, an improved estimate can be obtained by taking into account variations in tryptophan and tyrosine content of the protein to be quantitated ($\varepsilon^{1mg/mL} = 27.0 + 120 \times (A_{280}/A_{205})$). Absorbance at 205 nm can be used to quantitate dilute solutions, or for short path length applications. For example, continuous measurement in column chromatography, or for analysis of peptides where there are few, if any, aromatic amino acids. The disadvantage of this method is that some buffers and other components also absorb at 205 nm (Table 8.2).

Spectrofluorometric Measurement: Detecting Proteins by Intrinsic Fluorescence Emission

Protein concentration can also be determined by measuring the intrinsic fluorescence based on fluorescence emission by the aromatic amino acids tryptophan, tyrosine, and phenylalanine. Usually tryptophan fluorescence is measured as the indication of the fluorescence intensity of the protein sample solution. The concentration of the protein sample solution can be calculated from a calibration curve based on the fluorescence emission of standard solutions prepared from the purified protein. This assay can be used to quantitate protein solutions with concentrations of 5–50 μg/mL. Normally, the excitation wavelength is set to 280 nm and the emission wavelength to between 320 and 350 nm. If the exact emission wavelength is not known, determine it empirically by scanning the standard solution with the excitation wavelength set to 280 nm.

COLORIMETRIC PROTEIN ASSAY TECHNIQUES

All colorimetric protein assays require protein standard to estimate the concentration of a sample. The ideal protein standard to use in a quantitative assay is the exact same protein in a matched matrix/solution that has been assigned using a higher order method, for example AAA or gravimetric analysis. Gravimetric analysis is prone to errors due to the extensive dialysis and drying to remove water and salts from commercial preparations. In practice, there is not always a matched protein standard available. On the other hand, some commercially available standards may be suitable for use. The most commonly used standards are bovine serum albumin (BSA), bovine gamma globulins, and immunoglobulins (used for antibody quantitation). The use of a BSA standard can give misleading results in many assays, especially those methods that are sensitive to the protein sequence, that is, where the signal is generated by specific amino acids. Assays with a low protein sequence dependence will give better estimates when BSA calibration is compared to AAA assignment. This is because AAA assignment quantitates the amount of specific amino acids present following protein hydrolysis and separation; using peptide sequence information, the amount of target protein can then be calculated.

Biuret Reaction

The biuret reaction is based on the complex formation of cupric ions with proteins. In this reaction, copper sulfate is added to a protein solution in strong alkaline solution. A purplish-violet color is produced, resulting from complex formation between the cupric ions and the peptide bond. The biuret reaction with proteins is independent of the composition of the protein; therefore, protein composition is not a factor. However, protein purity and association state could influence the results obtained with the biuret reagent.

The biuret reaction is somewhat insensitive compared with the other methods of colorimetric protein determination. A "reverse biuret method," which is a modification of the biuret reaction, significantly increases sensitivity. In the "classic" biuret reaction, color is produced by the formation of a protein-copper-tartrate complex; in the "reverse" biuret reaction, color is generated by the reduction of excess cupric ions, not bound in the biuret complex, by ascorbic acid to cuprous ions, which are subsequently measured as a complex with bathocuproine. The amount of Cu^+-bathocuproine complex is inversely proportional to the protein concentration. The sensitivity of this reaction is greater than either the Lowry assay or the Bradford method. As with the direct biuret reaction, there is no dependence on protein composition. Solution constituents such as Tris buffer, ammonium ions, sucrose, primary amines, glycerol, and dextran can interfere with the biuret reaction.

Lowry Method

The Lowry protein assay is based on the biuret reaction with additional steps and reagents to increase the sensitivity of detection. In the biuret reaction, copper interacts with four nitrogen atoms of peptides to form a cuprous complex. Lowry adds phosphomolybdic/phosphotungstic acid also known as Folin-Ciocalteu reagent. This reagent interacts with the cuprous ions and the side chains of tyrosine, tryptophan, and cysteine to produce a blue-green color that can be detected between 650 nm and 750 nm. The protein detection range is 5–100 μg.

Although the Lowry method uses standards for calibration, which can be a source of error as the composition of the protein of interest may not necessarily match that of the protein standards, it is almost 100-fold more sensitive than determining absorbance at 280 nm. One disadvantage of the Lowry method is that many common substances, such as K^+, Mg^{2+}, NH_4^+, EDTA, Tris-HCl, carbohydrates, and reducing agents, interfere with the method. Furthermore, the Folin reagent is reactive for only a short period of time after addition. The method is complicated and requires more steps and reagents than the BCA or Bradford assays, and this method is destructive to proteins: once the protein sample has reacted with the dye, the protein cannot be used for other assays.

Bradford Method

The Bradford method uses the binding of Coomassie brilliant blue G-250 dye to proteins, which results in a dye-protein complex with increased molar absorbance for the determination of protein concentration. The Coomassie brilliant blue G-250 dye is protonated and is reddish/brown with an absorbance maximum of 465 nm at acidic pH. Under acidic conditions, the dye reacts primarily with arginine and to a lesser extent with lysine, histidine, tyrosine, tryptophan, and phenylalanine residues in proteins, producing a blue color with an absorbance maximum at 595 nm (the absorption range is between 575 nm and 615 nm), and 0.2–20 µg of protein can be detected. The method is the easiest and fastest of the protein determination methods.

The Bradford Coomassie Blue G-250 assay is fast, simple and sensitive. The Coomassie brilliant blue G-250 is stable for long periods of time. Furthermore, the volume of reagents can be reduced and the assay can be performed in a 96-well plate. However, the Bradford Coomassie brilliant blue G-250 dye stains cuvettes, although cuvettes can be cleaned by washing with a dilute SDS solution. Dye binding depends on the basic amino acid content that can vary between proteins. It is, however, useful as a general, sensitive, semiquantitative assay for proteins. With the selection of an appropriate standard, the assay can be both accurate and sensitive. It has been demonstrated that decreasing the acidity in the reaction by the addition of NaOH, the inclusion of Triton X-100, or the addition of SDS can decrease protein-to-protein variability.

Another issue for this method is that the concentrated protein solutions can form a precipitate upon contact with the dye reagent. If this is observed, the protein solution should be diluted to determine protein concentration. Furthermore, this method is also destructive to proteins, meaning that once the protein sample has reacted with the dye, the protein cannot be used for other assays.

Bicinchoninic Acid Assay (BCA) Method

The BCA assay method is based on the fact that the sodium salt of bicinchoninic acid reacts with the cuprous ion generated by the biuret reaction under alkaline conditions. The bicinchoninic acid cuprous complex forms a deep blue color that is read at 562 nm, and the detection range is 0.2–50 µg. The BCA reagent is stable under alkaline conditions, so it can be included in the biuret alkaline copper solution. The development of color in the BCA assay depends on time, temperature, and pH. The assay can be performed at room (ambient) temperature, but increases the reaction temperature can significantly reduces the time required for maximal color development and increases sensitivity. The assay is compatible with detergents, and therefore is better than both the Lowry and Coomassie dye assays.

BCA method has less protein/protein variability than the Bradford assay. The variation as a function of protein composition could be decreased by reaction at 60°C. Furthermore, the volume of reagents can be reduced, and it can be performed in 96-well plates. However, reducing reagents that reduce the cupric ions to cuprous ions interfere with the assay, and the chelating agents (e.g., EDTA) that chelate copper interfere as well. H_2O_2 can also interfere with the BCA assay. Phospholipid interferes with the BCA protein assay, resulting in artificially high values that reflect the interaction of phospholipid with the BCA reagent to yield a chromophore absorbing close to 562 nm. Like the Bradford method, this method is destructive to proteins. Therefore, once the protein sample has reacted with the copper, the protein cannot be used for other assays.

Standards and Assay Validation

Any assay for protein concentration must be validated to ensure an accurate value for the sample of interest. The selection of an appropriate standard is of critical importance because, with the possible exception of the biuret reaction, the protein assays described earlier all depend on the quality of the protein for the response. The selection of an appropriate standard together with a rigorous validation of the analytical process can solve the problems presented by the protein composition. The validation methodology must include measurements to ensure that the sample protein concentration is within the dynamic range of the assay. Furthermore, all sample types must be independently validated to assess the effect of protein composition as well as interfering or enhancing substances that may be contained in the sample.

FLUORESCENT DYE-BASED ASSAYS

Microplate Detection Method

Amine-labeling "derivatization" using various fluorescent probes is a common technique to quantitate amino acid mixtures in amino acid analysis. This technique can be used to quantitate proteins and peptides containing either lysine or a free N terminus, both of which need to be accessible to the dye. Upon reaction with amines, the dyes display a large increase in fluorescence that, for part of the dynamic range, will generate a linear response with increasing protein concentration. The range of this method is 0.05–25 μg. Three dyes used to quantitate proteins, or amino acids in a microplate format include o-phthalaldehyde (OPA), fluorescamine, and 3-(4 carboxybenzoyl)quinoline-2-carboxyaldehyde (CBQCA). Fluorescamine reacts directly with the amine functional group, whereas OPA and CBQCA require the addition of a thiol (2-mercaptoethanol) or cyanide.

Cuvette Detection Method

In a cuvette-based format, OPA reacts with primary amino acid (except cysteine) in the presence of 2-mercaptoethanol or 3-mercaptopropionic acid to form a highly fluorescent adduct. OPA also reacts with 4-amino-1-butanol and 4-aminobutane-1,3-diol produced from oxidation of proline and 4-hydroxyproline, respectively, in the presence of chloramine-T plus sodium borohydride at 60°C, or with S-carboxymethyl-cysteine formed from cysteine and iodoacetic acid at 25°C. Fluorescence of OPA derivatives is monitored at excitation and emission wavelengths of 340 nm and 455 nm, respectively. Detection limits are 50 fmol for amino acid.

The Limitations of Using Fluorescent Dyes

All three dyes offer improved sensitivity and dynamic range when compared with absorbance-based protein quantitation assays. OPA is generally preferred over fluorescamine because of its enhanced solubility and stability in aqueous buffers. The use of amine-derivatization agents for protein quantitation is limited, as the assay displays a large protein-to-protein variability due to variation in the number of lysine residues in proteins, requiring the need for a "matched" standard. Assay interference from glycine and amine containing buffers, ammonium ions, and thiols common in many biological-buffering systems limits the application of such assays. The reproducibility of the assay is dependent on the pH of the reaction. For example, protein samples that contain residual acids could reduce the rate of amine derivatization.

THE CHOICE OF MEASUREMENT FORMAT

Cuvette Measurement Method

The format used in the measurement depends on the throughput, sensitivity, and precision required of the assay. Traditionally, cuvettes have been used for the majority of spectrophotometric protein assays. Quartz cuvettes can be costly, therefore, glass cuvettes are preferred. However, both of these may have to be washed between measurements to remove dye and adsorbed protein. Disposable plastic cuvettes are available and can be used to increase the throughput where many samples have to be measured, or where the reagent is prone to sticking to the cuvette surface like in Bradford reagent. Staggering of sample analysis is especially important if the signal is not stable or does not run to completion within the time frame of the assay like in BCA or Lowry assays. Replacing the cuvette in the holder between each measurement because of cleaning, or the use of disposable cuvettes, can result in changes in alignment, resulting in significant changes in the amount of light reaching the detector. This is especially important if low-volume cuvettes are being used, where the transmission window is reduced in size.

Cuvette Cleaning Method

Care should be taken when handling and cleaning cuvettes. It is also important to prevent fingerprints from contaminating the transmitting surfaces. Cuvettes should be washed with either water or an appropriate solvent between runs and dried using a stream of nitrogen gas. If smearing of the transmitting surface is observed, the cuvette can be rewashed in water, ethanol, and finally acetone, or smears can be removed using ethanol and lint-less lens tissue. If protein deposition is a recurring problem, cuvettes can be soaked overnight in nitric acid and thoroughly washed before use.

Microplate Method

The use of microplate for protein assay can enhance speed and throughput and decrease sample and reagent usage. Many of the commercial fluorescent assays are specifically designed for plate formats. The plate reader format also offers the advantage of being able to read multiple samples within a short period (typically 25 s), reducing potential timing differences in reactions that do not go to completion or are unstable. Quartz 96-well plates tend to be expensive, difficult to clean, and prone to scratches that can affect light transmission. Since the cost decreases, disposable plastic plate can be the alternative.

Preparation of Samples for Measurement

Care should be taken in the preparation of protein assays in plate formats. The use of lower volume samples (down to 5 μL for some assays) can increase the relative pipetting errors of high-viscosity solutions. Well-to-well contamination should be avoided by using fresh pipette tips for each sample and reagent. Regular calibration of the instrument should be performed using either optical standards or solid-phase fluorescent standard plates to ensure equal transmission/light detection from all wells.

MOLECULAR SEPARATION OF PROTEIN MOLECULES THROUGH COLUMN CHROMATOGRAPHY

After the cells are broken and the cell extracts are released, proteins can then be further purified through various biochemical methods. The most common method for purifying proteins from other protein molecules within a given sample is column chromatography. Basically, this method uses a glass or plastic tube filled with resin that can separate proteins based on their physical properties as they flow through the column. At first, the protein sample is applied to the top of the column. As a buffer solution is flowed continuously through the column, proteins in the sample migrate through the column at different rate, depending on the nature of the matrix and the physical and the chemical properties of the proteins. Therefore, the targeted protein can be separated from others.

Size Exclusion Chromatography

This chromatography can separate proteins based on the size and shape on a gel filtration column (Fig. 8.3). The column matrix also known as resin consists of microscopic beads of inert material. The resin bead has many tiny pores like a whiffle ball (Fig. 8.4). In this type of column, small molecules, which is smaller than the cut-off threshold size, are more likely to go through the pore of the matrix, and thus are trapped in the resin and travel through the column more slowly. Larger molecules, on the other hand, move through the column more quickly because they cannot fit into the beads and are excluded from entering the pores in the beads. These larger molecules can only pass through the spaces between resin beads, so they travel a shorter distance overall. Thus, large molecules will emerge first from the column, and small molecules will emerge last.

Ion-Exchange Chromatography

Ion-exchange chromatography uses a resin to separate proteins according to their surface charges. This type of column contains a resin bearing either positively or negatively charged chemical groups. Resins containing positively charged groups attract negatively charged solutes and are referred to as anion-exchange resins (Table 8.3). Resins with negatively charged groups are cation exchangers. In low-salt solutions, proteins with a negative surface charge will bind more strongly

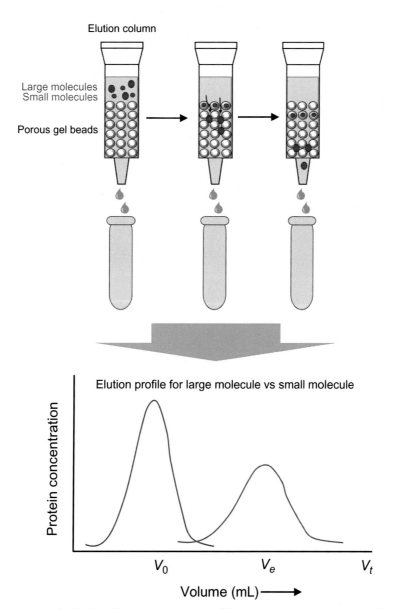

FIG. 8.3 Size exclusion chromatography in which small proteins (under the cutoff limit) enter pores of beads and take a longer route to exit the column than large proteins, which are excluded from porous beads and quickly exit column first.

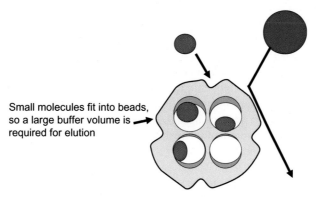

FIG. 8.4 Size exclusion chromatography column. Larger molecules are excluded from gel beads and emerge from the column sooner while smaller molecules must navigate the intricate network of pores and exclude later.

TABLE 8.3 Cation- and Anion-Exchange Resins Commonly Used for Biochemical Separations

Cation-Exchange Media	Structure
Strongly acidic, polystyrene resin	
Weakly acidic, carboxymethyl (CM) cellulose	
Weakly acidic, chelating, polystyrene resin (Chelex-100)	

Anion-Exchange Media	Structure
Strongly basic, polystyrene resin (Dowex-1)	
Weakly basic, diethylaminoethyl (DEAE) cellulose	

to positively charged anion-exchange columns (Fig. 8.5). Likewise, proteins with a positive surface charge will bind to negatively charged cation-exchange columns. Later, higher salt concentration buffer is applied to the column so it can compete with the resin for the bound proteins, and the bound proteins can then be eluted through the high-salt buffer.

Affinity Chromatography

Affinity chromatography uses the principle that the protein binds to a molecule for which it has specific affinity. This is because in most instances proteins carry out their biological activity through binding or complex formation with specific small molecules, or ligands. This small molecule can be immobilized through covalent attachment to the resin in a column (Fig. 8.6). When the sample goes through the column, the protein of interest, in displaying affinity for its ligand, become bound and immobilized itself. The protein of interest is thus removed from the mixture of sample. Finally, the protein is dissociated or eluted from the resin by the addition of high concentrations of free ligand in solution. Because this method relies on the biological specificity of the protein of interest, it is a very efficient procedure. A typical example is the resins coupled to an antibody that recognizes a specific protein, or it may contain an unreactive analogue of an enzyme's substrate. The power of affinity chromatography lies in the specificity of binding between the affinity reagent on the resin and the molecule to be purified. As such, it is possible to design an affinity chromatography procedure to purify a protein in a single step.

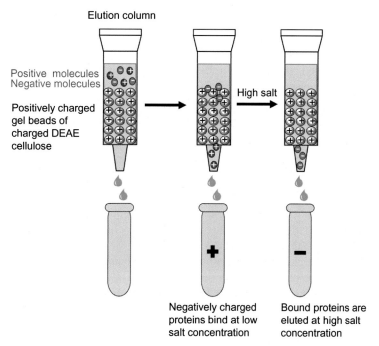

Elution column

Positive molecules
Negative molecules

Positively charged
gel beads of
charged DEAE
cellulose

High salt

Negatively charged
proteins bind at low
salt concentration

Bound proteins are
eluted at high salt
concentration

FIG. 8.5 Ion exchange chromatography separates molecules based on positive and negative charges. If the beads in column are positively charged, then negatively charged proteins will bind to column matrix at low salt as a result of ionic interactions. Proteins can then be induced to dissociate with high salt.

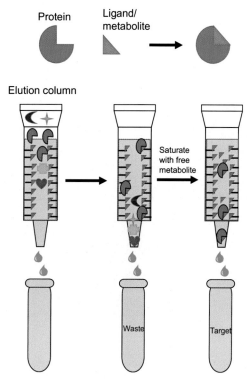

Protein

Ligand/
metabolite

Elution column

Saturate
with free
metabolite

Waste

Target

FIG. 8.6 Affinity chromatography to isolate molecules based upon ligand binding abilities.

Hydrophobic Interaction Chromatography

In this purification technique, proteins passed through a chromatographic column packed with a support resin to which the hydrophobic groups are covalently linked. Phenyl Sepharose, which contains a phenyl group, is commonly used in the column. Usually, the protein sample is prepared in high-salt buffer. When the protein sample goes through the column, proteins bind to the phenyl group by virtue of hydrophobic interactions. Proteins in a mixture can be differentially eluted from the phenyl group by using low salt concentration buffer or by adding solvents such as polyethylene glycol to the elution fluid.

MOLECULAR SEPARATION THROUGH ELECTROPHORESIS

Sodium Dodecyl Sulfate-Polyacrylamide Gel Electrophoresis (SDS-PAGE)

Prior to any downstream analysis, purity and integrity are the very first qualities that need to be assessed for any protein sample. This is routinely achieved by SDS-PAGE. This technique, associated with staining method, can detect bands of protein in a simple and relatively rapid manner (just a few hours).

The hydrophobic tail of SDS interacts strongly with polypeptide chains. The number of SDS molecules bound by a polypeptide is proportional to the length (the number of amino acid residues) of the polypeptide. Each sodium dodecyl sulfate contributes two negative charges (Fig. 8.7). Collectively, these charges overwhelm any intrinsic charges that the protein might have. Therefore, SDS has two advantages. (1) It coats all the polypeptides with negative charges, so they all electrophorese toward the anode. (2) It masks the natural charges of the subunits, so they electrophorese according to their molecular masses and not by their native charges. Small polypeptides fit easily through the pores in the gel, so they migrate rapidly. Large polypeptides migrate more slowly. SDS is also a detergent that disrupts protein tertiary structure. Both heat and reducing agent such as DTT and 2-mercaptoethanol can also denature the protein and break the covalent bonds between subunits.

After reduction and denaturation by SDS, proteins migrate in the gel according to their molecular mass, allowing to detect potential contaminants, proteolysis events, and so forth (Fig. 8.8A). The electrophoretic mobility of proteins upon SDS-PAGE is inversely proportional to the logarithm of the protein's molecular weight (Fig. 8.8B). SDS-PAGE is often used to determine the molecular weight of proteins.

FIG. 8.7 Chemical structure of sodium dodecylsulfate (SDS).

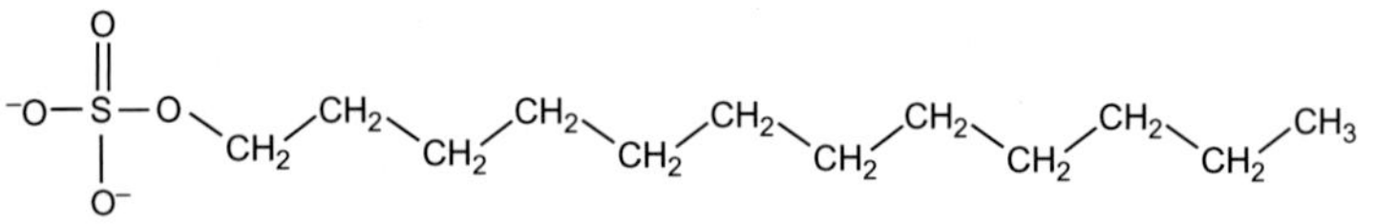

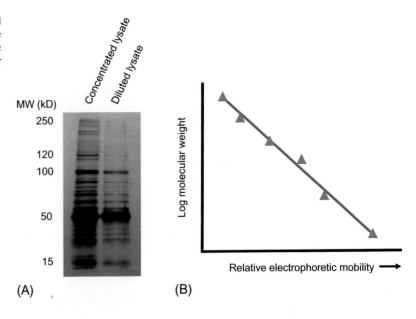

FIG. 8.8 (A) An example of SDS-polyacrylamide gel electrophoresis coupled with a silver stain to visualize polypeptides. (B) Relative SDS-PAGE electrophoretic mobility is inverse relative to the log of the molecular weights of the individual polypeptides.

Gel Staining After Electrophoresis: Colorimetric Staining Methods

Three high-sensitivity colorimetric staining methods can be used either directly after electrophoresis, including Coomassie blue staining, zinc-reverse staining, and silver staining. These can detect as low as 10-ng and 1-ng protein bands, respectively.

Staining with the organic dye Coomassie Brilliant Blue is the most frequently employed method for protein detection in SDS-PAGE gels. This anionic triphenylmethane dye is used in two modifications: Coomassie R-250 (Red tint) and Coomassie G-250 (Green tint), which has two additional methyl groups. In the presence of an acidic medium, these dyes stick to the amino groups of the proteins by electrostatic and hydrophobic interactions. The advantages of Coomassie staining methods are quantitative binding of the dye to proteins, low price, and good reproducibility. For mass spectrometric analysis, the dye can be removed from the gel with bicarbonate prior to tryptic digestion. When a protein is detectable with Coomassie Brilliant Blue, enough protein is present for appropriate mass spectrometry (MS) analysis. The disadvantages are the long staining times, the relatively low sensitivity, and the narrow dynamic range.

Zinc-reverse staining, also known as negative staining, uses imidazole and zinc salts for protein detection in electrophoresis gels. It is based on the precipitation of zinc imidazole in the gel, except in the zones where proteins are located. When zinc reverse staining is applied on a Coomassie Blue-stained gel, previously undetected bands can be spotted. This technique is rapid, simple, cheap, and reproducible and is compatible with MS analysis.

Silver staining is based on the binding of silver ions to the proteins followed by reduction to free silver, sensitization, and enhancement. If silver staining is used as a second staining, it is essential to fix the proteins in the gel with acidic alcohol prior to initial Coomassie Blue staining. Detecting proteins with silver staining is widely used, because the sensitivity is below 1 ng per spot, and the costs for reagents are relatively low. Although it is a multistep procedure, the results are available relatively quickly. Ideally, the spots are dark brown to black on a light beige background (Fig. 8.9). Two drawbacks of this technique are that proteins are differentially sensitive to silver staining and that the process may irreversibly modify them, preventing further analysis. In particular glutaraldehyde, which is generally used during the sensitization step, may interfere with protein analysis by MS due to the introduction of covalent cross-links. To circumvent this problem, a glutaraldehyde-free modified silver-staining protocol has been developed, which is compatible with both matrix-assisted laser desorption/ionization (MALDI) and electrospray ionization-MS.

Gel Staining After Electrophoresis: Fluorescence Staining Method

Fluorescence dyes give very wide linear dynamic ranges, over four orders of magnitude. Since fluorescent staining is an endpoint method, the results are highly reproducible. Several fluorescent dyes such as Nile Red, ruthenium(II) tris(bathophenantroline disulfonate) (RuBPS), SyPro, and Epicocconone, can also be used to reveal a few nanograms of proteins in gels. Although CyDyes can even reveal as low as 1 ng of protein, it is inconvenience because it has to be incorporated before gel electrophoresis. On the other hand, the sensitivities of SyPro Orange, Red, and Tangerine are similar to that of Coomassie Blue, SyPro Ruby can detect down to about 1 ng of protein in a spot. Deep Purple contains the fluorophore "epicocconone" from the fungus *Epicoccum nigrum* and is even more sensitive than SyPro Ruby, down to a few hundred picograms. Besides the high sensitivity, it does not create speckles in the background and is a natural product, which is easy to dispose of. Apart from Nile Red, these staining methods are compatible with subsequent MS analysis. However, their major disadvantage is that they require a fluorescence imager for visualization and that they are significantly more expensive than classical colorimetric dyes.

GEL-BASED PROTEOMICS: PROTEIN SEPARATION

SDS-PAGE gives very good resolution of polypeptides, but sometimes a mixture of polypeptides is so complex that we need an even better method to resolve them all. The field that uses techniques to resolve thousands of polypeptides is called proteomics. Proteomics tools have been exploited to address many molecular biological questions related to various healthy and disease states. Both gel-based proteomics, including two-dimensional gel electrophoresis (2DE) gel and 2DE fluorescence differential imaging gel electrophoresis (DIGE), and gel-free proteomics, such as liquid chromatography (LC) and capillary electrophoresis (CE), are useful methods to identify proteins. Currently, proteomics has been dominated by 2DE gel coupled with MS, as it is still the most reproducible and effective technology to separate overall proteins of microorganisms, cells, and heterogeneous tissues.

Two-Dimensional Gel Electrophoresis

Gel-based proteomics is the most popular and well-established technique for global protein separation and quantification. Through this technique, overall protein expression of tissues can be analyzed on a large scale, and it is a cheaper approach than gel-free proteomics. 2DE gel, MS, and bioinformatics tools are the key components of gel-based proteomics. The most important steps involved in the 2DE gel technique are summarized in Fig. 8.9. In 2DE, the complex protein samples are separated in two dimensions according to their net charge at different pH and their molecular weights determined by SDS-PAGE. Protein migration in a perpendicular direction provides a spot map of the proteins distributed in 2DE gel. The technique has great resolving power, making it possible to visualize >10,000 spots corresponding to >1000 proteins. After the separation, the next typical steps of 2DE gel technique are spot visualization, spot evaluation, and expression analysis. The final step is protein identification by MS.

First Dimension Electrophoresis: Conventional Isoelectric Focusing

In 2DE, the complex protein samples are separated in two dimensions according to their net charge at different pH, which is each protein's isoelectric focusing (IEF), and their molecular weights through SDS-PAGE. In the first step, the mixture of proteins is electrophoresed through a narrow tube gel containing molecules called ampholytes that set up a pH gradient

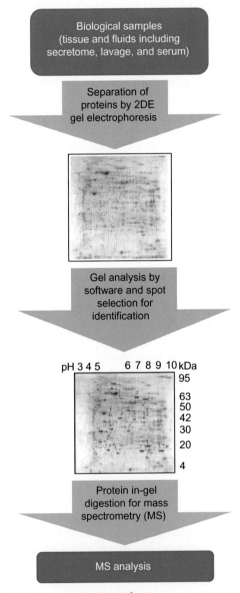

FIG. 8.9 Application of 2DE gel electrophoresis for mass spectrometry analyses.

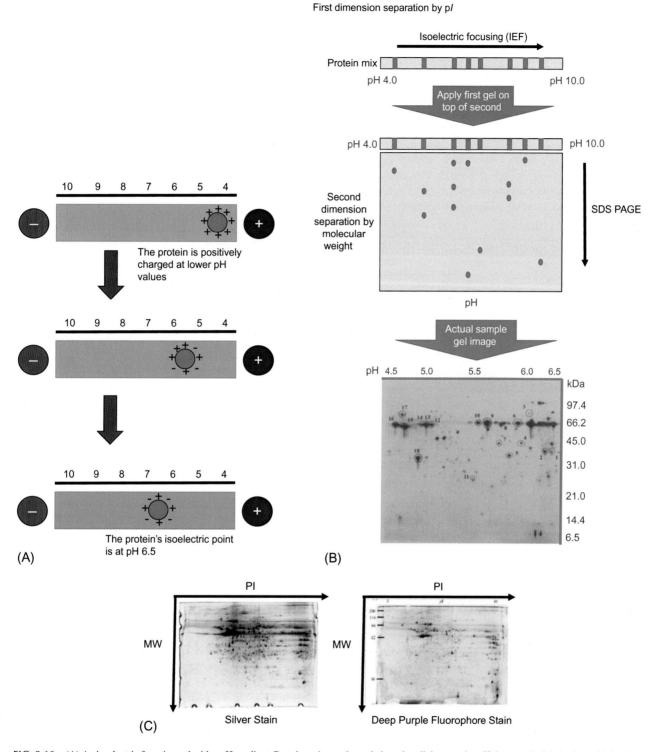

FIG. 8.10 (A) An isoelectric focusing gel with a pH gradient. Proteins migrate through the gel until they reach a pH that matched their p*I*, at which point they are no longer charged and therefore stop migration.(B) Two-dimensional gel electrophoresis separation of proteins by p*I* and molecular weight. (C) An example of two dimensional polyacrylamide gel electrophoresis of mouse colon protein stained by silver staining or deep purple flurophore dye.

from one end of the tube to the other. A negatively charged protein will electrophorese toward the anode. This is because the protein has an overall negative charge after treatment with heat, SDS, and reducing agents. As it migrates through a gradient of increasing pH, however, the protein's overall charge will decrease until the protein reaches the pH region that corresponds to its isoelectric point (p*I*) (Fig. 8.10). Since the p*I* is the pH at which the protein has no net charge, without

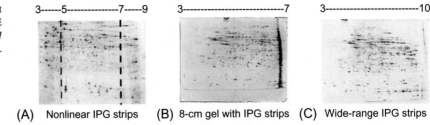

FIG. 8.11 Various immobilized pH gradient strips used in the first dimension step of 2-DE gels. (A) Gel with nonliner (NL) IPG strips (p*I* 5–7). (B) 8 cm gel with IPG strips (p*I* 3–7). (C) Wide-ranged (p*I* 3–10) IPG strip.

(A) Nonlinear IPG strips   (B) 8-cm gel with IPG strips   (C) Wide-range IPG strips

net charge, protein is no longer drawn toward the anode or the cathode, and thus it stops. IEF focuses proteins at their isoelectric point in the gel. Although this conventional method is easy to prepare and do not require much casting equipment, it has the disadvantage that the ampholytes have some susceptibility to flow toward the cathode, and this gradient flow usually causes a reduction in reproducibility.

First Dimension Electrophoresis: Immobilized pH Gradient (IPG)

An immobilized pH gradient strip (IPG) is an integrated part of a polyacrylamide gel matrix fixed on a plastic strip. Copolymerization of a set of nonamphoteric buffers with different chemical properties is included. Ready-made IPG strips are available with different lengths and p*I*. Usually, short IPG strips are used for fast screening while longer ones are used for maximal and comprehensive analysis. Various IPG gels are shown in Fig. 8.11. A commercial precast acrylamide gel matrix copolymerized with a pH gradient on a plastic strip results in a more stable pH value than the traditional ampholyte method. The IPG gel has the ability to avoid cationic accumulation and to produce a better-focused protein with less smearing, and it allows the protein molecules to move at different rates across the gel based on their charge and the volt hours setting, which determines speed pattern and reproducibility. The advantages of using IPG strips over ampholytes include reduced cathodic drift, higher mechanical strength as the strips are casted on a plastic backing that minimizes gel breakage, and higher protein loading capacity due to the sample loading method.

First Dimension Electrophoresis: Nonequilibrium pH Gel Electrophoresis

Nonequilibrium pH gel electrophoresis (NEPHGE) technique can resolve proteins with basic to extremely high p*I* (7.0–11.0) that cannot be separated by traditional methods. Compared to NEPHGE techniques, protein loss is higher in the IPG-based method, especially for basic proteins. The reproducibility of spots is slightly better in the NEPHGE-based method than in the IPG method. In general, about half of detected basic protein spots are not reproducible by IPG-based 2DE, whereas the NEPHGE-based method has excellent reproducibility in the basic gel zone. The reproducibility of acidic proteins is similar in both methods.

Second Dimension SDS-PAGE

The second step of 2DE separates proteins based on their molecular weight using a vertical electrophoretic device with either Laemmli buffer or Tris-Tricine buffer. Instead of loading protein sample within the wells, the first-dimension rehydrated strip is carefully placed on the top of the SDS-PAGE and sealed with agarose (Fig. 8.12). Here, the proteins are resolved according to their sizes by SDS-PAGE.

Postelectrophoresis

After 2DE, the proteins are detected by SDS-PAGE with small pore sizes, and the proteins are unfolded. Under this condition, it is easy to prevent the proteins from diffusing into and out of the matrix. Fixing with highly concentrated trichloroacetic acid is therefore not necessary. The gel can be stained as with the standard SDS-PAGE, as described in previous section.

In 2DE, proteins can be prelabeled with a fluorescent dye after the IEF step prior to SDS-PAGE or prior to IEF with monobromobimane. The spots can be directly scanned in the gel (while it is still in the cassette) when low-fluorescent glass is used. The fluorescent label exhibits high sensitivity and wide dynamic range of the signal.

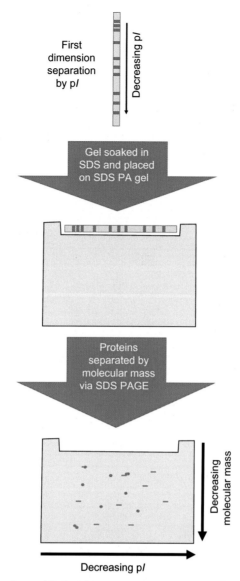

FIG. 8.12 The schematic process of the second step of 2DE which separates proteins based on their molecular weight.

The Analysis of 2DE Images

Analysis of a 2DE gel includes spot visualization, spot evaluation, and expression analysis through bioinformatics software. The key steps of the workflow of most of the automated methods are image quality control, image alignment, spot detection, automatic analysis, editing of spot detection, review the results, statistical analysis, calibration of spots against either a molecular weight ladder or known proteins, spot picking, and importing the protein ID. These automated methods are helpful for gel identification, particularly in quantitative proteomics (Fig. 8.13). The analysis identifies biomarkers by quantifying individual proteins and showing the separation between one or more protein spots on a scanned image of a 2DE gel. Additionally, spots can be matched between gels of similar samples—for example, proteomic differences between early and advanced stages of an illness. Finally, the last step of the analysis is protein identification by MS.

Modification of 2DE: Two-Dimensional Difference Gel Electrophoresis

The development of image technology has introduced the differential imaging gel electrophoresis (DIGE) technique. A quantum leap in 2D gel electrophoresis methodology uses size- and charge-matched cyanine dyes (CyDye, DIGE,

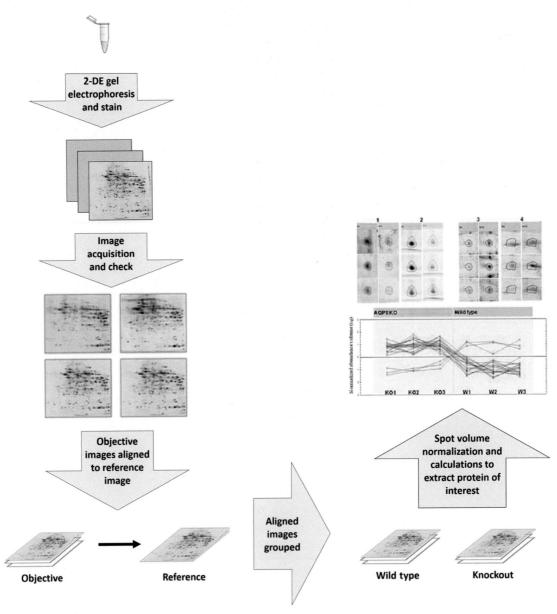

FIG. 8.13 Progenesis SameSpots (nonlinear dynamic) image analysis of a 2-DE gel electrophoresis.

Fluors) with different excitation and emission wavelengths as protein labels for different samples. This method can increase sensitivity and reproducibility of 2DE using multiplexed fluorescent dyes (e.g., Cy2, Cy3, and Cy5) to covalently label protein samples. 2D-DIGE can run more than one sample (maximum 3 samples) on a single gel at once to address the issue of gel-to-gel variability.

In the DIGE technique, different fluorescent cyanine (Cy) dyes are used for labeling proteins from different samples. After mixing these samples in equal ratio and running them together as one sample, the same protein from different samples migrates to the same position on the 2D gel, where it can be easily examined and differentiated by the different fluorophore-labeled dyes and imaged to calculate its abundance. The workflow of 2D-DIGE is shown in Fig. 8.14. Two important factors in 2D-DIGE can affect the sensitivity of the result: minimum labeling by attaching the dye to the free lysine residues, and saturation labeling of all cysteine residues. For the minimum labeling to the lysine residues, only 3%–5% of the total proteins will receive a label, which ensures that only singly labeled proteins will be detected. In this case the labeled proteins comigrate with the nonlabeled proteins in both separation directions, and the resulting 2D image looks like the one

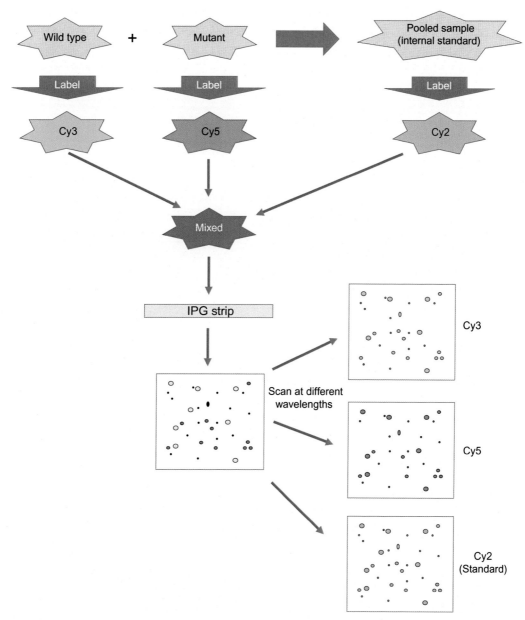

FIG. 8.14 Schematic of 2-D difference gel electrophoresis of two samples in one gel. When more samples are to be compared, more gels are run, every time including the internal standard, which is a mixture of pooled aliquots of all samples.

achieved with poststaining. Typically, 400 pmol/l of one of the three available cyanine dyes is added to 50 µg protein of each sample and the pooled standard. The sensitivity of detection is comparable to a sensitive silver staining method. Meanwhile, a considerable increase in sensitivity is obtained through saturation labeling of cysteine. 2D patterns resulting from as little as 1 µg total protein load can be visualized. Because all available cysteine residues are labeled, many multiply labeled proteins exist. The resulting spot patterns are different from those achieved with poststained or lysine labeled proteins. After scanning the gel at different wavelengths, the comigrated protein spots of different sample origins are codetected using dedicated software, which allows very fast and completely automatic pattern evaluation, thus avoiding any bias introduced by an operator. With the help of the pattern created by the internal standard, which is run in each gel, the patterns of different gels can easily be matched and the sample spot volumes can be normalized to the spot volume of the standard. This results in very accurate protein difference ratios.

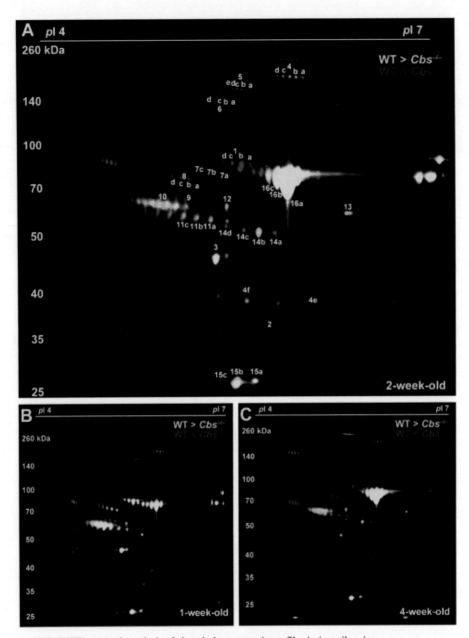

FIG. 8.15 An example of 2D DIGE proteomic analysis of altered plasma protein profiles in juvenile mice.

2D-DIGE is an important tool, especially for clinical laboratories involved in the determination of protein expression levels and disease biomarker discovery. Application of 2D-DIGE can reduce protein ratio errors because of low gel-to-gel variations, which is frequently occurred in the 2DE gel technique. When absolute biological variation between samples is the main objective, as in biomarker discovery, 2D-DIGE is one of the methods of choice. For example, Fig. 8.15 presents 2D-DIGE separation of plasma protein from wild-type mice and cystathionine β-synthase-deficient (Cbs −/−) mice, which is an animal model for homocystinuria with hepatic steatosis and juvenile semilethality.

SAMPLE BUFFER CONSTITUENTS FOR 2DE

Solubilization, disaggregation, denaturation, and reduction of proteins during or right after cell disruption is achieved by the use of chaotropic agents, detergents, reducing agents, buffers, and carrier ampholytes. These buffer constituents must be compatible with IEF, both electrically and chemically.

Chaotropic Agents

In a nondenaturing 2DE-urea, a neutral chaotropic agent, such as urea, can act as a denaturing agent in sample solutions at concentrations typically at 8–9 M. It effectively disrupts noncovalent and ionic bonds between amino acid residues. Urea-containing solutions may not be heated above 37°C to avoid degradation of urea to cyanate. Cyanate ions can react with the amine groups of proteins which can cause carbamoylation and remove the positive charge of the amine, thereby affect the p*I* of the respective proteins. The solubilizing power of urea-containing sample solutions can be dramatically increased by the addition of thiourea. Typically, thiourea is used at 2 M concentration in conjunction with 5–7 M urea.

Detergents

Hydrophobic interactions within a polypeptide chain or between proteins in protein complexes can be disrupted in the presence of detergents. The use of detergents can also increase solubility, especially of membrane proteins. Detergents operate synergistically with chaotropic agents during the solubilization process. This is because they prevent hydrophobic interactions between hydrophobic protein stretches exposed by the chaotropic agents. Detergents consist of a hydrophobic tail and a hydrophilic head moiety, which may be anionic, cationic, zwitterionic, or nonionic. To allow proteins to migrate according to their own charge during IEF, zwitterionic or nonionic detergents are preferred. Although nonionic detergents such as Nonidet P-40 or Triton X-100 can be used, the zwitterionic detergent CHAPS has much better solubilization, particularly for membrane proteins. The selection of the optimal detergent in a solubilization cocktail should take into account how the detergent interacts with high concentrations of urea.

SDS is excellent in its ability to efficiently and rapidly solubilize proteins. Although SDS is incompatible with IEF as an anionic detergent, it can be used in the initial preparation of concentrated protein samples. In these cases, another IEF-compatible detergent must be used in excess to disrupt the binding of SDS to protein. The IEF-compatible detergent: SDS concentration ratio should be at least 8:1 to avoid detrimental effects of SDS in IEF. In addition, its final concentration in the sample solution buffer should be below 0.25%.

Reducing Agents

Reducing agents cleave disulfide bond cross-links within and between protein subunits, thereby promoting protein unfolding and maintaining proteins in their fully reduced states. The compounds mostly used for 2D sample preparation are sulfhydryl reducing agents like DTT. A large excess of sulfhydryl reducing agent will shift the equilibrium of the oxidation/reduction reaction toward the fully reduced protein state. Therefore, sulfhydryl reducing agents are typically used in large excess. Typically, DTT is present in sample preparation cocktails at concentrations ranging from 20 mM to 100 mM to ensure reduction of the protein disulfide bonds. Therefore, it is necessary that DTT be evenly distributed over the whole length of the IPG strip during IEF in order to maintain the reduced state of all cysteines present in the protein. On the other hand, DTT is a weak acid and it can migrate toward the anode. As a consequence, some disulfide bonds will re-form, leading to precipitation of some disulfide rich-proteins. Artificial spots might be generated because of the formation of disulfide bridges. Phosphines such as tributylphosphine offer an alternative to thiols as reducing agents because they are uncharged and reduce cystines stoichiometrically at concentrations as low as 2 mM.

Another approach to avoid protein aggregation and precipitation in basic regions (above pH 8) of IPG gradients is to reduce proteins with tributylphosphine and then irreversibly alkylate with iodoacetamide (Fig. 8.16). Tributylphosphine is a more effective reducing agent than DTT because of it ability to increase protein solubility. This treatment, which is

Untreated sample   Treated sample

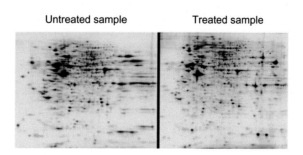

FIG. 8.16 HeLa cell extract separated by 2DE and stained to demonstrate the effect of reduction and alkylation treatment of protein samples (treated sample on right demonstrates better spot resolution than untreated sample on left).

performed after protein solubilization and before IEF, blocks protein sulfhydryls and prevents proteins from aggregating and precipitating due to oxidative cross-linking. This can ensure that proteins remain soluble throughout electrophoresis.

Carrier Ampholytes

Carrier ampholytes are usually included at a concentration of 0.5%–2% (v/v) in sample solutions for IPG strip focusing. They can reduce protein-matrix hydrophobic interactions, which tend to occur at the basic end of the IPG strip and lead to streaking caused by precipitation. Carrier ampholytes can also overcome detrimental effects from salt boundaries and help to compensate for insufficient salt in a sample. This is because even in the presence of detergents, certain samples may have stringent salt requirements to maintain the solubility of some proteins. Although IPG strips are tolerant to salts, salts should be in the sample only if they are absolutely required. During IEF, any salt will be removed, and proteins that need salt for solubility are then subject to precipitation. As such, salts forming strong acids and bases (e.g., NaCl, Na_2HPO_4) should be avoided, because strongly alkaline cationic and strongly acidic anionic boundaries are formed by the migration of the salt's ionic constituents. Salts formed from weak acids and bases such as Tris-acetate and Tris-glycine should be used instead.

Removal of Interfering Contaminants

Sample purity can determine success or failure of any protein analysis. Interfering substances are any contaminants that negatively impact IEF, SDS-PAGE, or both. Commonly seen contaminants include salts, detergents, small compounds, nucleic acids, lipids, and polysaccharides. Therefore, it is important to eliminate contaminants from the original biological sample during sample preparation. Four widely applied precipitation methods are trichloroacetic acid, acetone, chloroform/methanol and ammonium sulfate, and ultrafiltration. Protein precipitation with trichloroacetic acid and acetone as well as ultrafiltration results in efficient sample concentration and desalting. However, independent of precipitation and resolubilization method, protein recoveries in the liquid phase are rarely 100%. All protocols require careful optimization to ensure a true protein representation in the final 2D image.

GEL FREE PROTEOMICS TECHNIQUES: CAPILLARY ELECTROPHORESIS

The Principles of Capillary Electrophoresis

Capillary electrophoresis (CE) is a typical gel-free technique, with the advantage of superior separation efficiency, small sample consumption, short analysis time, and automatability. CE can readily be coupled online with MS. CE separates proteins, with or without prior denaturation, in slab gels or microfluidic channels. Protein separation in CE is based on a variety of properties, including their molecular mass (SDS-capillary gel electrophoresis [CGE]), their isoelectric point (capillary isoelectric focusing [CIEF]), or their electrophoretic mobility (capillary zone electrophoresis [CZE]).

In SDS-CGE, a capillary is filled with gel or viscous solution (to form molecular sieve) and the electroosmotic flow (EOF) is usually suppressed. EOF is a unique driving force in CE that is generated at the wall of a fused-silica capillary by the formation of an electrical double layer associated with the ionized silanol groups. Therefore, the analytes migrate solely based on their electrophoretic mobility. CGE can be accomplished using either a permanently coated gel (e.g., cross-linked polyacrylamide) or a dynamic coated gel (e.g., linear polyacrylamide and cellulose dextran). In the first approach, the gel is prepared and bonded inside the fused silica capillary by polymerization of monomer, whereas in the second approach, the gel is dissolved in aqueous solution to form molecular sieve and used as a separating medium.

With CZE, sample is introduced into a buffer-filled capillary either electrokinetically (with low voltage) or hydrodynamically (with pressure or suction). Both ends of the capillary and electrodes are then placed into a buffer solution that also contains the electrodes, and a high voltage is applied to the system. The applied voltage causes the analytes to migrate through the capillary and past a detector window, where information is collected and stored by an appropriate data acquisition system.

Microchip Electrophoresis

Microchip electrophoresis (ME) is a miniaturized form of conventional CE. In ME, the basic CE configuration has been miniaturized and transferred to a chip format. With ME, it is possible to integrate injection, separation, and detection on a single microchip with typical channel lengths of 5 cm to 15 cm. The separation mechanism for ME is similar to that of CE in that differential migration of analytes is governed by their effective ionic mobilities and the mobility of the EOF under an electric field (Fig. 8.17). In ME, the EOF can be generated on different materials rather than just silica. Compared to conventional CE the advantage to use ME include shorter analysis time, less sample/reagent consumption, and less

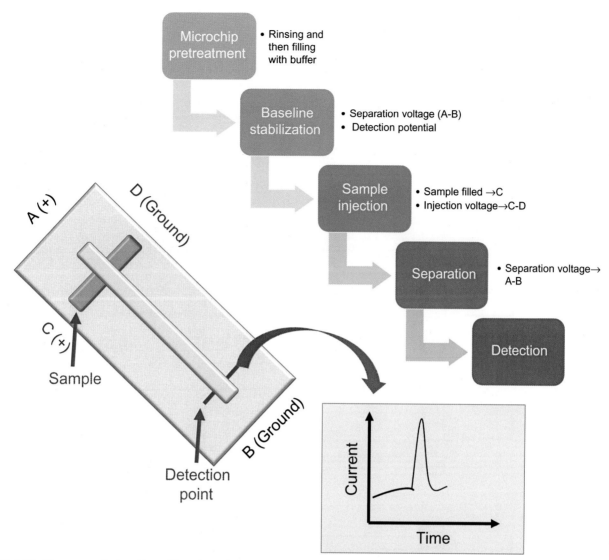

FIG. 8.17 The process of microchip capillary electrophoresis.

waste generation. Furthermore, ME requires lower voltages due to the shorter separation channels, and the associated instrumentation can be made small, leading to portable analytical systems. ME enables fast separation of protein, peptide, and single-cell analysis, which are normally present in small amounts in complex matrices.

GEL FREE PROTEOMICS TECHNIQUES: LIQUID CHROMATOGRAPHY

Liquid chromatography is the fundamental of protein separation science. It can detect molecules at the nanomolar level. Therefore, it is useful in proteomics and genome research. Many kinds of liquid chromatography, such as reversed-phase high performance liquid chromatography (HPLC), affinity HPLC, gel permeation HPLC, ligand-exchange HPLC, and capillary HPLC, have been used in proteomic research. The principles exploited in HPLC basically are the same as those used in common column chromatographic methods such as affinity chromatography, ion-exchange chromatography, or size-exclusion chromatography. The difference is that very high-resolution separations can be achieved quickly and with high sensitivity in HPLC using automated instrumentation.

Reversed-Phase HPLC

Reversed-phase HPLC is the most popular mode of chromatography because of its wide range of applications and the availability of various mobile and stationary phases. Stationary phases mostly comprise nonpolar alkyl hydrocarbons such as C8

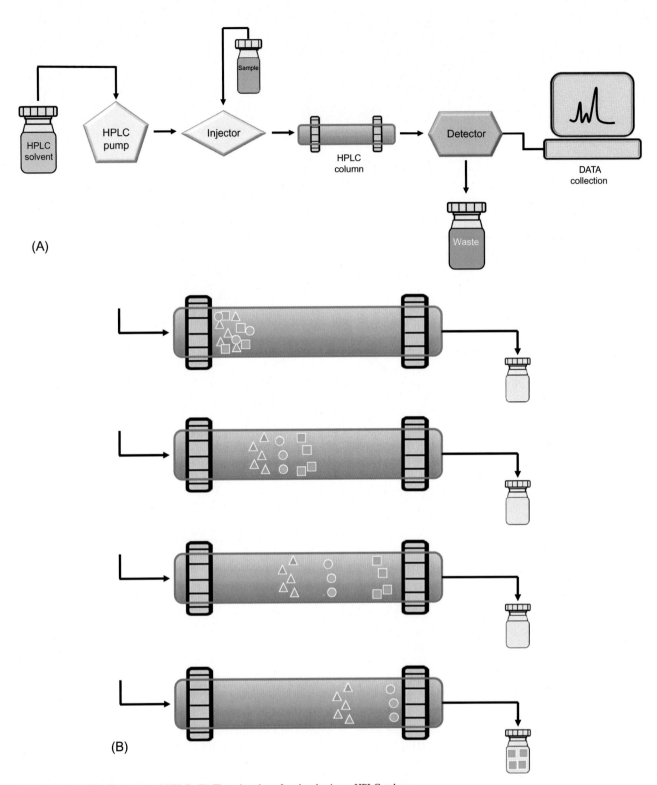

FIG. 8.18 (A) The flow chart of HPLC. (B) The migration of molecules in an HPLC column.

or C18 chains bound to silica or other inert supports (Fig. 8.18). C18 and C8 refer to the alkyl chain length of the bonded phase of the column. The chain length influences hydrophobicity of the sorbent phase and thus increases retention of ligands. C18 is often called the "**traditional reverse phase matrix**" because it has the highest degree of hydrophobicity. The reason why C-18 is more hydrophobic than other reverse phases is because length of the carbon chains are longer. C8 is used when shorter retention times are desired. Lower hydrophobicity means faster retention for nonpolar compounds, hence nonpolar compounds move down the column more readily with C8 than with C18. C8 is preferred over C18 if one is looking for a reverse phase matrix that has a lower degree of hydrophobicity. C18 columns actually can handle more than 60% of the applications in most HPLC labs. The mobile phase is polar, and the elution order is polar followed by less polar, and weakly polar or nonpolar compounds in the end.

Affinity HPLC

Affinity HPLC is a chromatographic method capable of separating biochemical mixtures of highly specific nature. It is possible to design a stationary phase that reversibly binds to a known subset of molecules just by combining affinity chromatography. This method exploits a well-known and well-defined property of analytes that can be used during purification process. The process can be considered as an entrapment, with the target molecule trapped on a stationary phase while the other molecules in solution are not trapped because they lack this property.

Gel-Permeation HPLC

Gel-permeation HPLC works on the principle of sizes of the compounds, which is the same as the size-exclusion chromatography. In this technique, large molecules are eluted first, followed by smaller molecules. Therefore, it is a method of choice for separation of biomolecules such as peptides, proteins, and enzymes. The stationary phase is a porous solid such as glass or silica, or a cross-linked gel that contains pores of appropriate dimensions to affect the separation desired. The separation and isolation of proteins can be done by using Sepharose chromatography followed by MS.

Ligand-Exchange HPLC

Ligand-exchange HPLC is an advanced version of reversed-phase-HPLC where the reversed-phase column is replaced by an ion-exchange column. It has been used widely for the analysis of all inorganic and organic ionic species. In ligand-exchange-HPLC, either anion- or cation-exchange column is used but, nowadays, mixed (anion and cation) columns are also available that improve the separation efficiency.

In cation-exchange chromatography, the stationary phase is usually composed of resins containing sulfonic acid groups or carboxylic acid groups of negative charge, and thus cationic molecules are attracted to the stationary phase by electrostatic interaction. In anion-exchange chromatography, the stationary phase is a resin, generally, containing primary or quaternary amine functional groups of positive charge, and thus these stationary phase groups attract solutes of negative charge.

Capillary Electrochromatography

A hybrid technique of HPLC and CE is capillary electrochromatography (CEC). It combines the high peak efficiency that is characteristic of electrically driven separations with high separation selectivity (Fig. 8.19). CEC can be carried out on wall coated open tubular capillaries or capillaries packed with particulate or monolithic silica or other inorganic materials as well as organic polymers. The chromatographic and electrophoretic mechanisms work simultaneously in CEC, and several combinations are possible.

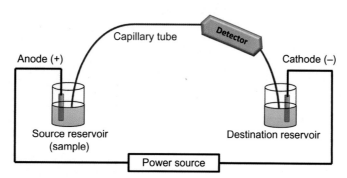

FIG. 8.19 A schematic of capillary electro-chromatography.

Comparison of Various Chromatographic Methods

All chromatographic techniques are important and useful for proteomics analysis. The choice of which technique to use depends on the type and nature of the proteins to be analyzed. For example, gel-permeation HPLC is useful for proteomic separation and identification because of the wide variation in the sizes of proteins. Ligand exchange is useful because proteins have charges, which may be exploited in this kind of chromatography. The comparison of each chromatographic method and their applications are listed in Table 8.4. Among various chromatographic methods used in proteomic analyses the order of application is reversed phase > gel permeation > ligand exchange > affinity. The most popular method

TABLE 8.4 Various Chromatographic Techniques for Proteomic Analyses

Proteomes	Columns	Mobile Phases
Reversed-phase high performance liquid chromatography		
Complex protein mix	C_{18} column	Acetonitrile-TFA
Cytoskeletal proteins & enzymes	PepMap C_{18}	–
Peptides of *Caenorhabditis elegans*	C_{18}	Acetonitrile-formic acid
S-transferase isoenzyme	Vydac 214TP C_4	Acetonitrile and water with TFA
Photosytem II antenna protein	Vydac C_4	Acetonitrile-water-TFA
Proteins of rat adipose cells	Zorbax 300 SB-C3	Acetonitrile-TFA
Complex protein mixture	Zorbax SB-C_{18}	Buffer-acetonitrile-FA Buffer-acetonitrile-TFA
GPI-Aps protein	Zorbax SB-C_{18}	Acetonitrile with acids HFBA (different combinations)
Phosphopeptides of rat kidney inner medullary collecting duct	Picofrit RP column	Acetonitrile-FA
Venoms of snakes	Lichrosphere RP100 C_{18}	Water-acetonitrile-TFA
Affinity high performance liquid chromatography		
Peptides of liver of mice	Magic C_{18} AQ	Water-acetonitrile-FA
Phosphopeptides of *Arabidopsis*	PepMap C_{18}	Water-acetonitrile-FA
Ligand-exchange high performance liquid chromatography		
Proteomic analysis of *E. coli*	218 TP 5415 Vydac C_{18} RP column	Acetonitrile-water-TFA
Pancreatic islet proteome	Polysulfoethyl A column	10 mM Ammonium formate buffer-water and acetonitrile
Membrane proteins of breast cancer MCF7 and BT474 cells	Polysulfoethyl A resin	–
C. elegans peptidomic analysis	Bio-SCX column	Water-acetonitrile-FA
Capillary electro-chromatography		
Protein of *Helicobacter pylori*	ReproSil-Pur C_{18} AQ	–
Tryptic peptide of fish	Phenomenex Jupiter C_{18}	Acetonitrile-TFA-formic acid
Synaptic proteomes of wild type mice	Polysulfoethyl A	Buffers-acetonitrile-10 mM KH_2PO_4
Proteomic analysis of *Escherichia coli*	TSKG3000SWxL	Water-50 mM KH_2PO_4-200 nM NaCl
Vacuum-dried peptides	Jupiter C_{18} RP capillary column	Water with TFA and FA, water and acetonitrile with TFA

FA, formic acid; *HFBA*, heptafluorobutyric acid; *TFA*, trifluoroacetic acid.

used in proteomic analyses is reversed phase-HPLC. This is because that it is well developed. There are many types of reversed-phase stationary phases available, which can be used for analyses of proteomes. Besides, the reversed-phase columns are able to work with a wide range of mobile phases, enhancing the application range of reversed-phase chromatography.

PROTEIN ANALYSIS: MASS SPECTROMETER

After the proteins are separated and quantified, the next step is to identify each protein in the sample. For example, individual spots are cut out of the gel and cleaved into peptides with proteolytic enzymes. These peptides can then be identified by MS. Mass spectrometers exploit the difference in the mass-to-charge (m/z) ratio of ionized atoms or molecules to separate them from each other. The basic operation of a mass spectrometer is to (1) evaporate and ionize molecules in a vacuum, creating gas-phase ions; (2) separate the ions in space and/or time based on their m/z ratios; and (3) measure the quantity of ions with specific m/z ratios. Different MS have been developed depending on the method to ionize proteins for analysis. The two most prominent MS methods for protein analysis are electrospray ionization MS (ESI-MS) and matrix-assisted laser desorption/ionization-time of flight MS (MALDI-TOF MS).

ESI-MS

A solution of macromolecules is sprayed in the form of fine droplets from a glass capillary under the influence of a strong electrical field (Fig. 8.20). The droplets pick up positive charges as they exit the capillary; evaporation of the solvent leaves multiply charged molecules. Individual protein molecules acquire about one positive charge per kilodalton. The MS spectrum of these charged molecules is a series of sharp peaks whose consecutive m/z values differ by the charge and mass of a single proton, leading to a spectrum of m/z ratios for a single protein species (Fig. 8.21). Computer analysis can convert these data into a single spectrum that has a peak at the correct protein mass. ESI-MS recently is evolved to the so-called nanospray ionization, while "tandem" MS is rapidly replaced by "cascades" of mass spectrometers using more sophisticated and sensitive detectors (such as orbit traps), and accommodating other intermediate "sequencing" instruments, like electron-transfer dissociation.

MALDI-TOF

In MALDI-TIF procedure, a peptide is placed on a matrix, which causes the peptide to form crystals (Fig. 8.22). A laser pulse is used to excite the chemical matrix, creating a microplasma that transfers the energy to protein molecules in the sample, ionizing them and ejecting them into the gas phase. An increase in voltage at matrix is used to shoot the ions toward

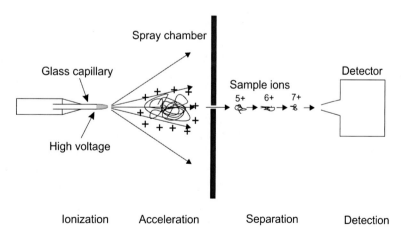

FIG. 8.20 The steps of electrospray ionization mass spectrometry (ESI-MS).

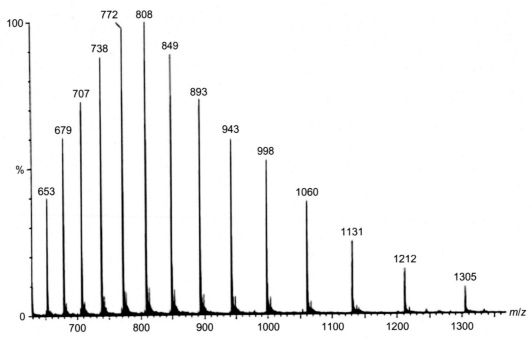

FIG. 8.21 A typical protein analysis via electrospray ionization mass spectrometry. This analysis represents the cluster ions of a highly purified horse myoglobin protein standard solution (5 μmol/L in 50% acetonitrile solution containing 0.2% formic acid).

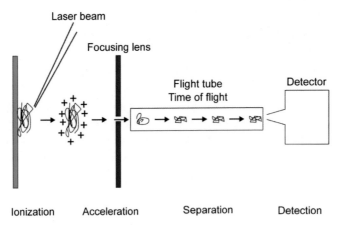

FIG. 8.22 MALDI-TOF mass spectrometry mechanism.

a detector. Protein molecules that have picked up a single proton can be selected by the MS for mass analysis. This is because the time it takes an ion to reach the detector depends on its mass. The higher the mass, the longer the time of flight of the ion. In a MALDI-TOF MS, the ion can also be deflected with an electrostatic reflector that also focuses the ion beam. Thus, one can determine the mass of the ions reaching a second detector with high precision, and these masses can reveal the exact chemical composition of the peptides. As such, MALDI-TOF MS is very sensitive and accurate for determining amino acid sequences.

Chapter 9

Amplification of Nucleic Acids

Chapter Outline

CONVENTIONAL AMPLIFICATION TECHNIQUE: THE POLYMERASE CHAIN REACTION

PCR is an in vitro enzymatic reaction to amplify a defined DNA region. A PCR reaction normally uses two oligonucleotide primers, which are hybridized to the $5'$ and $3'$ borders of the target sequence, and a DNA polymerase, which can extend the annealed primers by adding on deoxyribonucleoside triphosphates (dNTPs) to generate double-stranded products (Fig. 9.1). Subsequently, the copied DNA duplex is heat-denatured into single strand, and the next cycle starts with twice the number of DNA segments for copying. Therefore, the copied number of DNA segment ideally increases by twofold after a cycle of thermal treatment. PCR is useful as an end-point assay to examine the existence of target sequence of DNA in a sample.

Any region of any DNA molecule can be chosen. The only requirement is that the sequences at the borders of the region are known. The oligonucleotides, which act as primers for the DNA synthesis reactions, define the amplified region (Fig. 9.2). By raising and lowering the temperature of the reaction mixture, the two strands of the DNA product are separated and can serve as templates for the next round of annealing and extension, and the process is repeated. Amplification is usually carried out by *Taq* polymerase, which is thermostable DNA polymerase I enzyme from *Thermus aquaticus*. The thermostability of *Taq* polymerase is an essential requirement in PCR, as a PCR reaction will go through different temperature stages that may denature or disable the nonthermostable polymerase.

Diagnostic Molecular Biology. https://doi.org/10.1016/B978-0-12-802823-0.00009-2

FIG. 9.1 PCR amplification of defined region of DNA via multiple cycles of denaturation, primer annealing, and extension.

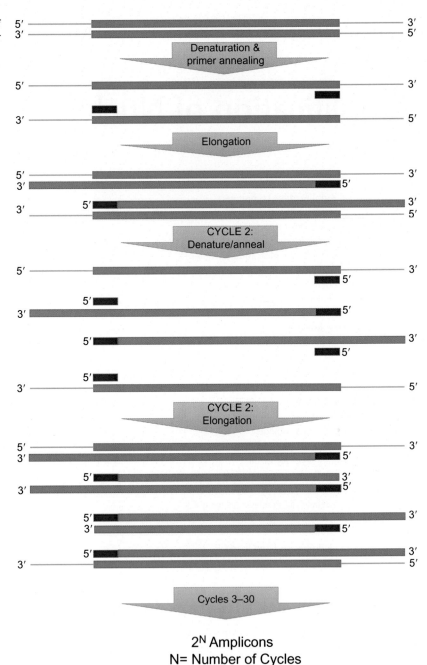

Three Steps of a PCR Reaction

In a typical PCR experiment, the target DNA is mixed with *Taq* polymerase, the two oligonucleotide primers, and a supply of dNTPs. The amount of target DNA can be very small because PCR is extremely sensitive and will work with just a single starting molecule. The reaction is started by heating the mixture to 94°C. At this temperature the hydrogen bonds that hold together the two polynucleotides of the double helix are broken, and the target DNA becomes denatured into single-stranded molecules (Fig. 9.3). The temperature is then decreased to 50–60°C, which is just below the annealing temperature. This results in some rejoining of the single strands of the target DNA, but also allows the primers to attach to their

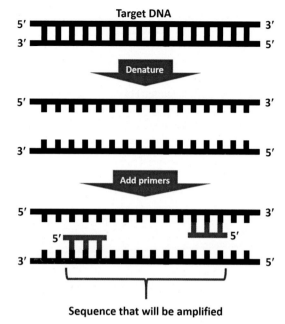

FIG. 9.2 Hybridization of oligonucleotide primers during the annealing step of PCR defines the region of amplification.

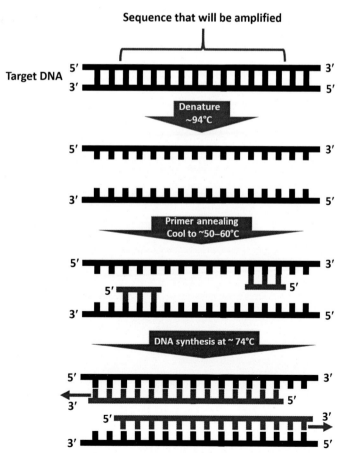

FIG. 9.3 A typical PCR reaction has three steps: denaturation, annealing, and extension.

complementary sequences. Normally, the temperature is about 2°C below the melting temperature. DNA synthesis can now begin when the temperature is raised to 72–74°C, just below the optimum for *Taq* polymerase.

In this first cycle of PCR, a set of "long products" is synthesized from each strand of the target DNA. These polynucleotides have identical 5′ ends but random 3′ ends, the latter representing positions where DNA synthesis terminates by chance (Fig. 9.2). The cycle of denaturation-annealing-synthesis is now repeated. The long products denature, and the four resulting strands are copied during the DNA synthesis stage. This gives four double-stranded molecules, two of which are identical to the long products from the first cycle and two of which are made entirely of new DNA. During the third cycle, the newly made two DNAs serve as the template to produce the correct amplification products, the 5′ and 3′ ends of which are both defined by the primer annealing positions. In subsequent cycles, the number of correct amplification products accumulates in an exponential fashion (doubling during each cycle) until one of the components of the reaction becomes depleted or until the end of a defined cycle. Thus, after 30 cycles, there will be more than 1 billion (2^{30}) short products derived from each starting molecule, and this is equivalent to several micrograms of PCR product from a few nanograms or less of target DNA.

Analysis of PCR Products

To evaluate whether a PCR reaction is successful, one has to examine the PCR products. The simplest and easiest procedure is to analyze the PCR products by agarose gel electrophoresis. Normally, sufficient DNA having been produced from the amplified fragment can be visible as a discrete band after staining with ethidium bromide. This may by itself provide useful information about the DNA region that has been amplified, or if the DNA yield is low, the product can be detected by Southern hybridization. If the expected band is present, the PCR reaction is successful. If the expected band is absent, or if additional bands are present, something has gone wrong and the experiment must be repeated. Alternatively, the PCR product can be examined by techniques such as DNA sequencing and cloning.

MODIFIED NUCLEIC ACID AMPLIFICATION TECHNIQUES

Reverse-Transcriptase Polymerase Chain Reaction

Reverse-transcriptase (RT) PCR (RT-PCR) was developed to amplify RNA targets. In this process, complementary DNA (cDNA) is first produced from RNA targets by reverse transcription, and then the cDNA is amplified by PCR. The reverse transcription is accomplished through the action of reverse transcriptase, an enzyme isolated from RNA virus. Reverse

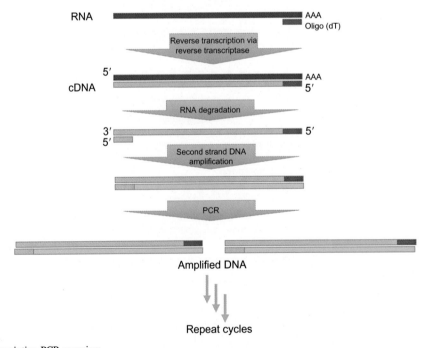

FIG. 9.4 Reverse transcription PCR overview.

transcriptase requires primers such as oligo dT primers or random hexamers to prime the synthesis of the initial DNA strand (Fig. 9.4). Oligo dT primers are 18-base-long single-stranded poly dT sequences that will prime cDNA synthesis from mRNA with poly A tails. Random hexamers or decamers are 6- or 10-base-long single-stranded oligonucleotides of random sequences. These primers will match and hybridize to random sites in the target RNA to prime cDNA synthesis. These primers can generate cDNA from all RNA in the specimen, and the last step is the amplification of cDNA by PCR.

Nested Polymerase Chain Reaction

Nested PCR was developed to increase both the sensitivity and specificity of PCR. This technique uses two pairs of amplification primers and two rounds of PCR (Fig. 9.5). Typically, one primer pair is used in the first round of the amplification of PCR of 15–30 cycles. The products of the first round of amplification are then subjected to a second round of amplification using the second set of primers. The second set of primers anneal to a sequence internal to the sequence amplified by the first primer set. The increased sensitivity arises from the high total cycle number, and the increased specificity arises from the annealing of the second primer set to sequences produced by the first round. The major concern for this method is the contamination that occurs during the transfer of the first-round product to the second tube for the second round of amplification. However, this can be avoided by physically separating the first- and second-round amplification mixtures with a layer of wax or oil.

Multiplex Polymerase Chain Reaction

In multiplex PCR, two or more primer sets designed for amplification of different targets are included in the same PCR reaction. Using this technique, more than one target sequence in a clinical specimen can be amplified in a single tube. As an extension to the practical use of PCR, this technique can save time and effort. The primers used in multiplex reactions must be selected carefully to have similar annealing temperatures and must be not complementary to each other. The amplicon sizes should be different enough to form distinct bands when visualized by gel electrophoresis. Multiplex PCR can be designed in either single-template PCR reaction that uses several sets of primers to amplify specific regions within a template, or multiple-template PCR reaction, which uses multiple templates and several primer sets in the same reaction tube (Fig. 9.6). Although the use of multiplex PCR can reduce costs and time to simultaneously detect two, three, or more pathogens in a specimen, multiplex PCR is more complicated to develop and often is less sensitive than single-primer-set PCR. The advantage of multiplex PCR is that a set of primers can be used as internal control, so that we can eliminate the possibility of false positives or negatives. Furthermore, multiplex PCR can save costly polymerase and template in short supply.

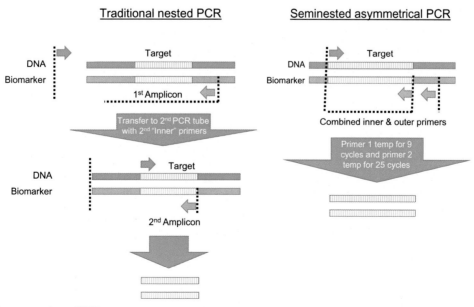

FIG. 9.5 Nested versus seminested PCR.

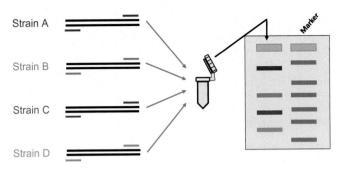

FIG. 9.6   Multiplex PCR overview.

Digital Polymerase Chain Reaction

PCR exponentially amplifies nucleic acids, and the number of amplification cycles and the amount of amplicon allows the quantitation of the starting quantity of targeted nucleic acid. However, many factors complicate this calculation, often creating uncertainties and inaccuracies, particularly when the starting concentration is low. Digital PCR overcomes these difficulties by transforming the exponential data from conventional PCR to digital signals that simply indicate whether or not amplification occurred. Digital PCR is accomplished by capturing or isolating each individual nucleic acid molecule present in a sample within many chambers, zones, or regions that are able to localize and concentrate the amplification product to detectable levels (Fig. 9.7). After PCR quantification, a count of the areas containing PCR product is a direct measure of the absolute quantity of nucleic acid in the sample. The capture or isolation of individual nucleic acid molecules may be done in capillaries, microemulsions, or arrays of miniaturized chambers, or on surfaces that bind nucleic acids. Digital PCR has many applications, including detection and quantification of low levels of pathogen sequences, expression of rare genetic sequences in single cells, and clonal amplification of nucleic acids for sequencing mixed nucleic acid samples.

PARAMETERS THAT AFFECT PCR

Both primer length and annealing temperature are the most important factors that can influence PCR result significantly. If the primers are designed correctly, the experiment results in amplification of a single DNA fragment, corresponding to the target region of the template molecule. If the primers are incorrectly designed, the experiment will fail, possibly because no amplification occurs, or possibly because the wrong fragment, or more than one fragment, is amplified.

The Size of the PCR Product

The first step to design the primer is to determine the target region that is for PCR amplification. Normally, the target sequences to be amplified should be kept below 3 kb. Fragments up to 10 kb can be amplified by standard PCR techniques, but the longer the fragment, the less efficient the amplification. Furthermore, it is more difficult to obtain consistent and accurate results in a longer amplification. However, many of today's PCR-based applications require longer read lengths,

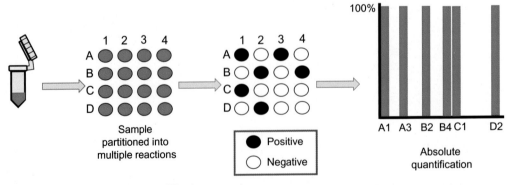

FIG. 9.7   Digital PCR can show whether or not amplification occurred.

greater fidelity, and higher yields than can be achieved with *Taq* DNA polymerase. For example, cloning and expression experiments, and cDNA analysis can be improved by using long and accurate amplification over routine amplification. Recent development has made the long and accurate amplification to be achieved by combining a thermostable polymerase with a second polymerase exhibiting a $3' \rightarrow 5'$ exonuclease activity. The exonuclease, or proofreading, activity repairs terminal mismatches and allows for greater read lengths and increased fidelity. As such, it is possible to generate amplicons up to 20 kb on complex, genomic templates or 40 kb on less complex DNA.

The Lengths of the Primers

Once the target region is determined, the next issue is the length of the primers. If the primers are too short, they might hybridize to nontarget sites and give undesired amplification products. For example, *Saccharomyces cerevisiae* total genomic DNA is used in a PCR experiment with a pair of primers five nucleotides in length (these are called "5-mers"). The likely result is that a number of different fragments will be amplified. This is because attachment sites for these primers are expected to occur, on average, once every $4 \times 4 \times 4 \times 4 \times 4 = 4^5 = 1024$ bp, giving approximately 11,788 possible sites in the 12,071,326 bp of nucleotide sequence that makes up the yeast genome. Therefore, it would be very unlikely that a pair of 8-mer primers would give a single, specific amplification product with yeast DNA (Fig. 9.8). On the other hand, if 17-mer primers are used in the PCR, the theoretical frequency is once every 17,179,869,184. The number is much larger than the genome size. The frequency is even smaller than the genome size. As such, a pair of 17-mer primers should therefore give a single, specific amplification product (see Fig. 9.8).

Although longer primers produce more specific PCR product than shorter primers, it is not possible to have very long primers. This is because the length of the primer influences the rate at which it hybridizes to the template DNA. Usually, long primers hybridize at a slower rate. The efficiency of the PCR is therefore reduced if the primers are too long. This is because the primers cannot complete hybridization to the template molecules in the time allowed during the reaction cycle, and also the annealing temperature increases as the length increases. Therefore, the length of primers is usually below 30 nucleotides.

Annealing Temperature

Each primer must be complementary to its template strand in order for hybridization to occur, and the 3' ends of the hybridized primers should point toward one another (Fig. 9.9).

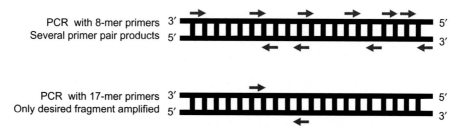

FIG. 9.8 Impact of primer length on the specificity of PCR.

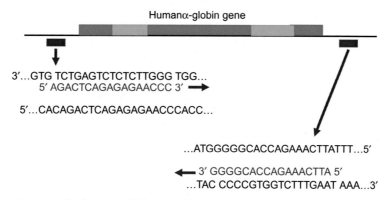

FIG. 9.9 An example of the primers targeting human α-globin gene with exons in *blue* and introns in *green*.

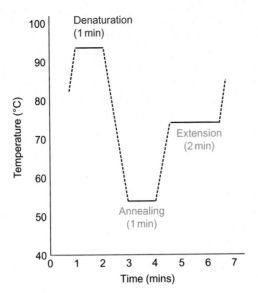

FIG. 9.10   Temperature profile for PCR stages.

For each cycle of a PCR reaction, the PCR reaction mixture is incubated between three temperatures (Fig. 9.10):

1. The denaturation temperature, usually 94°C, which breaks the hydrogen bonds between base pairs and releases single-stranded DNA to act as templates for the next round of DNA synthesis;
2. The hybridization or annealing temperature, usually between 50°C and 60°C, at which the primers bind to the templates;
3. The extension temperature, at which DNA synthesis occurs. This is usually set at between 72°C and 74°C, just below the optimum for *Taq* polymerase.

Both denaturation and extension temperature are usually constant in any PCR reaction. However, the annealing temperature is dependent on the primers sequences and is the critical factor for a successful PCR reaction. This is because the primer sequences determine the annealing temperature and this temperature can affect the specificity of the PCR reaction. If the annealing temperature is too high, no hybridization takes place, and this is high-stringency condition. Therefore, under this condition, the primers and templates remain dissociated (Fig. 9.11). On the other hand, if the annealing temperature is too low, mismatched hybrids are produced. Under this low-stringency condition, the number of potential hybridization sites for each primer is greatly increased, and amplification is more likely to occur at nontarget sites in the template molecule.

The ideal annealing temperature must be low enough to enable hybridization between primer and template, but high enough to prevent mismatched hybrids from forming. Because the annealing temperature determines the hybridization

FIG. 9.11   The effect of temperature on primer annealing in PCR.

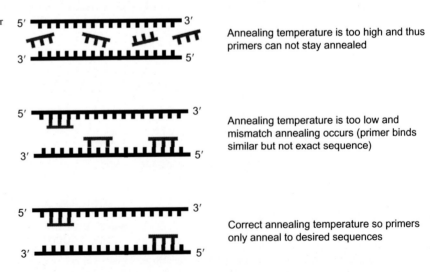

Annealing temperature is too high and thus primers can not stay annealed

Annealing temperature is too low and mismatch annealing occurs (primer binds similar but not exact sequence)

Correct annealing temperature so primers only anneal to desired sequences between primers and the templates, this temperature is determined by estimating the melting temperature (T_m) of the primer-template hybrid. When a DNA-containing solution is heated enough, the noncovalent forces that hold the two DNA strands together weaken and finally break. When this happens, the two DNA strands come apart in a process known as DNA denaturation, or DNA melting. The temperature at which the DNA strands are half correctly base-paired hybrid dissociates/denatured is called the T_m. Normally, the annealing temperature is 1–2°C below T_m in a PCR reaction and this temperature should be low enough to allow the correct primer-template hybrid to form, but too high for a hybrid with a single mismatch to be stable.

The T_m can be determined experimentally but is more usually calculated from the formula

$$T_m = 81.5°C + 16.6(\log M) + 0.41(\%GC) - 0.61(\%form) - 500/L$$

in which M is the concentration of potassium chloride in the reaction, which is around 25–50 mM; %GC is the percentage of G and C nucleotides in the primer sequence; %form is the percentage of the formaldehyde in the reaction; and L is the length of the primer sequence. Alternatively, the T_m can be estimated by the simple formula

$$T_m = (4 \times [G+C]) + (2 \times [A+T])°C$$

in which [G + C] is the number of G and C nucleotides in the primer sequence, and [A + T] is the number of A and T nucleotides.

Because T_m is determined by the primer sequence, it is a normal practice that the two primers in a PCR reaction should be designed to have similar T_m. The GC content of a DNA has a significant effect on its T_m: the higher a DNA's GC content, the higher its T_m. This is because G-C pairs form three hydrogen bonds, whereas A-T pairs have only two. Two strands of DNA rich in G and C will hold to each other more tightly than those of AT-rich DNA. To have similar T_m, it is therefore critical to have similar G+C contents in the primer sequence.

QUANTITATIVE PCR

Real-time PCR or quantitative PCR (qPCR) is the technology for the quantification of nucleic acids. This method is a way of quantifying the amplification of a DNA as it occurs, which means it detects the accumulation of amplicon after each thermal cycle in real time. "Real time" implies that data collection and analysis occur as a reaction proceeds. The initial concentration of target sequence can be determined by a comparison with the amplification kinetics between the target and an internal control with known copy number (Fig. 9.12). Thus, qPCR can be used in gene expression analysis and in routine DNA quantification, and is becoming very popular in research and diagnostics of various fields. Because of its prominence, it is replacing conventional PCR. For example, several new qPCR detection chemistries have been developed for the quantification and genotyping of pathogens, gene expression, methylated DNA and microRNA analysis, validation of microarray data, allelic discrimination and genotyping (detection of mutations, analysis of single-nucleotide polymorphisms (SNPs) and microsatellites, identification of chromosomal alterations), validation of drug therapy efficacy, forensic studies, and quantification of genetically modified organisms.

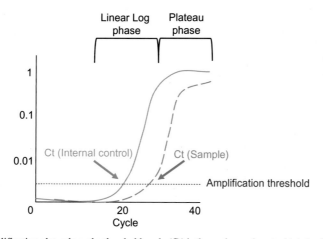

FIG. 9.12 An example of PCR amplification plot, where the threshold cycle (Ct) is the cycle number at which the fluorescence crosses the amplification threshold. The initial concentration of target sequence can be determined by a comparison with the Ct between the target and an internal control with known copy number. The number of PCR cycles required to exceed the background fluorescence is called the Ct.

The Comparison Between qPCR and Conventional PCR

The main advantage of qPCR over the traditional PCR assays is that the starting DNA concentration can be determined with accuracy and high sensitivity. Thus, the obtained results from a qPCR can be either qualitative (showing the presence or absence of the DNA sequence of interest) or quantitative (showing the starting quantity of the DNA sequence of interest). In contrast, conventional PCR is, at best, semiquantitative. This is because conventional PCR can only evaluate the end product of a PCR reaction, which can be qualitatively but not quantitatively accurate. Therefore, the quantity of the starting materials will not be distinguished as long as they reached the maximum end-product quantity. Moreover, the amplification reactions are run and data are analyzed in a real-time fashion in qPCR, eliminating the need for postamplification manipulation and therefore reducing opportunities for contamination.

Real-Time PCR Instruments

Basically, the qPCR instrument consists of a thermal cycler with an integrated excitation light source (a lamp, laser, or LED [light-emitting diode]), a fluorescence detection system or fluorimeter, and software that displays the recorded fluorescence data as a DNA amplification curve. Since all real-time PCR instruments monitor sample fluorescence during thermal cycling, it is necessary to add a dye- or fluorophore-labeled probe to the reaction mixture.

Real-Time PCR Detection Chemistry

In real-time PCR, the amplification reaction is monitored in "real time." The number of PCR cycles required to exceed the background fluorescence is called the cycle threshold (C_T) (Fig. 9.12). In the linear part of the PCR amplification, the C_T is inversely correlated with the log of the number of nucleic acid target sequence copies initially present in the sample, and this approach can thus be used in a quantitative manner (Fig. 9.13).

There are mainly two types of DNA analysis in qPCR. One is the ds DNA binding dye chemistry for both specific and nonspecific detection of amplified products, and the other is the fluorophore-linked probes for specific PCR product detection. For the nonspecific detection, DNA binding dyes are simply added to the PCR mix, and the level of amplification could be monitored by observing the florescent properties of the reaction vessel. Because intercalating dyes bind to double-stranded DNA (dsDNA), they give an increase in fluorescence proportional to the amount of dsDNA present. When using intercalating dyes, the observed fluorescence may originate from the specific and/or the unspecific products. This is because these dyes bind to all dsDNA present in the reaction. To avoid the problems associated with nonspecific PCR amplification products being detected, one or more fluorescent probes can be added to the reaction mix. The probe(s) are designed to be complementary to a sequence motif in the target amplicon, and a PCR product will only be detected if the amplified sequence is complementary to the annealing probe.

Double-stranded DNA binding dyes have the advantage of being compatible with multiple primer sets and are often substantially cheaper than acquiring specific probes for an equivalent number of gene targets. However, the specificity required for accurate quantification is dependent solely on the sequence-specific primers. Sequence-specific probes provide greater specificity, while also allowing multiple gene targets to be analyzed in the same reaction if different fluorescent reporters are attached to each probe.

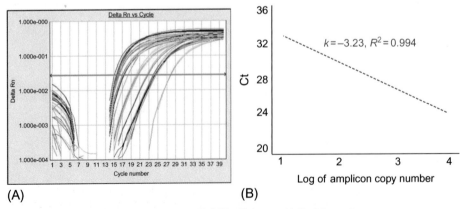

FIG. 9.13   (A) A typical example of qPCR amplification curve. (B) PCR efficiency with TaqMan probe.

DNA INTERCALATING DYES FOR qPCR

The first available fluorescent DNA dyes for real-time PCR is ethidium bromide which were used by simply adding ethidium bromide to the PCR mix and monitoring the amplification via the fluorescent properties of the reaction volume. SYBR Green, an asymmetrical cyanine dye (2-[N-(3-dimethylaminopropyl)-N-propylamino]-4-[2,3-dihydro-3-methyl-(benzo-1,3-thiazol-2-yl)-methylidene]-1-phenylquinolinium), is another intercalating dye with higher sensitivity than ethidium bromide. It has two positive charges under standard PCR reaction conditions contributing to its high dsDNA binding affinity. The resulting DNA-dye complex absorbs blue light ($\lambda_{max} = 497$ nm) and emits green light ($\lambda_{max} = 520$ nm). Therefore, the dye's excitation and emission spectra match well with the optical settings in most instruments and are largely compatible with the chemistry of PCR.

Principles of the Intercalating Dyes

SYBR Green and the third-generation DNA binding dye EvaGreen, like all other intercalating dyes, bind any double-stranded DNA (Fig. 9.14). When such a dye binds to the minor groove of dsDNA, its fluorescence is increased and can be measured in the extension phase of each cycle of qPCR. Because of the nonspecific binding, nonspecific products and primer-dimers can be formed during the PCR process. To distinguish the difference, one has to assume that different PCR products will have different melting temperatures to separate nonspecific from specific amplicons.

It is usually a standard procedure that a melting (dissociation) curve analysis is performed to check the specificity of the amplified fragments. This analysis consists of applying heat to the sample (from 50°C to 95°C) and monitoring the fluorescence emission during the process. The temperature of DNA denaturation is shown as a sharp drop in the fluorescence signal due to dissociation of the dye (Fig. 9.15). Nonspecific products and primer-dimers are denatured at lower temperatures than the specific products. Instead of one peak, there are usually two peaks in a melting curve analysis. Therefore, monitoring the fluorescence properties of the PCR amplification products during a melting phase can help to ensure that

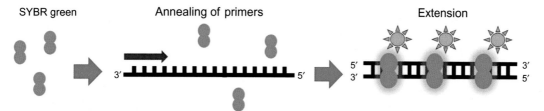

FIG. 9.14   SYBR *green* fluorescence mechanism.

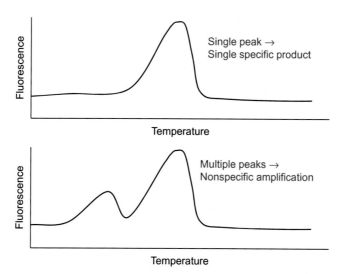

FIG. 9.15   Example of how melting curves can demonstrate the specificity of a qPCR reaction.

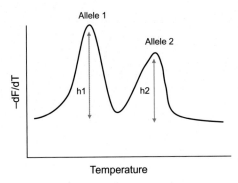

FIG. 9.16 Principle of high resolution melting curve analysis for detection of allelic amplification.

nonspecific amplification has not taken place. The DNA intercalating dyes can be used to detect either single or two or more different DNA sequences in a single PCR reaction (multiplex assays).

Comparison Between SYBR Green and EvaGreen

The costs of employing DNA binding dyes in qPCR are much lower than those of methods requiring fluorescent probes described in the next section. Despite its popularity, SYBR Green presents some limitations, including limited dye stability and dye-dependent PCR inhibition. For example, one of the problems is that only a relatively low SYBR Green concentration (e.g., <0.5 μM) can be used in the reaction. This is due to the fact that high concentration of SYBR Green can inhibit PCR and promote mispriming. Low SYBR Green concentration not only may compromise PCR signal strength but also may make DNA melt curve data unreliable due to the so-called dye redistribution phenomenon. The unreliability is the main reason why SYBR Green is unsuitable for high-resolution melt curve analysis (HRM), which is now gaining popularity because of the high specificity it offers and its potential to be used in diagnostic applications such as allelic detection (Fig. 9.16). In addition, SYBR Green is relatively unstable, especially under the alkaline conditions of Tris buffer during storage. Finally, although SYBR Green is only weakly mutagenic by itself, it has been reported to be a strong mutation enhancer by possibly impairing the natural DNA repair mechanism in cells.

On the other hand, EvaGreen offers several advantages, such as being less inhibitory to PCR than SYBR Green, and it can be used under saturating conditions to generate greater fluorescent signals (Fig. 9.17). The relatively low PCR inhibition of EvaGreen can be exploited for several benefits. First, the dye may be used at a relatively high concentration (i.e., 1.34 μM), which should provide a robust PCR signal as well as a strong and sharp DNA melt peak. Second, the use of a relatively high dye concentration in qPCR eliminates dye redistribution problem, which is a major problem for SYBR Green during post-PCR DNA melting curve analysis. This makes EvaGreen suitable for HRM application in a closed-tube format. As a result, EvaGreen is more suitable for multiplex real-time PCR. EvaGreen-based multiplex real-time PCR assays provided better peak resolution. Third, EvaGreen may be suitable for qPCR using a fast cycling protocol. Finally, EvaGreen is spectrally compatible with existing instruments and very stable.

FLUOROPHORE-LABELED OLIGONUCLEOTIDES FOR qPCR

The Principle of Fluorophore-Labeled Oligonucleotides

Primer-based chemistry assays are more reliable than those based on intercalating dyes. In primer-based chemistry, a dye does not have to be added directly to the reaction mixture because a fluorescently labeled PCR primer is used instead. These fluorescently labeled PCR primers are called fluorophores. They are small fluorescent molecules that are attached to oligonucleotides in order to function as probes in qPCR technology. These assays are fairly straightforward to design.

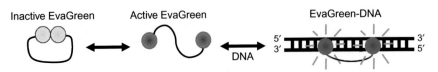

FIG. 9.17 EvaGreen dye's "release-on-demand" mechanism of binding dsDNA.

Types of Fluorophores Used in qPCR

There are two types of fluorophores: donor/reporter and acceptor/quencher. When a donor fluorophore absorbs energy from light, it rises to an excited state. The process of returning to the ground state is driven by the emission of energy as fluorescence. This emitted light from the donor has a lower energy and frequency and a longer wavelength than the absorbed light and can be transferred to an acceptor fluorophore. If both fluorophores are within a specific distance, usually 10 to 100 Å, the transfer of excited-state energy from a reporter to a quencher is denoted as fluorescence resonance energy transfer (FRET). When the donor is separated from the quencher, it can lead to a fluorescent signal proportional to the amount of amplified PCR product (Fig. 9.18). Fluorescent oligonucleotides are usually classified as one of (i) primer-probes, (ii) probes, or (iii) analogues of nucleic acids. This depends what kind of oligonucleotides the fluorophores are attached to.

Fluorophore-Labeled Oligonucleotides: Primer-Probe Chemistries

Primer-probes are oligonucleotides that combine a primer and probe in a single molecule. As such, a dye does not have to be added directly to the reaction mixture. These assays are relatively inexpensive to run and also fairly straightforward to design. Fluorescence emitted from primer-probes is detected and measured during the denaturation or extension phase of the qPCR, depending on the type of primer-probe used. The use of these primer-probes can sometimes lead to amplification of unspecific products or dimer-primers during the PCR reaction. Therefore, melting curve analysis to determine the efficiency of the reaction is always recommended for this application. Primer-probes can be classified into three groups: hairpin, cyclicon, and Angler primer-probes.

Hairpin Primer-Probes

Hairpin primer-probes are single-stranded oligonucleotides with four features. One feature is a hairpin secondary structure, in which the loop of the structure specifically binds to the target DNA. This probe contains a short tail sequence of six nucleotides (CG) at the 5'-end of the probe complementary to the 3'-end region. Hairpin primer-probes include Scorpions, Amplifluor, and Light-Upon-eXtension (LUX).

In the Scorpion primer-probes, the hairpin structure has a reporter at the 5'-end and an internal quencher at the 3'-end. The 3'-end of the hairpin is attached to the 5'-end of the primer by a HEG (hexaethylene glycol) blocker, which prevents primer extension by the polymerase (Fig. 9.19). In solution, the reporter and quencher are in close proximity, and energy transfer via FRET-quenching is produced. After binding of the primer-probe to the target DNA, the polymerase copies the sequence of nucleotides from the 3'-end of the primer. In the next denaturation step, the specific sequence of the probe binds to the complementary region within the same strand of newly amplified DNA. This hybridization opens the hairpin structure. As a result, the reporter is separated from the quencher, and this separation leads to a fluorescent signal proportional to the amount of amplified PCR product. Therefore, the Scorpion primer-probe combines the binding and detection mechanisms in the same molecule, making it an inexpensive system. In Scorpion applications, the hairpin structure prevents the formation of primer-dimers and nonspecific PCR amplification products. The use of stems can minimize background signals as the unincorporated primer-probes are switched off. Scorpion primer-probes can be used in single and multiplex formats for pathogen detection, viral/bacterial load quantitation, genotyping, Single Nucleotide Polymorphism (SNP) allelic discrimination, and mutation detection. Sometimes the addition of a nucleic acid analogue to a Scorpion primer-probe containing reaction can provide greater accuracy in SNP detection and allele discrimination.

The reporter of Amplifluor primer-probes is located at the 5'-end and the internal quencher is linked at the 3'-end of the hairpin. The 3'-end acts as a PCR primer (Fig. 9.20). When the primer-probe is not bound to the templates, the hairpin structure is intact and the reporter transfers energy to the quencher via FRET quenching. DNA amplification occurs after binding of the primer-probe to the target sequence. In the next step of denaturation, reporter and quencher are separated and, as a result, the emitted fluorescence of the donor is measured by the fluorimeter. Thus, the principle of the Amplifluor primer-probe is very similar to that of the Scorpion primer-probe. The use of this primer-probe has the same advantage as the Scorpion primer-probe. Amplifluor primer-probes can be used in single and multiplex formats for pathogen detection, viral/bacterial load quantitation, genotyping, allelic discrimination, mutation detection, SNP detection, and genetically modified organism (GMO) detection.

In the LUX technology, one of the primers is labeled with a fluorophore close to the 3'-end that is quenched by the hairpin structure of the primer. The 3'-end contains a single reporter located in the guanosine-rich region of the primary

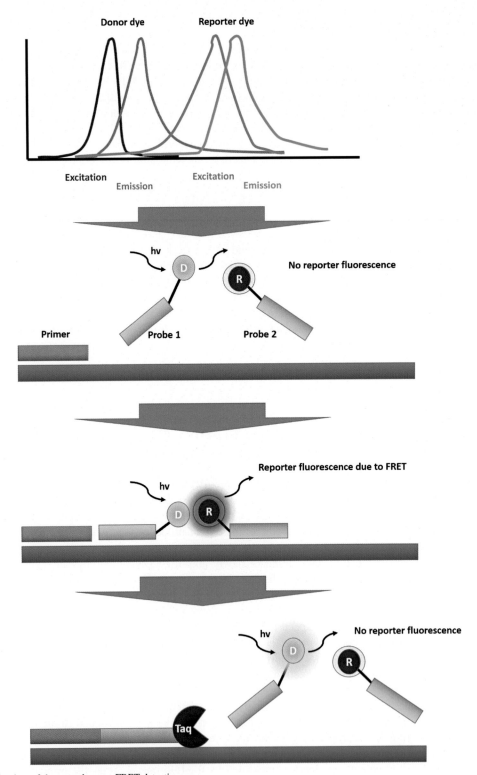

FIG. 9.18   Mechanism of donor and report FRET detection.

sequence. The LUX primer-probes do not require the presence of an internal quencher (Fig. 9.21). The hairpin structure confers the ability to decrease the fluorescence signal when the primer-probe is free. On integration of labeled primer into a PCR product, its fluorescence increases up to eightfold because of extension of the hairpin structure. Fluorescence is measured during the extension phase. The LUX primer-probes can be used in single and multiplex formats for pathogen

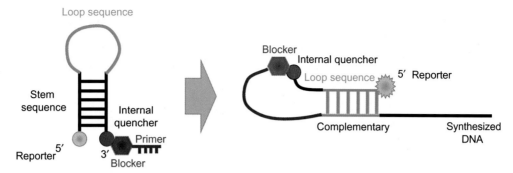

FIG. 9.19 Scorpion primer-probe system.

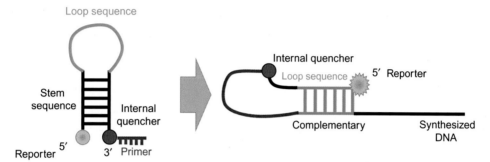

FIG. 9.20 Amplifluor primer-probe system.

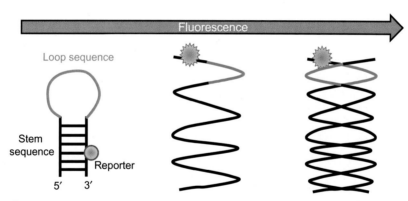

FIG. 9.21 LUX primer-probe system.

detection, viral/bacterial load quantitation, genotyping, allelic discrimination, mutation detection, SNP detection and gene expression analysis, and GMO detection.

Cyclicon Primer-Probes

Cyclicons contain a long primer-probe (complementary to the target DNA sequence) and a short modified oligo attached through 5'-5'ends, which binds to six to eight nucleotides at the reporter-containing modified 3'-end of the primer-probe. A quencher is placed on a thymine base at the 5'-position in the primer-probe sequence. This binding forms a cyclic structure with two 3'-ends (Fig. 9.22). In the absence of the target sequence, reporter and quencher molecules are in close proximity and energy transfer occurs via FRET quenching. The binding of cyclicon probes to template DNA opens up the cyclic structure and leads to extension of the 3'-end primer-probe by DNA polymerase without any interference from the quencher. The 3'-end of the modified oligo does not bind to the target DNA, and its 3'-end is blocked by a reporter.

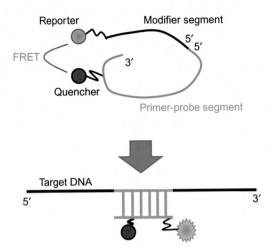

FIG. 9.22 Cyclicons primer-probe system.

Thus, the 3′-end of the modified oligo is not extendible. The separation between donor and acceptor molecules results in emission of fluorescence, which is measured during the extension phase.

The integrated primer-probe structure of cyclicons is an important benefit for DNA detection in qPCR systems. It allows the use of shorter oligonucleotides, reducing the costs of the assay, simplifies the reaction set up and avoids unnecessary carry-over contaminations. On the other hand, the linkage between the long primer-probe and the short oligo can also be through their 3′-3′ ends. In this case, cyclicons would function as probes similar to TaqMan probes and Molecular Beacons (described in the next section). Cyclicons can be used in single and multiplex qPCR for pathogen detection, viral/bacterial load quantitation, genotyping, allelic discrimination, mutation detection, and SNP detection. Cyclicons can also be directly fixed to solid supports on chips for high-throughput screening in solid-phase PCR.

Angler Primer-Probes

Angler primer-probes contain a DNA sequence identical to that of the target, which is bound to a reverse primer through a hex-ethylene glycol (HEG) linker. It has an acceptor fluorescent moiety at its 5′-end. SYBR Gold DNA intercalating dye is employed in the assay as the donor fluorescent moiety (Fig. 9.23). In solution, the primer-probe does not emit fluorescence because there is no donor fluorescent moiety close enough for FRET. When the Angler primer-probe binds to its target DNA during the annealing step, DNA polymerase starts the extension of the 3′-end reverse primer. Subsequently, during the denaturation phase, the specific sequence of the probe binds to the complementary region of newly amplified DNA, producing a dsDNA fragment in which SYBR Gold dye can be intercalated to generate fluorescence. Therefore, the emitted fluorescence is measured during the denaturation step in each cycle.

The combination of a dsDNA intercalating agent and a primer-probe in real-time PCR allows nonspecific (SYBR Gold) and specific (Angler primer-probe) amplified products to be distinguished without performing melting curves. The fluorescent signal from nonspecific intercalation of the SYBR Gold dye is different from the signal generated by the specific binding of the Angler primer-probe. The Angler primer-probe can be used in single or multiplex formats for rapid detection

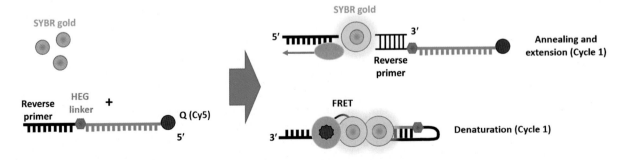

FIG. 9.23 Angler primer-probe system.

of DNA, in studies of gene expression, allelic discrimination, genotyping, SNP detection, identification and quantitation of infectious organisms, and screening of environmental and biological samples.

FLUOROPHORE-LABELED OLIGONUCLEOTIDES: PROBES-BASED CHEMISTRY

An extra fluorescence probe, which is complementary to a target sequence lying between the PCR primers, can be used to increase PCR sepcificity. Each probe has a reporter fluorophore covalently attached to one end and a quencher attached to the other. As long as both dyes stay in close proximity, the signal is quenched. The signal is released only when dyes become physically separated. There are two types of probes: hydrolysis and hybridization probes. The original TaqMan probe design is a typical hydrolysis probe. Because of their increased melting temperatures, these probes can be designed to be shorter and have higher hybridization specificity. Molecular beacons, on the other hand, are hybridization probes. They consist of a sequence-specific loop region flanked by two inverted repeats that form a hairpin structure. To bind to a complementary target sequence, the beacon unfolds, leading to separation of the fluorophore from the quencher and an increase in fluorescence.

Hydrolysis Probes

For the hydrolysis probes, the mechanism relies on the $5' \rightarrow 3'$ exonuclease activity of *Taq* polymerase degrades the bound probe during amplification. This also prevents performing a melting curve analysis. In this system, the fluorescence is measured at the end of the extension phase and is proportional to the amount of amplified specific product.

TaqMan probes are hydrolysis probes and are oligonucleotides containing a donor fluorescent moiety at the 5'-end and an acceptor fluorescent moiety at the 3'-end that quenches the fluorescence emitted from the donor molecule due to their close proximity. The TaqMan probe is designed to bind to a specific region of the target DNA and is located between two PCR primers (Fig. 9.24). In solution, the fluorescent signal from the donor fluorophore is suppressed by the acceptor fluorophore. During the extension phase, following the hybridization of the TaqMan probe to its complementary sequence within the PCR target, the bound hydrolysis probe is degraded by the $5' \rightarrow 3'$-exonuclease activity of DNA polymerase. Therefore, the reporter dye is separated from the quencher, and its fluorescence intensity increases. This process is repeated in each cycle without interfering with the exponential synthesis of the PCR products. The design and synthesis of TaqMan probes are easy. However, if TaqMan probes are not well designed, primer-dimers might be formed during PCR reaction. They can be used in single and multiplex formats for virus detection, viral/bacterial load quantitation, gene expression, microarray validation, allelic discrimination, mutation detection, SNP detection, and GMO detection.

The original TaqMan probe can be extended to include ligand such as minor groove binding (MGB) to improve target DNA-binding specificity and sensitivity. MGB ligand is conjugated to TaqMan probes through either 3' or 5' ends, and MGB ligands are small molecule tripeptides that form a noncovalent union with the minor groove of dsDNA. This type of ligand selectively binds to AT-rich sequences, favoring the inclusion of aromatic rings by van der Waals and electrostatic interactions. This interaction produces minimal distortion in the phosphodiester backbone but greatly stabilizes the DNA structure. As shown in Fig. 9.25, FRET quenching occurs when the random coiling form of the MGB-conjugated probe brings the nonfluorescent quencher and the fluorophore reporter together. The probe is straightened out when it binds to its target, causing an increase in the fluorescent signal. The highly stable interaction between the MGB probe and the target increases the T_m of the probe and prevents the amplification of nonspecific products. The MGB probe can be used in single and multiplex formats for pathogen detection, viral/bacterial load quantitation, gene expression, microarray validation, allelic discrimination, mutation detection, SNP detection, GMO detection, and forensic analysis.

FIG. 9.24 Taqman probes system.

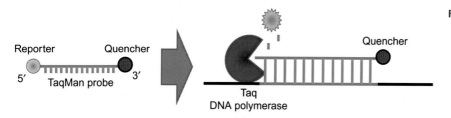

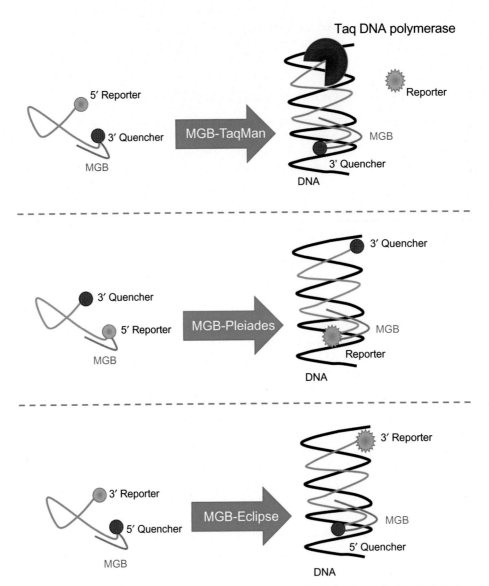

FIG. 9.25 The binding of MGB-conjugated probes to the target DNA. Three probes including MGB-TaqMan hydrolysis probe, MGB-Pleiades hybridization probe and MGB-Eclipse hybridization probe are shown.

Hybridization Probes

The fluorescence emitted by binding hybridization probes can be measured during either the annealing or the extension phase. The use of these probes allows amplified fragments to be analyzed by performing melting curves, and this is the main advantage over hydrolysis probes. The amount of fluorescent signal detected is directly proportional to the amount of the target amplified during the qPCR reaction. The typical hybridization probes are Hybprobe and Molecular Beacon probes.

Hybprobe, also known as FRET probe, consists of a pair of oligonucleotides binding to adjacent target DNA sequences. One probe carries a reporter fluorophore at its 3′-end and the other probe contains a quencher at its 5′-end and a phosphate group attached to its 3′-end to prevent DNA amplification (Fig. 9.26). The sequences of the probes are designed to hybridize to the target DNA sequences in a head-to-tail orientation so that the two fluorophores are in close proximity. During the annealing stage, in which the probes are adjacently bound, the quencher emits fluorescence because it has been previously excited by the energy released from the reporter (Fig. 9.26). The advantage of this probe is that their design and synthesis, as well as the optimization of the PCR reaction conditions, is quick and easy. They can be used in the multiplex

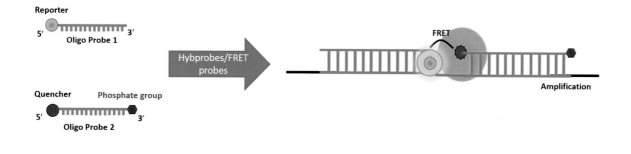

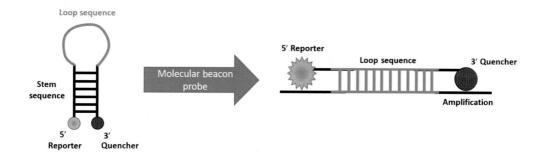

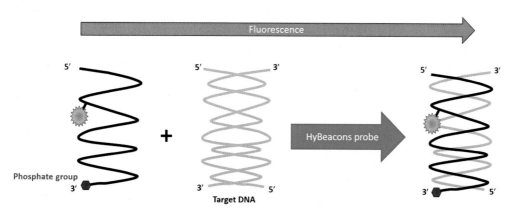

FIG. 9.26 The FRET probe system.

format for pathogen detection, viral/bacterial load quantitation, microarray validation, genotyping, allelic discrimination, mutation detection, and SNP detection.

Molecular Beacon probes are single-stranded hairpin shaped oligonucleotide probes. The probes are usually divided into two parts: (i) a loop, a fragment of 18–30 bp complementary to the target DNA sequence; and (ii) a stem, which is formed by two complementary sequences of 5–7 bp located at each end of the probe. The 5′-end of the probe has a fluorescent reporter, and a nonfluorescent quencher is attached to the 3′-end, which absorbs the emitted fluorescence from the reporter when the Molecular Beacon probe is in closed form (Fig. 9.26). During the annealing phase, this probe unfolds and binds to the target. Fluorescence is emitted because the reporter is no longer quenched. This fluorescent signal is proportional to the amount of amplified PCR product. If the Molecular Beacon probe and target DNA sequences are not perfectly complementary, there will be no emission of fluorescence because the hairpin structure prevails over the hybridization. The binding specificity of Molecular Beacon probes is higher than that of fluorescent oligonucleotides because they are able to form a hairpin stem. Therefore, the use of such probes allows discrimination between target DNA sequences that differ in a single nucleotide. They can be used in single and multiplex formats for pathogen detection, viral/bacterial load quantitation, genotyping, allelic discrimination, mutation detection, SNP detection, mRNA analysis in living cells, and GMO detection.

FLUOROPHORE-LABELED OLIGONUCLEOTIDES: NUCLEIC ACID ANALOGUES

Nucleic acid analogues are compounds that are analogous (structurally similar) to naturally occurring RNA and DNA. An analogue may have alterations in its phosphate backbone, pentose sugar (either ribose or deoxyribose), or nitrogenous bases. Normally, the analogues incorporate all of the advantages of native DNA but are more stable in biological fluids and have increased affinity for complementary nucleic acid targets. Some typical analogues are peptide nucleic acids (PNAs), locked nucleic acids (LNAs), Zip Nucleic Acids (ZNAs), and nonnatural bases: isoguanine (iso-dG) and 5′-methylisocytosine (iso-dC).

PNAs

The sugar-phosphate backbone of PNAs has been replaced by a peptide of N-(2-aminoethyl)-glycine units linked to the nitrogenous bases by methylene carbonyl, so PNAs are electrically neutral DNA analogues. PNA hybridizes to complementary oligonucleotides as the natural DNA obeying the Watson-Crick rules (Fig. 9.27A). PNAs can interact with either dsDNA or RNA with higher affinity and greater specificity than conventional oligonucleotides. This binding takes place by strand displacement rather than by triple helix formation. PNAs can be attached to a molecule of thiazole orange or a fluorophore for qPCR reactions, and they can also be introduced to primer-probes or probes and follow the method of action of conventional probes. PNA-containing probes are more resistant to nucleases and proteases and can interact with DNA at lower salt concentration than standard probes/primer-probes.

LNA

LNAs are DNA or RNA sequences in A conformation that contain one or more modified nucleotides. More specifically, a bicyclic ring is formed through a methylene bridge between atoms 2′-O and 4′-C in the ribose ring (Fig. 9.27B). LNA nucleotides can be introduced into primer-probes and probes such as Molecular Beacon and TaqMan probes. LNA-containing primer-probes or probes exhibit the same mode of action as that of conventional primer-probes or probes. Like the PNA system, LNA probes are resistant to degradation by nucleases. LNA nucleotides are often used in combination with nonmodified DNA/RNA nucleotides to increase the thermal stability of the probe, resulting in a high specificity for their target sequences.

ZNA

ZNAs are a novel type of synthetic modified oligonucleotide through conjugating cationic moieties (Z units), such as derivatives of spermine, to an oligonucleotide (Fig. 9.27C). The modification can increase their affinity for the target. during the annealing phase, ZNA oligonucleotides are attracted toward the nucleic acid strands because of their polycationic nature, starting their scanning of DNA sequences. Next, hybridization takes place by zipping up when the ZNA oligonucleotide finds its complementary sequence (Fig. 9.27D). ZNA oligonucleotides display an exceptionally high affinity for their targets, mainly due to the presence of the Z units, which enhance the interaction with the DNA target.

Nonnatural Bases: Plexor Primers

Two modified bases, isoguanine (Iso-dG) and 5′-methylisocytosine (Iso-dC), which generate novel base pairings, have been successfully designed to allow protein recognition and site-specific enzymatic incorporation. In Plexor reactions, one PCR primer contains an Iso-dC residue and a fluorescent reporter label at the 5′-end, whereas the second one is an unlabeled oligonucleotide that carries standard nucleotides. In this system, Iso-dG nucleotides, covalently coupled to a quencher, are added into the qPCR reaction (Fig. 9.28). During the amplification phase, the incorporation of Iso-dG nucleotides brings the quencher and reporter into close proximity, producing the quenching of the fluorescent signal released from the labeled primer. In this system, the decrease in initial fluorescence is proportional to the starting amount of target.

ISOTHERMAL AMPLIFICATION TECHNIQUES

The limitation of PCR application in molecular diagnosis is the requirement of a thermocycler to separate the two DNA strands. Thermal cyclers are usually costly, and not portable. Therefore, there is a growing demand for developing a similar method with no requirement for thermocycling that could be used in point-of-care testing. A wide variety of isothermal

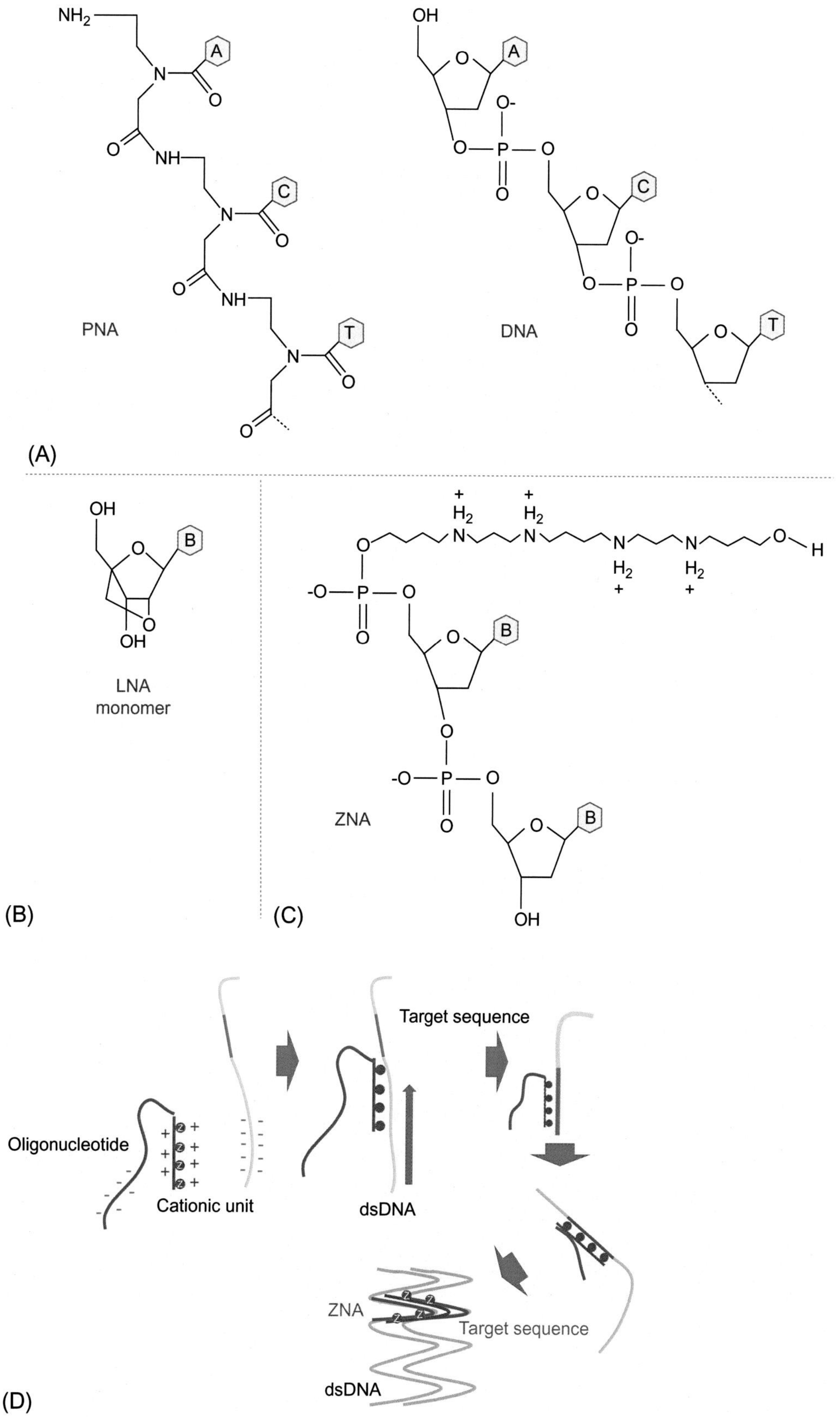

FIG. 9.27 Structure of nucleic acid analogues. (A) PNA; (B) LNA; (C) ZNA; (D) ZNA hybridization process.

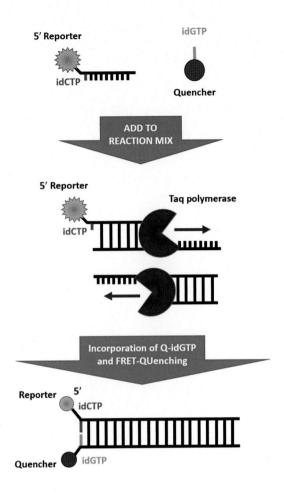

FIG. 9.28 Plexor probe system.

amplification techniques have been developed without using thermocyclers. These non-PCR-based methods have been developed according to some new findings in molecular biology of DNA/RNA synthesis by polymerase and assisting proteins in vivo under isothermal conditions for nucleic acid amplification.

Nucleic Acid Sequence-Based Amplification (NASBA)

Transcription-based amplification includes transcription-mediated amplification (TMA) and nucleic acid sequence-based amplification (NASBA). Both are isothermal amplification reactions mimicking retroviral RNA replication. Both are specific for target RNA sequences and have been gaining popularity because they have been shown to have a wide range of applications for pathogen detection in clinical, environmental, and food samples. Both are also convenient to use with commercially available kits. As an alternative to RT-PCR for RNA amplification, NASBA and TMA have the advantage of not requiring thermal cycling. These two techniques use the function of an RNA polymerase to make RNA from a promoter engineered in the primer region, and a reverse transcriptase, to produce DNA from the RNA templates. For TMA, the reverse transcriptase itself degrades the initial RNA template as it synthesizes its complementary DNA. For NASBA, this RNA amplification technology has been further improved by introducing a third enzymatic activity, RNase H, to remove the RNA from cDNA without the heat-denaturing step (Fig. 9.29). During the reaction, the forward primer hybridizes to any target RNA present in a sample. The enzymes reverse transcriptase and RNase H along with the reverse primer then produce a dsDNA with the target sequence and a T7 promoter. T7 DNA-dependent RNA polymerase uses this dsDNA to produce many RNA strands complementary to the original target RNA. After this initial phase of NASBA, each newly synthesized RNA can be copied in a cyclical phase, resulting in an exponential amplification of RNA complementary to the target. Thus, the thermocycling step has been eliminated, generating an isothermal amplification method named self-sustained sequence replication. The end products of NASBA and TMA can be detected using gel electrophoresis, fluorescent probes, and colorimetric assay.

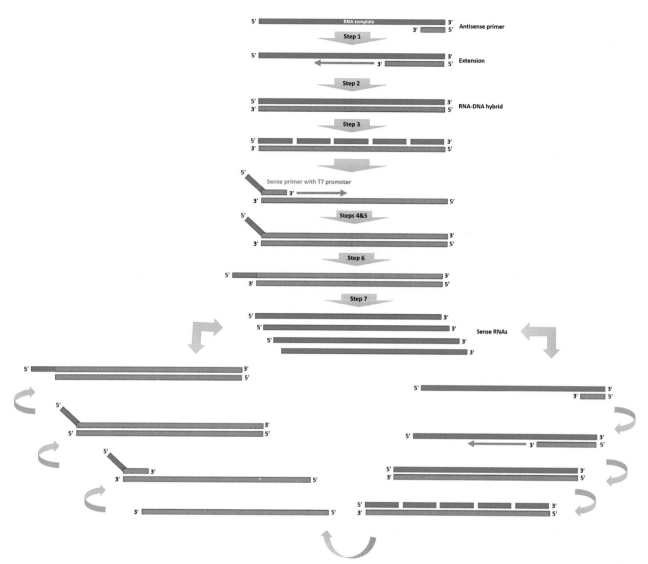

FIG. 9.29 The mechanism of nucleic acid sequence-based amplification (NASBA).

Signal-Mediated Amplification of RNA Technology (SMART)

The SMART technology is based on the formation of a three-way junction (3WJ) structure. It consists of two single-stranded oligonucleotide probes. Each probe includes one region that can hybridize to the target at adjacent positions and another, much shorter, region that hybridizes to the other probe. The two probes are annealed to each other in the presence of the specific target, thus forming a 3WJ (Fig. 9.30). Following 3WJ formation, *Bst* DNA polymerase extends the short (extension) probe by copying the opposing template probe to produce a double stranded T7 RNA polymerase promoter sequence. The formed promoter allows T7 RNA polymerase to generate multiple copies of an RNA amplicon, therefore being produced only when a specific target is present to allow 3WJ formation. Each RNA amplicon may itself be amplified by binding to a second template oligonucleotide (probe for amplification) and is extended by DNA polymerase to generate a double-stranded promoter. This leads to transcription that increases the RNA amplicons. The RNA amplicons can be detected by an enzyme-linked oligosorbent assay or in real-time format. The method actually relies on signal amplification and does not require thermal cycling or involve the copying of target sequences. The assay generates a signal that is highly target dependent and is appropriate for the detection of DNA or RNA targets.

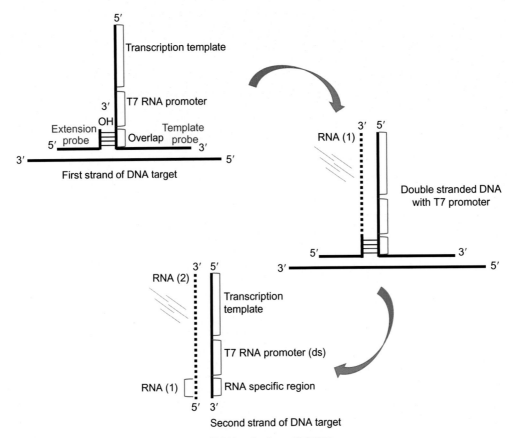

FIG. 9.30 The mechanism of signal mediated amplification of RNA technology (SMART).

Strand Displacement Amplification (SDA)

Strand displacement amplification (SDA) uses the activity of a restriction endonuclease and an exonuclease-deficient strand displacing DNA polymerase to generate copies of a target DNA sequence. SDA combines the ability of a restriction endonuclease to nick the unmodified strand of its target DNA and the action of an exonuclease-deficient DNA polymerase to extend the 3′ end at the nick and displace the downstream DNA strand. The displaced strand serves as a template for an antisense reaction and vice versa, resulting in exponential amplification of the target DNA (Fig. 9.31). After an initial heat denaturation to separate dsDNA, the extension of SDA primers containing restriction sites and their displacement by extension of flanking bumper primers produce double-stranded target sequences flanked by nickable restriction sites. Briefly, four primers (B1, B2, S1, and S2) present in excess and bind the target strands at positions flanking the sequence to be amplified. Primers S1 and S2 have *Hinc*II recognition sequences (5′-GTTGAC-3′) located 5′ to the target complementary sequences. The four primers are simultaneously extended by exo⁻ Klenow (exonuclease-deficient DNA polymerase) using dGTP, dCTP, TTP, and dATP. Extension of B1 displaces the S1 primer extension product, S1-ext. Likewise, extension of B2 displaces S2-ext. B2 and S2 bind to displaced S1-ext. B1 and S1 bind to displaced S2-ext. Extension and displacement reactions on templates S1-ext and S2-ext produce two fragments with a hemiphosphorothioate *Hinc*II at each end and two longer fragments with a hemiphosphorothioate *Hinc*II site at just one end. *Hinc*II nicking and exo⁻ Klenow extension/displacement reactions initiate at these four fragments, automatically entering the SDA reaction cycle. These reaction steps continuously cycle during the course of amplification. Because two SDA primers (S1 and S2) present in excess, the 3′-end of S1 binds to the 3′-end of the displaced target strand T1, forming a duplex with 5′-overhangs. Likewise, S2 binds T2. The 5′-overhangs of S1 and S2 contain the *Hinc*II recognition sequence (5′- GTTGAC-3′). exo⁻ Klenow extends the 3′-ends of the duplexes using dGTP, dCTP, TTP, and dATP, which produces hemiphosphorothioate recognition sites on S1:T1 and S2:T2. *Hinc*II nicks the unmodified primer strands of the hemiphosphorothioate recognition sites, leaving intact the modified complementary strands. exo⁻ Klenow extends the 3′-end at the nick on S1:T1 and displaces the downstream strand that is equivalent to T2. Likewise, extension at the nick on S2:T2 results in

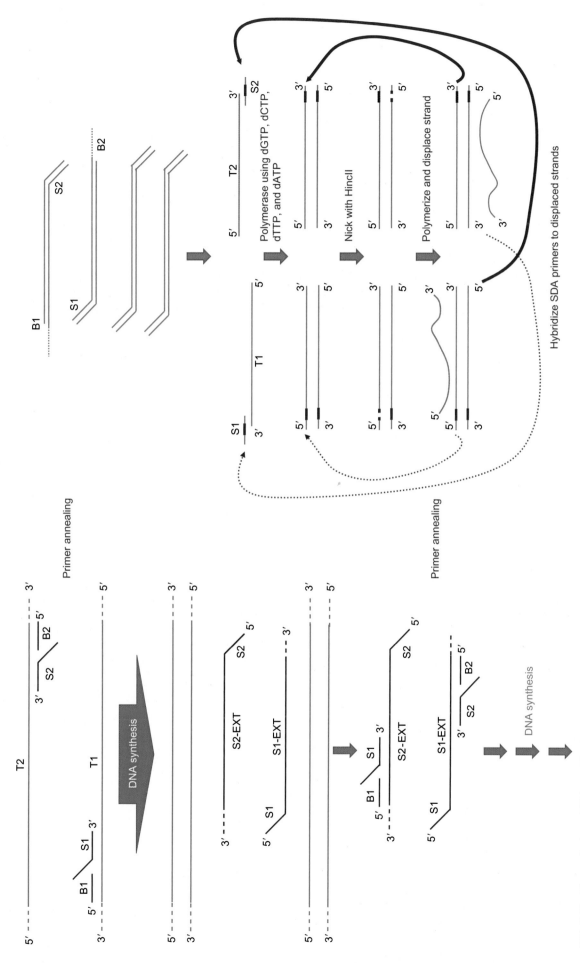

FIG. 9.31 The mechanism of SDA amplification process.

displacement of T1. Nicking and polymerization/displacement steps cycle continuously on S1:T1 and S2:T2 because extension at a nick regenerates a nickable *Hinc*II recognition site. Target amplification is exponential because strands displaced from S1:T1 serve as target for S2 while strands displaced from S2:T2 serve as target for S1. In this approach, each ssDNA product can be copied for an exponential 10^{10}-fold amplification of the target. SDA amplicon can be detected in real time, and the reaction has been used in assays to detect infectious diseases, including a widely used kit for *Chlamydia trachomatis* and *Neisseria gonorrhoeae*.

Rolling Circle Amplification

Rolling circle amplification (RCA) uses a strand-displacing DNA polymerase to continuously amplify a circular DNA template at a constant low temperature, producing a long DNA molecule with tandem repeats of the circular template. The RCA reaction involves numerous rounds of isothermal enzymatic synthesis in which Φ29 DNA polymerase extends a circle-hybridized ssDNA primer by continuously progressing around the circular ssDNA probe of several dozen nucleotides to replicate its sequence over and over again (Fig. 9.32). After completing DNA synthesis in the primer loop, the strand displacement ability continues to synthesize a new strand by displacing the previously formed dsDNA and eventually produces a long ssDNA amplicon with repeated sequence of the original target. This reaction is widely used for diagnostic purposes in direct or indirect detection of different DNA/RNA, protein, and other biomarkers via a set of various bimolecular recognition events.

Loop-Mediated Isothermal Amplification (LAMP)

The loop-mediated isothermal amplification (LAMP) method requires a set of four specifically designed primers that can recognize six distinct regions on the target DNA and a DNA polymerase with strand displacement activity to amplify a target 1 billion times in less than 1 h at a temperature around 60°C. The amplification products are stem-loop DNA structures with several inverted repeats of the target and cauliflower-like structures with multiple loops (Fig. 9.33). By combining processes for simplicity, rapidity, and precision, the LAMP method could be employed in a wide range of applications, such as point-of-care testing, genetic testing in clinical settings, and rapid testing of food products and environmental samples for a variety of pathogens including viruses and bacteria. The LAMP method can also amplify RNA with the addition of reverse transcriptase. The synthesis of large amounts of DNA in a short time creates the by-product pyrophosphate, which is produced during the reaction and forms a white precipitate that can be used to detect positive reactions both visually by observing the precipitate and quantitatively by measuring the turbidity change for real-time analysis.

Several aspects of the LAMP reaction differ from those of other amplification methods. First, only a single type of enzyme is required, and the amplification can be carried out at a constant temperature. Secondly, LAMP uses six gene regions for amplification. The fundamental characteristics of the inner primer provide the amplification with a specificity that is much greater than that observed in other methods. The amplification process is usually completed within 1 h, with sensitivity similar to that of nested PCR. The rapid, sequential progression of the LAMP amplification reaction contributes to its high amplification efficiency. In addition, small quantities of a gene can be amplified within a short time. Furthermore,

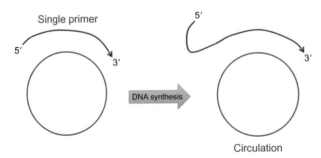

FIG. 9.32 Schematic of rolling circle amplification (RCA) process. The RCA reaction use a free DNA minicircle as the template and a single primer. The reaction is initiated by the hybridization of a linear single stranded DNA to a specific DNA minicircle. Therefore, the amplification products generally exhibit a wide, essentially continuous distribution over length and are normally seen in gel-electrophoresis images as a broad of smear of stranded nature of high-molecular-weight DNAs.

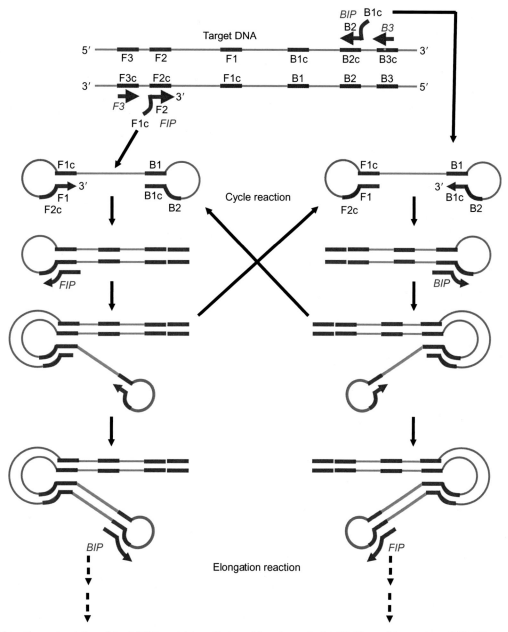

FIG. 9.33 Schematic representation of the LAMP mechanism. Design of the primers for LAMP. Forward inner primer (FIP): F2 sequence with F1c sequence at the 5Gend. Backward inner primer (BIP): B2 sequence with B1c sequence at the 5' end. Outer primers are designed at the regions of F3 and B3. The LAMP method employs a single strand of DNA shaped like a dumbbell with loops at both ends. This starting material is used to initiate a cycle of amplification reactions. DNA having an inverse structure relative to the starting material is produced, and the starting material is formed again by the same reaction. This cycle produces amplified DNA products that are connected to an inverted repeat structure at the amplified region. The amplified products again pass through repeated elongation reactions, which generate amplified DNA products of various stem lengths.

this method can be used to amplify a target RNA sequence. In this case, a one-step amplification—the same as the amplification of DNA—can be performed by the simultaneous addition of a reverse transcriptase enzyme because the reverse transcriptase also exhibits strand displacement activity. Simple detection methods, such as turbidity or fluorescence, can be employed because of the high specificity and large output of the amplification products. The turbidity detection method uses the turbidity of magnesium pyrophosphate, a by-product of DNA synthesis, as an indicator, whereas the fluorescence detection method uses fluorescent chelation reagents. Both methods allow for real-time detection and visual inspection. In addition, detection using normal DNA probes is also possible.

Isothermal Multiple Displacement Amplification

Isothermal multiple displacement amplification (IMDA) is based on strand displacement replication of the nucleic acid sequences by multiple primers. One of the IMDA methods is multiple strand displacement amplification. In this method, two sets of primers are used, a right set and a left set (Fig. 9.34). The primers in the right set are complementary to one strand of the nucleic acid molecule to be amplified, and the primers in the left set are complementary to the opposite strand. The 5′ ends of primers in both sets are distal to the nucleic acid sequence of interest when the primers have hybridized to the nucleic acid sequence molecule to be amplified. Amplification proceeds by replication initiated at each primer and continuing through the nucleic acid sequence of interest. A key feature of this method is the displacement of intervening primers during replication by the polymerase. This method has been further developed for whole-genome amplification by using a random set of primers to randomly prime a sample of genomic nucleic acid. In this way, multiple overlapping copies of the entire genome can be synthesized in a short time.

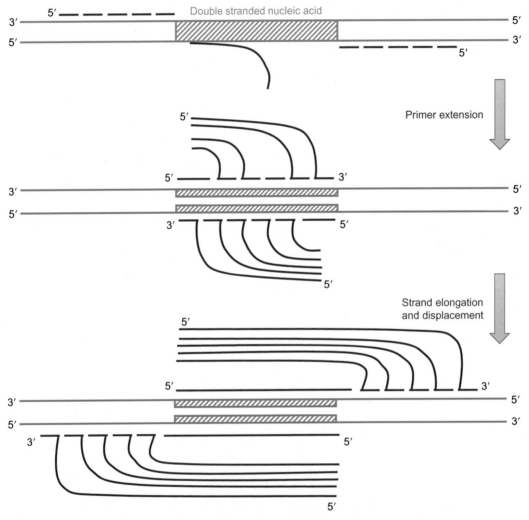

FIG. 9.34 Schematic representation of Isothermal Multiple Displacement Amplification (IMDA) mechanism. This is an example of isothermal multiple strand displacement amplification. At the top is a double-stranded nucleic acid molecule which contains a nucleic acid of interest *(hatched area)*. Hybridized to the nucleic acid molecules are right and left sets of primers. The middle panel is the multiple strands of replicated nucleic acid being elongated from each primer. The polymerase at the end of each elongating strand displaces the elongating strand of the primer ahead of it. The bottom panel is the multiple strands of replicated nucleic acid further elongated. Also shown are the next sets of primers, which hybridize to their complementary sites on the newly replicated strands. The newly replicated strands are made available for hybridization to the primers through displacement by the polymerase elongating the following strand.

Single Primer Isothermal Amplification

Single primer isothermal amplification (SPIA) uses a single chimeric primer for amplification of DNA and RNA. For DNA amplification, it employs a single, target-specific chimeric primer composed of deoxyribonucleotides at the 3' end and ribonucleotides at its 5' end, RNase H, and a DNA polymerase with a strong strand displacement activity. Amplification is initiated by hybridizing the chimeric primer to a complementary sequence in the target DNA molecule. DNA polymerase initiates primer extension of the hybridized primer and extends along the target DNA strand. Following initiation of the primer extension step, the 5' RNA portion of the extended primer (RNA-DNA hybrid) is cleaved by RNase H, thus freeing part of the primer-binding site on the target DNA strand from binding of a new chimeric primer. The newly bound primer competes with the previous primer extension product for binding to the complementary DNA target sequence and is stabilized by binding of DNA polymerase and displaces the 5' end of the previous extension product. As replication is again initiated by primer extension, RNase H cleavage of the 5' RNA portion of the newly extended primer again frees part of the primer binding site for subsequent primer binding, and the replication cycle is repeated. This amplification technique can be used for global genomic DNA amplification and for the amplification of specific genomic sequences and synthetic oligonucleotide DNA targets.

For single primer isothermal amplification for RNA (Ribo-SPIA), it is suitable for global and target-specific RNA amplification (Fig. 9.35). This technology can perform linear, isothermal amplification of the mRNA species in a total RNA population. Replication is initiated and repeated up to 10,000 times off of each original transcript. Therefore, this process can be used for amplification of large populations of nucleic acid species that are limited in biological samples, as commonly encountered in clinical research.

Helicase-Dependent Amplification (HDA)

Helicase-dependent amplification (HDA) is based on the natural mechanism of the DNA replication fork. HDA employs DNA helicase to denature dsDNA instead of conventional heat denaturation step at 95°C in PCR. The separated nucleotides by helicase are protected by single-stranded binding protein, allowing the binding of primers to initiate elongation by polymerase at a constant temperature of 60–65°C without any thermal steps, unlike both SDA and RCA methods, which require an initial heat denaturation step even though the rest of the amplification reaction is isothermal. The newly synthesized dsDNA products are then used as substrate for the next round of the chain reaction, which results in exponential amplification of the target sequence as a function of incubation period. This technique has been used to detect HIV-1 in human plasma and *Clostridium difficile* in fecal samples. The simple reaction, when also combined with other advancements such as the ability to detect RNA with reverse transcription, gives HDA much potential for use in a portable nucleic acid detection system.

In HDA, basically, duplex DNAs normally are unwound by a DNA helicase in the presence of ATP, and the displaced DNA strands are coated by single-stranded binding proteins (SSBs; step 1 in Fig. 9.36). Two sequence-specific primers anneal to the 3'-end of each ssDNA template, and exonuclease-deficient DNA polymerases produce dsDNA by extending the primers annealed to the target DNA (step 2 in Fig. 9.36). Completion of DNA synthesis produces another copy of the dsDNA template (step 3 in Fig. 9.36). The two newly synthesized dsDNAs are used as substrates by the DNA helicase and enter the next round of the reaction (step 4 in Fig. 9.36). Thus, exponential amplification of the selected target DNA sequence is possible by simultaneous chain reactions. This process allows multiple cycles of replication to be performed at a single incubation temperature, eliminating the need for thermocycler. The HDA amplicons can be detected using gel electrophoresis, qPCR, and enzyme-linked immunosorbent assay.

Circular helicase-dependent amplification (cHDA) is used for amplifying nucleic acids from a circular DNA template and is based on the T7 replication machinery without initial heat denaturation (Fig. 9.37). In this system, a circular DNA template is used to amplify a very long DNA fragment by performing a helicase-dependent and strand-displacement reaction with the T7 helicase and 3'→5' exonuclease-deficient T7 DNA polymerase (T7 Sequenase). Using circular DNA as a template, DNA amplification is accomplished through helicase-dependent RCA, producing multiple copies of a specific product defined by two primers and concatemers of circular DNA. In this method, the T7 replisome extends a primer annealed to the complementary region in the single-stranded circular template DNA (step 1). After one round of rolling-circle synthesis of DNA, the 5'-end of the newly synthesized DNA strand is displaced by the T7 replisome (step 2). The rolling-circle synthesis of DNA is continued, and the displaced strand provides multiple sites for annealing of the reverse primer, which is extended by the T7 replisome (step 3). The synthesized ssDNAs that are extended from the reverse primers are released because of the activity of strand displacement by the T7 replisome, which provides annealing sites for the forward primers (step 4). The T7 replisome then extends the forward primers annealed to the released ssDNAs,

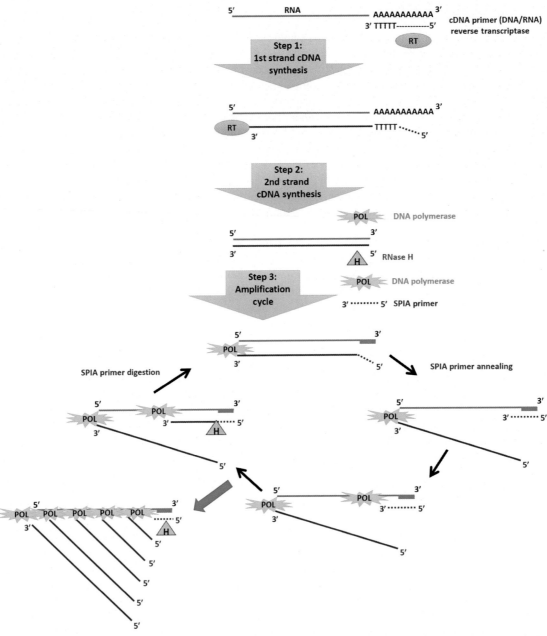

FIG. 9.35 Schematic representation of the 3-initiated Ribo-SPIA process. Step 1, first strand cDNA is produced using a unique DNA/RNA chimeric primer and reverse transcriptase. Step 2, the RNA template is partially degraded in a heating step which also serves to denature the reverse transcriptase. DNA polymerase is added to the reaction mixture to carry out second strand cDNA synthesis along the first strand cDNA product from the first step. The DNA polymerase elongates the product along the RNA portion of the chimeric primers, by its RNA-dependent DNA polymerase activity, thus forming a double-stranded cDNA with a unique RNA/DNA heterodouplex at one end. This unique product serves as a substrate for the subsequent SPIA DNA amplification step. Step 3 amplification is initiated by adding RNase H, DNA polymerase, and a second chimeric DNA/RNA heterodouplex at the end of the double stranded cDNA. This exposes 3 single strands at one end of the second strand cDNA whose unique sequence is complementary to the SPIA primer. The SPIA primer binds to this site, and the primer is extended by a strand displacing DNA polymerase. As soon as this extension begins, RNase H can again digest the RNA portion of the primer at the 5 end of the new strand, thus revealing another priming site for the RNA portion of a new primer molecule to bind. This continuous and isothermal cycle of degradation the RNA/DNA heterodouplexes, annealing of new SPIA primer, cDNA extension, and strand displacement continues in a linear fashion, producing microgram quantities of anti-sense cDNA.

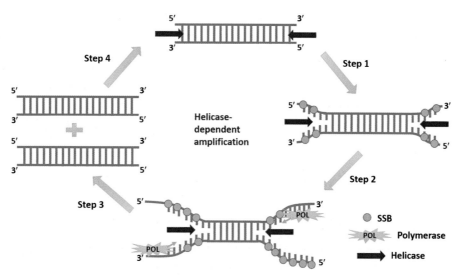

Step 4

Step 1

Helicase-
dependent
amplification

Step 2

Step 3

POL

POL

○ SSB

POL Polymerase

➡ Helicase

POL

FIG. 9.36   Schematic diagram of the helicase-dependent DNA amplification. Helicases unwind dsDNA and SSB proteins bind to exposed ssDNA. Subsequently, DNA polymerases start synthesizing the complementary strand from the bound primers, and the cycles repeat continuously.

generating dsDNA products defined by two primers and concatemers of the circular DNA template (step 5). The T7 replisome continues this process with the released ssDNA produced in step 5 for the next round of strand displacement synthesis. The process can be carried out at one temperature (25°C) for the entire process.

Primase-based whole-genome amplification (pWGA) uses the dual activities of T7 gp4 as helicase and primase to denature dsDNA templates and generate primers without addition of DNA primers and thermocycling. After T7 gp4 denatures the dsDNA template and synthesizes short RNA primers, the T7 DNA polymerase subsequently extends the strand. Overall, this technique is quite close to the cHDA method described earlier (Fig. 9.38). Using this method for DNA amplification, microgram quantities of DNA product could be obtained from nanogram quantities of input DNA within 1 h. More importantly, elimination of primer annealing reduces the risk of amplification bias due to uneven annealing of the added primers.

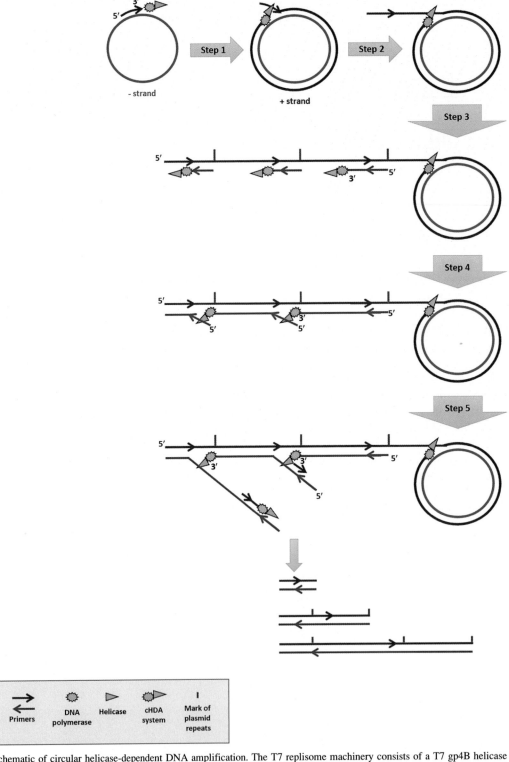

FIG. 9.37 Schematic of circular helicase-dependent DNA amplification. The T7 replisome machinery consists of a T7 gp4B helicase and T7 DNA polymerase gp5/trx complex. Primer extension and strand displacement produces a concatamer of the circular template DNA. Multiple reverse primers anneal to the concatamer and are extended by the T7 DNA polymerase. The helicase/DNA polymerase complex displaces the nontemplate strands, which provide complementary sites for forward primers to anneal. Duplex DNAs are produced by the T7 replisome after the release of ssDNAs for the next round of strand displacement synthesis.

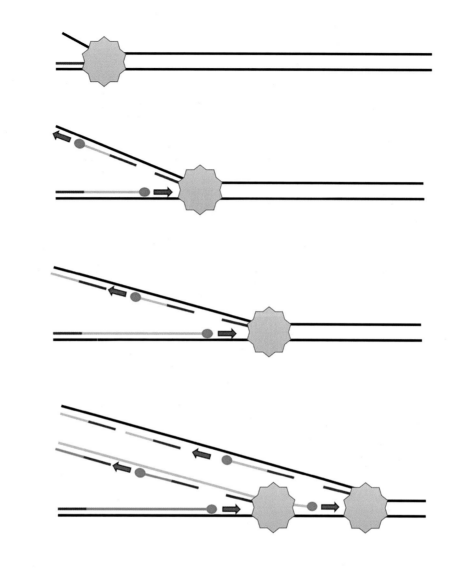

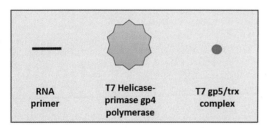

RNA primer   T7 Helicase-primase gp4 polymerase   T7 gp5/trx complex

FIG. 9.38 Mechanism of primase-based whole genome amplification. The T7 helicase-primase gp4 denatures the dsDNA template and synthesizes primers. Primers are extended by the T7 DNA polymerase gp5/trx complex, resulting in DNA replication in both strands. Newly synthesized DNA is displaced and serves as a template for whole-genome amplification.

Chapter 10

Characterization of Nucleic Acids and Proteins

Chapter Outline

NUCLEASES

Nucleases degrade DNA molecules by breaking the phosphodiester bonds that link one nucleotide to the next in a DNA strand. There are two different kinds of nuclease: exonuclease and endonuclease (Fig. 10.1). An endonuclease can hydrolyze internal bonds within a polynucleotide chain, whereas exonucleases remove nucleotides one at a time from the end of a DNA molecule. Nucleases may be specific for DNA or RNA, such as DNases or RNases, respectively, or even be specific for a DNA/RNA hybrid, such as RNase H, which cleaves the RNA strand of a DNA-RNA hybrid. Therefore, the specificity of nucleases varies dramatically.

When a nuclease hydrolyzes an ester bond in a phosphodiester linkage, it will have specificity for either of the two ester bonds, generating either $5'$ nucleotides or $3'$ nucleotides (Fig. 10.2). An exonuclease may either attack a polynucleotide chain from the $5'$ end and hydrolyze $5'$ to $3'$ or attack from the $3'$ end and hydrolyze $3'$ to $5'$. Similarly, an endonuclease can attack the phosphodiester bond from the 5' end or from the 3' end of the linkage.

For the strand preference, nucleases may be specific for single-strand nucleotide chain, double-helix stands, or both. For example, some exonucleases such as *Bal*31 can remove nucleotides from both strands of a double-stranded molecule (Fig. 10.3A). On the other hand, other enzymes such as exonuclease III degrade just one strand of a double-stranded molecule, leaving single-stranded DNA as the product (Fig. 10.3B). Similarly, some endonucleases such as S1 endonuclease only cleaves single strands (Fig. 10.4A), whereas deoxyribonuclease I (DNase I) cuts both single- and double-stranded

Diagnostic Molecular Biology. https://doi.org/10.1016/B978-0-12-802823-0.00010-9

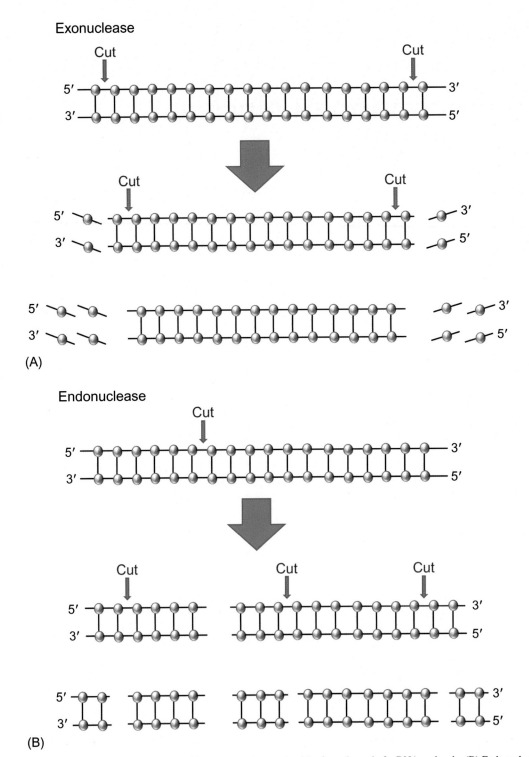

FIG. 10.1   Two different types of nucleases. (A) Exonuclease can remove nucleotides from the end of a DNA molecule. (B) Endonuclease can break internal phosphodiester bonds.

molecules (Fig. 10.4B). Since DNase I is nonspecific in that it attacks DNA at any internal phosphodiester bond, the end result of prolonged DNase I action is a mixture of mononucleotides and very short oligonucleotides. A restriction endonuclease can recognize a specific nucleotide sequence and cleave the DNA molecules internally (Fig. 10.4C). For single-strand molecules such as RNA, ribonuclease can rapidly degrade RNA molecules into ribonucleotide subunits.

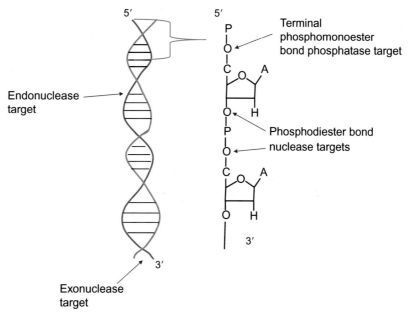

FIG. 10.2 The mechanism of nuclease hydrolysis.

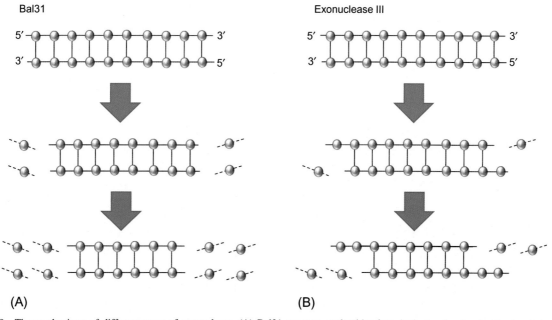

FIG. 10.3 The mechanisms of different types of exonuclease. (A) Bal31 removes nucleotides from both strands of a double-stranded molecule. (B) Exonuclease III removes nucleotides from the 3′ terminus.

Endonucleases for Cutting DNA: Restriction Endonucleases

The initial observation that led to the eventual discovery of restriction endonucleases was made in the early 1950s, when it was shown that some strains of bacteria are immune to bacteriophage infection, a host defense mechanism. Restriction occurs because the bacterium produces an enzyme that degrades the phage DNA before it has time to replicate and direct synthesis of new phage particles (Fig. 10.5A). The bacterium's own DNA is protected from attack because it carries additional methyl groups that prevent the degradative enzyme action (Fig. 10.5B). These degradative enzymes are called restriction endonucleases and are synthesized by many species of bacteria.

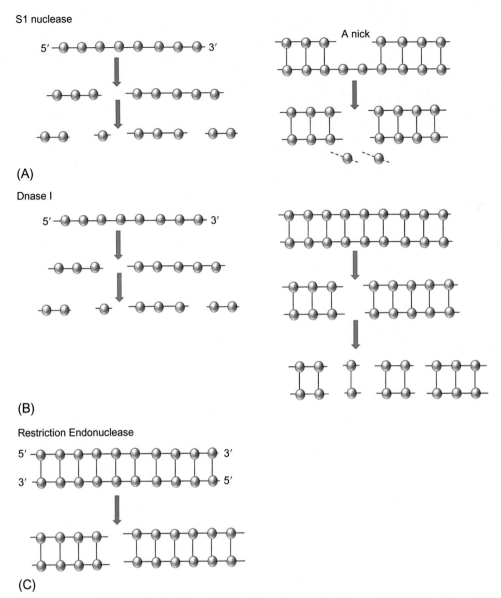

FIG. 10.4   The mechanisms of different types of endonuclease. (A) S1 nuclease cleaves only single-stranded DNA, including single-stranded nicks in mainly double-stranded molecules. (B) DNase I cleaves both single- and double-stranded DNA. (C) A restriction endonuclease cleaves double-stranded DNA, but only at a limited number of sites.

There are three different classes of restriction endonuclease, and each is distinguished by the mode of action. Types I and III are rather complex and have only a limited role in practical biotechnology applications. Type II restriction endonucleases, on the other hand, are the cutting enzymes that are so important in laboratory and clinical analysis.

Conventionally, the name of restriction endonucleases refers to the type II restriction endonucleases. This is because each enzyme has a specific recognition sequence at which it cuts a DNA molecule. Type I and III enzymes differ from type II enzymes in that the recognition site and cleavage site are different. With a type I enzyme, the cleavage site can be up to 1000 bp away from the recognition site. Type III enzymes have closer cleavage sites, usually 20 to 30 bp away. Type II restriction endonuclease is important because the recognition site and the cleavage site are the same. These recognition/cutting sites range in length from 4 to 8 bp. The sites are typically inverse palindromic, which means reading the same forward and backward on complementary strands (Fig. 10.6). Many restriction endonucleases recognize hexanucleotide target sites. There are also examples of restriction endonucleases with degenerate recognition sequences, meaning that they cut DNA at any one of a family of related sites. For example, *Hin*fI (*Haemophilus influenzae* strain Rf) recognizes GANTC, so cuts at GAATC, GATTC, GAGTC, and GACTC.

Phage DNA cleaved by bacterial restriction endonucleases

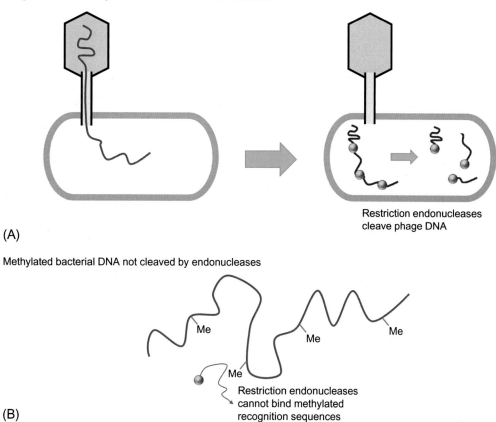

Restriction endonucleases
cleave phage DNA

(A)

Methylated bacterial DNA not cleaved by endonucleases

Me

Me

Me

Me

Restriction endonucleases
cannot bind methylated
recognition sequences

(B)

FIG. 10.5 The restriction endonuclease can protect bacterial cell from phage infection: (A) phage DNA is cleaved, but (B) bacterial DNA is not. This is because of the bacterial DNA is methylated.

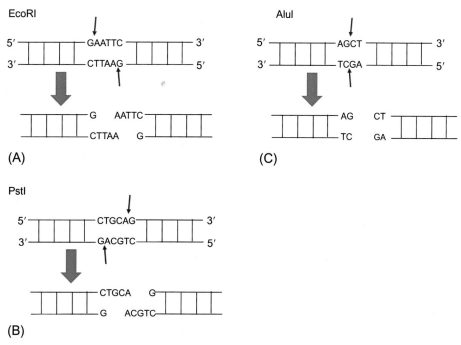

FIG. 10.6 Restriction endonucleases can cut DNA in the middle of the recognition sites to produce a blunt end or cut DNA staggeringly to produce sticky/cohesive ends. (A) EcoRI cleaves its recognition site and makes a 5′ overhang; (B) PstI cleaves its recognition site and makes a 3′ overhang; (C) cleaves its recognition site and makes a blunt end cut.

Different Sources of Type II Restriction Endonucleases

Restriction enzymes isolated from different bacteria may recognize and cut DNA at the same site. This type of enzymes is known as isoschizomers. Typical examples of isoschizomers are *Bsp*EI from a *Bacillus* species and *Acc*III from *Acinetobacter calcoaceticus*. They both bind the same DNA sequence and cut at the same sites. On the other hand, some restriction enzymes recognize and bind to the same sequence of DNA but cleave at different positions, producing different single-stranded extensions. These are neoschizomers, such as *Nar*I from *Nocardia argentinensis* and *Sfo*I from *Serratia fonticola*. They bind the same DNA sequence but cut at different sites. Isocaudomers are restriction endonucleases that produce the same nucleotide extensions but have different recognition sites. The examples are *Nco*I from *Nocardia corallina* and *Pag*I from *Pseudomonas alcaligenes*. They bind different DNA sequences but produce the same sticky ends. In some cases, a restriction endonuclease will cleave a sequence only if one of the nucleotides in the recognition site is methylated.

Restriction-Modification System

Almost all restriction endonucleases are paired with methylases that recognize and methylate the same DNA sites. The two enzymes, both restriction endonucleases and methylases, are collectively called a restriction-modification system (R-M) system. After methylation, DNA sites are protected from the cleavage by the restriction endonucleases. During the replication, one strand of the daughter duplex is a newly made strand and is unmethylated. Because the parental strand is still methylated, this half-methylation (hemimethylation) is enough to protect the DNA duplex against cleavage by the endonucleases (Fig. 10.7). Therefore, the methylase has time to find the site and methylate the other strand, producing fully methylated DNA. This is a host defense mechanism in bacterial cells to protect their own DNA from restriction enzyme degradation.

The Frequency of Recognition Sequences in a DNA Molecule

The number of recognition sequences for a particular restriction endonuclease in a DNA molecule of known length can be calculated mathematically. A tetranucleotide sequence should occur once every $4 \times 4 \times 4 \times 4 = 4^4 = 256$ nucleotides, and a hexanucleotide once every $4^6 = 4096$ nucleotides. These calculations assume that the nucleotides are placed in a random fashion and also that the four different nucleotides are present in equal proportions. Therefore, if one enzyme has a hexanucleotide recognition sequence, it will find the site in every 4096 nucleotides. However, this is not the case in reality. For example, the λ DNA molecule, at 49 kb, should contain about 12 sites for a restriction endonuclease with a hexanucleotide recognition sequence. However, it has been shown that there are six sites for *Bgl*II, five for *Bam*HI, and only two for *Sal*I. This is a reflection of the fact restriction sites are generally not evenly spaced out along the λ DNA molecule.

Analysis of Restriction Digested Fragments

Before sequencing a large stretch of DNA, some preliminary mapping is usually done to locate the cutting site and to examine the sizes of fragments. A restriction digest can be performed in a microcentrifuge tube in the presence of all necessary components, including template DNA, restriction enzyme, and Mg^{2+} at the right conditions. A restriction digest results in a number of DNA fragments, the sizes of which depend on the exact positions of the recognition sequences for the endonuclease in the original molecule. These restriction digested DNA fragments can be analyzed by gel electrophoresis, in which the restriction fragments can be separated on the basis of their size. A map based on the restriction sites determined by gel electrophoresis is a restriction map, which represents a linear sequence of the sites at which particular restriction enzymes find their targets. A typical example of a restriction map is shown in Fig. 10.8. Here, the relative locations of the recognition sites for restriction enzyme A and B, respectively, are shown (Fig. 10.8). To identify a specific DNA region, these digested fragments can also be analyzed by Southern blot, as discussed in a later section.

POLYMERASES

DNA polymerases are enzymes that synthesize a new strand of DNA complementary to an existing DNA or RNA template (Fig. 10.9). Most polymerases can function only if the template possesses a double-stranded region that acts as a primer for initiation of polymerization.

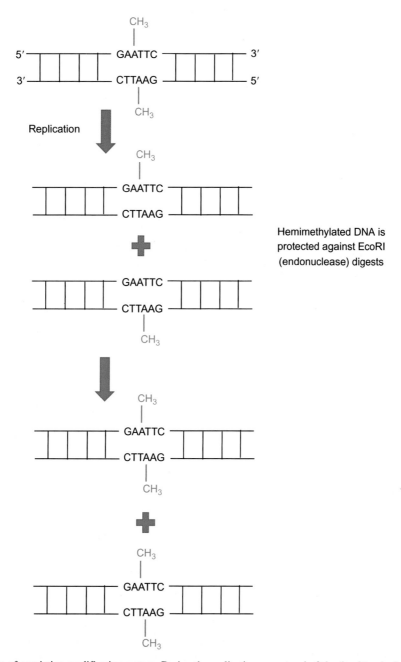

FIG. 10.7 The mechanism of restriction-modification system. During the replication, one strand of the daughter duplex is a newly made strand and is unmethylated. Because the parental strand is still methylated, this half-methylation (hemimethylation) is enough to protect the DNA duplex against cleavage by the endonucleases. Therefore, the methylase has time to find the site and methylate the other strand, producing fully methylated DNA.

FIG. 10.8 A restriction map represents a linear sequence of the sites at which particular restriction enzymes find their targets.

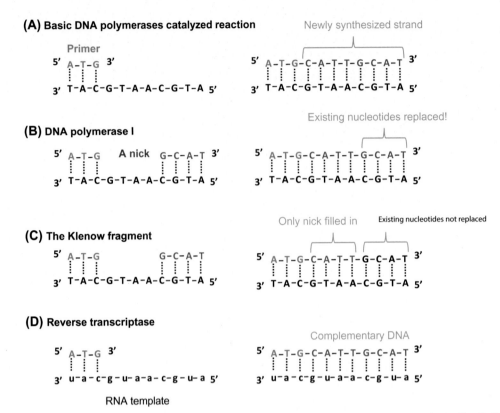

FIG. 10.9 The reactions catalyzed by DNA polymerases. (A) The basic reaction: a new DNA strand is synthesized in the 5' to 3' direction. (B) DNA polymerase I, which initially fills in nicks but then continues to synthesize a new strand, degrading the existing one as it proceeds. (C) The Klenow fragment, which only fills in nicks. (D) Reverse transcriptase, which uses a template of RNA.

DNA Polymerase I

Four types of DNA polymerase are used routinely in laboratory analysis. The first is DNA polymerase I, which is usually prepared from *E. coli*. DNA polymerase I has DNA polymerase activity, so this enzyme can attach to a short single-stranded region (or nick) in a mainly double-stranded DNA molecule, and then synthesizes a completely new strand (Fig. 10.9B). On the other hand, it also has two different exonuclease activities: a 3'→5' and a 5'→3' exonuclease activity. The 3'→5' activity is important in proofreading newly synthesized DNA. The 5'→3' exonuclease activity allow DNA polymerase I to degrade a strand ahead of the advancing polymerase, so it can remove and replace a strand all in one pass of the polymerase. DNA polymerase I is therefore an example of an enzyme with a dual activity: DNA polymerization and DNA degradation.

Klenow Fragment

Mild proteolytic treatment of DNA polymerase I produces two polypeptides: a large fragment (the Klenow fragment), which has the polymerase and 3'→5' exonuclease activity; and a small fragment with the 5'→3' exonuclease activity. As such, a Klenow fragment can synthesize a complementary DNA strand on a single-stranded template also known as a nick region. Since it has no 5'→3' activity, it cannot continue the synthesis once the nick is filled in (Fig. 10.9C). The major application of the Klenow fragment is to perform DNA end-filling or DNA sequencing.

Taq DNA polymerase

The *Taq* DNA polymerase used in the PCR is the DNA polymerase I enzyme of the bacterium *Thermus aquaticus*. This organism lives in hot springs, and many of its enzymes, including the *Taq* DNA polymerase, are thermostable. Therefore, they are resistant to denaturation by heat treatment. This is the special feature of *Taq* DNA polymerase that makes it suitable for PCR, because if it were not thermostable, it would be inactivated when the temperature of the reaction was raised to 94° C to denature the DNA.

Reverse Transcriptase

Another type of DNA polymerase that is important in clinical analysis is reverse transcriptase, an enzyme involved in the replication of several kinds of virus. Reverse transcriptase is unique in that it uses RNA as a template not DNA through reverse transcriptase PCR (RT-PCR) (Fig. 10.9D). The ability of this enzyme is to synthesize a DNA strand complementary to an RNA template, and this newly synthesized DNA is called complementary DNA (cDNA). Reverse transcriptase is used to evaluate the amount of RNA. As such, it can be used to establish the expression profile, which is important to clinical evaluation of the change in gene expression pattern.

DNA MODIFYING ENZYMES AND LIGASES

DNA Modifying Enzymes

There are numerous enzymes that modify DNA molecules by addition or removal of specific chemical groups. For example, is alkaline phosphatase (from *E. coli*, calf intestinal tissue, or arctic shrimp) removes the phosphate group present at the 5′ terminus of a DNA molecule (Fig. 10.10). Polynucleotide kinase (from *E. coli* infected with T4 phage), which has

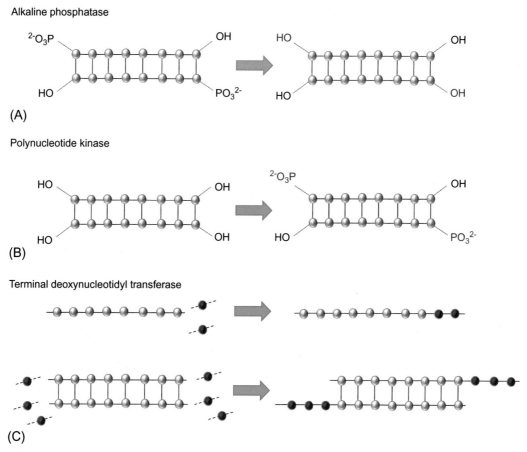

FIG. 10.10 The reactions catalyzed by DNA-modifying enzymes. (A) Alkaline phosphatase, which removes 5′-phosphate groups. (B) Polynucleotide kinase, which attaches 5′-phosphate groups. (C) Terminal deoxynucleotidyl transferase, which attaches deoxyribonucleotides to the 3′ termini of poly-nucleotides in either single- or double-stranded molecules.

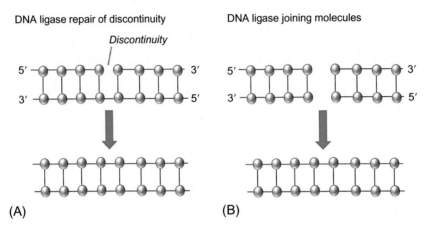

FIG. 10.11 The two reactions catalyzed by DNA ligase. (A) Repair of a discontinuity—a missing phosphodiester bond in one strand of a double-stranded molecule. (B) Joining two molecules together.

the reverse effect to alkaline phosphatase, adds phosphate groups onto free 5′ termini (Fig. 10.10). Terminal deoxynucleotidyl transferase (from calf thymus tissue) adds one or more deoxyribonucleotides onto the 3′ terminus of a DNA molecule (Fig. 10.10).

DNA Ligase

In the cell, the function of DNA ligase is to repair single-stranded breaks ("discontinuities") that arise in double-stranded DNA molecules during DNA replication or during DNA damage repair. DNA ligases from most organisms can also join together two individual fragments of double-stranded DNA (Fig. 10.11). In the test tube, purified DNA ligases can also join together individual DNA molecules or the two ends of the same DNA molecule. The chemical reaction involved in ligating two molecules is exactly the same as discontinuity repair, except that two phosphodiester bonds must be made, one for each strand.

THE APPLICATION OF NUCLEIC ACID HYBRIDIZATION

Hybridization is the formation of hydrogen bonds between two complementary strands of nucleic acids. This is a direct consequence of the stable double-stranded structure of nucleic acid under physiological conditions. This is due to the fact that the binding between separate, complementary nucleic acids is both reversible and base sequence-specific. During the annealing process, both nucleic acid strands are not labeled with any isotopes or fluorescence. However, if one strand is labeled, this labeled strand is referred to as a probe and the process is called hybridization. This is because the hybrid molecule is formed between a labeled and unlabeled strand. A hybridization assay is the hybridization reaction that is used to analyze the nucleic acid content of an unknown sample. Typical hybridization assays include Southern hybridization, Northern hybridization, dot/blot hybridization, microarray, and fluorescent in situ hybridization. Here, we discuss Southern hybridization, Northern hybridization, and array-based hybridization, which includes dot/blot hybridization and microarray. Fluorescent in situ hybridization will be discussed in Chapter 13.

Southern Hybridization

Southern blotting allows the detection of a given DNA sequence in a complex mixture of DNA sequences. It can be used to identify homologous sequences in genomic DNA, or to facilitate gene mapping through restriction mapping of genes or in the detection of restriction fragment length polymorphisms.

The first step of Southern hybridization is to purify genomic DNA from eukaryotic cells or bacteria followed by one or a few specific restriction enzyme digestion. Enzyme digestions will produce thousands of fragments of genomic DNA. Each specific enzyme will produce different digest results. Subsequently, these DNA fragments are separated by agarose gel electrophoresis, which separates the fragments according to size, the smaller DNA

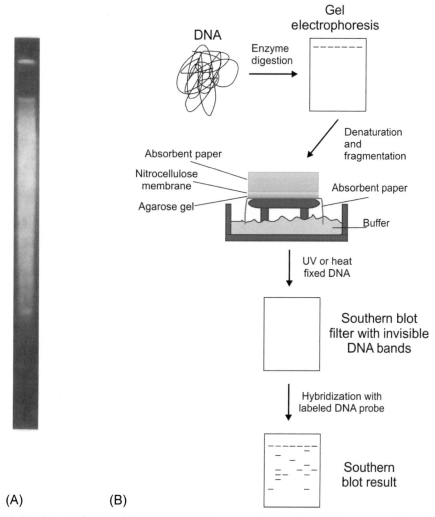

FIG. 10.12 Southern hybridization. (A) Genomic DNA is digested with one or a few specific restriction enzymes followed by gel electrophoresis. (B) The schematic process of Southern hybridization.

fragments migrating farthest in the electric field. If those fragments are visualized by staining, the gel should show a series of smear bands without any discrete, distinguishable bands (Fig. 10.12A). After gel electrophoresis, DNA fragments are further fragmented to smaller than 1 kb size and denatured by mild alkali solution followed by transfer (Fig. 10.12B). Finally, a labeled probe is added to hybridize with the fragments on the membrane for detection and identification.

Parameters for Southern Hybridization

In classical Southern analysis, various parameters are modified to achieve optimal resolution in the desired range of DNA length. These important parameters depend on the resolution of the agarose gel, which include the percentage of the agarose gel used and the voltage applied in gel electrophoresis. For the analysis of genomic DNA, 0.7%–1.2% agarose gels are used. Gels below 0.7% are difficult to handle, while transfer efficiencies with high-percentage gels (>1.2%) are poor in the analysis of single-copy genes where the amount of target DNA is the limiting factor. Table 10.1 is an example of guidelines for the percentage of agarose gels used to achieve an optimal resolution in the 0.2–20 kbp range. For the voltage applied in gel electrophoresis, high voltage combined with short run times gives optimal size separation of DNA of 10–20 kbp in length, whereas low voltage combined with long run times gives good resolution between 1 and 10 kbp (Fig. 10.13).

TABLE 10.1 Optimal Resolution of DNA of Varying Length in Agarose Gels

Size of DNA Fragment (kbp)	Agarose Concentration (w/v%)	Bromophenol Blue Dye Migration in TAE Buffer
0.1–0.5	4% agarose blend	35 bp
0.5–1.0	3% agarose blend	Not determined
0.4–6	1.2% agarose	400 bp
1–20	0.7% agarose	700 bp

TAE, tris-acetate-EDTA.

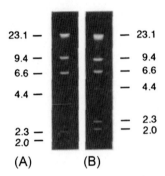

FIG. 10.13 Effect of voltage conditions on size separation of ds DNA. (A) 28 V/37 mA for 16 h; (B) 70 V/ 97 mA for 5.25 h. All other conditions were identical.

For restriction digestion, restriction endonucleases cleave DNA only at specific cleavage sites. Each enzyme produces a characteristic number of fragments from a single specific length of DNA. Therefore, allelic variants of a particular gene can be detected if the differences in the nucleic acid sequence include the cleavage sites. This is because digestion by the appropriate restriction enzyme will yield different fragments from allelic genes (Fig. 10.14). As such, mutations may result in either the loss of cleavage sites or the introduction of new, additional cleavage sites. Such differences in the fragments produced from one gene are referred to as restriction fragment length polymorphisms.

Several parameters are critical to achieve an optimal transfer of DNA from a gel to a membrane support. The most critical step is the depurination of DNA molecules. The depurination by 0.2 M hydrochloric acid upon denaturation by sodium hydroxide leads to nicks in the DNA strands, resulting in a breakdown of long DNA fragments into shorter pieces of single-stranded DNA. These shorter fragments are more efficiently transferred in the blotting. After a subsequent neutralization step, the DNA can be transferred onto a membrane support.

For the physical transfer of DNA onto the membrane support, two different transfer methods have been developed. The classical Southern transfer is ascending capillary transfer (Fig. 10.15). In this setup, the agarose gel is placed on top of a solid supporter in the presence of salt (10 × saline–sodium citrate [SSC] buffer). Nylon membrane is place on top of the agarose gel, and then a stack of blotting paper is placed on top of the nylon membrane. This arrangement allows efficient diffusion of DNA onto the nylon membrane. On the other hand, descending capillary transfer uses the gravitational flow of the transfer buffer with the help of vacuum to allow more rapid and reproducible blotting of DNA. Nylon membrane is used is because it has high DNA/RNA binding capacity and strong mechanical strength on stropping and reprobing.

The last step of transfer is to fix DNA onto the nylon membrane. This can be achieved by baking the membrane at 80°C for 2 h or by UV light irradiation. Both methods are equally effective for fixation of DNA to a nylon membrane. The baked or UV-fixed Southern blot can be stored for prolonged periods of time at room temperature.

Northern Hybridization

Northern blotting is very similar to Southern blotting. The only different is that Northern blotting allows detection of a given RNA molecules in a mixture of heterogeneous RNA, instead of DNA as in the Southern blot. Northern blotting uses DNA

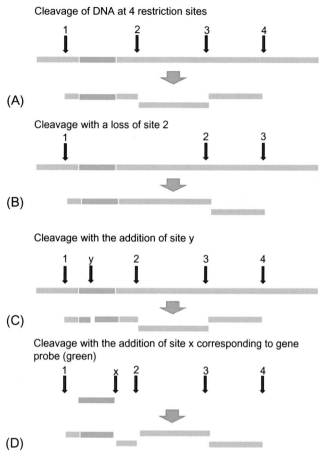

Cleavage of DNA at 4 restriction sites

(A)

Cleavage with a loss of site 2

(B)

Cleavage with the addition of site y

(C)

Cleavage with the addition of site x corresponding to gene probe (green)

(D)

FIG. 10.14 Allelic variants of a particular gene can be detected if the differences in the nucleic acid sequence include the cleavage sites, because digestion by the appropriate restriction enzyme will yield different fragments from allelic genes.

probes to hybridize with complementary RNA sequences. It is an ideal tool to study the products of gene transcription. In contrast to electrophoresis of DNA, electrophoresis of RNA typically requires more caution, as RNA is more prone to degradation. The reagents and the apparatus used to prepare RNA and to conduct RNA electrophoresis must be kept free from ubiquitous RNase. This can be achieved by treatment with diethyl pyrocarbonate (DEPC).

Because single-stranded, native RNA tends to form secondary structure, it has to be denatured before electrophoresis and kept in a denatured state during electrophoresis. Before electrophoresis, denaturation is achieved by heating the sample to 55°C in the presence of formaldehyde and formamide. During electrophoresis of RNA, addition of formaldehyde to the gel prevents reformation of secondary structure. Ethidium bromide is not usually recommended to add in electrophoresis, as it will always decrease the hybridization signal compared with unstained RNA. A typical Northern hybridization protocol is shown in Table 10.2. After electrophoresis, the separated RNA can be transferred onto a nylon membrane as in the Southern blot, through capillary blotting or vacuum blotting. Subsequently, RNA can be fixed by UV light irradiation or baking at 80°C.

Restriction Fragment Length Polymorphism (RFLP) Analysis

The availability of a variety of restriction endonuclease enzymes that cleave DNA at specific sites has made it possible to identify the presence of polymorphic regions in the isolated fragments. Restriction fragment length polymorphism (RFLP) results from a variable number of tandem repeats (VNTR) in a short DNA segment. These VNTR sequences can uniquely specify an individual and, as such, are used in DNA fingerprinting and in paternity testing. Therefore, RFLP is a difference in homologous DNA sequences that can be detected by the presence of fragments of different lengths after digestion of the DNA samples in question with specific restriction endonucleases. RFLP, as a molecular marker, is specific to a single clone/restriction enzyme combination.

Electrophorese *Bam*HI-restricted DNA

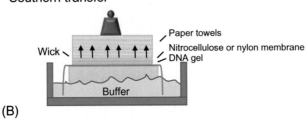

(A)

Southern transfer

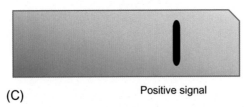

(B)

Result of hybridization probing

(C)

FIG. 10.15 Ascending capillary transfer—transfer of DNA to membrane in Southern hybridization.

TABLE 10.2 Northern Blot Denaturing, Staining, and Electrophoretic Steps

1) RNA sample loaded onto gel as follows:

 – 5.5 μL RNA (5 μg) in water
 – 19 μL premix (1.3 × MOPS, 3 M formaldehyde, 64% formamide)
 – 5 μL loading solution (1 mM EDTA pH 8, 0.25% bromophenol blue, 0.25% xylene cyanol, 50% glycerol)
 – 1.0 μL EtBr (0.5 mg/mL stock)
 – TOTAL: 31 μL

2) RNA denatured with 15 min of 55°C heating

3) Denaturation quenched with ice; sample then loaded onto a 1.2% agarose-1.1% formaldehyde gel. The RNA is subjected to electrophoretic separation at 70 V for 3.5 h with recirculation of buffer

EtBR, ethidium bromide; *MOPS*, 3-(*N*-morpholino)propanesulfonic acid.

RFLP analysis is performed when genomic DNA is collected and is digested with a specific restriction enzyme followed by gel electrophoresis. Subsequently, the gel is subject to Southern blot analysis. RFLP may be found close to a disease gene, and, as such, can be used as a genetic disease marker. Certain criteria need to be fulfilled, however, for RFLP to be useful as a genetic disease marker, such as its closeness to the disease gene.

Isolation of sufficient DNA for RFLP analysis is sometimes time consuming and labor intensive. However, PCR can be used to amplify very small amounts of DNA, usually in 2–3 h, to the levels required for RFLP analysis. Therefore, more samples can be analyzed in a shorter time. This approach is known as cleaved amplified polymorphic sequence (CAPS) assay.

Amplified Fragment Length Polymorphism (AFLP)

AFLP is another fingerprinting technique that has been extensively used in evolutionary, population genetics, and epidemiological studies. The main feature of AFLP is that it permits the simultaneous evaluation of hundreds to thousands of different DNA regions distributed randomly throughout the genome without prior sequence knowledge. This makes the implementation of AFLP particularly useful in nonmodel species which have no complete genome sequences available and where other types of genome-wide markers such as single nucleotide polymorphisms (SNPs) and microsatellites are difficult to obtain.

The first step of the AFLP basic protocol is the generation of restriction fragments by using two different restriction endonucleases, one with a rare cutting frequency and a second with a high cutting frequency (e.g., *Eco*RI: 5′-G | AATTC-3′; and *Mse*I: 5′-T | TAA-3′). The digested DNA fragments are ligated to double-stranded oligonucleotide adapters that contain a known core sequence of approximately 20 nucleotides. These adapters also contain enzyme-specific cohesive ends complementary to the overhangs left after enzymatic digestion. The second step is the preselective amplification. Here, the sequences of the ligated adapters serve as primer binding sites for PCR amplification of all the fragments derived from cutting with both enzymes, which are called heterosite restriction fragments. As such, only fragments with ends for both enzymes will effectively amplify. This step provides a higher amount of template DNA for subsequent rounds of more selective AFLP reactions.

The third step in the generation of usable AFLP profiles is the selective amplification using the pre-selective PCR product. The primers used for this purpose contain one to three extra nucleotides at the 3′ end. This new PCR reaction amplify defined sets of the *Eco*RI-*Mse*I fragments, producing fingerprints with manageable number of fragments that may be visualized either by conventional denaturing polyacrylamide gel electrophoresis (PAGE) or by capillary electrophoresis (CE). For PAGE detection, fragments may be labeled using fluorescent or radioactive nucleotides or primers, whereas detection by CE relies mostly on fluorescent PCR primers. The number of extra nucleotides at the 3′ end of the selective PCR primers determines the complexity of the resulting fingerprint, requiring a careful optimization step to define the best selective primer combinations. Once the best primers are selected, the study may proceed with a larger number of samples, as resulting fragment patterns are very reproducible and the technique is robust.

AFLP is highly reproducible and robust because it combines the specificity of RFLP with the sensitivity of the PCR. AFLP methods rapidly generate hundreds of highly replicable markers from DNA of any organism. Therefore, they allow high-resolution genotyping of fingerprinting quality. The time and cost efficiency, replicability, and resolution of AFLPs are superior to those of RFLP. Because of their high replicability and ease of use, AFLP markers have emerged as a major new type of genetic marker with broad application in systematics, pathotyping, population genetics, DNA fingerprinting, and quantitative trait locus mapping.

DETECTION METHODS

Probe

For both Southern and Northern blots, the detection of signals is similar. In general, a labeled, denatured single-stranded probe is hybridized to membrane-bound DNA or RNA. The probe must be labeled with a radioactive or other type of marker, denatured by heating, and applied to the membrane in a solution of chemicals that promote nucleic acid hybridization. Many parameters have been proved to influence the hybridization efficiency (Table 10.3). After a period to allow hybridization to take place, the nylon membrane is washed to remove unbound probe, dried, and the label detected in order to identify the specific DNA/RNA fragments to which the probe has become bound.

The idea of nucleic acid hybridization is that any two single-stranded nucleic acid molecules have the potential to form base pairs with one another. If only a small number of individual interstrand bonds are formed, the resulting hybrid structure are unstable (Fig. 10.16A). However, if the polynucleotides are complementary, extensive base pairing can occur to form a stable double-stranded molecule, which is stabilized by the hydrogen bonds between the two complementary strands of nucleic acids (Fig. 10.16B). This can occur not only between single-stranded DNA molecules to form the DNA double helix, but also between a pair of single-stranded RNA molecules or between combinations of one DNA strand and one RNA strand (Fig. 10.16C).

RFLP Probe

An RFLP probe is a labeled DNA sequence that hybridizes with one or more fragments of the digested DNA sample after they have been separated by gel electrophoresis, thus revealing a unique blotting pattern characteristic of a specific

TABLE 10.3 Factors That Influence Hybrid Stability and Hybridization Rate

Factor	Influence
A. Hybrid Stability	
Ionic strength	Between 0.01 and 0.4 M NaCl, melting temp increases 16.6°C for each 10-fold increase in monovalent cations
Base composition	In NaCl solutions, GC base pairs are more stable than AT base pairs
Destabilizing agents	Every 1% of formamide decreases the melting temp by around 0.6°C for a DNA-DNA hybrid. 6 M urea decreases the melting temp around 30°C
Mismatched base pairs	For every 1% of mismatch, melting temp decreases by 1°C
Duplex length	No detectable effect with probes greater than 500 bp
B. Hybridization Rate	
Temperature	For DNA-DNA hybrids: max rate at 20–25°C below melting temp For DNA-RNA hybrids: max rate at 10–15°C below melting temp
Ionic strength	1.5 M Na$^+$ for optimal hybridization
Destabilizing agents	No effect when formamide is at 50%, but when at higher or lower conc, hybridization rate is reduced
Mismatched base pairs	Hybridization rate decreases by a factor of 2 for each 10% of mismatching
Duplex length	Hybridization rate is directly proportionate to duplex length
Viscosity	Rate of membrane hybridization is increased with increased viscosity
Probe complexity	Hybridization rate increased with an increase in repetitive sequences
Base composition	No detectable effect
pH	No detectable effect between pH 5 and pH 9

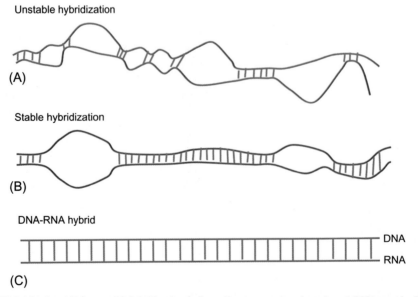

Unstable hybridization

(A)

Stable hybridization

(B)

DNA-RNA hybrid

DNA

RNA

(C)

FIG. 10.16 Nucleic acid hybridization. (A) An unstable hybrid molecule formed between two non-homologous DNA strands. (B) A stable hybrid formed between two complementary strands. (C) A DNA-RNA hybrid, such as may be formed between a gene and its transcript.

genotype at a specific locus. Short, single- or low-copy genomic DNA or cDNA clones are typically used as RFLP probes. The RFLP probes are frequently used in genome mapping and in variation analysis including genotyping, forensics, paternity tests, hereditary disease diagnostics, and so forth.

Labeling With a Radioactive Marker

A DNA molecule is usually labeled by incorporating nucleotides that carry a radioactive isotope of phosphorus, ^{32}P (Fig. 10.17A). Several methods are available, including nick translation, end filling, and random priming.

Most purified samples of DNA contain some nicked molecules; therefore, DNA polymerase I can be attached to the DNA and catalyze a strand replacement reaction as nick translation (Fig. 10.17B). This reaction requires a supply of nucleotides, one of which is radioactively labeled with ^{32}P-modified deoxynucleoside triphosphate. During the synthesis, the DNA molecule will become labeled as the radiolabeled deoxynucleotides are attached to the newly synthesized strand. Nick translation can be used to label any DNA molecule but might under some circumstances also cause DNA cleavage.

End filling is a gentler method than nick translation and rarely causes breakage of the DNA, but unfortunately can only be used to label DNA molecules that have sticky ends. The enzyme used is the Klenow fragment, which "fills in" a sticky end by synthesizing the complementary strand (Fig. 10.17C). As with nick translation, the end filling reaction is carried out in the presence of labeled nucleotides.

Random priming results in a probe with higher activity and therefore able to detect smaller amounts of membrane-bound DNA. However, the resultant probe is usually shorter than the nick translation labeled probe (Fig. 10.17D). The denatured DNA is mixed with a set of hexameric oligonucleotides of random sequence. By chance, these random hexamers will contain a few molecules that will base pair with the probe and prime new DNA synthesis. The Klenow fragment is used as this enzyme lacks the nuclease activity of DNA polymerase I and so only fills in the gaps between adjacent primers. Labeled nucleotides are incorporated into the new DNA that is synthesized. One important advantage for random priming is that it allows the labeling of short dsDNA fragments that are inefficient substrate for nick translation.

Once probes are labeled, the DNA probe is denatured and is used for hybridization. The specificity of the hybridization signal in Southern and Northern blotting is mainly determined by the hybridization temperature (Fig. 10.18A). The higher the hybridization temperature, the more stringent. After hybridization, the location of the bound probe is detected by autoradiography. In this method, a sheet of X-ray-sensitive photographic film is placed over the membrane. The radioactive DNA exposes the film, which is developed to reveal the positions of the fragment to which the probe has hybridized (Fig. 10.18B). Sometimes an intensifying screen can be used to enhance the sensitivity of autoradiography. This is a screen coated with a compound that fluoresces when it is excited by an electron at low temperature. This intensifying screen can be used directly with X-ray film or can be read by a phosphorimager.

Labeling With a Nonradioactive Marker

The advantage of radioactive labeling methods is their sensitivity, but they are starting to fall out of favor, partly because of the hazard to the researcher and partly because of the problems associated with disposal of radioactive waste. It is also because the development of the high sensitive nonradioactive tracers. Many different methods have been developed recently. The first makes use of deoxyuridine triphosphate (dUTP) nucleotides modified by reaction with biotin, an organic molecule that has a high affinity for a protein called avidin. After hybridization, the positions of the bound biotinylated probe can be determined by washing with avidin coupled to a fluorescent marker (Fig. 10.19). This method is as sensitive as radioactive probing and is becoming increasingly popular.

The same is true for a second procedure for nonradioactive hybridization probing, in which the probe DNA is complexed with the enzyme horseradish peroxidase and is detected through the enzyme's ability to degrade luminol with the emission of chemiluminescence. The signal can be recorded on normal photographic film in a manner analogous to autoradiography.

ARRAY-BASED HYBRIDIZATION

Dot Blot and Slot Blot Hybridization

The principle of this assay is that multiple samples are immobilized in a geometric array on a nitrocellulose or nylon membrane. When samples are applied by hand, the shape is more random and is called a blot (Fig. 10.20). If a commercially available manifold is used to apply the samples on the membrane with the aid of suction, the sample shape is regular and is

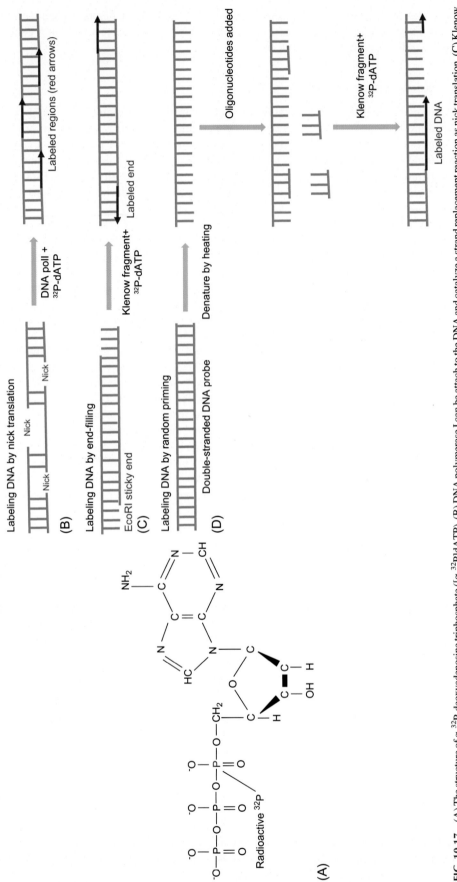

FIG. 10.17 (A) The structure of α-³²P-deoxyadenosine triphosphate ((α-³²P]dATP). (B) DNA polymerase I can be attach to the DNA and catalyze a strand replacement reaction as nick translation. (C) Klenow fragment fills in at the sticky end by synthesizing the complementary strand. (D) Klenow fragment is used in Random priming.

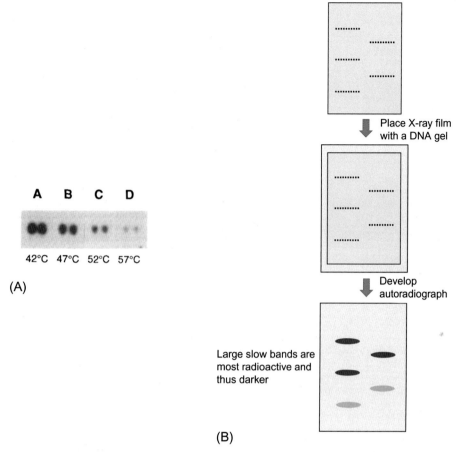

FIG. 10.18   (A) Effect of the hybridization temperature on the final Northern hybridization signal. Total RNA from HeLa cells (5 μg per lane) was separated on a 1.2% agarose-1.1% formaldehyde gel. After vacuum transfer onto a Nylon membrane, the lanes were cut apart and hybridized at varying temperatures but otherwise standard conditions to a rat glyceraldehyde phosphate dehydrogenase (GAPDH) probe that is 85% identical to human GAPDH. All hybridized blots were washed under the conditions. (B) Autoradiography. After hybridization, the location of the bound probe is detected by autoradiography. In this method, a sheet of X-ray-sensitive photographic film is placed over the membrane. The radioactive DNA exposes the film, which is developed to reveal the positions of the fragment to which the probe has hybridized. In this case, the large, slowly migrating bands are the most radioactive, so the bands on the autoradiograph that correspond to them are the darkest.

called dot or slot blot. The solid matrix allows multiple samples to be processed through all steps of the assay simultaneously. Samples are usually applied to the membrane using a manifold attached to a suction device. Dot and slot blots differ only in the geometry of the blot.

For this assay, samples can be completely pure or unpurified. Pure samples provide more specific results, whereas unpurified samples may have background from nonspecific binding. Interpretation of the results of a dot blot or a slot blot hybridization is relatively straightforward. If hybridization has occurred, a signal is generated in the specific spot. Therefore, a simple yes or no interpretation is usually given. No information is available about the size of the hybridizing fragments.

Macroarrays

Macroarrays have been applied to formats in which areas of probe localization are large enough to be visualized without magnification. A macroarray is a reverse dot/slot blot of up to several thousand targets on nitrocellulose or nylon membranes. Radioactive or chemiluminescent signals are typically used to detect hybridization of a labeled sample to the target probes on the membrane. Those probes are usually deposited onto the membrane by printing or dot blotting, then dried and stored for future use. Because the size of the probes, the density of macroarrays is much lower than that of microarrays. As such, a macroarray is limited by the area of the membrane and the specimen requirements.

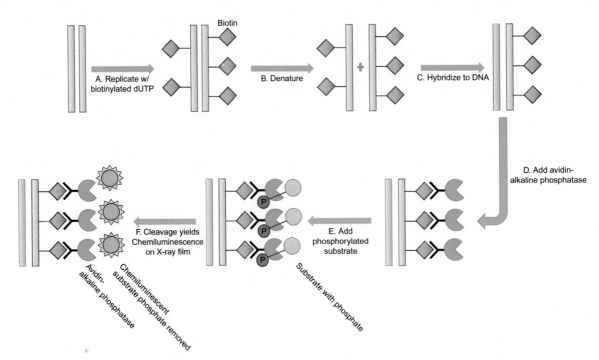

FIG. 10.19 Detecting nucleic acids with a nonradioactive probe. This sort of technique is usually indirect; detecting a nucleic acid of interest by hybridization to a labeled probe that can, in turn, be detected by virtue of its ability to produce a colored or light-emitting substance. In this example, the following steps are executed. (A) Replicate the probe DNA in the presence of dUTP that ls tagged with the vitamin biotin *(yellow)*. This generates biotinylated probe DNA. (B) Denature this probe, and (C) hybridize it to the DNA to be detected *(blue)*. (D) Mix the hybrids with a bifunctional reagent containing both avidin and the enzyme alkaline phosphatase *(light yellow)*. The avldin binds tightly and specifically to the biotin in the probe DNA. (E) Add a phosphorylated compound that will become chemiluminescent as soon as its phosphate group is removed. (F) The alkaline phosphatase enzymes attached to the probe cleave the phosphates from these substrate molecules, rendering them chemiluminescent (light-emitting). The light emitted from the chemiluminescent substrate can be detected with an X-ray film.

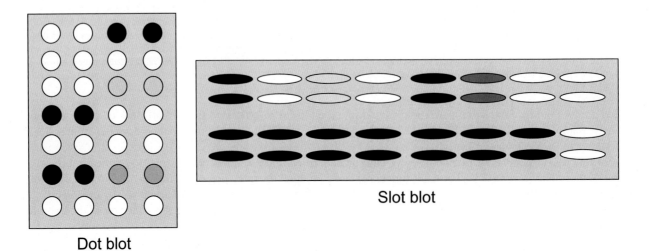

Dot blot

Slot blot

FIG. 10.20 An example of a dot blot *(left)* and a slot blot *(right)* analysis. For the dot blot, the target is spotted in duplicate, side by side. The *last* two rows of spots contain positive and negative control, followed by a blank with no target. For the slot blot, the *top* two rows of the slot blot gel on the left represent positive and negative control followed by four samples spotted in duplicate blank with no target are in the last four spots on the right. The *bottom* two rows represent a loading or normalization control that is often useful to confirm that equal amounts of DNA or RNA were spotted for each test sample.

Microarrays

Microarray technology is a variation of the dot/slot blot in which the dotted material is arranged in a regular gridlike pattern with each feature reduced to a very small size so that hundreds to thousands of probes can be placed on one solid surface, such as a glass microscope slide. This technology is an analysis of nucleic acid on a genome-wide scale.

Microarray technology usually refers to both microarrays and GeneChips technologies, which are referred to as delivery and synthesis technologies, respectively. Microarrays are robotically printed sets of PCR products or conventionally synthesized oligonucleotides and this is known as the delivery method, whereas Affymetrix GeneChips are high-density arrays of oligonucleotides synthesized in situ using light-directed chemistry and this is known as the synthesis method.

The most common application of microarray technology is transcript profiling, which is the gene-by-gene determination of differences in transcript abundance between two mRNA preparations. Microarrays and chips accomplish this in somewhat different ways. In the case of microarrays, a pair of mRNA samples (e.g., a control and an experimental sample) are independently copied as labeled cDNAs by reverse transcriptase and distinguishable fluorochromes, mixed and then hybridized to spotted microarrays. In the case of chips, the control and experimental cDNAs are usually synthesized with a single fluorochrome, and each labeled cDNA is hybridized to separate arrays. In either case, fluorescence intensities are determined for each gene, and ratios are computed.

Globally, the magnitude of these log-transformed ratios takes the form of a bell-shaped curve, with a relatively small number of expression ratios exceeding some threshold and receiving most attention. Fig. 10.21 shows one design for microarray analysis of gene expression.

Microarray-Manufacturing Technology

Microarray-manufacturing technologies fall into two main categories: synthesis and delivery. In the synthesis approaches, microarrays are prepared in a stepwise fashion by the in situ synthesis of nucleic acids and other biopolymers from biochemical building blocks. With each round of synthesis, nucleotides are added to growing chains until the desired length is achieved. The delivery technologies, by contrast, use the exogenous deposition of pre-prepared biochemical substances for chip fabrication. Molecules such as cDNAs are amplified by PCR and purified, and small quantities are deposited onto known locations using a variety of delivery technologies. A considerable disadvantage of the delivery approach versus the in situ synthesis approach is that amplification of the DNA by PCR must be carried out for each gene of interest. This is relatively straightforward when vector-specific primers can be used to amplify cDNA clones, but when a clone is not in hand, or the clone contains undesirable repeat elements, PCR from genomic DNA or RT-PCR from mRNA with gene-specific primers must be carried out. Approaches to reduce the cost of oligonucleotide synthesis and improve success of PCR enable this to be done on a large scale.

The key technical parameters for evaluating both the synthesis and delivery technologies include microarray density and design, biochemical composition and versatility, reproducibility, throughput, quality, cost, and ease of prototyping. Three types of advanced technologies have emerged in automated microarray production (Fig. 10.22). One novel synthesis technology combines photolithography technology from the semiconductor industry with DNA-synthetic chemistry to enable high-density oligonucleotide microarray manufacture (Fig. 10.22A). A key advantage of this approach over nonsynthetic methods is that photoprotected versions of the four DNA building blocks allow chips to be manufactured directly from sequence databases, thereby removing the uncertain and burdensome aspects of sample handling and tracking. Furthermore, the use of synthetic reagents minimizes chip-to-chip variation by ensuring a high degree of precision in each coupling cycle. One disadvantage of this approach is, however, the need for photomasks, which are expensive and time-consuming to design and build. Affymetrix chips is a typical example of this technology and it can accommodate up to 400,000 groups of oligonucleotides or features in an area of 1.6 cm^2.

The second technology is the mechanical microspotting approach. Microspotting, a miniaturized version of earlier DNA spotting techniques, encompasses a family of related deposition technologies that enable automated microarray production by printing small quantities of premade biochemical substances onto solid surfaces (Fig. 10.22B). Printing is accomplished by direct surface contact between the printing substrate and a delivery mechanism that contains an array of tweezers, pins, or capillaries that serve to transfer the biochemical samples to the surface. Advantages of the microspotting technologies include ease of prototyping and therefore rapid implementation, low cost, and versatility. One disadvantage of microspotting is that each sample must be synthesized, purified, and stored prior to microarray fabrication. The microspotted microarrays contain as many as 10,000 groups of cDNAs in an area of ~3.6 cm^2. As such, microspotting is unlikely to produce the densities of photolithography.

A third group of microarray technologies, the "drop-on-demand" delivery approaches, provide another way to manufacture microarrays (Fig. 10.22C). The most advanced of these approaches are adaptations of ink-jetting technologies, which use piezoelectric and other forms of propulsion to transfer biochemical substances from miniature nozzles to solid surfaces. Similar to the microspotting approaches, drop-on-demand technologies allow high-density gridding of virtually

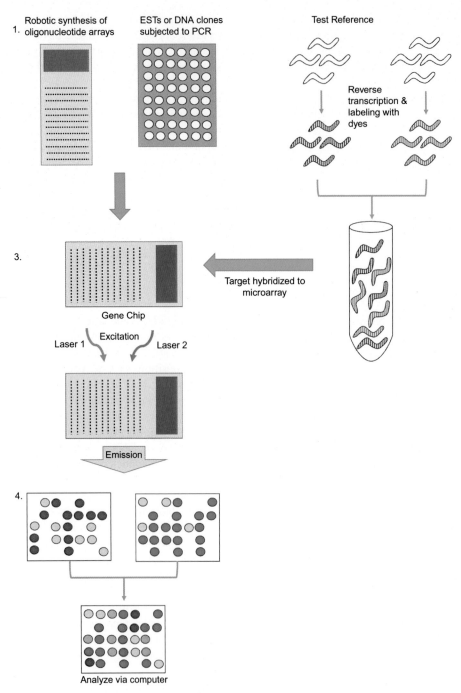

FIG. 10.21 The schematic process of DNA microarray analysis. (1) Random or sequence specific oligonucleotides array can be synthesized and printed on coated glass microscope slides to create gene chip. (2) Control and experimental RNA and cDNA sample preparation. (3) The two batches of cDNA are pooled and hybridized to gene chip. (4) Laser excitation of the hybridized gene chip for data collection.

any biomolecule of interest, including cDNAs, genomic DNAs, antibodies, and small molecules. Because ink jetting does not require direct surface contact, piezoelectric delivery is theoretically amenable to very high throughput. Piezoelectric-based delivery of phosphoramidite reagents has recently been used for the manufacture of high-density oligonucleotide microarrays. The successful application of ink jetting in a gene-expression setting (Fig. 10.23) demonstrates the immediate utility of this technology for genome analysis.

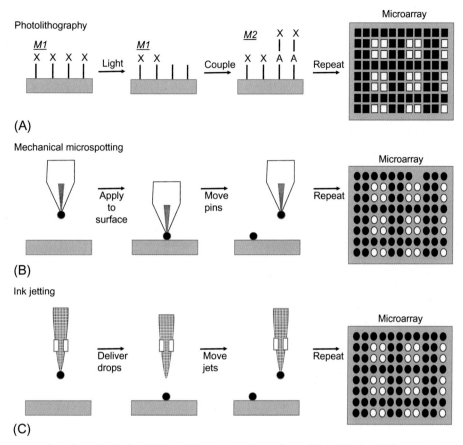

FIG. 10.22 Microarray manufacturing technologies. (A) Photolithography: a glass wafer modified with photolabile protecting groups (X) is selectively activated for DNA synthesis by shining light through a photomask (M1). The wafer is then flooded with a photoprotected DNA base (A–X), resulting in spatially defined coupling on the chip surface. A second photomask (M2) is used to deprotect defined regions of the wafer. Repeated deprotection and coupling cycles enable the preparation of high-density oligonucleotide microarrays. (B) Mechanical microspotting: a biochemical sample is loaded into a spotting pin by capillary action, and a small volume is transferred to a solid surface by physical contact between the pin and the solid substrate. After the first spotting cycle, the pin is washed and a second sample is loaded and deposited to an adjacent address. Robotic control systems and multiplexed printheads allow automated microarray fabrication. (C) Ink jetting: a biochemical sample is loaded into a miniature nozzle equipped with a piezoelectric fitting *(rectangles)*, and an electrical current is used to expel a precise amount of liquid from the jet onto the substrate. After the first jetting step, the jet is washed, and a second sample is loaded and deposited to an adjacent address. A repeated series of cycles with multiple jets enables rapid microarray production.

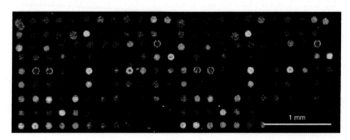

FIG. 10.23 An example of gene-expression monitoring with an ink-jetted microarray. This is a fluorescent scan of a high-density microarray printed. Coupling of the cDNAs to the chip surface occurs via a succinimidyl-ester-displacement reaction. The fluorescent sample was prepared by biotin incorporation into antisense RNA, followed by secondary labelling with Cy-5 conjugated streptavidin. Fluorescent intensities, represented in a pseudocolor scale, reflect gene-expression levels.

Sample Processing and Detection

A critical step in sample preparation is the isolation of mRNA from cells or tissues of interest. Because the expression pattern in a cell can change quite rapidly following perturbations such as heat shock or activation with lipopolysaccharide or other reagents, it is essential that the material be stored in the desired state, and all subsequent steps for isolation of the mRNA be carried out with the utmost care and speed. Laser capture microdissection has been used to separate single cells of interest from the surrounding tissue to improve sample purity.

Hybridization of the labeled sample which is known as probe is ideally sensitive so that low-abundance genes are detected, and specific so that targets hybridize only to the desired gene in the complex probe mixture. Consideration of the melting temperature equation for polynucleotides dictates that reactions performed in $4-5 \times$ SSC are performed at $60-65°C$. Hybridization reactions employing 50% formamide are typically performed at $42°C$, as the duplex melting temperature is lowered by $0.63°C$ per percentage formamide for DNA-DNA hybrids. This ensures adequate stringency and reduces the temperature-induced detachment of noncovalently attached target molecules. The fluorescent dyes in the probes also lower the duplex melting temperature so hybridization and washing conditions need not be as stringent as with nonfluorescently labeled probes. Salmon sperm DNA, poly[dA], tRNA, and SDS are added to the hybridization to eliminate nonspecific hybridization due to repetitive sequences.

The most common detection method is laser-induced fluorescence, detected using confocal optics. One scanner reads a 1-cm square chip in a single color at a resolution of $11.25 \mu m$ in $15 min$. Unlabeled probe can be detected using surface plasma resonance. Hybridization can be monitored in real time using a charge-coupled device imaging camera. Synthesis of arrays on optical fiber bundles allows easy and sensitive reading. Another real-time detection method is the use of the evanescent wave effect excites only fluorophores that are bound to the surface, thereby allowing real-time hybridization detection on microarrays without washing.

WESTERN BLOT

Separation of proteins from a complex mixture can be done by electrophoretic procedures, separating the protein molecules according to size or isoelectric point. Gel electrophoresis and electrofocusing have high resolving power, but the separated proteins, particularly in polyacrylamide gels, are difficult to access by macromolecular reagents. To identify individual proteins, it is necessary to render the proteins accessible for the identifying ligand. This can be achieved by transferring/blotting the proteins from the gel onto the surface of an adsorbent membrane. This technique is called Western blotting, analogous to Southern and Northern blotting of nucleic acids.

The blotted proteins form a replica of the gel, and they have been used in various ways. Antibodies are widely used to detect particular proteins which is known as the immunoblotting, usually with the aid of a second, enzyme-labeled antibody, or enzyme-labeled protein A or protein G. Biotin labels in combination with enzyme-labeled streptavidin are also frequently used. Most complementary binding reactions can be exploited, but, for those requiring intact native conformation of the separated protein, the electrophoresis must be carried out under nondenaturing conditions. Blotted proteins can function as ligands in the affinity purification of small amounts of proteins such as antibodies and can act as substrates for various protein-modifying reactions such as phosphorylation. In addition, portions of blots containing a protein of interest may be subjected to gas phase sequencing or be used as an immunogen either in vivo or in vitro.

Western blotting is currently widely used in clinical research. Individual proteins from a protein mixture such as plasma or other body fluids can be identified and then "rescued" and further studied by removing the labeling antibody, making the technique a powerful tool. It is gaining widespread application particularly for clinical diagnosis into the acquired immunodeficiency syndrome. For example, this method allows serum samples from patients to be screened for antibodies against a variety of viral antigens. It uses the antibodies in the serum as "probes" to show the presence of known viral antigens that have been run on an electrophoretic strip. By the use of standard, that is, positive-antisera, antigens from different sources can be studied and cross reactivity identified.

Gel Electrophoresis for Protein Separation

Two different types of electrophoresis have been developed for different purpose. The most widely used gel electrophoresis method is SDS-PAGE as described previously. In this method, proteins are fully denatured with SDS and heat in the presence of β-mercaptoethanol. Therefore, proteins are uniformly negatively charged and thus separated according to size by the sieving effect of the gel.

To separate native proteins, nondenaturing or native gels are used, where the proteins are dissolved in a suitable buffer without denaturing agents. As a gel, steep (3%–20%) acrylamide gradient may be used. The proteins are separated according to size and charge. Protein-protein or protein-nucleic acid interactions stay intact. A special form of native gel electrophoresis is isoelectric focusing (IEF), where native proteins are separated in a gel containing a pH gradient. The protein will finally be accumulated at their isoelectric point. Native electrophoresis works well for hydrophilic proteins. Hydrophobic or amphiphilic proteins, however, require detergents to stay in solution during separation.

Methods for Transferring Protein onto Membranes

Proteins can be transferred to a membrane either by diffusion (passive transfer), aided by capillary action or vacuum, or by electroblotting or electrotransfer, where the proteins are driven by an electric field.

Passive transfers of proteins out of the gel and onto the surface of a membrane can be achieved by simply posing a membrane sheet on one or both of the gel surfaces. Transfer occurs by diffusion and the proteins tend to accumulate on the membrane because of its high affinity for proteins. The transfer is not efficient with polyacrylamide gels but work well with agarose gels that have larger pores. This approach is particularly useful when the agarose gels are cast on a plastic backing sheet. To increase the speed of transfer, one can create a steady flow of buffer through the gel and the membrane, for example, by the capillary action of a stack of filter paper put on top of the membrane. The capillary action can also be improved by the vacuum action.

The transfer of proteins separated by IEF in agarose or in acrylamide gels is limited. The focused proteins are not in solution and they are at their isoelectric point, which means that they are not charged. Therefore, blotting to a membrane has to be by passive transfer and not by electrotransfer. If IEF is followed by SDS-PAGE, the separated proteins may be subjected to electroblotting.

Electrotransfer forces the proteins out of the gel and onto the membrane by an electric field applied across the gel and the superimposed membrane. This is achieved by putting the gel-membrane sandwich in a suitable holder and immersing it in a tank filled with buffer and fitted with two plate electrodes. Alternatively, stacks of filter paper, wetted with transfer buffer are placed on both sides of a gel-membrane sandwich. To increase transfer efficiency, discontinuous buffer systems are possible. For example, the filter paper stack on the anodic side of the membrane-gel sandwich may contain another buffer than the stack on the cathodic side. One or several such assemblies, put on top of each other and separated by a dialysis membrane, are then placed between plate electrodes. This so called semidry blotting requires no buffer tank and no cooling system. Furthermore, several gels can be processed simultaneously and the transfer time is shorter than with tank blotting.

Electroblotting is fast and gives a sharp replica of the gel. The strong electric force, however, can cause blow through. This means that some proteins may not be retained by the membrane. The effect increases proportionally with pore size of the membrane, the strength of the electric field, and decreasing size of the protein. Thus, for a given gel-membrane combination, careful adjustment of the electric field to lower values leaving enough time for the protein to bind to the membrane is essential.

Types of Membranes

Three types of widely used membranes used for protein blotting are nitrocellulose, charged or uncharged nylon and polyvinylidene difluoride (PVDF) (Table 10.4). The most commonly used membrane for protein blotting is nitrocellulose; it has satisfactory capacity, gives low backgrounds but is fragile. Nylon has much superior mechanical strength than nitrocellulose. Its protein binding capacity is good and it can be obtained as positively charged membrane that even increases the binding and retention capacity. PVDF seems to be the membrane of choice for many applications of Western blotting. It combines high protein binding capacity with excellent mechanical resistance and good staining properties. It is a

TABLE 10.4 Comparison of Membranes Used in Western Blotting

Membrane	Strength	Relative Binding; Protein Capacity	Staining
Nitrocellulose (NC)	Poor	1; 0.5	Good
Nylon	Good	0.6–0.8; 0.9	Poor
Polyvinylidene difluoride (PVDF)	Good	0.7; 1	Good

TABLE 10.5 Comparison of Protein Stains

Stain (Concentration)	Solvent	Destain	Membrane	Relative Sensitivity
Amido black (0.1%–0.5%)	10%–50% methanol, 2%–5% acetic acid	50% methanol, 5% acetic acid	NC, PVDF	1
Colloidal gold	–	Rinse in distilled water	NC, PVDF	1–28
Coomassie (0.1%)	50% methanol, 10% acetic acid	50% methanol, 10% acetic acid	PVDF	0.5
India ink (0.05%–0.1%)	0.3% Tween 20 or 0.5% Triton X-100	Rinse in PBS	NC, PVDF	1.25
Ponceau red (0.2%)	Distilled water	Water or 5% acetic acid	NC, PVDF	0.1
Transillumination	20% methanol	–	PVDF	1

NC, nitrocellulose; PBS, phosphate-buffered saline; PVDF, polyvinylidene difluoride.

hydrophobic membrane that must be made wet before use by briefly immersing it in methanol, and the membrane must not be allowed to dry during the whole blotting and detection procedure.

Before the protein sample on the blot is further analyzed, it is advisable to test if the electrophoretic separation worked properly, if the transfer onto the blot was satisfactory and if the blot spans the necessary range of molecular weight. This can conveniently be monitored by running in one lane of the gel with pre-stained molecular weight standard. Furthermore, a variety of dyes with different sensitivities can be used to stain all proteins transferred onto nitrocellulose and PVDF membranes (Table 10.5), whereas nylon is not suitable for total protein stain. Amino black and India ink are standard stains of similar sensitivity. Ponceau is the stain with the lowest sensitivity, however, it is convenient for many applications as it is reversible by briefly immersing the stained blot in 0.1 N NaOH. Protein bands on PVDF become translucent on immersion in methanol and thus can be visualized on a lightbox.

Detection of Proteins

Before probing blotted proteins with antibody, it is nearly always essential to block the unused binding sites on the membrane. This is done by incubating with a protein or protein mixture that is inert in the subsequent probing and detection reactions, or with detergents such as Tween-20 (Table 10.6). Ovalbumin/gelatin gives optimal blocking with lowest background and without impairing immunoreactivity of the blotted proteins. Although ovalbumin gives the best result, fat free milk powder has proved to be an economical and effective blocker in most systems. Blocking with Tween-20 is very simple but has a tendency to give elevated background. Furthermore, it may mask or detach immunoreactive proteins in some occasions.

Immunodetection of specific proteins in Western blots is usually done by the indirect method. This involves a poly- or monoclonal antibody (Ab, MAb) directed against the protein on the blot (primary antibody) and a labeled antispecies antibody (secondary Ab) or labeled protein A or G. The primary Ab should not only have a high binding affinity but also match the antigen configuration. This is usually no problem with polyclonal Ab, as they consist of a mixture of different Ab. On the other hand, MAb may be very specific in recognizing protein configuration. In a native gel, the MAb can recognize its epitope in the native state by either conformational or sequential MAb. If proteins are denatured by SDS-PAGE, the

TABLE 10.6 Comparison of Blocking Agents

Blocking Agent	Blocking Solution	Time/Temperature
Ovalbumin/gelatin	3% Ovalbumin, 0.25% gelatin in 100 mM Tris-HCl, pH 7.4	60 min at 40°C
Milk powder	5% nonfat powdered milk in 100 mM Tris-HCl, pH 7.4, 150 mM NaCl	30 min at room temp
Tween-20	0.3% Tween-20 in PBS	30 min at 37°C

PBS, phosphate-buffered saline.

TABLE 10.7 Comparison of Antibody Labels for Protein Detection

Label	Visualization by	Relative Sensitivity
Colloidal gold	Color	5
	Silver enhancement	50
Enzyme peroxidase	Color: 4-chloro-1-napthol	1
	3,3′-diaminobenzidine	1
	Chemiluminescence	50 (–500)
Phosphatase	Color: 5-bromo-4-chloro-3-indolyl phosphate and nitroblue tetrazolium	5
	Naphthol phosphate and fast red	5
	Chemiluminescence	50

MAb has to recognize a sequential epitope. Such MAb, however, may have problem to recognize a configuration. Proteins A and G are more universal than antispecies Ab, but protein A and protein G does not recognize IgG of all species or all IgG subtypes of a given species. Furthermore, it is hard to demonstrate convincingly a general superiority of one or the other.

A variety of secondary Ab labeling techniques have been developed, but enzyme as label for secondary Ab is commonly used. The most commonly used enzymes are horseradish peroxidase (HRP) and alkaline phosphatase (AP). For Western blots, chromogens which produce a colored precipitate rather than a colored solution are used. Table 10.7 lists two general purpose chromogens for each of the two enzymes.

The sensitivity of an enzyme detection system can be enhanced with bridging Ab, which recognizes the primary Ab as well as an antienzyme Ab, raised in the same species as the primary Ab. The antienzyme Ab then in turns binds the enzyme. This method can be performed with HRP or with AP. Other modifications consist of biotin-labeled secondary Ab followed by an avidin-biotin-AP complex. The purpose of these modifications is to increase the number of enzyme molecules per binding site of the primary Ab.

Chemiluminescence (CL) system provides a higher sensitivity than the color production system. The principle is that the enzyme activates a substrate for AP and cyclic diacylhydrazide for HRP, and then emits light. The signal is recorded on a photographic film. Advantage is that quantitative determinations of the detected protein are possible by densitometry of the exposed and developed film. Drawback of this method is that the necessity to find the right exposure time of the film for each blot.

Detection of radioactivity on blots can be necessary if the blotted proteins are radioactively labeled or if a radioactive ligand—for example, ^{125}I-labeled Ab—is used (Fig. 10.24). Radioactivity is most easily detected and recorded by exposing the blot on an X-ray film. The drawback of this method is the disposal of the radioactive materials.

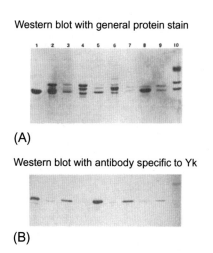

Western blot with general protein stain

(A)

Western blot with antibody specific to Yk

(B)

FIG. 10.24 An example of Western blotting: (A) 10 samples separated by electrophoresis and different proteins identified with a general protein stain; (B) the specific polypeptide Yk picked out by radioactively labeled antibody to Yk.

Any ligand that interacts firmly enough with its target to withstand the manipulation involved may serve to detect a specific protein on a blot. There are many examples for such interactions, among them the detection of glycosylated proteins by directly labeled or indirectly detected lectins. Two other fast expanding fields are DNA- or RNA-binding proteins and protein-protein interactions. For DNA- or RNA- binding proteins, one application is that SDS is removed during blotting so that the protein can renature. Subsequently, the blotted proteins are incubated under appropriate salt conditions with suitable labeled DNA or RNA. The nucleic acid bound to the proteins is then detected by its radioactivity. Likewise, proteins also need to be renatured for detecting protein-protein interaction. The protein counterpart interacting with the blotted protein may be radioactively labeled or may be traced immunologically for visualization of the protein-protein interaction.

Western blotting has proved invaluable in protein chemistry applied to such areas as enzymology and oncology: for example, to find different isoenzymes and their pattern of distribution in diseased and healthy tissue. Similarly, identification and localization of tumor markers within neoplastic material can easily be carried out with the appropriate antisera.

Enzyme-Linked Immunosorbent Assay (ELISA)

ELISA is based on the concept of antigen-antibody reactions, representing the chemical interaction between antibodies produced by the B cells of leukocytes and antigens. This specific immune response plays an important role in protecting the body from invaders such as pathogens and toxins. Therefore, by exploiting this reaction, ELISA permits the highly sensitive and selective quantitative/qualitative analysis of antigens, including proteins, peptides, nucleic acids, hormones, herbicides, and plant secondary metabolites. To detect these molecules, an antigen or antibody is labeled using enzymes, the so-called enzyme immunoassay, in which alkaline phosphatase (ALP), horseradish peroxidase (HRP), and β-galactosidase are commonly used. The antigen in the fluid phase is immobilized on a solid support, such as a microtiter plate constituted of rigid polystyrene, polyvinyl chloride, and polypropylene. Subsequently, the antigen is allowed to react with a specific antibody, which is detected by an enzyme-labeled secondary antibody. The development of color using a chromogenic substrate corresponds to the presence of the antigen.

During the color developing stage, ALP hydrolyzes *p*-nitrophenyl phosphate to produce *p*-nitrophenol, which can be detected at 405 nm (yellow color), and HRP catalyzes the conversion of chromogenic substrates, into colored products. By using chemiluminescent substrates such as chloro-5-substituted adamantyl-1,2-dioxetane phosphate and luminol for ALP and HRP, respectively, and fluorogenic substrates such as 4-methylumbelliferyl galactoside and nitrophenyl galactoside for β-galactosidase, even more sensitive detection can be achieved. These enzyme-substrate reactions are typically completed within 30–60 min, and the reaction stops with the addition of an appropriate solution, such as sodium hydroxide, hydrochloric acid, sulfuric acid, sodium carbonate, or sodium azide, for individual reactions. Finally, colored or fluorescent products are detected using a microtiter plate reader.

ELISA exhibits several advantages. For example, (i) it is a simple procedure; (ii) it has high specificity and sensitivity, because of an antigen-antibody reaction; (iii) it has high efficiency, as simultaneous analyses can be performed without complicated sample pretreatment; (iv) it is generally safe and eco-friendly, because radioactive substances and large amounts of organic solvents are not required; and (v) it is cost-effective, as low-cost reagents are used. However, ELISA is labor-intensive and expensive to prepare antibody because it is a sophisticated technique, and expensive culture cell media are required to obtain a specific antibody. Furthermore, there is a high possibility of false-positive or -negative ELISA results because of insufficient blocking of the surface of microtiter plate immobilized with antigen, and because antibodies are unstable.

Chapter 11

Techniques in Sequencing

Chapter Outline

CHAIN-TERMINATION DNA SEQUENCING TECHNIQUE

Principle of Dideoxynucleotide Procedure

Chain-termination DNA sequencing, also called the dideoxynucleotide procedure, is based on the principle that during DNA synthesis, addition of a nucleotide triphosphate requires a free hydroxyl group on the $3'$ carbon of the sugar of the last nucleotide of the growing DNA strand (Fig. 11.1). However, if a synthetic dideoxynucleotide that lacks a hydroxyl group at the $3'$ carbon of the sugar moiety is incorporated at the end of the growing chain, DNA synthesis stops because a phosphodiester bond cannot be formed with the next incoming nucleotide. As a result, DNA synthesis is terminated and this is the defining feature of the dideoxynucleotide DNA sequencing method.

Procedure

The starting material for a chain-termination sequencing experiment is a preparation of identical single-stranded DNA molecules. The first step is to anneal a short oligonucleotide (17–24 nucleotides) to the same position on each molecule. This oligonucleotide subsequently acts as the primer for synthesis of a new DNA strand that is complementary to the template (Fig. 11.2A). The strand synthesis reaction is catalyzed by the Klenow fragment of DNA polymerase.

Sanger's method performs this DNA synthesis in four separate tubes. Each tube requires a small amount of each of four dideoxynucleotides (ddNTPs—ddATP, ddCTP, ddGTP, and ddTTP), and the four deoxyribonucleotide triphosphates (dNTPs—dATP, dCTP, dGTP, and dTTP) as substrates. Klenow enzyme does not discriminate between deoxy- and dideoxynucleotides, but once incorporated, a dideoxynucleotide blocks further elongation because it lacks the $3'$-hydroxyl group needed to form a connection with the next nucleotide. As such, an excess of normal deoxynucleotides must be used, with just enough dideoxynucleotide to stop DNA strand extension once in a while at random. Because of random

Diagnostic Molecular Biology. https://doi.org/10.1016/B978-0-12-802823-0.00011-0

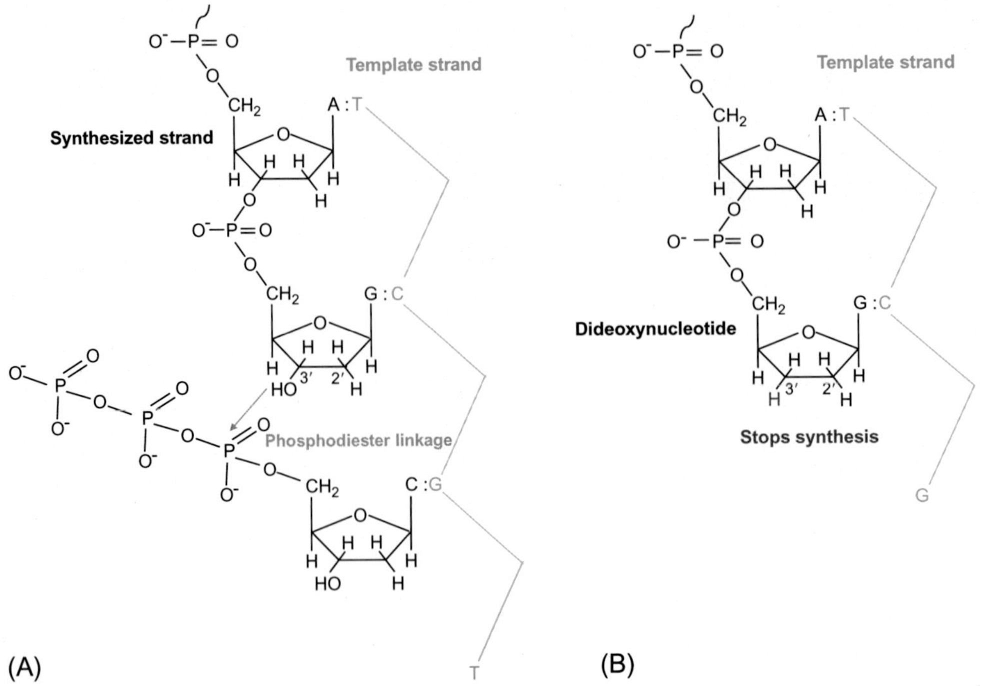

FIG. 11.1 The principle of chain termination of DNA synthesis. The 3′ OH (A) is replaced by an H (B), and this results in the termination of DNA synthesis.

termination of DNA synthesis, the reaction will produce different lengths of DNA strands. The result is a series of fragments of different lengths in each. For example, in tube 1 where ddATP is added, all the fragments end in A; in tube 2 where ddGTP is added, all the fragments end in G; in tube 3 where ddCTP is added, all fragments end in C; and in tube 4 where ddTTP is added, all fragments end in T (Fig. 11.2B). Subsequently, all four reaction mixtures are electrophoresed under denaturing conditions, so all DNAs are single-stranded. Because radioactive dATP is included in the reaction, autoradiography is performed to visualize the DNA fragments, which appear as horizontal bands on an X-ray film (Fig. 11.2C). As a result, the DNA sequence can be read through each band appeared on each tube.

DNA Polymerase Used in Chain-Termination Sequencing

In the original method for chain-termination sequencing, the Klenow polymerase was used as the sequencing enzyme. Because the Klenow polymerase has low processivity, it can only synthesize a relatively short DNA strand before dissociating from the template. This limits the length of sequence that can be obtained from a single experiment to about 250 bp. To improve this short synthesis problem, most sequencing today makes use of a more specialized enzyme, such as Sequenase which is a modified version of the DNA polymerase encoded by bacteriophage T7. Sequenase has high processivity and no exonuclease activity and so is ideal for chain-termination sequencing, enabling sequences of up to 750 bp to be obtained in a single experiment.

Template for Chain-Termination Sequencing

The template for a chain-termination experiment is a single-stranded version of the DNA molecule to be sequenced. One way of obtaining single-stranded DNA is to use an M13 vector. However, M13 vector is only good to use for DNA fragments that are shorter than 3 kb. Plasmid vectors are the alternatives, but some means is needed of converting the double-stranded plasmid into a single-stranded form. One way to convert the plasmid vectors to single-stranded DNA is to use alkali or boiling denaturation. This is a common method for obtaining template DNA for DNA sequencing. However, a shortcoming is that it can be difficult to prepare plasmid DNA that is not contaminated with small quantities of bacterial DNA and RNA, and these contaminated materials can act as spurious templates or primers in the DNA sequencing experiment. Another approach is to clone the DNA into a phagemid, a plasmid vector that contains an M13 origin of replication and which can therefore be obtained as both double- and single-stranded DNA versions. Phagemids can be used with fragments up to 10 kb or more.

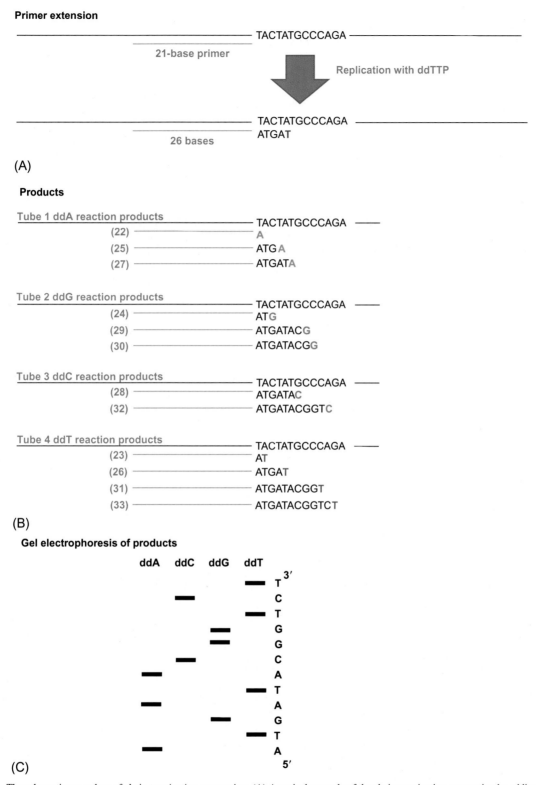

FIG. 11.2 The schematic procedure of chain termination sequencing. (A) A typical example of the chain-termination sequencing by adding ddTTP to stop the reaction. (B) The chain-termination sequencing is performing in four separate tubes and each tube contain one of the four dideoxynucleotides and the four deoxyribonucleotide triphosphates. The numbers in front of each nucleotide chain represent the length of that corresponding nucleotide chain in each reaction. (C) Gel electrophoresis of the chain termination sequencing to separate every nucleotide chains in each tube and reveal the sequence.

Automated DNA Sequencing

Automated sequencing has been developed to sequence a really large amount of DNA. This procedure uses the principle of the Sanger chain-termination method. Instead of labeling dATP in the original Sanger method, each of the dideoxynucleotides used in the reaction is labeled with a different fluorescent marker.

After the extension reactions and chain termination are completed, to work out the DNA sequence, the mixture is loaded into a well of a polyacrylamide slab gel, or into a tube of a capillary gel system, and electrophoresis is carried out to separate the molecules according to their lengths. After separation, the molecules are run past a fluorescent detector capable of discriminating the labels attached to the dideoxynucleotides (Fig. 11.3). The detector therefore determines if each molecule ends in an A, C, G, or T. The sequence can be printed out for examination, or entered directly into a storage device for future analysis. The entire dideoxynucleotide sequencing process has been automated to increase the rate of acquisition of DNA sequence data. This is essential for large-scale sequencing projects such as those involving whole prokaryotic or eukaryotic genomes.

FIRST HIGH-THROUGHPUT SEQUENCING: PYROSEQUENCING

Pyrosequencing was the first of the high-throughput DNA sequencing (also known as next-generation sequencing [NGS]) technologies to be made commercially available and has contributed to the rapid output of large amounts of sequence data by the scientific community. The advantage with pyrosequencing is that it can be automated in a massively parallel manner that enables hundreds of thousands of sequences to be obtained at once, perhaps as much as 1000 million base pairs (Mb) in a single run. A sequence is therefore produced much more quickly than is possible by the chain-termination method.

Principle of Pyrosequencing

The basis of pyrosequencing is the detection of pyrophosphate that is released during DNA synthesis. It requires the preparation of identical single-stranded DNA molecules as the starting material. These can be obtained by alkali denaturation of PCR products or, more rarely, recombinant plasmid molecules. After attachment of the primer, the template is copied by a DNA polymerase (usually a Klenow fragment) in a straightforward manner without added dideoxynucleotides. When a DNA strand is extended by DNA polymerase, the α-phosphate attached to the 5′ carbon of the sugar of an incoming deoxynucleotide triphosphate forms a phosphodiester bond with the 3′ hydroxyl group of the last nucleotide of the growing strand. The terminal β- and γ-phosphates of the incoming nucleotide are released as a unit known as pyrophosphate (Fig. 11.4A). The release of the pyrophosphate correlates with the incorporation of a specific nucleotide in the growing DNA strand. In this way, the addition of a deoxynucleotide to the end of the growing strand is detectable through the release of a molecule of pyrophosphate in real time, which can be converted by the enzyme sulfurylase into a flash of chemiluminescence (Fig. 11.4B).

The pyrophosphate, released from the incorporation of a nucleotide, combines with adenosine-5′-phosphosulfate in the presence of the enzyme ATP sulfurylase to form adenosine triphosphate (ATP). In turn, ATP drives the conversion of luciferin to oxyluciferin by the enzyme luciferase, a reaction that generates light. Detection of light after each cycle of nucleotide addition and enzymatic reactions indicates the incorporation of a complementary nucleotide. Thus, light emission is proportional to the amount of pyrophosphate produced, which is directly proportional to the number of nucleotides added. Whether a given dispensed dNTP can be incorporated or not, apyrase catalyzes the degradation of excess dNTPs before the next dNTP is dispensed. Therefore, pyrosequencing does not require electrophoresis or any other fragment separation procedure and so is more rapid than chain-termination sequencing.

In pyrosequencing, each deoxynucleotide is added separately, one after the other, with a nucleotidase enzyme also present in the reaction mixture so that if a deoxynucleotide is not incorporated into the polynucleotide, it is rapidly degraded before the next one is added (Fig. 11.5). This procedure makes it possible to follow the order in which the deoxynucleotides are incorporated into the growing strand. If all four deoxynucleotides were added at once, the flashes of light would be seen all the time and it would be difficult to determine which deoxynucleotide is added. Therefore, this technique sounds complicated, but it simply requires that a repetitive series of additions be made to the reaction mixture, precisely the type of procedure that is easily automated.

Advantage of Pyrosequencing

Although it is only able to generate up to 150 bp in a single experiment, the advantage of pyrosequencing is that it can be automated in a massively parallel manner that enables hundreds of thousands of sequences to be obtained at once, perhaps as much as 1000 Mb in a single run. Sequence is therefore produced much more quickly than is possible by the chain-termination method.

Primer extension

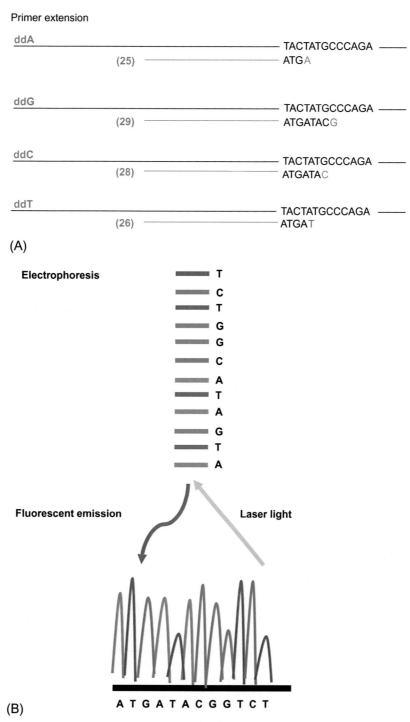

(A)

(B)

FIG. 11.3 The process of automated sequencing. The procedure is very similar to Sanger's chain termination. (A) Each of the dideoxynucleotides used in the reaction is labeled with a different fluorescent marker. The chain-termination sequencing is performing in one tube and this tube contain four fluorescent labeled dideoxynucleotides. The numbers in front of each nucleotide chain represent the length of that corresponding nucleotide chain in each reaction. (B) After the extension reactions, the mixture is loaded into a well of a polyacrylamide slab gel, or into a tube of a capillary gel system, and electrophoresis is carried out to separate the molecules according to their lengths. After separation, the molecules are run past a fluorescent detector capable of discriminating between the labels attached to the dideoxynucleotides.

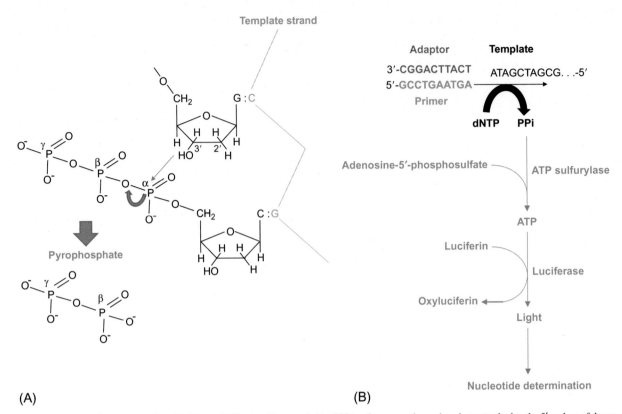

FIG. 11.4 Principle of pyrosequencing. (A) When a DNA strand is extended by DNA polymerase, the α-phosphate attached to the 5′ carbon of the sugar of an incoming deoxynucleotide triphosphate forms a phosphodiester bond with the 3′ hydroxyl group of the last nucleotide of the growing strand. The terminal β- and γ-phosphates of the added nucleotide are cleaved off as a unit known as phosphate. (B) The molecule of pyrophosphate can be converted by the enzyme sulfurylase into a flash of chemiluminescence.

TECHNIQUES TO SEQUENCE A GENOME

Sequencing the whole genome not only can contribute to our understanding of the nature of an organism but also can help to unravel the complexity of human disease or identify microorganisms that may cause a disease. The first DNA genome to be sequenced was bacteriophage φX174 (5375 bp). This was quickly followed by sequences for SV40 virus (5243 bp) and pBR322 (4363 bp). Gradually sequencing was applied to larger molecules such as the human mitochondrial genome (16.6 kb) and bacteriophage λ (49 kb). The pioneering projects today are the massive genome initiatives, each aimed at obtaining the nucleotide sequence of the entire genome of a particular organism. For example, the first chromosome sequence, for chromosome III of the yeast *Saccharomyces cerevisiae*, was published in 1992, and the entire yeast genome was completed in 1996. There are now complete genome sequences for many organisms.

The total size of a fairly typical bacterial genome is 4,000,000 bp and the human genome is 3,200,000,000 bp. However, a single pyrosequencing yields up to 150 bp. Clearly a large number of sequencing experiments must be carried out in order to determine the sequence of an entire genome. Although the generation of sufficient sequence data can be done in an automated system, the problem that arises is the need to assemble the thousands or perhaps millions of individual sequences into a contiguous genome sequence. Two different strategies have been developed for sequence assembly (Fig. 11.6). The shotgun approach breaks down genome randomly into short fragments. The resulting sequences are examined for overlaps, and these are used to build up the contiguous genome sequence. The second technique is the clone contig approach, which involves a presequencing phase during which a series of overlapping clones is identified. This contiguous series is called a contig. Each piece of cloned DNA is then sequenced, and this sequence placed at its appropriate position on the contig map in order to gradually build up the overlapping genome sequence.

Shotgun Cloning Strategy

A shotgun library is constructed by fragmenting genomic DNA and inserting the fragments into a universal vector to generate a sequencing template. To obtain random, overlapping fragments, the genomic DNA is usually sheared and the

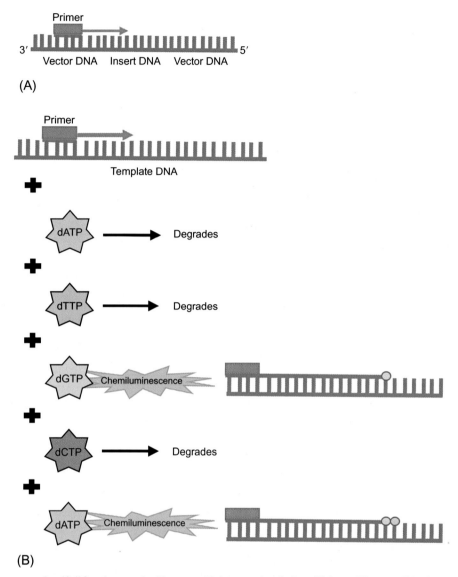

FIG. 11.5 (A) In pyrosequencing, if all four deoxynucleotides were added at once, then flashes of light would be seen all the time, and no useful sequence information would be obtained. (B) Each deoxynucleotide is, therefore, added separately, one after the other, with a nucleotidase enzyme also present in the reaction mixture so that if a deoxynucleotide is not incorporated into the polynucleotide, then it is rapidly degraded before the next one is added. This procedure makes it possible to follow the order in which the deoxynucleotides are incorporated into the growing strand.

shearing conditions are optimized to obtain uniformly sized DNA fragments. The fragmentation usually produces either a 5′ or a 3′ single-stranded end. For the 5′ single-stranded end, polymerase is used to fill in the 3′ recessed end of the complementary strand, and for the 3' single-stranded end, exonuclease is used to remove the protruding 3′ end. Subsequently, these DNA fragments are ligated to the vector DNAs randomly. The resulting recombinant DNA molecules are known as the library, which is introduced into *Escherichia coli* separately for amplification and act as sequencing template. Primers that anneal to the complementary vector sequences flanking the insert are used to obtain the sequence of both ends of the cloned DNA fragment using the dideoxynucleotide method.

The next step for the shotgun approach is to identify overlaps between all the individual sequences that are generated. This identification process must be accurate and unambiguous so that the correct genome sequence is obtained. An error in identifying a pair of overlapping sequences could lead to the genome sequence becoming scrambled, or parts being left out entirely. The probability of making mistakes increases with larger genome sizes, so the shotgun approach has been used mainly with the smaller bacterial genomes.

Shotgun sequencing has been successful with many bacterial genomes. This is because these genomes are small and they have very few or none repetitive DNA sequence. Therefore, the computational requirements for finding sequence

FIG. 11.6 Genome sequence assemble. In the shotgun approach, the genome is randomly broken into short fragments, and the resulting sequences are examined for overlaps and are used to build up the contiguous genome sequence. The clone contig approach involves a presequencing phase during which a series of overlapping clones is identified. This contiguous series is called a contig. Each piece of cloned DNA is then sequenced, and this sequence is placed at its appropriate position on the contig map in order to gradually build up the overlapping genome sequence.

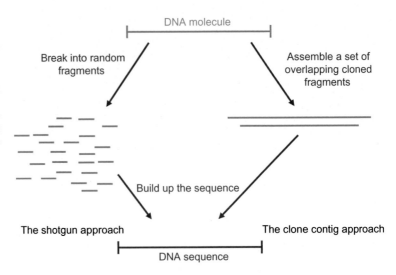

overlaps are not too great. The repetitive DNA sequence usually causes problems for the shotgun approach. This is because when sequences are assembled, those that lie partly or wholly within one repeat element might accidentally be assigned an overlap with the identical sequence present in a different repeat element (Fig. 11.7). This could lead to a part of the genome sequence being placed at the incorrect position or left out entirely. For this reason, it has generally been thought that shotgun sequencing is inappropriate for eukaryotic genomes, as these have many repeat elements.

The Clone Contig Approach

The clone contig approach does not suffer from the limitations of shotgun sequencing and so can provide an accurate sequence of a large genome that contains repetitive DNA. Its drawback is that the additional time and effort is needed to construct the overlapping series of cloned DNA fragments. This is because that the genome is broken down into several segments of up to 1.5 Mb, and each with a known position on the genome map before sequencing is carried out. This can be achieved by partial restriction followed by cloning into a high-capacity vector such as a Bacterial Artificial Chromosome (BAC) or a Yeast Artificial Chromosome (YAC). A clone contig is built up by identifying clones containing overlapping fragments, which are then individually sequenced by the shotgun method. Contig assembly is an important step in genome assembly. For mapping, overlapping clones are assembled to sequence that overlap. The basic principle of sequence assembly is that the overlapping clones share the same nucleotide sequence. Each fragment is cloned in a vector and sequenced from both ends to produce a sequence length of approximately 600–700 bp. The sequence from both ends of DNA fragment is called a pair end. The sequence from both ends of insert DNA and the distance between them is approximately known. Each fragment sequence is compared with other similar fragment. The next step is to cluster the fragments which share similarities. Subsequently, all fragments are subject to sequencing and the genome sequence built up step by step (Fig. 11.8). The cloned fragments should be as long as possible in order to minimize the total number needed to cover the entire genome. A high-capacity vector is therefore necessary.

ASSEMBLY APPROACH FOR CLONE CONTIGS—CLONE FINGERPRINTING TECHNIQUES

Clone Contig Assembly by Chromosome Walking

One technique that can be used to assemble a clone contig is chromosome walking. Chromosome walking was originally used to move relatively short distances along DNA molecules, using clone libraries prepared with λ or cosmid vectors. The most straightforward approach is to use the clone at random from the library, labeled, and used as a hybridization probe against all the other clones in the library (Fig. 11.9A). Those clones that give hybridization signals are ones that overlap with the probe. One of these overlapping clones is now labeled, and a second round of probing is carried out. More hybridization signals are seen, some of these indicating additional overlaps (Fig. 11.9B). Gradually the clone contig is built up in a step-by-step fashion. However, this is a laborious process and is only attempted when the contig is for a short chromosome

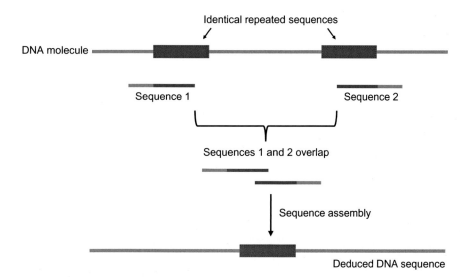

FIG. 11.7 A possible problem with the shotgun sequencing approach. It is possible that an incorrect overlap is made between two sequence reads that both end within repeat sequences. This can lead to the loss of a segment of the DNA in between the two sequences.

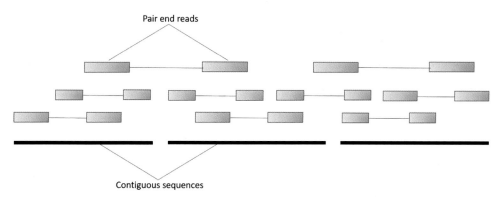

FIG. 11.8 Genome sequence assembly by the clone contig approach. Paired end reads are sequence data generated from both ends of a DNA fragment. Using a computer program, a large quantity of reads can be generated and assembled into longer contiguous sequences (contigs).

and so involves relatively few clones, or when the aim is to close one or more small gaps between contigs that have been built up by more rapid methods. Another problem that arises is that if the probe contains a genome-wide repeat sequence then it will hybridize not only to overlapping clones but also to nonoverlapping clones whose inserts also contain copies of the repeat.

Rapid Methods for Clone Contig Assembly

The weakness of chromosome walking is that it begins at a fixed starting point and builds up the clone contig step by step from that fixed point. It is a slow process and it is rarely possible to assemble contigs of more than 15–20 clones. The more rapid techniques for clone contig assembly do not use a fixed starting point. Instead, they aim to identify pairs of overlapping clones. The idea is that when enough overlapping pairs have been identified, the contig is revealed (Fig. 11.10). The various techniques that can be used to identify overlaps are collectively known as clone fingerprinting, which is based on the identification of sequence features that are shared by a pair of clones. Clone fingerprinting provides information on the physical structure of a cloned DNA fragment.

The simplest approach is to digest each clone with a variety of restriction endonucleases to separate the products in an agarose gel and to look for pairs of clones that share restriction fragments of the same size, excluding those fragments that derive from the vector rather than the inserted DNA. If two clones contain overlapping inserts then their restriction fingerprints will have bands in common, as both will contain fragments derived from the overlap region.

FIG. 11.9 Clone contig assembled by chromosome walking. In the beginning, a clone is selected at random from the library, labeled, and used as a hybridization probe against all the other clones in the library. Those clones that give hybridization signals are ones that overlap with the probe. One of these overlapping clones is now labeled, and a second round of probing carried out. More hybridization signals are seen, some of these indicating additional overlaps. Gradually, the clone contig is built up in a step-by-step fashion. For example, Oligonucleotides 2 and 5 both hybridize to the same clone in the genomic library, indicating that Contigs I and III are adjacent.

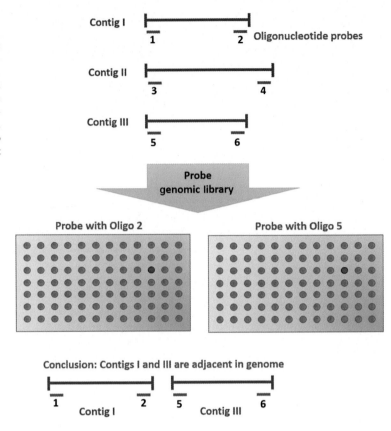

FIG. 11.10 Building up a series of overlapping clones using a clone fingerprinting technique.

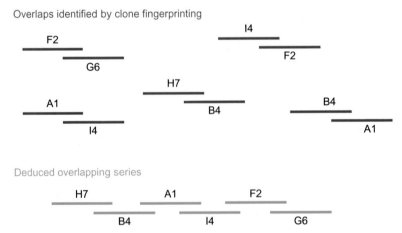

Another approach is to perform repetitive DNA PCR, or interspersed repeat element PCR (IRE-PCR). This type of PCR uses primers that are designed to anneal within repetitive DNA sequences and direct amplification of the DNA between adjacent repeats (Fig. 11.11). Because genome-wide repeat sequences are not evenly spaced in a genome, the sizes of the products obtained after repetitive DNA PCR can be used as a fingerprint in comparisons with other clones, in order to identify potential overlaps. If a pair of clones gives PCR products of the same size, they must contain repeats that are identically spaced, possibly because the cloned DNA fragments overlap.

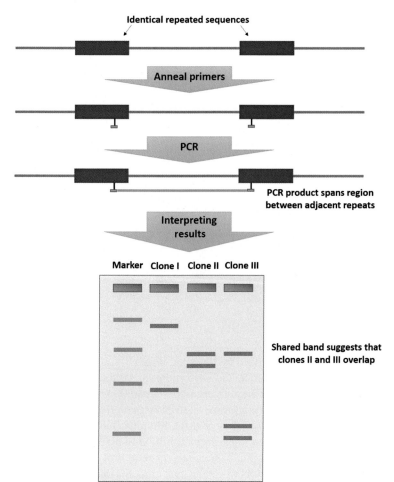

Identical repeated sequences

Anneal primers

PCR

PCR product spans region between adjacent repeats

Interpreting results

Marker   Clone I   Clone II   Clone III

Shared band suggests that clones II and III overlap

FIG. 11.11 Repetitive DNA PCR interspersed repeat element PCR (IRE-PCR). This type of PCR uses primers that are designed to anneal within repetitive DNA sequences and direct amplification of the DNA between adjacent repeats. Repeats of a particular type are distributed fairly randomly in a eukaryotic genome, with varying distances between them, so a variety of product sizes is obtained when these primers are used with clones of eukaryotic DNA. If a pair of clones gives PCR products of the same size, they must contain repeats that are identically spaced, possibly because the cloned DNA fragments overlap.

Clone Contig Assembly by Sequence Tagged Site Content Analysis

A third way to assemble a clone contig is to search for pairs of clones that contain a specific DNA sequence that occurs at just one position in the genome under study. If two clones contain this DAN sequence which is called a sequence tagged site (STS), then clearly they must overlap (Fig. 11.12). Often an STS has been sequenced in an earlier project. As the sequence is known, a pair of PCR primers can be designed that are specific for that STS and then used to identify which members of a clone library contain the STS. The STS does not have to be a gene and can be any short piece of DNA sequence, the only requirement being that it occurs just once in the genome.

HIGH-THROUGHPUT NGS STRATEGIES

Although both shotgun and clone contig approaches have been used successfully to obtain the sequences of many whole genomes, the preparation of clone libraries is costly and time-consuming for routing sequencing of large amounts of genomic DNA. To circumvent these problems, high-throughput NGS strategies have been developed. For NGS analysis, cell-free methods is used to generate a library of genomic DNA fragments in a dense array on the surface of a glass slide or in picoliter volume wells of a multiwall plate. This minimizes the volume of reagents for the sequencing reactions and enables hundreds of millions of sequences to be acquired simultaneously. The first NGS technologies to appear, referred to as second-generation sequencing (SGS), relied on cycles of the termination of DNA polymerization and recording of the incorporated nucleotides in each cycle.

Since 2005, at least four SGS platforms have been developed, including the Roche 454 GS FLX system, Applied Bio-systems SOLiD (supported oligonucleotide ligation and detection), Life Technologies Ion Personal Genome Machine

FIG. 11.12 Clone contig assembly by sequence tagged content. If two clones contain a specific DNA sequence that occurs at just one position in the genome under study, then they must overlap. A sequence of this type is called a sequence tagged site (STS). A pair of PCR primers can be designed that are specific for a gene that has been sequenced in an earlier project, and then used to identify which members of a clone library contain the gene.

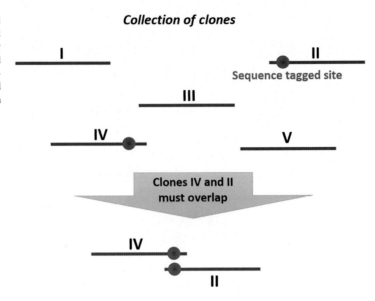

(PGM)/MiSeq systems, and Solexa GA (Genome Analyzer)/HiSeq developed by Illumina. While SOLiD uses sequencing by ligation (no DNA polymerase is required), all other systems employ sequencing by synthesis. All these NGS platforms conduct either bridge amplification (Illumina platforms) or emulsion PCR amplification for preparation of "clusters" of the same templates for sequencing. The read lengths are variable among these platforms: up to 75 bp (paired end), 300 bp (overlapping paired end), 400 bp (bidirectional), and 700 bp (paired end) produced by SOLiD, Illumina, Ion Torrent, and 454 systems, respectively. Furthermore, the number of reads per NGS run can range from 1 million to 5000 million with the machine running time varying from 8 h to 11 days, depending on the platform. In the relatively short time frame since 2005, NGS has fundamentally altered genomics research and allowed investigators to conduct experiments that were previously not technically feasible or affordable.

Fundamentals of NGS Platforms

NGS platforms share a common technological feature: massively parallel sequencing of clonally amplified or single DNA molecules that are spatially separated in a flow cell. In NGS, sequencing is performed by repeated cycles of polymerase-mediated nucleotide extensions or, in one format, by iterative cycles of oligonucleotide ligation. As a massively parallel process, NGS generates hundreds of megabases to gigabases of nucleotide sequence output in a single instrument run, depending on the platform.

Massively Parallel Pyrosequencing: Roche/454 Life Sciences

The 454 technology is derived from the technological convergence of pyrosequencing and emulsion PCR and is termed massively parallel pyrosequencing. The principle of pyrophosphate detection was described previously. This technique has been commercialized for the analysis of 96 samples in parallel in a microtiter plate. The first next-generation system developed by 454 Life Sciences was introduced in 2005. In 2007, the second version of the 454 instrument, the GS FLX, was developed. Sharing the same core technology as the first version, the GS FLX flow cell is referred to as a "picotiter well" plate. In its newest configuration, approximately 3.4×10^6 picoliter-scale sequencing-reaction wells are etched into the plate surface, and the well walls have a metal coating to improve signal-to-noise discrimination.

For sequencing, a library of template DNA is prepared by fragmentation into several hundred base pairs in length through sonication. Subsequently, these fragments are end-repaired and ligated to adapter oligonucleotides (Fig. 11.13). The library is then diluted to single-molecule concentration, denatured, and hybridized to individual beads containing sequences complementary to adapter oligonucleotides. The beads are compartmentalized into water-in-oil microvesicles, where clonal expansion of single DNA molecules bound to the beads occurs during emulsion PCR.

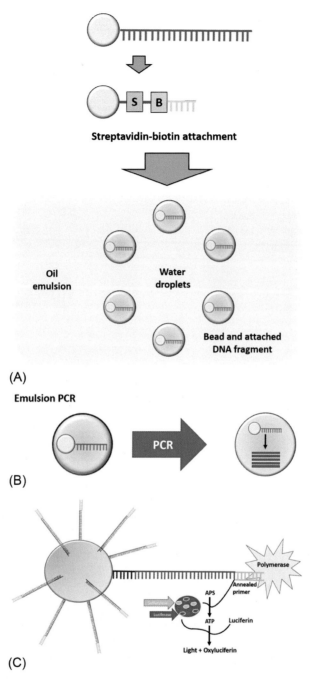

Streptavidin-biotin attachment

Oil
emulsion

Water
droplets

Bead and attached
DNA fragment

(A)

Emulsion PCR

PCR

(B)

Polymerase

Annealed
primer

APS

Sulfurylase

Luciferase

ATP   Luciferin

Light + Oxyluciferin

(C)

FIG. 11.13   The principle of massively parallel pyrosequencing. (A) Each DNA fragment is attached to a single bead via a streptavidin-biotin linkage. Immobilization of the DNA fragments by base pairing to metallic beads. Individual beads, with their attached DNA fragments, are placed within water droplets in an oil-water emulsion. (B) The beads are compartmentalized into water-in-oil microvesicles, where clonal expansion of single DNA molecules bound to the beads occurs during emulsion PCR. (C) Schematic illustration of the pyrosequencing reaction which occurs on nucleotide incorporation to report sequencing-by-synthesis.

Emulsion PCR is carried out for fragment amplification, with water droplets containing one bead and PCR reagents immersed in oil. The amplification is necessary to obtain sufficient light signal intensity for reliable detection in the sequencing-by-synthesis reaction steps.

For emulsion PCR, fragments of the source's genomic DNA are first ligated at each end to two different adaptors. The adaptors have specific sequences for binding PCR primers and for binding sequencing primers (Fig. 11.14). PCR primers that are complementary to a sequence on one of the adaptors are bound to DNA capture beads. After ligation of the adaptors,

FIG. 11.14 The principle of Emulsion PCR. Genomic DNA fragments are digested and ligated to two different adaptors that have sequences for capturing the fragments on the surface of beads and for binding PCR and sequencing primers. The fragments are denatured and mixed with beads under condition that favor binding of one DNA molecule per bead. The DNA-bound beads are mixed with PCR components and oil to create a PCR microreactor in an oil globule. Emulsion PCR results in millions of copies of the genomic DNA fragment attached to the bead within the globule. At the end of the PCR cycles, the beads are released from the emulsion, the DNA is denatured, and the beads are deposited into the wells of a multi-well plate. The bead-deposition process maximizes the number of wells that contain a single amplified library bead (avoiding more than one single stranded DNA library bead per well). The single-stranded genomic DNA fragment attached to the beads serve as template for pyrosequencing or sequencing by ligation.

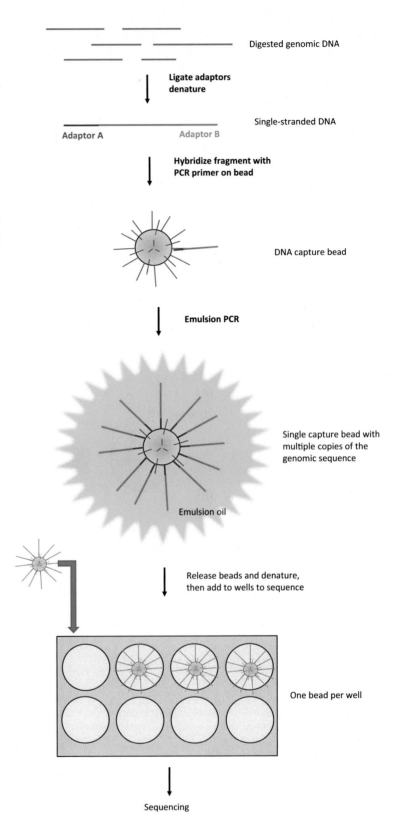

the genomic DNA fragments are melted and the single-stranded molecules anneal to the beads through complementary base-pairing. Each DNA capture bead carries more than 10^7 primer molecules. For amplification by emulsion PCR, the DNA capture beads carrying the hybridized DNA templates are mixed with the PCR components and oil to create a water-in-oil emulsion. Each oil globule is a separate reaction chamber, and the amplification products that remain bound to the bead by hybridization with the primes are contained within the globule.

The emulsion is disrupted after amplification, and the beads containing clonally amplified template DNA are enriched. The beads are again separated by limiting dilution, deposited into individual picotiter-plate wells, and combined with sequencing enzymes. Each bead with its one amplified fragment is placed at the top end of an etched fiber in an optical fiber chip, created from glass fiber bundles. The individual glass fibers are excellent light guides, with the other end facing a sensitive charge-coupled device (CCD) camera, enabling positional detection of emitted light. Each bead thus sits on an addressable position in the light guide chip, containing several hundred thousand fibers with attached beads.

Subsequently, the picotiter plate functions as a flow cell wherein iterative pyrosequencing is performed by successive flow addition of the four dNTPs in the presence of polymerase enzyme and primer. One unlabeled nucleotide only is supplied to the reaction mixture to all beads on the chip, so that synthesis of the complementary strand can start. Incorporation of a following base by the polymerase enzyme in the growing chain releases a pyrophosphate group, which can be detected as emitted light. Knowing the identity of the nucleotide supplied in each step, the presence of a light signal indicates the next base incorporated into the sequence of the growing DNA strand. Therefore, the luminescence is transmitted through the fiber-optic plate and recorded on a CCD camera. With the flow of each dNTP reagent, a linear sequence output can be generated. With the newest chemistry, termed "Titanium," a single GS FLX run generates approximately 1×10^6 sequence reads, with read lengths of ≥ 400 bases yielding up to 500 Mb of sequence. The advantage of the 454 technology is thus the longer read length, which facilitates de novo assembly of genomes.

Sequencing-by-Synthesis: Illumina/Solexa

The Solexa sequencing platform was commercialized in 2006. The principle is on the basis of sequencing-by-synthesis chemistry, with novel reversible terminator nucleotides for the four bases each labeled with a different fluorescent dye, and a special DNA polymerase enzyme able to incorporate them. Reversible chain terminators are synthetic nucleotides. All four deoxynucleotides can be added to the reaction at the same time which can determine the DNA sequence rapidly. Reversible chain terminators can be controlled to ensure that only a single nucleotide is incorporated during each cycle and each of the four deoxynucleotide triphosphates can be recognized. In line with these two important modifications, each chain terminator carries a chemical blocking group at the 3' carbon of the sugar moiety to prevent addition of more than one nucleotide during each round of sequencing, and a different fluorescent dye is added to each of the four nucleotides to enable identification of the incorporated nucleotide (Fig. 11.15A). The fluorophore is added at a position that does not interfere with either base-pairing or phosphodiester bond formation. Similar to dideoxynucleotide procedure and pyrosequencing, DNA polymerase is used to catalyze the addition of the modified nucleotides to an oligonucleotide primers specified by the DNA template sequence (Fig. 11.15B). After recording fluorescent emission, the fluorescent dye and the 3' blocking group are removed. After the removal of the blocking group, the 3' hydroxyl group of the sugar is restored so the subsequent addition of another nucleotide is possible. Generally, 50–100 nucleotides per run are generated. Because of its lower cost, high throughput, and accuracy, the Solexa sequencing platform has become the first choice sequencing technology across many fields of research and medical diagnostics.

The Genome Analyzer uses a flow cell consisting of an optically transparent slide with eight individual lanes, on the surfaces of which are bound oligonucleotide anchors (Fig. 11.16). Template DNA is fragmented into lengths of several hundred base pairs and end-repaired to generate 5'-phosphorylated blunt ends. The polymerase activity of Klenow fragment is used to add a single A base to the 3' end of the blunt phosphorylated DNA fragments. This addition helps the DNA fragments for ligation to oligonucleotide adapters because of an overhang of a single T base at their 3' end. This can increase ligation efficiency. The adapter oligonucleotides are complementary to the flow-cell anchors. Under limiting-dilution conditions, adapter-modified, single-stranded template DNA is added to the flow cell and immobilized by hybridization to the anchors.

Subsequently, DNA templates are amplified in the flow cell by "bridge" amplification. This specific amplification relies on captured DNA strands "arching" over and hybridizing to an adjacent anchor oligonucleotide. In the mixture containing the PCR amplification reagents, the adapters on the surface act as primers for the PCR amplification. After several PCR cycles, random clusters of about 1000 copies of single-stranded DNA fragments, also known as DNA "polonies," are created on the surface. Approximately 50×10^6 separate clusters can be generated per flow cell. The sequencing and DNA synthesis reaction mixture is supplied onto the surface and contains primers, four reversible terminator nucleotides

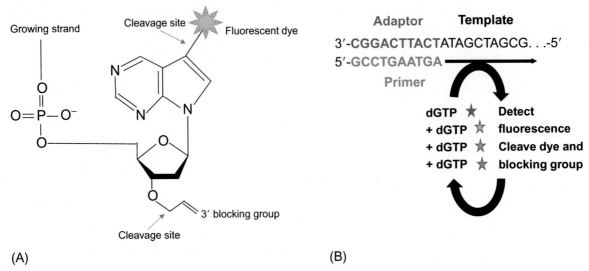

FIG. 11.15 The principle of sequencing using reversible chain terminators. (A) Reversible chain terminators are modified nucleotides that have a removable blocking group on the oxygen of the 3′ position of the sugar to prevent addition of more than one nucleotide per sequencing cycle. To enable identification, each of the four nucleotides carries different fluorescent dye. (B) An adaptor sequence is added to the 3′ end of the DNA sequencing template that provides a binding site for a sequencing primer. All four modified nucleotides are added at the same time. Once the fluorescence is detected, both the dye and the 3′ blocking group are cleaved before the next cycle.

each labeled with a different fluorescent dye, and the DNA polymerase. For sequencing, the clusters are denatured, and a subsequent chemical cleavage reaction and wash leave only forward strands for single-end sequencing. Sequencing of the forward strands is initiated by hybridizing a primer complementary to the adapter sequences, which is followed by addition of polymerase and a mixture of four differently colored fluorescent reversible dye terminators. This iterative, sequencing-by-synthesis process requires approximately 2.5 days to generate read lengths of 36 bases. With 50×10^6 clusters per flow cell, the overall sequence output is >1 billion base pairs (Gb) per analytical run.

The Genome Analyzer II is a revised platform and has optical modifications enabling analysis of higher cluster densities. A paired-end module for the sequencer was introduced, and the new instrument triples the output per paired-end run from 1 to 3 Gb. The system generates at least 1.5 Gb of single-read data per run, and at least 3 Gb of data in a paired-end run, recording data from more than 50 million reads per flow cell. The run time for a 36-cycle run was decreased to 2 days for a single-read run, and 4 days for a paired-end run.

Sequencing by Ligation: Applied Biosystems/SOLiD

The SOLiD (Supported Oligonucleotide Ligation and Detection) System 2.0 platform is a short-read sequencing technology based on ligation. Sequencing by ligation extends the DNA strand by ligation of short oligonucleotide through DNA ligase in a template-dependent fashion. This technique is completely different form both pyrosequencing and sequencing using reversible terminators because these two techniques extend the growing DNA strand by a single base during each cycle. For sequencing by ligation, the oligonucleotides are eight nucleotides (octamers) in length with two known nucleotides at the 3' (query) end, any nucleotides in the next three (degenerate) positions, and a sequence that is common (universal) to all of the oligonucleotides at the 5' end (Fig. 11.17A). A set of 16 different oligonucleotides representing all possible combinations of two nucleotides in the query position is used. Each oligonucleotide is tagged at the 5' end with a different fluorescent dye that corresponds to the query nucleotide composition. On the other hand, a short nucleotide adaptor is joined to the 5' ends of the DNA templates that are to be sequenced (Fig. 11.17B). This adaptor is complementary to a particular primer. After denaturation of the template, this particular primer binds to the adaptor sequence. In this way, a free 5' phosphate is available for the 3' hydroxyl end of the hybridized octamer. Ligation reaction occurs in the presence of the T4 DNA ligase. After washing away of the nonligated octamers and other octamer, the identity of the dinucleotides in the query position is determined by the distinctive fluorescent signal and is recorded. Subsequently, the fluorescent dye is removed after fluorescence emission by cleaving the terminal universal nucleotides. As such, cleavage provides a free end for ligation of another octamer in the next ligation cycle. The successive cycles of ligation can extend the complementary strand and identify the next dinucleotides with three nucleotides gap from the previous

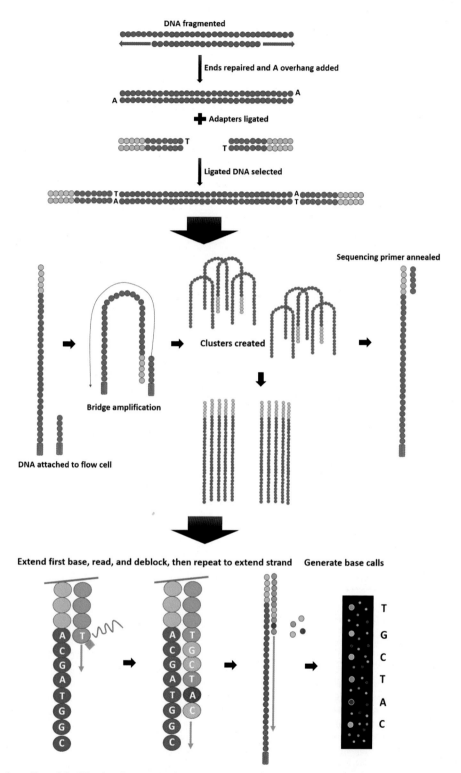

FIG. 11.16 Schematic outline of the Illumina Genome Analyzer workflow. Similar fragmentation and adapter ligation steps take place (I), before applying the library onto the solid surface of a flow cell. Attached DNA fragments form "bridge" molecules which are subsequently amplified via an isothermal amplification process, leading to a cluster of identical fragments that are subsequently denatured for sequencing primer annealing (II). Amplified DNA fragments are subjected to sequencing-by-synthesis using 3′ blocked labelled nucleotides (III).

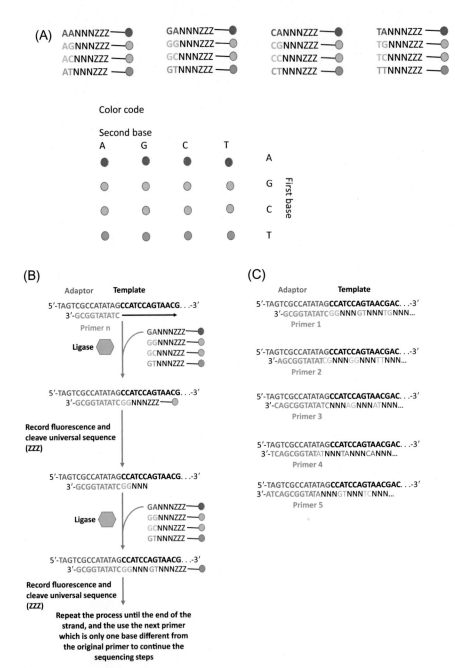

FIG. 11.17 Sequencing by ligation. (A) Pool of octamers. A set of 16 different octamers representing all possible combinations of two nucleotides in the query position and the color code are shown. Each probe is an octamer, which consists of (3'-to-5'direction) 2 probe-specific bases followed by 6 degenerate bases (nnnzzz) with one of 4 fluorescent labels linked to the 5' end. (B) To determine a sequence by ligation, an adaptor is first added to the 5' end of the template DNA molecule. The adaptor provides a binding site for a sequencing primer. Pools of octamers are added, and an octamer that contains the dinucleotide that are complementary to the template will hybridize. After the unbound octamers are washed away, the fluorescent signal from the ligated octamer is recorded to identify the dinucleotides in the query position followed by the removal of the fluorescent dye. The cycle is repeated until the end of the strand. Complete extension of primer through the first round of ligation. The product extended from original primer (*n*) is denatured from the adapter/template, and the second round of sequencing is performed with primer (*n* − 1). With the use of progressively offset primers, in this example (*n* − 1), adapter bases are sequenced, and this known sequence is used in conjunction with the color-space coding for determining the template sequence by deconvolution.

dinucleotides. The length of the sequence read is determined by the number of the ligation cycles. To determine the nucleotide sequence of the intervening region between the identified dinucleotides, the primer and the octamers are removed from the template strand. The process is repeated using the new primer that is set back on the template DNA by one nucleotide from the first primer (Fig. 11.17C). Therefore, the entire primer extension process by serial ligation requires five different primers and each primer is offset from the previous primer by one nucleotide. In this way, the nucleotide is each position in the template DNA is identified twice in separate reactions, which increases the accuracy of the sequence determination. For SOLiD system, samples preparation shares similarities with the 454 technology in that DNA fragments are ligated to oligonucleotide adapters, attached to beads, and clonally amplified by emulsion PCR. In this technique, a water droplet in oil emulsion contains the amplification reagents and only one fragment bound per bead; DNA fragments on the beads are amplified by the emulsion PCR. Beads with clonally amplified template are immobilized onto a derivatized-glass flow-cell surface, and sequencing is begun by annealing a primer oligonucleotide complementary to the adapter at the adapter-template junction in the presence of ligation mixture.

After annealing, a ligation step is performed, followed by wash removal of unbound probe. Fluorescence signals are optically collected before cleavage of the ligated probes, and a wash is performed to remove the fluorescence and regenerate the 5′ phosphate group. In the subsequent sequencing steps, interrogation probes are ligated to the 5′ phosphate group of the preceding pentamer. Seven cycles of ligation, referred to as a "round," are performed to extend the first primer.

By this approach, each template nucleotide is sequenced twice. A 6-day instrument run generates sequence read lengths of 35 bases. Because each base is determined with a different fluorescent label, error rate is reduced. Sequences can be determined in parallel for more than 50 million bead clusters, resulting in a very high throughput of the order of gigabases per run. Placing two flow-cell slides in the instrument per analytical run produces a combined output of 4 Gb of sequence or greater. Unextended strands are capped before the ligation to mitigate signal deterioration due to dephasing. Capping coupled with high-fidelity ligation chemistry and interrogation of each nucleotide base twice during independent ligation cycles yields a sequence consensus accuracy of 99.9% for a known target at a 15-fold sequence coverage over sequence reads of 25 nucleotides.

Sequencing by Semiconductor Technology: Compact Personal Genome Machine (PGM) Sequencer

Ion Personal Genome Machine (PGM) and MiSeq were launched by Ion Torrent and Illumina. They are both small in size and feature fast turnover rates but limited data throughput. They are targeted to clinical applications and small labs.

Ion PGM was released by Ion Torrent at the end of 2010. PGM uses semiconductor sequencing technology. When a nucleotide is incorporated into the DNA molecules by the polymerase, a proton is released. By detecting the change in pH, PGM recognized whether the nucleotide was added or not. Each time the chip was flooded with one nucleotide after another, if it was not the correct nucleotide, no voltage would be found; if two nucleotides are added, double voltage is detected. PGM is the first commercial sequencing machine that does not require fluorescence and camera scanning, resulting in higher speed, lower cost, and smaller instrument size. Currently, it enables 200-bp reads in 2 h, and the sample preparation time is less than 6 h for eight samples in parallel. An exemplary application of the Ion Torrent PGM sequencer is the identification of microbial pathogens. PGM has the potential of offering a fast but limited-throughput sequencer when there is an outbreak of new disease.

THE THIRD-GENERATION SEQUENCER

With the increasing use and new modifications in next-generation sequencing, third-generation sequencing is coming out with new insights into sequencing. Third-generation sequencing has two main characteristics. First, PCR is not needed before sequencing, which shortens DNA preparation time. Second, the signal is captured in real time, which means that it is monitored during the enzymatic reaction of adding nucleotide in the complementary strand. Single-molecule sequencing is the third-generation sequencing method, and it simplifies sample preparation, reduces sample mass requirements, and eliminates amplification of DNA templates.

Because the properties of different single-molecule sequencing technologies vary so much from each other and from other generations of sequencing technologies, it is important to understand those properties and their impact on experimental design and output. The advantages of single-molecule sequencing are the ability to resequence the same molecule multiple times for improved accuracy and the ability to sequence molecules that cannot be readily amplified because of extremes of GC content, secondary structure, or other reasons. Some single-molecule sequencing also has long read length and quantitative superiority. A comparison of relative performance for different generation sequencing is shown in Table 11.1. Fig. 11.18 shows the basic principles of three common third-generation sequencing technologies.

TABLE 11.1 Sequencing Platform Performance Summaries

Method	Generation	Read Length (bp)	Single Pass Error Rate (%)	# Reads per Run	Time per Run	Cost per Million Bases ($)
Sanger ABI 370x1	1st	600–1000	0.001	96	0.5–3 h	500
Ion Torrent	2nd	200	1	8.2×10^7	2–4 h	0.1
454 (Roche) GS FLX +	2nd	700	1	1×10^6	23 h	8.57
Illumina HiSeq 2500 (High Output)	2nd	2×125	0.1	8×10^9 (paired)	7–60 h	0.03
Illumina HiSeq 2500 (Rapid Run)	2nd	2×250	0.1	1.2×10^9 (paired)	1–6 days	0.04
SOLiD 5500x1	2nd	2×60	5	8×10^8	6 days	0.11
PacBio RS II: P6-C4	3rd	$1.0–1.5 \times 10^4$ on average	13	$3.5–7.5 \times 10^4$	0.5–4 h	0.4–0.8
Oxford Nanopore MinION	3rd	$12–5 \times 10^3$ on average	38	$1.1–4.7 \times 10^4$	50 h	6.44–17.90

The Helicos Single-Molecule Sequencing Device, HeliScope

In the Helicos Biosciences single molecule sequencing device. The DNA fragments are hybridized to primers covalently anchored in random positions on a glass cover slip in a flow cell. The primer, polymerase enzyme, and labeled nucleotides are added to the glass support. The next base incorporated into the synthesized strand is determined by analysis of the emitted light signal, in the sequencing-by-synthesis technique. This system also analyzes many millions of single DNA fragments simultaneously, resulting in sequence throughput in the gigabase range.

In this system, the fluorescent "Virtual Terminator" nucleotide prevents the incorporation of any subsequent nucleotide until the nucleotide dye moiety is cleaved. The images from each cycle are assembled to generate an overall set of sequence reads. On a standard run, 120 cycles of nucleotide addition and detection are carried out. Roughly about a billion molecules can be followed simultaneously in this approach. Because there are two 25-channel flow cells in a standard run, 50 different samples can be sequenced simultaneously. Sample requirements are the simplest of all technologies: only subnanogram amounts are necessary, and very poor-quality DNA, including degraded or modified DNA, can be sequenced. Average read lengths are relatively short (about 35 nt) with raw individual nucleotide error rates currently about 3% to 5%, occurring randomly throughout the sequence reads and predominantly in the form of a "dark base" or deletion error, which is accounted for in the alignment algorithm. With 30 X coverage level, accurate consensus sequences are generated regardless of error rates within this range. Single-molecule systems have a much more even coverage and thus do not require as much depth for complete detection of heterozygotes.

The Helicos Sequencer system can also sequence RNA molecules directly. This can avoid the many artifacts associated with reverse transcriptase and providing unparalleled quantitative accuracy for RNA expression measurements. The very high read count per sample allows precise expression measurements to be made with either RNA or cDNA, a feature not yet possible with other single-molecule technologies. As with many single-molecule systems, repeated reads of the same molecule can markedly improve the error rate and also allow detection of very rare variants in a mixed sample. For example, with repeat sequencing of the same molecule, the error rate can be driven sufficiently low that mutations in heterogeneous samples such as tumors can be readily detected. Because of the minimal sample preparation needs, the ability to use exceptionally small starting quantities, and the high read count, the Helicos Sequencer system is ideal for quantitative applications such as chromatin immunoprecipitation (ChIP), RNA expression, and copy number variation, and situations in which sample quantity is limiting or degraded.

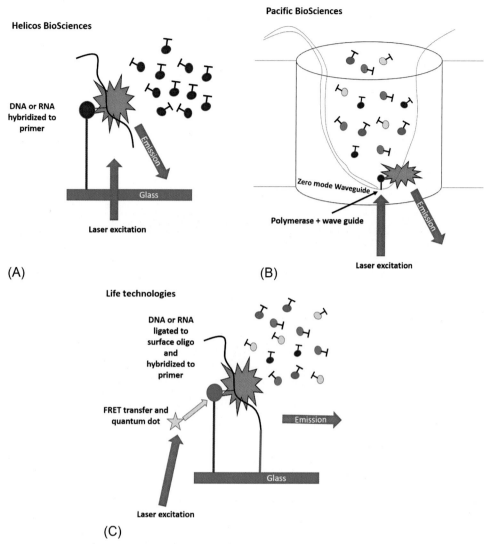

Helicos BioSciences

DNA or RNA
hybridized to
primer

Glass

Laser excitation

(A)

Pacific BioSciences

Zero mode Waveguide

Polymerase + wave guide

Emission

Laser excitation

(B)

Life technologies

DNA or RNA
ligated to
surface oligo
and
hybridized to
primer

FRET transfer and
quantum dot

Emission

Glass

Laser excitation

(C)

FIG. 11.18 The principle of single-molecule sequencers. The three most advanced single-molecule sequencing systems all carry out sequencing-by-synthesis using laser excitation to generate a fluorescent signal from labeled nucleotides, which is then detected using a camera. In the Helicos BioSciences system, single nucleotides, each with a fluorescent dye attached to the base, are sequentially added. In the Pacific Biosciences and Life Technologies systems, four different nucleotides, each with a different color dye attached to the phosphates, are continuously added. Background fluorescence is minimized differently in the three systems. (A) Helicos uses total internal reflectance fluorescence (TIRF) to create a narrow evanescent field of light in which the intensity of the light decays exponentially away from the glass surface. Only dyes that are in the TIRF evanescent field can fluoresce. (B) Pacific Biosciences uses a zero-mode waveguide (ZMW), which limits illumination to a narrow region near the bottom of the well containing the polymerase. Only dyes near the opening of the ZMW can fluoresce. (C) Life Technologies uses fluorescence resonance energy transfer (FRET) between the initially absorbing quantum dot on the polymerase and the emitting dye on the nucleotide. Only dyes close to the polymerase-attached quantum dot can be excited by FRET and then fluoresce.

Pacific Biosciences

Pacific Biosciences (PacBio) has developed another sequencing-by-synthesis approach using fluorescently labeled nucleotides. In Pacific Biosciences (PacBio) sequencing-by-synthesis system, the cell consists of millions of zero-mode waveguides (ZMWs), embedded with only one set of enzymes and a DNA template that can be detected during the whole process. The template, called a SMRTbell, is a closed, single-stranded circular DNA that is created by ligating hairpin adaptors to both ends of a target double-stranded DNA molecule (Fig. 11.19). When a sample of SMRTbell is loaded to a chip called a SMRT cell (Fig. 11.20), a SMRTbell diffuses into a ZMW, which provides the smallest available volume for light detection. All four potential nucleotides are included in the reaction, each labeled with a different color of fluorescent dye so that they can be distinguished from each other. Each nucleotide has a characteristic incorporation time that

FIG. 11.19 SMRT bell template. Hairpin adaptors *(green)* are ligated to the end of a double stranded DNA molecule *(blue and purple)*. This ligation forms a closed circle. The polymerase *(gray)* is anchored to the bottom of a ZMW and incorporates bases into the read strand *(orange)*.

FIG. 11.20 Picture of a SMRT cell. Each SMRT cell contains 150,000 ZMWs. Approximately 35,000–75,000 of these wells produce a read in a run lasting 0.5–4h, resulting in 0.5–1 Gb of sequence. *ZMW*, zero-mode waveguide.

can further aid in improving base calls. During the reaction, the enzyme will incorporate the nucleotide into the complementary strand and cleave off the fluorescent dye previously linked with the nucleotide. Then the camera inside the machine will capture signal in a movie format in real-time observation (Fig. 11.21). This will give out not only the fluorescent signal but also the signal difference over time, which may be useful for the prediction of structural variance in the sequence, which is especially helpful in epigenetic studies such as DNA methylation.

Direct RNA sequencing with this system is problematic. This because nucleotides bind repeatedly to the reverse transcriptase before nucleotide incorporation, thereby giving false signals with multiple insertions that prevent determination of a meaningful sequence. In addition, the low read count of this system will limit it to the identification of common mRNA isoforms rather than quantitative expression profiling or complete transcriptome coverage.

This system has several advantages. First, the sample preparation is very fast; it takes 4 to 6h instead of days. Also, it does not require PCR in the preparation step, which reduces the bias and error caused by PCR. Second, the turnover rate is also fast; runs are finished within a day. Third, the average read length is 1300bp, which is longer than that of any second-generation sequencing technology. Although the throughput is lower than that of a second-generation sequencer, this technology is quite useful for clinical laboratories, especially for microbiology research to assemble genomes, and to assess the analysis of structural variation, haplotyping, metagenomics, and identification of splicing isoforms.

Life Technologies Method

Life Technologies system uses the fluorescence resonance energy transfer (FRET)-based single-molecule sequencing-by-synthesis technology. It consists of a quantum dot-labeled polymerase that synthesizes DNA using four distinctly labeled

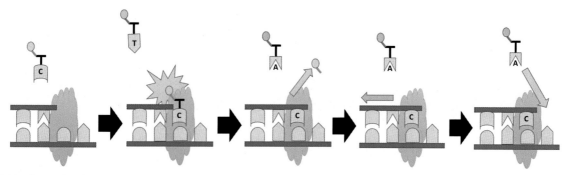

FIG. 11.21 Sequencing via light pulses. Each of the four nucleotides is labeled with a different fluorescent dye (indicated in *red*, *yellow*, *green*, and *blue*, respectively for G, C, T, and A) so that they have distinct emission spectrums. As a nucleotide is held in the detection volume by the polymerase, a light pulse is produced that identifies the base. (1) A fluorescently labeled nucleotide associates with the template in the active site of the polymerase. (2) The fluorescence output of the color corresponding to the incorporated base (*yellow* for base C as an example here) is elevated. (3) The dyelinker-pyrophosphate product is cleaved from the nucleotide and diffuses out of the ZMW, ending the fluorescence pulse. (4) The polymerase translocates to the next position. (5) The next nucleotide associates with the template in the active site of the polymerase, initiating the next fluorescence pulse, which corresponds to Base A here.

nucleotides in a real-time system. Quantum dots, which are fluorescent semiconducting nanoparticles, have an advantage over fluorescent dyes in that they are much brighter and less susceptible to bleaching, although they are also much larger and more susceptible to blinking. The genomic sample to be sequenced is ligated to a surface-attached oligonucleotide of defined sequence and then read by extension of a primer complementary to the surface oligonucleotide. When a fluorescently labeled nucleotide binds to the polymerase, it interacts with the quantum dot, causing an alteration in the fluorescence of both the nucleotide and the quantum dot. The quantum dot signal drops, whereas a signal from the dye-labeled phosphate on each nucleotide rises at a characteristic wavelength. The real-time sequence is captured for each extending primer. Because each sequence is bound to the surface, it can be reprimed and sequenced again for improved accuracy. The applications of this approach are in genome assembly, structural variation, haplotyping, and metagenomics.

Nanopore Sequencing

Nanopore sequencing is another method of third-generation sequencing. A nanopore is a tiny biopore with a diameter in nanoscale, which can be found in a protein channel embedded on a lipid bilayer that facilitates ion exchange. Because of the biological role of nanopore, any particle movement can disrupt the voltage across the channel. The core concept of nanopore sequencing involves putting a thread of single-stranded DNA across an α-hemolysin (αHL) pore, a 33-kD protein isolated from *Staphylococcus aureus*, undergoes self-assembly to form a heptameric transmembrane channel. It can tolerate extraordinary voltage up to 100 mV with current 100 pA. This unique property supports its role as a building block of nanopore.

In nanopore sequencing, an ionic flow is applied continuously. Current disruption is simply detected by standard electrophysiological techniques. Readout relies on the size difference between all deoxyribonucleoside monophosphates (dNMPs). Thus, for a given dNMP, the characteristic current modulation is shown for discrimination. Ionic current is resumed after the trapped nucleotide is entirely squeezed out.

Nanopore sequencing possesses a number of useful advantages over existing commercialized next-generation sequencing technologies. First, it potentially reaches long read lengths >5 kbp with speeds of 1 bp/ns. Second, detection of bases is fluorescent tag-free. Third, nanopore sequencing is less sensitive to temperature throughout the sequencing reaction and reliable outcome can be maintained. Fourth, instead of sequencing DNA during polymerization, single DNA strands are sequenced through nanopore by means of DNA strand depolymerization. Hence, hands-on time for sample preparation such as cloning and amplification steps can be shortened significantly.

NGS DATA ANALYSIS

NGS experiments generate unprecedented volumes of data. Typically, NGS data begin as large sets of tiled fluorescence or luminescence images of the flow-cell surface recorded after each iterative sequencing step (Fig. 11.22). The read lengths of those fragments depend on the type of NGS platform and can be in the range of 25–450 base pairs. While the reads are much shorter than those created by Sanger sequencing, NGS has a higher throughput. Data volumes generated during single runs

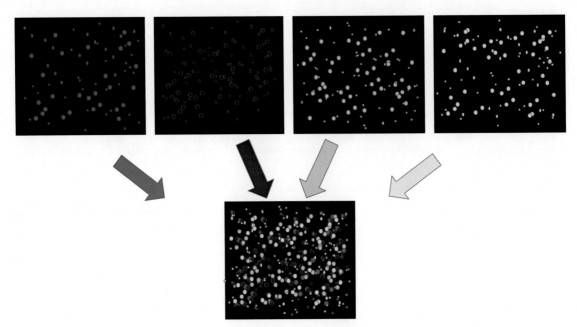

FIG. 11.22 NGS data analysis. Here is an example of a pseudocolor image from the Illumina flow cell. Each fluorescence signal originates from a clonally amplified template cluster. *Top panel* illustrates four emission wavelengths of fluorescent labels d Applied Biosystems SOLiD sequencing depicted in *red, green, blue,* and *yellow.* Images are processed to identify individual clusters and to remove noise or interference. The *lower panel* is a composite image of the four fluorescence channels.

of the 454 GS FLX, Illumina, and SOLiD instruments are typically 15 GB, 1 TB, and 15 TB, respectively. This requires improved algorithms that are able to process such huge amounts of raw data. These present challenges and opportunities for data management, storage, and, most importantly, analysis in a resource-intensive data-pipeline system.

Conversion of Raw Data to Base Sequences

The raw data generated by NGS consists of fluorescence signals. These signals need to be converted into base sequences by platform-specific base calling-algorithms provided by the manufacturer. Therefore, the main processing feature of the data pipeline is the computationally intensive conversion of image data into sequence reads, known as base calling. First, individual beads or clusters are identified and localized in an image series. Image parameters such as intensity, background, and noise are then used in a platform-dependent algorithm to generate read sequences and error probability-related quality scores for each base. The base calls can be generated by the platform-specific data-pipeline software. Some alternative base-calling programs consider the incorporation of ambiguous bases into reads, improved removal of poor-quality bases from read ends, and the use of data sets for software training. Incorporation of these features can reduce read error and improve alignment, especially as platforms are pushed to generate longer reads.

The quality values calculated during NGS base calling provide important information for alignment, assembly, and variant analysis. The NGS error rates estimated by quality values depend on several factors, including signal-to-noise levels, cross talk from nearby beads or clusters, and dephasing. Substantial effort has been made to understand and improve the accuracy of quality scores and the underlying error sources, including inaccuracies in homopolymer run lengths on the 454 platform and base-substitution error biases with the Illumina format.

Alignment and Assembly

Alignment and assembly are substantially difficult for NGS because of the shorter reads lengths. Several factors need to be considered when choosing an appropriate sequencing platform with its associate read length. One limitation of short-read alignment and assembly is the inability to uniquely align large portions of a read set when the read length becomes too short. Similarly, the number of uniquely aligned reads is reduced when aligning to larger, more complex genomes or reference sequences because of a higher probability of repetitive sequences. Unique alignment or assembly is reduced not only by the presence of repeat sequences but also by shared homologies within closely related gene families and pseudogenes.

Nonunique read alignment is handled in software by read distribution between multiple alignment positions or leaving alignment gaps.

NGS Accuracy

Error rates for individual NGS reads are higher compared to conventional sequencing. Accuracy in NGS is achieved by sequencing a given region multiple times, enabled by the massively parallel process, with each sequence contributing to "coverage" depth. Through this process, a "consensus" sequence is derived. To assemble, align, and analyze NGS data requires an adequate number of overlapping reads, or coverage. In practice, coverage across a sequenced region is variable and is influenced by factors such as differential ligation of adapters to template sequences and differential amplification during clonal template generation. Beyond sequence errors, inadequate coverage can cause failure to detect actual nucleotide variation, leading to false-negative results for heterozygotes. Coverages of less than 20- to 30-fold begin to reduce the accuracy of single-nucleotide polymorphism calls in data on the 454 platform. For the Illumina system, higher coverage depths (50- to 60-fold) can improve short-read alignment, assembly, and accuracy, although coverage in the 20- to 30-fold range may be sufficient for certain resequencing applications. Coverage gaps can occur when sequences are not aligned. Alignment of repetitive sequences in repeat regions of a target sequence can also affect the apparent coverage. Reads that align equally well at multiple sites can be randomly distributed to the sites or in some cases discarded, depending on the alignment software. In de novo-assembly software, reads with ambiguous alignments are typically discarded, yielding multiple aligned read groups, or contigs, with no information regarding relative order.

NGS Software

NGS software features vary with the application and in general may include alignment, de novo assembly, alignment viewing, and variant-discovery programs. Software packages available for alignment and assembly to a reference sequence include Zoom, MAQ, Mosaik, SOAP, and SHRiMP, which supports SOLiD color-space analysis. Software for de novo assembly includes Edina, EULER-SR, SHARCGS, SSAKE, Velvet, and SOAPdenovo. Recently released commercial software for alignment and de novo assembly includes packages from DNAStar, SoftGenetics, and CLC bio that feature data viewers that allow the user to see read alignments, coverage depth, genome annotations, and variant analysis. Fig. 11.23 presents examples of NGS data viewed in two different software systems.

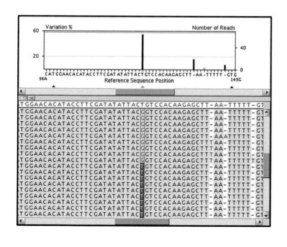

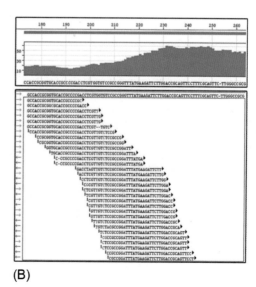

(A) (B)

FIG. 11.23 Examples of NGS data viewed in two different software systems. (A) Roche Amplicon Variant Analyzer software displaying an example of DNA sequence. *Lower pane* shows reference sequence *(green)* above 18 of 68 aligned reads. Column highlighted in *yellow* and *blue* shows a heterozygous single-nucleotide polymorphism (SNP). Single T/A insertions *(red)* may represent errors. *Upper pane* shows percent variation from reference *(vertical bars)* and coverage *(pale blue line)*. (B) DNAStar SeqMan Pro software displaying Illumina data. *Lower pane* shows reference sequence above aligned reads. *Green* and *red arrows* show direction of sequencing; base calls at variance with reference are indicated *(red)*. Three columns in agreement *(red)* indicate presumptive SNPs. Other bases in *red* may be errors. *Upper pane* shows read coverage, with relative alignment positions above the graph.

The Outlook for NGS Technologies

The NGS technologies has tremendously reduced time and cost requirements for large genome sequencing. Therefore, NGS technologies became very interesting for de novo sequencing of eukaryotic genomes. The initial generation of genomic sequences enables a detailed genetic analysis of a particular organism and the analysis of genomic diversity among different individuals. NGS is also used for whole-genome or targeted resequencing, which is currently one of the broadest applications of high-throughput sequencing. Therefore, NGS has markedly accelerated multiple areas of genomics research, enabling experiments that previously were not technically feasible or affordable.

From the impact that NGS has made at the basic research level, we anticipate its translation into molecular diagnostics. As NGS technologies have made it possible to generate high-resolution genomic data much more efficiently, researchers in many fields have turned to NGS to identify particular features of the genome that contribute to specific phenotypes, evolutionary history, epidemiology and clinical diagnosis. It is no doubt that future cutting-edge development in NGS will improve genome-scale technologies and generation of genomic resources, datasets, and computational tools. All these developments will eventually continue to improve clinical diagnosis and disease research for sustained and broadly applicable scientific innovation.

Chapter 12

Genome and Transcriptome Analysis

Chapter Outline

GENOME-WIDE SAMPLING SEQUENCING FOR GENETIC POLYMORPHISMS

The development of NGS has significantly improved the process to diagnose and predict diseases. However, predicting disease phenotypes from genetic sequences is still challenging. A single gene can be associated with multiple disease phenotypes while a single disease phenotype can be caused by mutations in multiple genes. Importantly, mutations do not have identical effects on individuals due to the individual variation in interaction between genes, proteins, metabolites and environmental factors. The complete set of (physical) interactions between molecules, such as genes, proteins and metabolites is known as the interactome. Interactome can translate the phenotypic effects determined by both genotypes and environmental factors (Fig. 12.1). Therefore, analysis to link genetic polymorphisms to the specific disease is very useful to predict diseases.

The techniques that have a subset of a genome for sampling and sequencing is referred as genome-wide sampling sequencing (GWSS) methods. GWSS comprise a collection of reduced genome complexity sequencing, reduced genome representation sequencing, and selective genome target sequencing. This technique is suitable for detecting genetic polymorphisms, particularly single nucleotide polymorphisms (SNPs). SNPs are the predominant forms of sequence. For example, approximately 150 million SNPs have been discovered in the human genome (see dbSNP database at the National Center for Biotechnology Information, http://www.ncbi.nlm.nih.gov/SNP/). Since the whole genome

Diagnostic Molecular Biology. https://doi.org/10.1016/B978-0-12-802823-0.00012-2

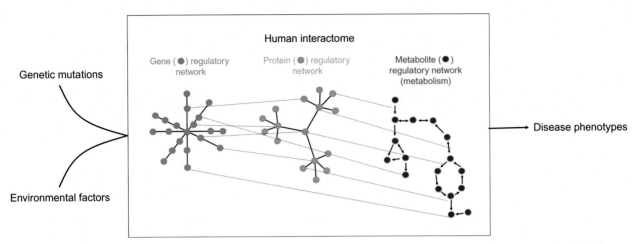

FIG. 12.1 Genetic mutations and environmental effects can perturb the human interactome (which is a complex network constituted by the gene regulatory network, protein interaction network, and metabolism) and thus lead to disease phenotypes.

sequencing/resequencing for SNP genotyping is technically unnecessary and is cost-prohibitive, GWSS is an alternative strategy to discover and genotype genetic variants in a cost-effective manner.

Traditionally, GWSS methods involve three common steps: DNA digestion with restriction enzymes, ligation with adapters that fit into the relevant sequencing platforms, and PCR amplification to increase the yield of library products for sequencing. The genome sampling process depends on the use of rare-cutter enzymes, size selection of products, or selective amplification of products. Generally speaking, GWSS libraries can be constructed by digesting DNA with one or two restriction enzymes and analyzing the subsequent library with Illumina or Ion Torrent sequencing platforms.

Currently, available GWSS methods have evolved mainly from reduced representation (library) sequencing (RRS or RRLS), complexity reduction of polymorphism sequencing (CRoPS™), restriction site associated DNA sequencing (RAD-seq), and genotyping by sequencing (GBS) methods. These methods can be classified into four categories: GWSS without size selection, GWSS with semi-size selection, GWSS with size selection, and GWSS with selective amplification.

GWSS Without Size Selection

Both GBS and genome reducing and sequencing (GGRS) methods are designed to prepare libraries without size selection. The main steps in these methods include digestion of DNA with restriction enzymes, simultaneous ligation of adapters, and selective amplification of products with PCR using primers complementary to ligated adapter sequences (Fig. 12.2). Of these methods, only GGRS employs selection of 300- to 500-bp PCR products for sequencing. Size selection at library preparation stage probably does not significantly affect the sequencing outcome in terms of genome complexity when compared to both GBS and two-enzyme GBS methods. The main difference of these methods relies in the design of their adapters.

Sample preparation with the GBS method uses a barcode adapter and a common adapter, which allow for multiplexing of 48 or 96 samples. The "barcode" adapter terminates with a 4- to 8-bp barcode on the 3′ end of its top stand and a 3 bp overhang on the 5′ end of its bottom strand that is complementary to the "sticky" end generated by *Ape*KI (CWG) (Fig. 12.3A). The sequences of the two oligonucleotides comprising the barcode adapter are: 5′-ACACTCTTTCCCTACACGACGCTCT TCCGATCTxxxx and 5′-CWGyyyyAGATCGGAAGAGCGTCGTGTAGGGAAAGAGTGT, where "xxxx" and "yyyy" denote the barcode and barcode complement and sequences, respectively (Fig. 12.3B). The second, or "common," adapter has only an *Ape*KI-compatible sticky end: 5′-CWGAGATCGGAAGAGCGGTTCAGCAGGAATGCCGAG and 5′-CTCGGCATTCCTGCTGAACCGCTCTTCCGATCT. Adapters were designed so that the ApeKI recognition site did not occur in any adapter sequence and was not regenerated after ligation to genomic DNA. Adapter design also allows for either single-end or paired-end sequencing on the Illumina NGS platforms.

For the sample preparation with two-enzyme GBS, a conventional adapter is designed for ligation to one cohesive end of the fragmented DNA, while a Y-type adapter is designed for ligation to the other end of the DNA fragment. The combination of a rare-cutting enzyme and a second common-cutting enzyme will produce a digest largely consisting of fragments with a rare cut-site and a common cut-site, or fragments with two common cut-sites.

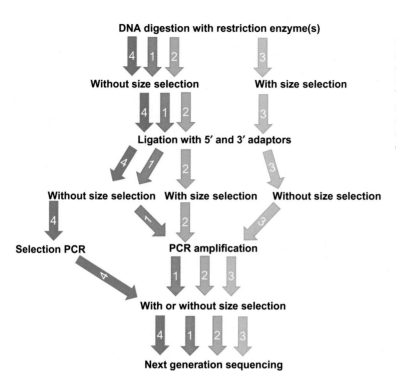

FIG. 12.2 Classification of genome-wide sampling sequencing (GWSS) methods and the process to prepare samples. (1) GWSS without size selection after DNA digestion and ligation. Size selection may be employed after PCR amplification. (2) GWSS with semi-size selection. These methods perform size selection after ligation with adapters. Some methods may combine random shearing with size selection and then ligate with the second adapter. (3) GWSS with size selection immediately after DNA digestion. (4) GWSS with selective amplification.

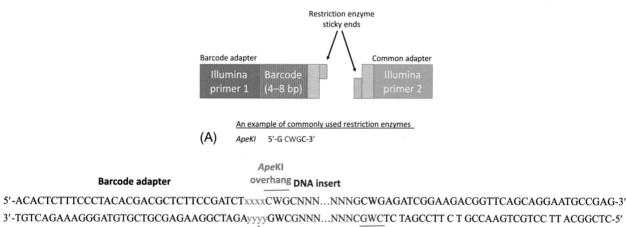

FIG. 12.3 (A) General format of genotyping by sequencing (GBS) adapters and enzymes. (B) An example of GBS double-stranded barcode and common adapter sequences. Adapters are shown ligated to ApeKI-cut genomic DNA. Positions of the barcode sequence and ApeKI overhangs are shown relative to the insert DNA.

In Fig. 12.4, a rare-cutting restriction enzyme, *Pst*I (CTGCAG), is chosen. The second enzyme used, *Msp*I (CCGG), has a more common recognition site. Barcoded forward adapters are designed with the *Pst*I restriction overhang while the reverse adapter matches the *Msp*I overhang. To eliminate amplification of the more common *Msp*I-*Msp*I fragments, the Y-adapter that contains an exact match (but no complement) to the reverse primer is used in PCR amplification. During the first cycle of PCR, amplification proceeds only from primer annealing to the forward adapter. Binding sites for the reverse primer are created only during the first round of PCR by extension from forward primers on the other end of the same fragment

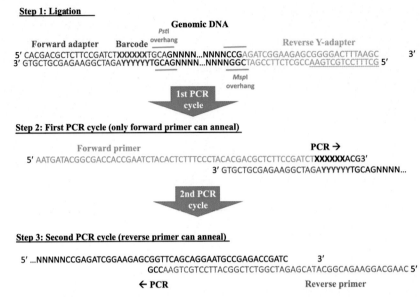

FIG. 12.4 An example of adapter design, PCR amplification of fragments for two-enzyme GBS. Step 1: The ligation product of a genomic DNA fragment (*black*) containing a *Pst*I restriction site and a *Msp*I restriction site. The forward adapter (*purple*) binds to a *Pst*I generated overhang. The 4–8 bp barcode for this adapter is in bold with "X." The *Msp*I generated overhang corresponds to the reverse Y-adapter (*yellow*). The unpaired tail of the Y-adapter is underlined. Step 2: During the first round of PCR only the forward primer (*green*) can anneal. PCR synthesis of the complementary strand proceeds to the end of the fragment synthesizing the compliment of the Y-adapter tail. Step 3: During the second round of PCR the reverse primer (*pink*) can anneal to the newly synthesized compliment of the Y-adapter tail. This PCR reaction then proceeds to fill in the compliment of the forward adapter/primer on the other end of the same fragment.

(Fig. 12.4). This design allows amplification of only *Pst*I-*Msp*I fragments and produces a uniform library. Under this circumstance, all fragments are Forward Adapter—genomic DNA—Reverse Adapter. PCR amplification with a short extension time (<30s) enriches for shorter fragments suitable for bridge-amplification on the Illumina flow-cell. This design ensures specific amplification of only fragments produced by digestion with the two restriction enzymes.

The GGRS approach has a similar procedure as GBS but it requires usage of only one Y-type adapter for library preparation. Two ends of digested resulting fragments are ligated to an identical barcode-adapter, instead of one end ligated common-adapter and the other end ligated barcode-adapter. Compared to GBS, the GGRS approach further improves the procedure to keep the simplicity, the rapidity, and the reproducibility. In the GGRS approach, fragments are selected by gel electrophores based on the genomic properties of an outbred population, which decreased cost. One set of adapter-barcodes is designed to meet the requirements of depth and coverage to attain greater genotype accuracy in GGRS. Further, GGRS method merged removing-adapter steps and PCR clean-up steps into the last PCR gel-purification steps to reduce variation of fragment number between individuals.

These methods rely on five- or six-cutter enzymes to prepare sequencing libraries with reduced genome complexity. In theory, DNA fragments too long to be amplified by polymerase would be excluded from the library. Overall, libraries prepared with GWSS without size-selection methods require specific sets of customized barcoded adapters for different enzymes or enzyme combinations. These methods are relatively simple, robust, and time- and cost-efficient.

GWSS With Semi-Size Selection

Some GWSS that use semi-size selection during library preparation include restriction-site associated DNA sequencing (RAD-seq), paired-end reduced representation libraries (paired-end RPLs), double-digest RAD-seq (DDRAD-seq) and flexible and scalable GBS. RAD-seq is an emerging method for SNP detection in genomes and it can identify polymorphic variants adjacent to restriction enzyme digestion sites. The RAD-seq approach can also be used in association mapping, population genetics inferences, genetic mapping, and in estimation of allele frequencies. RAD-seq differs from RNA-seq in that nontranscribed loci also are sequenced, thus increasing an opportunity to broaden the known SNPs.

RAD genome fragments share a unique sequence similarity: a sequence anchored by the restriction enzyme cleavage site and a variable sequence end generated from a shearing step during library construction (Fig. 12.5). When RAD is coupled with paired-end sequencing approaches available on NGS platforms, the opposite ends of the RAD fragment

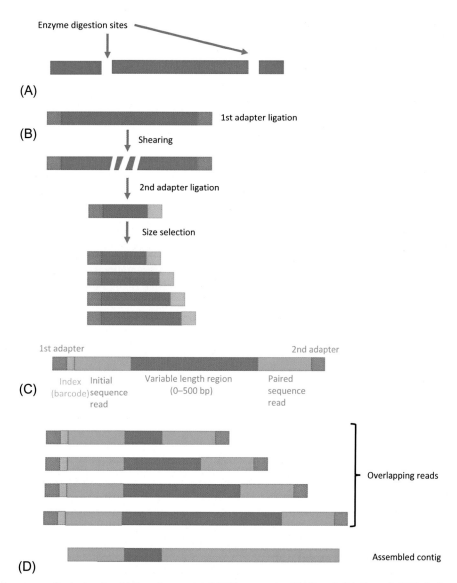

FIG. 12.5 Schematic process of paired-end restriction-site associated DNA sequencing (RAD-seq). (A) Genomic DNA is digested with a restriction endonuclease. (B) After ligation with a primary adapter, the fragments are sheared, then ligated with a secondary adapter. (C) A composite mixture of variable length fragments is recovered from each restriction enzyme digestion site. These fragments are size selected, amplified, and sequenced on a next-generation DNA sequencing platform using paired-end chemistry. (D) Development of the genomic assembly around each digestion site is then completed bioinformatically.

are linked in *cis* and the fragment can be interrogated. Mate-pairs with identical single-read sequences then can be readily assembled into contigs spanning hundreds of base pairs.

For the library construction processes, a DNA fragment size selection step immediately after restriction digestion is not necessary (Fig. 12.2). Instead, size selection is performed in ensuing steps of the library preparation process. Fragment size selection is performed twice when libraries are prepared with RAD-seq and paired-end RPLs methods. After the first adapter is ligated to DNA fragments, the ligated products are randomly sheared. Adapter-ligated restriction fragments are sheared to a size suitable for Illumina sequencing (typically 300–700 bp), and sheared fragments containing restriction site. Next, a size selection step is performed, followed by end repair, dA tailing, and ligation with the primer (Y adapter design). After PCR amplification, size selection is repeated, and products are subsequently sequenced.

In general, the RAD-seq technique is the most popular of reduced-representation library sequencing methods. It allows reduction of the complexity of genomes leading to deep sequence coverage of the fragments adjacent to the restriction site, subsequently leading to detection of SNPs. It has several advantages over restriction fragment length polymorphisms (RFLPs), amplified fragment length polymorphisms (AFLPs), and random amplified polymorphic DNA (RAPD) because of its ability to identify, verify, and score markers concurrently. The RAD-seq technique is suitable for organisms without an existing reference genome.

Both GWSS methods without size selection and with semi-size selection perform random shearing of ligated products and this approach can improve the possibility that large fragments are sequenced. Random shearing, however, can affect the position of the reads, which might complicate read mapping, particularly when the paired-end sequencing strategy is used. When libraries are prepared with the DDRAD-seq and flexible and scalable GBS procedures, size selection is conducted after ligation with both 5′ and 3′ adapters. The flexible and scalable GBS technique uses restriction enzymes to produce DNA fragments with blunt ends, and these blunt ends do not require end-repair. Instead, dA-tailing is used for ligation to universal Illumina adapters. If the ligation step is not optimized, size selection of ligated products might lead to inefficient genome representation reduction. As such, the reduced genome complexity of libraries prepared with GWSS methods that do not employ size selection or use only semi-size selection might not be significantly different if DNA is digested with the same enzyme(s).

GWSS With Size Selection

GWSS library preparation with size selection methods include reduced representation shotgun sequencing (RRS), reduced representation libraries (RRLs), paired-end reduced representation libraries (paired-end RPLs), GBS with one enzyme digestion, GBS with two enzyme digestion, and type IIB endonucleases restriction-site associated DNA. The library preparations for these procedures are almost identical: DNA digestion, size selection, adapter ligation, PCR amplification, and sequencing (Fig. 12.2). Only the original RRS method used the traditional cloning technique. For the size selection step, DNA is digested with a specific enzyme followed by gel electrophoresis on polyacrylamide gel to separate digested fragments. DNA fragments of 100–200 bp in length were sliced out of the gel, followed by elution and precipitation for library preparation using the NGS platform.

GWSS With Selective Amplification

The CRoPS™ method, which is based on the AFLP (amplified fragment length polymorphisms) technique, is an example of GWSS with selective amplification methods. This technology involves digestion of genomic DNA with a pair of enzymes, followed by ligation with adapters and PCR amplification (pre-amplification and selective-amplification). The amplified products are separated on a gel for traditional AFLP analysis or sequenced using an NGS platform. Another method is called scalable GBS. This method uses longer 3′ primers that contained three parts: the entire common adapter, the 3′ restriction site, and an extension of one or two bases into the insert fragments.

These methods do not involve size selection but employ selective amplification of products for sequencing. The whole exome sequencing (WES) method usually starts with construction of a whole genome shotgun library. Fragments that range in size from 150–250 bp are collected after gel electrophoresis, and exome capture arrays are designed based on Refseq mRNA sequences. The shotgun library then is hybridized to the exome arrays, and exon region related products are captured. The hybridized products are recovered from the slides and amplified for sequencing. Although WES method selectively sequences 2%–3% of the genome, this portion represents more than 98% of the exon regions.

GWSS Challenges

GWSS methods can reduce genome complexities, reduce genome representations, or select genome targets compared to whole genome sequencing/resequencing. On the other hand, GWSS methods have their drawbacks, such as inconsistency in the number of reads per sample library, and the number of sites/target per individual and the number of reads per site/target. Consequently, missing data and uneven distribution of markers along each chromosome are unavoidable. As such, library coverage, genotype calling accuracy, and read mapping rate need to be improved so that these problems can be solved.

Preparation of a GWSS library is relatively simple, but the process needs to be optimized further. Some strategies can improve the library construction process, including one sample-one library approach, which can minimize inconsistency in the number of reads per sample, good efficiencies of primers for amplification, and the gel-free size selection with beads, which can reduce site variation. To improve genotype calling accuracy, GWSS methods need to use enzymes with recognition sequences that are rarely mutated in the genome and that are insensitive/resistant to Dam (methylated by the dam methylase), Dcm (methylated by the dcm methylase), and CpG (methylated by the CpG methylase) blocks, thus producing millions of NGS reads that can be used to maximize the discovery of codominant markers for linkage mapping or genome-wide association studies. Furthermore, these enzymes must produce an even distribution of fragment sizes and cannot generate any concentrated bands after DNA digestion.

Ligation is one of the critical steps involved in the preparation of GWSS libraries. Many ligated products are fabricated in the reaction, which eventually lead to artifacts among the sequenced reads. The separation of ligated artifacts is important to improve the read mapping rate. The library construction process can be improved if one considers the design of adapters and primers, selection of restriction enzymes, and optimization of PCR conditions. As such, a unique set of targets with dense distribution and even coverage across all individual samples can be amplified and sequenced.

GENOME-WIDE ASSOCIATION STUDIES

Identification of SNPs

Genome-wide association studies (GWAS) analyze a level of association between common genetic variants and phenotypic traits. GWAS have identified a great number of common SNPs that statistically are associated with complex disease phenotypes. Fig. 12.6 demonstrates the association study design that compares the frequency of an allelic variants between a case and control sample. Having controlled for potential confounders between the samples (e.g., ethnic composition, relatedness, or similar factors that might influence genetic makeup), a significant difference in variant frequencies between cases and controls signals the presence of a potential risk or protective allele. An improved understanding of human genomic variation coupled with advances in genotyping technology and statistical methods have enabled association testing on an unbiased, genome-wide scale.

During the last two decades, human GWAS have used to reveal genetic risk factors for countless common disorders with complex genetic etiologies, including most of the major causes of morbidity and mortality in the developed world. So far,

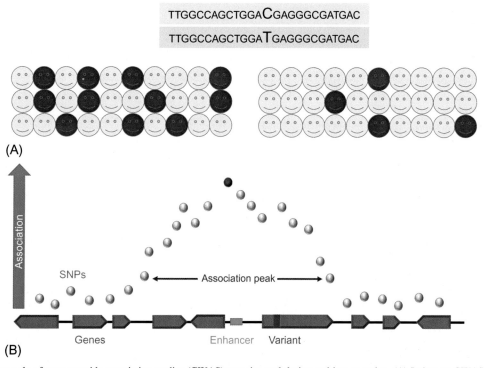

FIG. 12.6 An example of genome-wide association studies (GWAS) experimental design and interpretation. (A) In human GWAS, subjects with a disease or other trait of interest (cases) are compared with a cohort of unaffected individuals (controls). Genotyping array is used to profile genome-wide genetic variation, and the resulting frequencies of single nucleotide polymorphisms (SNPs) are evaluated between the groups. In the schematic, a C/T SNP is revealed. (B) An example of association test through statistical methods. The association test is generated based on the results of statistical tests for differences in SNP frequencies versus genomic position, superimposed on the annotated human genome reference. An association signal typically results in a "peak," The SNP showing the strongest association *(red)* falls within an intergenic region (sequence between genes). In this example, a number of candidate causal genes *(pink)* are shown. Three genes fall directly under the association peak; however, the top SNP is within an annotated enhancer element *(blue)*. It is also possible that a candidate, a rare exonic variant *(purple)* responsible for the regional association has not been directly assayed.

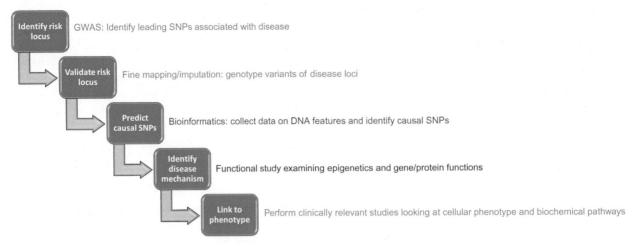

FIG. 12.7 Schematic workflow for identification of putative causal variants and the genes they affect. (1) GWAS is a hypothesis-free method for identifying SNPs correlating with disease risk. (2) Dense, targeted genotyping arrays such as Immunochip can be used for both replication of GWAS loci and genetic fine-mapping of all variants in disease-associated loci, further narrowing down the association signal. (3) Bioinformatics can then be used to both locate and functionally annotate SNPs in linkage disequilibrium (LD) with the lead SNPs. In noncoding regions, SNPs coinciding with regulatory features such as histone modifications or transcription factor binding sites are most likely to have a functional effect. (4) Appropriate functional experimental techniques can then be used to investigate genotype-specific protein interactions (ChIP), DNA conformation (3C), and gene expression (eQTL and reporter gene assays) in disease-relevant cell types. (5) SNPs associated with disease that coincide with coding regions of genes require bespoke experimental confirmation dependent on both their position within gene and function of protein. Once a causal variant affecting gene function has been identified, its effect on relevant biochemical pathways and resultant disease phenotype can be investigated.

The National Human Genome Research Institute Catalog of Published GWAS (March 2019; http://www.ebi.ac.uk/gwas/) reports 126,603 unique SNP-trait associations, based on aggregated results from 3841 publications.

GWAS also identified SNPs associated with drug efficacy and toxicity, boosting the development of pharmacogenomics and guiding individualized therapies. GWAS-associated variants usually require further extensive interrogation for a full interpretation of the data because of the number of highly correlated genetic variants in linkage dysequilibrium (LD) that might be causal. It has shown that only 5% of lead GWAS SNPs are likely to be causal and tend to lie an average distance of 14 kb from the probable causal SNP. Thus, the first task after GWAS is to perform dense genotyping, resequencing, or imputation, to test the association of all variants in LD with the lead variant and gain a detailed picture of potentially causal variants. Fig. 12.7 is a workflow for identification of putative causal variants and the genes they affect.

Integration of Epigenetic Effects

Epigenetics play a key role in the cross talk between environment and genome. This is because epigenetic marks can be used to explain the role of the environment in disease development. An organism's genome can be modified by various chemical compounds or species in the biological system, and this change can lead to changes in gene expression. These modifications are known as the epigenome. Major epigenetic alterations include DNA methylation, histone modification, and chromatin remodeling. High-throughput sequencing techniques with analysis of the pattern of DNA methylation and chromatin modifications are used for methylation analysis.

Because of epigenetic modifications, cells can exhibit different phenotypes in response to various environmental factors, such as nutritional changes and oxidative stress. This ability is phenotypic plasticity, whose abnormality is linked to diseases, such as cancers, neurodegenerative disorders, and autoimmune disorders. By integrating GWAS SNPs with epigenetic annotations, the DNA sequence variation can be linked better with the diseases. Therefore, a map of epigenomic variations shows how inter-individual differences in DNA sequences can be linked to phenotypic variations which can, in turn, lead to diseases.

Gaining a clear understanding about how variations in DNA sequences and variations in epigenetic markings contribute to the development of complex diseases is challenging. After the human epigenome has been mapped fully, identification of the effects of all deleterious environmental exposures according to duration of exposure and time period will be a complex undertaking and likely will require collaborative epidemiologic studies. In general, information deriving from GWAS in epigenomics provide possible etiological pathways rather than the exact molecular mechanisms underlying diseases.

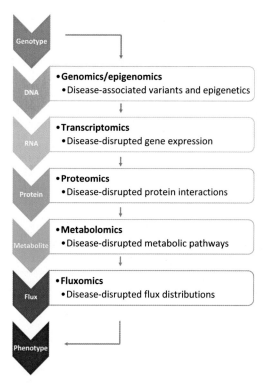

FIG. 12.8 DNA, RNA, protein, metabolite, and flux can bridge the genotype-phenotype relationship hierarchically. These molecules are profiled in the genomics, epigenomics, transcriptomics, proteomics, metabolomics, and fluxomics, respectively. Bioinformatics and systems biology approaches try to translate these omics data sets into unified knowledge. From genomics and epigenomics, one can identify the disease-associated genetic/epigenetic alterations. From transcriptomics, proteomics, metabolomics, and fluxomics, one can identify the genes, proteins, pathways, and the flux distributions involved in disease pathogenesis.

Although dramatic progress has been made in epigenomics research, there is still a gap between epigenomic knowledge and clinical application. To fill such gap, an accurate understanding of the genotype-phenotype relationship, which is hierarchically bridged by DNA, RNA, protein, metabolite, and flux, must be developed (Fig.12.8). It is no doubt that we need to combine omics data sets to understand the accurate relationship. For the remainder of the chapter, we will discuss the transcriptome, proteomics, and metablomic.

TRANSCRIPTOME SEQUENCING

Advantages to Using Whole Transcriptome Analysis

Transcriptome can provide the real-time gene expression information inside cells. Understanding transcriptome dynamics and patterns can deliver new insights into mechanisms controlling many biological events and processes, such as the cell cycle and mitosis, nuclear reprogramming, and stem cell biology, organogenesis, and tissue remodeling during metamorphosis and regeneration, and innate response to pathogens and environmental challenges. Microarray methods have been used widely to monitor global gene expression changes and have been the standard in expression profiling studies for most of the last two decades. On the other hand, the rapid development of NGS has made sequencing-based expression profiling emerging as a cost-effective alternative to microarrays. As a sequence-based expression profile, whole transcriptome analysis aims to capture both coding and noncoding RNA and quantifying gene expression heterogeneity in cells, tissues, organs, and even a whole body. This analysis can provide the first steps toward functional characterization and annotation of genes/genomes; builds blueprints for reconstruction of genetic interaction networks to understand cellular functions, growth/development, and biological systems; produces molecular fingerprints of disease processes and prognoses to pinpoint potential targets for drug discovery and diagnostics, and offers opportunities to examine the relationship between host and pathogen for novel strategies that can be used for therapeutic and prophylactic intervention.

Whole Transcriptome Analysis With Sequencing Methods and Technologies

Whole transcriptome analysis with sequencing methods and technologies generally is classified into four categories: whole transcriptome shotgun sequencing (WTSS), whole transcriptome target/tag sequencing with restriction digestion, whole transcriptome target/tag sequencing without restriction digestion, and other developments. These methods and technologies differ in library preparation, such as what serves as the library starting material, such as total RNAs, polyA+ RNAs or polyA- RNAs; how adapters or linkers are added to both 5′ and 3′ ends of RNA molecules, such as through cDNA synthesis or ligation; and how products are amplified for sequencing, such as via cloning or PCR amplification.

TRANSCRIPTOME SEQUENCING: WHOLE TRANSCRIPTOME SHOTGUN SEQUENCING—EST (EXPRESSED SEQUENCE TAG) SEQUENCING PROCEDURE

Source of EST Library

ESTs are single-pass reads derived from cDNA libraries with random selection. Therefore, an EST is a short sub-sequence of a cDNA sequence and the cDNA used for ESTs library are typically individual clones from a cDNA library. EST libraries can be prepared from single or multiple sources of tissues, organs, or cell types to meet various goals. ESTs also have been used to examine exclusively expressed genes, co-expressed genes, housekeeping genes, and genes that are differentially expressed because of different conditions or diseases.

Preparation of EST Library

EST sequencing as well as full-length cDNA and 3'-directed cDNA sequencing are all cloning based approaches using Sanger sequencing. Since most eukaryotic mRNAs have a poly(A) tail at their 3′ ends, EST library preparation often uses oligo(dT) primers (typically 20 nucleotides in length) to initiate reverse transcription. During the process, a switching primer also is incorporated in the first-strand cDNA synthesis to form DNA-RNA hybrids. Subsequently, the product is amplified with primers derived from the known adapter/linker sequences, followed by restriction digestion, fractionation, and cloning (Fig. 12.9A). Cloning involves ligation with vectors and transformation into *E. coli* for replication. Finally, the clones are picked randomly for Sanger sequencing. The conventional full-length cDNA library preparation involves five steps: first-strand cDNA synthesis using plasmid (pUC19) primers that have an overhang of dTs; extension of first-strand cDNA by adding an oligo(dC) tail using terminal deoxynucleotidyl transferase; digestion with *Hind*III to produce a sticky end for next step; ligation with oligo(dG) DNA linkers to form circular DNAs; followed by second-strand synthesis and subsequent transformation into *E. coli* for cloning collection and sequencing (Fig. 12.9B). Preparation of 3'-directed cDNA uses only pUC-19 based vector primers (Fig. 12.9C). First-strand cDNA is synthesized in the same manner as for full-length cDNA described above, but without addition of oligo(dC) tails. After second-strand cDNA synthesis, the constructs are cleaved with both *Bam*HI and *Mbo*I and ligated for plasmid re-circularization. The products are introduced into *E. coli* for clone selection and sequencing (Fig. 12.9C). The 3'-directed cDNA library proportionally represents the original mRNA population. In particular, the uniqueness of 3′ sequences allows for gene assignments and provides signaturesfor global profiling of gene expression.

TRANSCRIPTOME SEQUENCING: WHOLE TRANSCRIPTOME SHOTGUN SEQUENCING—RNA-SEQ PROCEDURE

The Principle and the Procedure

The whole transcriptome shotgun sequencing also can be achieved by using randomly primed cDNA and massively parallel short-read sequencing on an Illumina Genome Analyzer I. The initial steps for library preparation are similar to those described previously for EST library preparation (Fig. 12.9A). Basically, both modified oligo(dT) and template-switching primers can be used to produce full-length single-stranded cDNAs containing the complete 5′ end of the mRNA and universal priming sequences for end-to-end PCR amplification. The amplified products are fragmented and size-selected for 100–300 bp, which are then end-repaired, dA-tailed, and ligated with the Illumina sequencing adapters. PCR is performed again using Illumina's genomic DNA primer set for cluster generation and sequencing on Illumina Genome Analyzer I. RNA-seq can detect transcript levels, and reveal splicing isoforms and expressed polymorphisms. RNA-seq reads also can

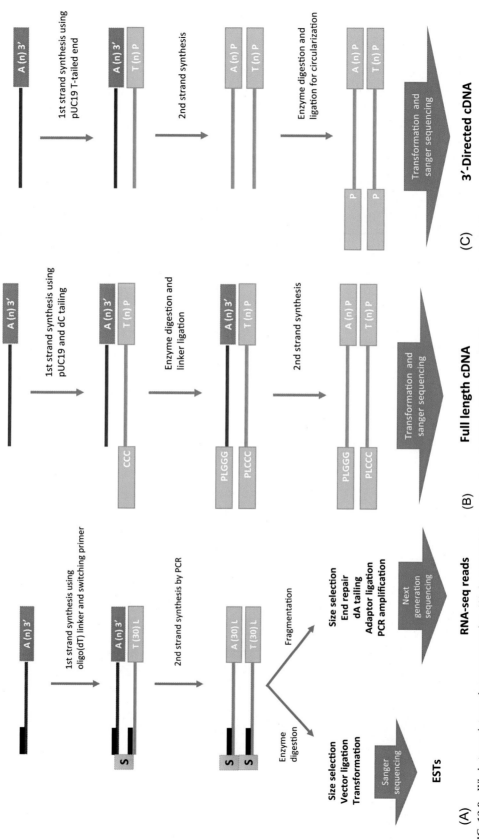

FIG. 12.9 Whole transcriptome shotgun sequencing. (A) the preparation procedure for both EST and RNA-seq, (B) library preparations for full-length cDNA, (C) library preparation for 3′-directed cDNA sequencing. EST, full-length cDNA, and 3′-directed cDNA are all cloning-based approaches using Sanger sequencing so that the adaptors are provided by the cloning vectors. However, the adaptors used in RNA-seq library preparation depend on the next generation sequencing platforms. L, Linker; S, Switching primer; P, pUC19.

be used for genome-wide profiling of polyA sites for discovery of alternative polyadenylation. Therefore, the whole transcriptome shotgun sequencing technique, with RNA-seq procedure has shaped the landscape of whole transcriptome profiling dramatically.

Comparison Between EST and RNA-seq

The construction of libraries for both EST and RNA-seq procedures are similar, but sequences are produced on different platforms. ESTs usually are determined by Sanger sequencing, which requires a cloning step to produce identical templates for the sequencing reaction. RNA-seq, however, is performed in a massive parallel manner on next generation sequencing platforms. Therefore, the advantages of RNA-seq are high throughputs, large data outputs, and relatively low costs. All of these have led RNA-seq to dominate the field, almost replacing conventional EST sequencing in recent years.

Analysis of RNA-seq

The analysis of RNA-Seq (Fig. 12.10) consists of four fundamental analysis steps, providing that an already sequenced reference genome or transcriptome is available for the target organism. Raw image data have to be converted into shortread sequences, which are subsequently aligned to the reference genome or transcriptome. The number of mapped reads is

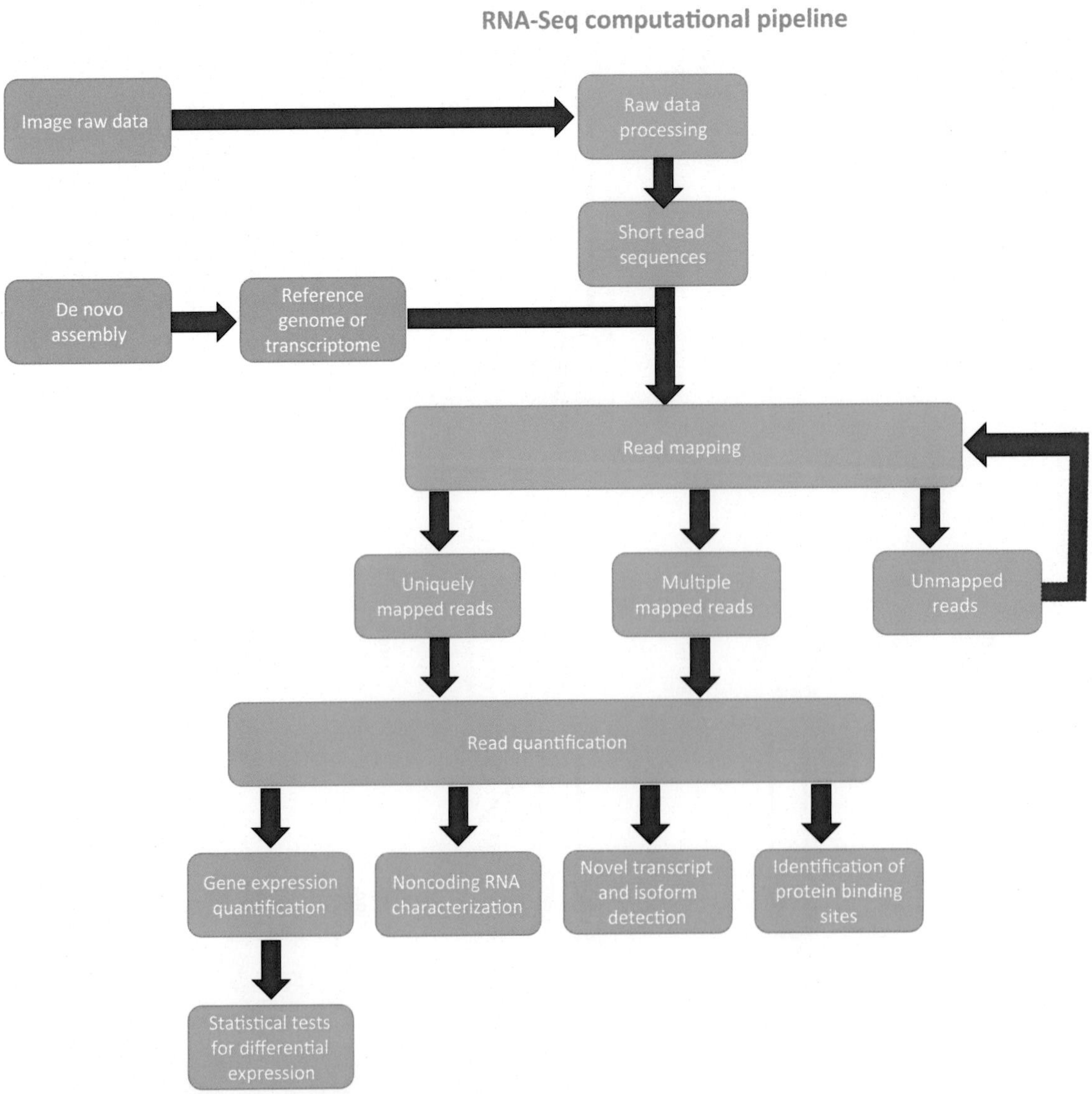

FIG. 12.10 A schematic diagram of computational pipeline for RNA-Seq data.

counted, and the gene expression level is calculated by peak calling algorithms. Finally, using statistical tests, differential gene expression is determined. A variety of bioinformatic tools can be combined to obtain an appropriate analysis system optimized to their requirements.

For RNA-Seq analysis of eukaryotes, cDNA reads preferably are aligned to a reference transcriptome. Since these data are still rarely available so far, the genomic sequences are commonly used instead. Therefore, spliced read mapping software is required to analyze the genomic intron-exon structure by splitting unmapped reads and aligning the read fragments independently. To estimate the gene expression level using RNA-seq reads, which are mapped to a particular gene, bioinformatic tools need to be used to count the number of reads in a window of defined size. By moving the window along the whole sequence, an expression profile is generated. Subsequently, these expression scores have to be normalized. The gene length influences the number of reads mapped to it and depends on the sequencing depth (total number of sequenced reads). Normalization of read counts enables the comparison of expression level between different genes as well as different experiments.

Challenges of RNA-seq Analysis

There are several challenges associated with RNA-seq analysis. The methodological challenge includes fragmentation bias, length bias, and transcriptome composition bias. For the fragmentation bias, when RNA is fragmented, library preparation favors the internal transcript body, while depleting the transcript ends by producing shorter fragments. The short 5' and 3' ends fragment can get lost easily during the size selection steps of library preparation. As a consequence, RNA-seq lacks uniform coverage for the whole gene region, implying that RNA-seq data cannot be used to accurately determine both transcription start and end sites. The number of reads per gene should depend on expression abundance, transcript lengths, and degradation process. When genes are expressed at a similar level, longer transcripts would produce more reads than shorter ones, resulting in gene length bias. Defects in nonsense-mediated decay pathways can decrease the decay of the aberrant RNA molecules, influencing the number of reads for a specific mRNA. Transcriptome composition bias occurs when one or a few transcripts in a given sample are expressed at extremely high levels, thereby downplaying the number of reads collected for other transcripts. To correct these biases, bioinformatics tools need to be introduced to adjust gene expression level. As a result, these corrections and adjustments can improve the outcomes significantly, as the adjusted data are highly correlated with qRT-PCR validation, overlap with microarray data, or are more consistent with known biology.

RNA-seq often requires 10–20 times more reads than a typical tag or target sequencing method, thus large data processing create challenges in bioinformatics analysis and computational methods. Although RNA-seq can provide reads for annotation of known genes, assembly of novel transcripts, and compilation of potential splicing forms within a gene or transcript, RNA-seq cannot sufficiently detect genes/ transcripts with low levels of expression.

WHOLE TRANSCRIPTOME TAG/TARGET SEQUENCING WITH RESTRICTION DIGESTION— SAGE (SERIAL ANALYSIS OF GENE EXPRESSION)

SAGE is a snapshot of the transcript in biological samples. Originally, magnetic beads coated with polyT tails are used to capture polyA + RNA, which then are converted into cDNA using reverse transcriptase. Subsequently, cDNA molecules are digested with anchoring enzymes and then ligated with adapters containing the tagging enzyme site that subsequently can be digested to produce small tags (Fig. 12.11). Two tags are combined into a di-tag by ligation. The di-tags are glued together to form long concatamers, which subsequently are cloned and copied millions of times and sequenced. The data are processed to count the small sequence tags for transcriptome analysis. Because it is difficult to assign the short lengths of tags to genes/transcripts, the LongSAGE (*Mme*I as the tagging enzyme) and SuperSAGE (*Eco*P15I as the tagging enzyme) methods have improved the technique by increasing tag size up to 21–26 bp in length. A method very similar to SAGE, called TALEST, or "tandem arrayed ligation of expressed sequence tags," employs an oligonucleotide adapter containing a type IIs restriction enzyme site to facilitate the generation of short (16 bp) ESTs of fixed position in the mRNA. This process involved cloning and sequencing and it does not require PCR at all stages of the assay (Fig. 12.11).

Increase of the Number of Tags Can Simplify SAGE Analysis Procedures

To further simplify SAGE analysis procedures, it is necessary to increase the number of tags sequenced per transcriptome/ library using high-throughput sequencing platforms. A novel sequencing approach, called the massively parallel signature sequencing (MPSS) method, combines nongel-based signature sequencing with *in vitro* cloning of millions of templates on

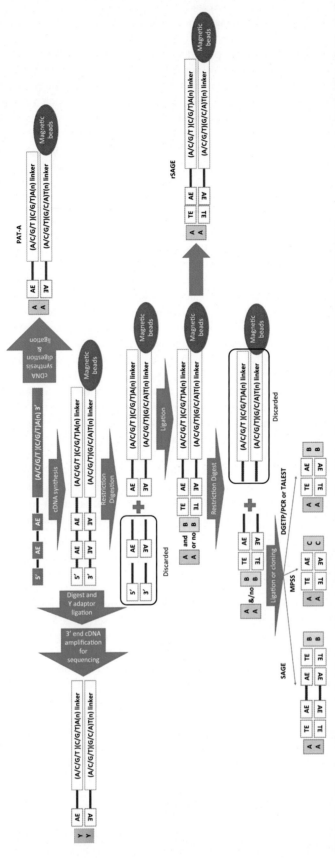

FIG. 12.11 The schematic process of whole transcriptome target/tag sequencing with restriction digestion. SAGE, MPSS, DGETP, and TALEST share many common steps in library preparation: both anchoring (AE) and tagging enzymes (TE) are used but adaptors are added in slightly different ways. In PAT-A (polyA tags), rSAGE (reverse serial analysis of gene expression) and 3′ end cDNA amplification, only the anchor enzyme cut site is used for 5′ adaptor ligation, while the 3′ adaptor/linker is combined with reverse transcription. For 3′ most target/tag collection, only 3′ end cDNA amplification uses specific primers, while the remaining methods rely on magnetic bead selection. AE, Anchoring Enzyme; TE, Tagging Enzyme; A, B, and Y, Adaptors.

separate 5 micron diameter microbeads. A 17-base sequence was generated for each transcript using enzymes *Dpn*II and *Bbv*I, followed by cloning and sequencing on beads (Fig. 12.11). MPSS can generate more than one million signature sequences (tags). As such, these tags provide enough sequence depth to identify low-expressed transcripts with high accuracy. The MPSS design led to development of several NGS platforms, such as the Roche/454 FLX, the Illumina/Solexa Genome Analyzer, the Applied Biosystems SOLiDTM System, the Helicos HeliscopeTM, and Pacific Biosciences SMRT instruments. Therefore, the traditional SAGE method has been adapted easily into NGS platforms, such as DGETP (digital gene expression tag profiling) on the Illumina/Solexa Genome Analyzers. Without formation of di-tags and concatamers, the tags are simply "sandwiched" by the Illumina GEX Adapters 1 and 2 (50 and 30 adapters) for amplification and sequencing (Fig. 12.11). SuperSAGE also has been successfully integrated with NGS as high-throughput SuperSAGE with Illumina Genome Analyzers and the Applied Biosystems SOLiDTM System.

Elongation of Tag/Target Sizes in SAGE Analysis

The tag length produced by SAGE-related methods can be increased from 10 to 26 bp by use of different tagging enzymes. However, it is still a challenging task to assign tags to known transcripts. To overcome this problem, at least three methods have been developed to collect long tags or targets. These are 3′ end cDNA amplification, rSAGE (reverse SAGE) and PATs (polyA tags) using restriction digestion. The 3′ end cDNA amplification method uses a 2-base anchored oligo(dT) primer with a heel (like a linker sequence) for first-strand cDNA synthesis, followed by the second-strand cDNA synthesis. The cDNA products then are digested with restriction enzymes and ligated to a Y-shaped adapter, which blocks amplification of the Y-Y ligated products. Therefore, only 3′ ends can be amplified for sequencing and profiling.

The rSAGE method uses primers containing 64 nucleotides (30 Ts included) as linkers for reverse transcription to synthesize cDNA molecules followed by digestion with *Nla*III and ligation with 5′ adapters. Both 5′ adapter and 3′ linker sequences are used to design primers that amplify the long tags or targets for sequencing.

In the PAT method using restriction digestion, switching primers containing enzyme cut sites are used in reverse transcription along with linker. The cDNA products then are digested with *Nla*III or *Tai*I and ligated with new adapters that have overhangs complementary to the enzyme cut sites. The remaining steps for PATs are the same as those for rSAGE, but PATs only need targets with length between 100 and 600 bp for sequencing.

Tag size and data amount are two major drawbacks associated with the conventional SAGE method. On the other hand, rSAGE and PATs using restriction digestion can extend the tag/targets sizes up to a few hundred bp in length, depending on the sequencing platforms. The long tags or targets make it easier and more accurate to assign them to genes/transcripts.

Problem With Restriction Digestion and the Solution

Whole transcriptome tag/target sequencing with restriction digestion always faces problems with the restriction enzymes themselves. First, none of the enzymes can make whole transcriptome analysis possible. This is mainly because a transcript can be cut by one enzyme, but lack the recognition site for another. Second, when an enzyme cuts only at the polyA junction site, the traditional SAGE method collects tags with all As, while rSAGE will not collect targets for these genes/transcripts. Therefore, these transcripts will be lost in the transcriptome analysis. Third, SAGE and its derivatives generally focus on collection of tags or targets associated with 3' most cut sites of transcripts. As such, rSAGE will collect targets of various lengths. When PCR is performed to enrich the targets for sequencing, the short fragments might be favored for amplification, and thus result in length bias with false abundance.

To overcome the limitations related to enzymes themselves, a combined set of enzymes can be used to conduct multiple rSAGE analysis (Fig. 12.12). Multiple rSAGE processes involve mRNA extraction, reverse transcription, enzyme digestion, bead collection, adapter ligation, PCR amplification, and NGS with sequence fragments of sizes ranging from 150 to 450 bp. Using this method, a combination of four enzymes: *Tsp*509I (AATT), *Nla*III (CATG), *Msp*I (CCGG), and *Dpn*II (GATC) can cover 99.82 percent of the *X. tropicalis* transcriptome. Both *Nla*III and *Dpn*II have been used heavily in tag-based RNA-seq analyses. The addition of *Tsp*509I (AATT) and *Msp*I (CCGG) to the enzyme combination can help produce tags for AT-rich and CG-rich mRNA sequences, respectively.

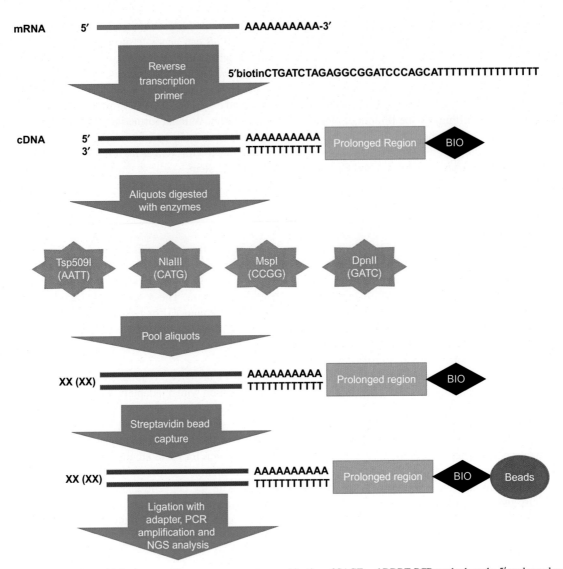

FIG. 12.12 SAGE-seq with multiple enzymes. The process represents a combination of SAGE and DDRT-PCR methods as the 5′ prolonged region was adapted from the former, while the 3′ prolonged region from the latter technique, respectively. Adaptors for the 5′ prolonged regions are designed according to the enzyme cut sites: *Tsp*509I (AATT), *Nla*III (CATG), *Msp*I (CCGG), and *Dpn*II (GATC).

WHOLE TRANSCRIPTOME TAG/TARGET SEQUENCING WITHOUT RESTRICTION DIGESTION

Transcriptome Profiling of 3′-Ends With Enrichment of PolyA+ RNA

RNA fragmentation, 3PC (3′Poly(A) site mapping using cDNA circulation), and 3′ READS (3′ region extraction and deep sequencing) are typical methods in PAT (Fig. 12.13). A common step involved in these methods is fragmentation of total RNA or polyA+ RNA followed by purification of fragmented polyA+ RNA using the Life Technologies oligo(dT) magnetic beads in 3PC, New England Biolabs oligo(dT) magnetic beads in RNA fragmentation, or Sigma CU5T45 coated beads in 3′ READ. The polyA+ containing fragments are used for reverse transcription in both 3PC and PATs. In 3′ READS, the enriched polyA+ fragments are first ligated to 5′ and 3′ adapters and then reverse transcribed to cDNA. All three methods use PCR to amplify products for deep sequencing.

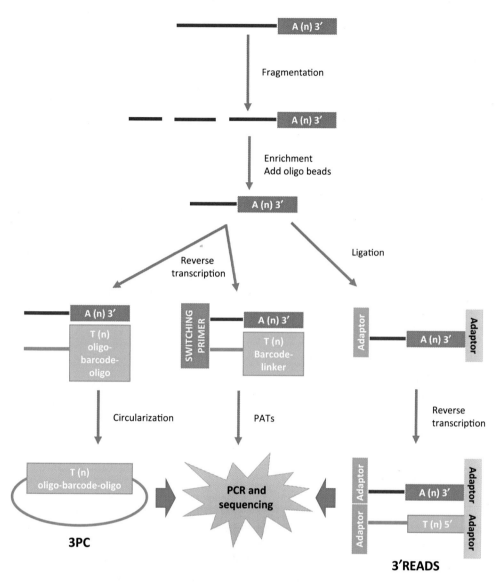

FIG. 12.13 Whole transcriptome target/tag sequencing without restriction digestion—Transcriptome profiling of 3'-ends with enrichment of polyA+ RNA. 3PC (3'Poly(A) site mapping using cDNA circulation), PATs, and 3'READS represent methods with enrichment of fragmented polyA+ RNA to conduct polyA site sequencing. These methods share common steps including size selection, PCR amplification, and next generation sequencing.

Transcriptome Profiling of 3'-Ends With Enrichment of PolyA+ cDNA

In the 3' T-fill method, total RNA is fragmented and first-strand cDNA is synthesized by reverse transcription with a biotinylated oligo (Fig. 12.14). The first-strand cDNA is treated with RNase H and second-strand cDNA is synthesized using DNA polymerase I. The double-stranded cDNA molecules are enriched for polyA+ cDNA using Dynabeads (Life Technologies), followed by dA tailing with ligation to a 5' end adapter. Eighteen cycles of PCR are performed, and the products then are size-selected by gel electrophoresis for deep sequencing.

The EXPRSS (Expression Profiling through Random Sheared cDNA Tag Sequencing) method uses total RNA as a template to synthesize first-strand cDNA with oligo(dT) primers containing the P7 sequence of the Illumina flow cell (Fig. 12.15). Second-strand cDNA synthesis is based on a traditional protocol. The double stranded cDNA products are sheared physically using Covaris Adaptive Focused Acoustics to a target size of 200 bp followed by end repair, dA tailing,

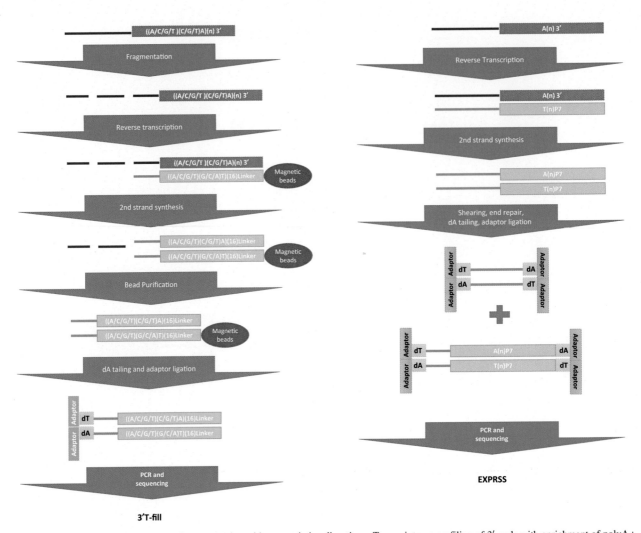

FIG. 12.14 Whole transcriptome target/tag sequencing without restriction digestion—Transcriptome profiling of 3′-ends with enrichment of polyA+ cDNA. 3′T-fill and EXPRSS represent methods with enrichment of fragmented polyA+ cDNA to conduct polyA site sequencing. These methods share common steps including size selection, PCR amplification, and next-generation sequencing.

ligation with Y-shaped adapters, and size selection with an agarose gel. The Y-shaped adapters allow amplification of only fragments derived from the 3′ end primer, and thus enrich polyA+ cDNA for sequencing. EXPRSS generates one tag per transcript at a relatively defined position from the 3′ end of a gene, ensuring no length-based data transformation, and enabling straightforward statistical analysis. The advantage is that the use of sonication for fragmenting cDNA allowed to sequence tags from all expressed genes, unlike restriction enzyme-based methods, which require the enzyme recognition site in the gene to get a sequence tag.

Transcriptome Profiling of 3′-Ends With Custom Primers Containing Oligo(dT) at 3′ End for Sequencing

PAS-seq (PolyA site sequencing) (Fig.12.16) and PolyA-seq (polyA sequencing) (Fig. 12.17) have been developed to profile 3′ ends of transcripts using oligo(dT) containing primers for sequencing. PAS-seq uses purified poly(A+) RNAs that are fragmented and reverse transcribed into cDNA using both an oligo(dT) primer and a switching primer. To sequence the products on the Illumina platform, both PE 1.0 and PE 2.0 primers are used for the first round of PCR using only three cycles. The amplified products are size-selected (200–300 bp) and used for the second round of PCR with 15 cycles. The products are purified and submitted for sequencing using a custom primer containing oligo (dT$_{20}$) at the 3′ end. Library

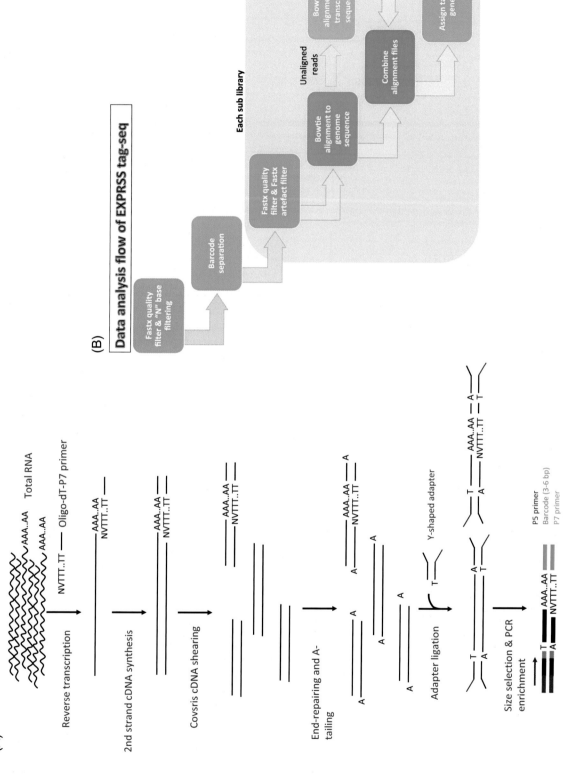

FIG. 12.15 Schematic diagram representing EXPRSS (Expression Profiling through Random Sheared cDNA Tag Sequencing) Tag-seq. (A) Library preparation. (B) Data analysis pipeline.

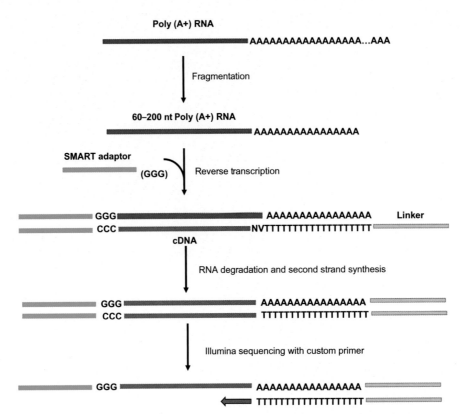

FIG. 12.16   PAS-Seq. The procedure of PAS-Seq is described in detail in text. The purple boxes represent RNAs and yellow, blue, and pink boxes represent linkers, adaptor, and cDNAs. Two degenerate nucleotides are present in the anchored oligo(dT) primer; "V" represents A/C/G and "N" represents A/T/C/G.

FIG. 12.17 Schematic overview of Poly-A-sequencing (PolyA-seq). Input was polyA+ selected RNA *(green)*. Reverse transcription using U1T10VN (10 thymidines, then a random base other than a thymidine, then a random base) is followed by RNase H treatment to degrade RNA. Second-strand synthesis using U2-N6 (random hexamers) is achieved through a random-primed Klenow extension. U1 and U2 have sequence complementarity to Illumina-specific adapters, which are added through PCR. This yields DNA libraries that can be directly sequenced.

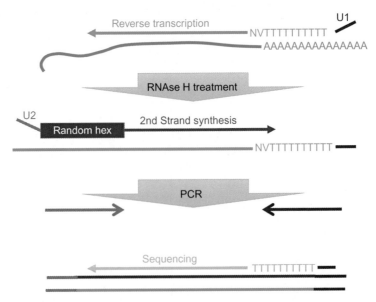

preparation for PolyA-seq begins with reverse transcription of unfragmented poly(A+) RNA molecules using an oligo. The first-strand cDNA is treated with RNase H and second-strand cDNA is synthesized using N7 random primers plus a 5′ heel of 10 bp. A total of 32 cycles of PCR is performed to amplify the cDNA products for sequencing using a custom primer containing oligo (dT$_{10}$) at 3′ end.

Whole Transcriptome Analysis With Sequencing: Profiling of Nonpolyadenylated RNAs

Some techniques have been developed for the whole transcriptomes that contain transcripts without polyA tails, or polyA-RNA molecules. These nonpolyadenylated RNAs include ribosomal RNAs, other small RNAs, replication-dependent histone RNAs, and long non-coding RNAs. Unlike eukaryotic mRNA, most bacterial mRNAs do not have a polyA+ tail, and hence cannot be isolated from other RNA sources by hybridization to immobilized oligo d(T). Furthermore, bacterial RNA preparations usually contain up to 80% rRNA and tRNA, and bacterial mRNA often has a short half-life and can be highly unstable.

Fig. 12.18 outlines the basic steps involved in generating cDNA libraries from ribosomal RNA for NGS sequencing. A common RNA adapter is ligated to the 3′ ends of all RNA transcripts, followed by removal of 18S and 28S ribosomal RNAs using biotinylated ribosomal specific probes and removal of other small RNAs using size selection. The common

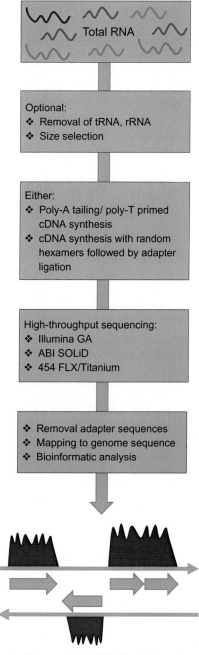

FIG. 12.18 Schematic diagram of the basic steps involved in generating cDNA libraries from ribosomal RNA for NGS sequencing. The starting material is a mix of RNA, followed by optional subtraction of tRNA and rRNA, generation of cDNA libraries, sequencing, bioinformatics, and interpretation of cDNA sequencing read histograms.

adapter ligated to the 3′ end of RNAs then is used to synthesize cDNAs, followed by digestion with *Nla*III for addition of the 5′ end adapter. The products that contained both 5′ and 3′ end adapters are amplified and sequenced from their 3′ ends using the Roche 454 system. The relatively long reads produced by this platform are required for this method because short reads with only As are useless in the analysis. Alternatively, the polyA+ RNAs can be removed using the oligo-dT beads. The remaining polyA-RNAs then are analyzed by a routine RNA sequencing method.

WHOLE TRANSCRIPTOME ANALYSIS WITH SEQUENCING

Profiling of Circular RNAs

In cells, different RNA molecules sometimes are joined together by splicing reactions (trans-splicing). Circular RNAs (circRNAs) are a large class of noncoding RNAs that form covalently closed RNA circles. Plant circRNAs encode subviral agents. Exonic circRNA and intronic circRNA are two types of circular RNA that can be detected by different RNA-seq strategies. In unicellular organisms, circRNAs mostly stem from self-splicing introns of preribosomal RNA but also can arise from protein-coding genes in archaea. In animals, the spliceosome links the 5′ and downstream 3′ ends of exons within the same transcript. In general, RNase R treatment degrades linear RNA molecular and Y-structure RNAs so that only circular RNAs are retained in the reaction. Special algorithms based on existing transcript models or based on genomic sequences can be used to identify exonic circRNAs. A polyA- enrichment step can be added to reduce noisy data and directly identify abundant intronic circular RNAs.

Profiling of RNA Methylation

RNA methylation is a common phenomenon in both eukaryotes and prokaryotes. To examine cytosine methylation within RNAs, RNA bisulfite sequencing combined with deep sequencing is often used. Basically, RNAs are treated with bisulfite to convert methylated cytosine to uracil. The conversion of cytosine residues can be detected by PCR-based methods. RNA sequencing of PCR amplicons and/or of cloned PCR products (DNA bisulfite sequencing) allows the characterization of RNA methylation patterns in their native sequence context. Fig. 12.19 is an example of the reverse transcription of bisulfite-treated tRNA. Besides chemical treatment, immunoprecipitation to detect RNA cytosine methylation also can be used. For example, both mammalian cytosine RNA methyltransferases (m5CRMTs) and the cytosine analog 5-azacytidin (5-aza-C) are used in an Aza-immunoprecipitation (Aza-IP) methodology to form stable m5C-RMT-RNA linkages. The products then can be immunoprecipitated and subjected to NGS sequencing. Similarly, RNA methyltransferase Nsun2 also can be used to develop the methylation individual-nucleotide-resolution crosslinking and immunoprecipitation method to detect cytosine methylation in RNA species.

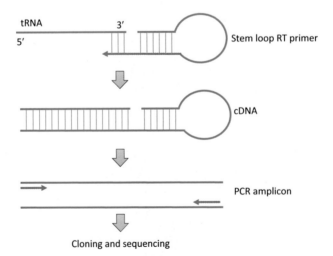

FIG. 12.19 Outline of the basic strategy to analyze tRNA for m5C methylation. Bisulfite-treated tRNAs are reverse transcribed using a tRNA 3′-sequence-specific stem-loop primer, amplified with primers binding only to deaminated sequences at the 5′ end, followed by standard cloning and sequencing.

Profiling of 5′-Ends With Enrichment of 5′G-CAP

The cap analysis of gene expression (CAGE) using Sanger sequencing to map transcription start sites have been converted to NGS known as deep CAGE, PEAT [paired-end analysis of TSSs (transcription start sites)], nanoCAGE and CAGEscan methodologies. The first step of these NGS techniques usually involves treatment of RNAs with bacterial alkaline phosphatase (BAP) to degrade the phosphate group of RNA without 5′G-cap. Tobacco acid pyrophosphatase (TAP) then is used to hydrolyze the phosphodiester bond of the 5′ triphosphate of an mRNA molecule and generate mRNA molecules with one phosphate group at the 5′ end. Subsequently, the BAP-TAP-treated RNAs are ligated with sequence adapters. As a result, only RNAs with cap structures can be sequenced. Deep CAGE expression profiling is unique among high-throughput expression profiling methods because the 5′ end of the CAGE tag identifies the corresponding transcription start site.

THE RNA STRUCTUROME

RNA has a significant role in nearly every process in living cells, including transcription, RNA processing, and translation. Moreover, RNA can sense biomolecules (e.g., proteins, RNAs, or DNAs), ligands, temperature, and mutations, all of which can modulate RNA structure (Fig. 12.20). The single-stranded (ss) nature of RNA provides the plasticity needed for it to fold into diverse secondary structures (such as hairpins or three-way junctions) and tertiary structures (such as pseudoknots and G-quadruplexes) that govern its functional roles, ranging from regulation of gene expression to ligand sensing and enzymatic functions. Therefore, elucidating RNA structure can provide profound insights into living systems.

Commonly used chemical probes for the assessment of RNA secondary and tertiary structure include dimethyl sulfate (DMS), 2-methylnicotinic acid imida-zolide (NAI) and 1-methyl-7-nitroisatoic anhydride (1M7). These probes can readily be applied in vivo. This enables interrogation of in vivo RNA structure and its change upon exposure to various conditions.

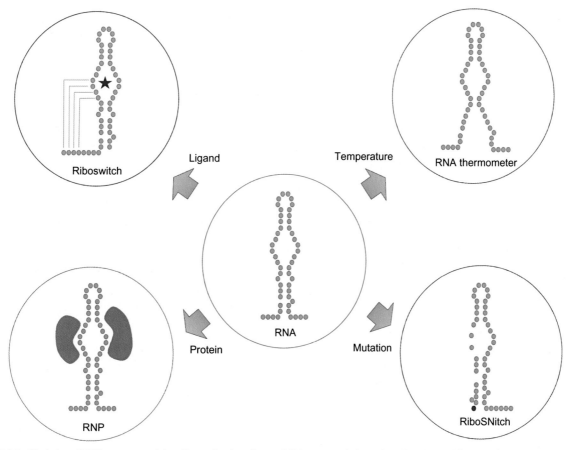

FIG. 12.20 Variation of RNA structure and the effects of various factors. RNA structure is dynamic and can be sensitive to various cellular factors. The type of factors that an RNA structure is sensitive to categorizes it into a particular class, such as riboswitches (sensitive to ligands), ribonucleoproteins (RNPs, sensitive to proteins), RNA thermometers (sensitive to temperature), and riboSNitches (sensitive to mutation).

Transcriptome-Wide RNA Structural Probing In Vivo

The classical method to read out RNA structure probing data is gel electrophoresis, but it is low throughput and generally provides information only on approximately 150 nucleotides (nt) of one given transcript. Capillary electrophoresis (CE) has a somewhat improved throughput, which has extended the length amenable to analysis to approximately 500 nt.

Initial high-throughput transcriptome-wide RNA structural probing approaches were conducted in vitro and those approaches, which enable the analysis of thousands of in vitro RNA structures in a single experiment, couple RNase probing with NGS, and differ mainly in the choice of RNase probe (Table 12.1). Three methods that use DMS to interrogate in vivo RNA structures transcriptome-wide have been developed: Structure-seq, DMS-sequencing (DMS-seq), and modification-sequencing (Mod-seq) (Fig. 12.21). All three methods have similar sample preparation procedure, including in vivo DMS treatment, reverse transcription, and ligation to enable PCR amplification before NGS. Although these methods have similar preparation process, they differ in the number of chemical modifications per molecule, reverse transcription priming, ligation methodology, and RNA population bias.

Process for In Vivo RNA Structures Transcriptome-Wide Techniques

To generate a cDNA library for sequencing, the DMS-modified RNAs must be converted to cDNA and fused with NGS adapters on each side. For Structure-seq, random hexamer (N_6) reverse transcription is conducted to generate first-strand cDNAs of varying lengths, together with part of an NGS adapter on one side, whereas in DMS-seq and Mod-seq, random RNA hydrolysis is first performed to fragment the RNA sample. The fragmented RNAs, which bear $5'$ hydroxyl and $2',3'$ cyclic phosphate termini in DMS-seq and Mod-seq, are then $3'$ dephosphorylated. For Mod-seq, a $5'$ phosphorylation of the RNA is performed. Resultant RNA fragments are subjected to $3'$ RNA adapter ligation to provide a known common

TABLE 12.1 NGS Methods of Interrogating RNA Structures

Methods	Application to Date	System	Probe
PARS, PARTE	In vitro	Saccharomyces cerevisiae Human lymphoblastoid cells	Rnase S1 (ssRNA) Rnase V1 (dsRNA)
Frag-seq	In vitro	Mus musculus (KH2 embryonic stem cells)	Rnase P1 (ssRNA)
Ss/dsRNA-seq	In vitro	Arabidopsis thaliana Drosophila melanogaster Caenorhabditis elegans	Rnase I (ssRNA) Rnase V1 (dsRNA)
SHAPE-seq 1.0, 2.0	In vitro	Synthetic	1M7
MAP-seq	In vitro	Synthetic	1M7 DMS CMCT
HRF-seq	In vitro	Escherichia coli	Hydroxyl radical
ChemMod-seq	In vitro	S. cerevisiae	DMS 1M7
SHAPE-MaP	In vitro	E. coli HIV-1	1M7
RING-MaP	In vitro	Synthetic	DMS
CIRS-seq	In vitro	M. musculus (E14 embryonic stem cells)	DMS CMCT
Structure-seq	In vitro	A. thaliana	DMS
DMS-seq	In vitro, in vivo	S. cerevisiae Human (K562/fibroblast cells)	DMS
Mod-seq	In vitro	S. cerevisiae	DMS

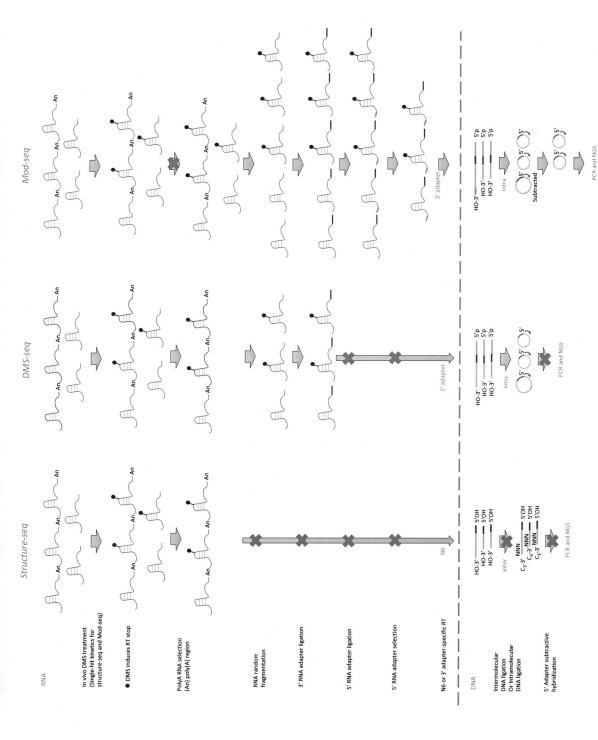

FIG. 12.21 The schematic workflow for structure-seq, dimethyl sulfate sequencing (DMS-seq), and modification-sequencing (Mod-seq). The steps that are not performed in a particular method are marked as "X." DMS modification is marked with *black oval*, which will lead to DMS-induced reverse transcriptase stop. The "An" in the poly(A) RNA selection step depicts the poly(A) tail in RNA. The "N" in the DNA ligation step denotes degenerate base. (A) For structure-seq (*red*), RNA is treated with DMS in vivo and poly(A) RNA is selected. Subsequently, random hexamer (N6) reverse transcription (RT) is used to generate cDNA. Intermolecular single-stranded DNA (ssDNA) ligation is then performed to ligate the cDNA (*green-black*) and a ssDNA linker with NNN on its 5′ end (*gray*). The ligated cDNA is amplified by PCR and submitted for next-generation sequencing (NGS). (B) For DMS-seq (*cyan*), RNA is treated with DMS in vivo and poly(A) RNA is selected. Then, random fragmentation by Zn²⁺-mediated hydrolysis is used to generate RNA fragments. A 3′ RNA ligation is performed to ligate the RNA fragment with a 3′ RNA adapter (*black*). A 3′ adapter-specific RT is performed on the ligated RNA to generate cDNA. Intramolecular circular DNA ligation is then performed, followed by PCR and NGS. (C) For Mod-seq (*purple*), RNA is treated with DMS in vivo and total RNA is used for the library preparation. RNA is subjected to random fragmentation by Zn²⁺-mediated hydrolysis to generate RNA fragments. A 3′ RNA ligation is performed to ligate the RNA fragment with a 3′ RNA adapter (*orange*) to the RNA from the previous step (RNA fragment plus 3′ RNA adapter), followed by a 5′ RNA adapter selection via biotinylated DNA oligonucleotide that is antisense to the 5′ adapter to select for successfully ligated RNAs. A 3′ adapter-specific RT is performed on the ligated RNA to generate cDNA. A 5′ adapter subtractive hybridization using biotinylated DNA oligonucleotide that senses the 5′ adapter is performed to remove cDNAs generated from unmodified RNA. Intramolecular circular DNA ligation is then performed, followed by PCR and NGS.

sequence for primer annealing to initiate reverse transcription. In Mod-seq, a 5′ RNA adapter ligation is also performed, followed by a 5′ adapter selection step via biotinylated oligonucleotide to select for successfully ligated RNA products. The 5′ RNA adapter ligation in Mod-seq is also important for the 5′ adapter subtractive hybridization step. Overall, Structure-seq has significantly less processing steps at the RNA level compared with DMS-seq and Mod-seq, which reduces the chance of undesired RNA degradation.

All three methods used Circligase for cDNA ligation. Structure-seq uses an intermolecular linear DNA ligation, whereas DMS-seq and Mod-seq use intramolecular circular DNA ligation. Circligase, similar to T4 RNA ligase, is known to have sequence bias. In all methods, after the processing steps, the cDNAs are subjected to PCR and NGS, and raw sequencing reads are generated for use in computational analysis.

RNA structurome data provide intriguing insights into roles of RNA structure in cellular processes, including translation, splicing, polyadenylation, miRNA-mediated regulation, transcript stability, and transcript localization. RNA structure is dynamic and known to be sensitive to cellular conditions. Development of transcriptome-wide structure probing of RNA has led to entirely new insights into RNA structure and function.

METABOLOMICS FOR GENOME-WIDE ASSOCIATION STUDIES: LINKING THE GENOME TO THE METABOLOME

The Principle of Metabolomics

Metabolites and small molecules are organized in biochemical pathways and in a wider metabolic network, which is itself dependent on various genetic and signaling networks for its regulation. Numerous biochemical methods to measure the concentrations of specific metabolites in human body fluids or tissue samples are available and partially used in diagnoses. Metabolomics is the emerging field of measuring ideally all small molecules (metabolites below 1500 Da) in a biological sample in one single experiment. All small molecules present in a sample are termed metabolomes. Metabolomics reveals global insights into intermediate phenotypes (changes in metabolite levels) not depicted by other diagnostic approaches (Fig. 12.22). The metabolites are being considered not only as biological end points but as a driving force in the pathophysiology of human disease.

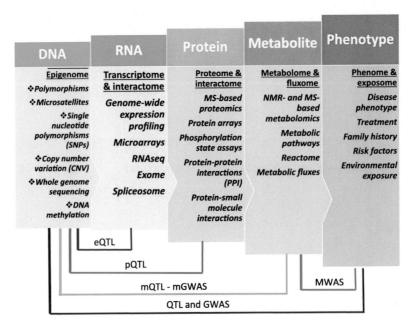

QTL: Quantitative trait loci
eQTL: Gene expression quantitative trait loci
pQTL: Protein quantitative trait loci
mQTL: Metabotype quantitative trait loci
mGWAS: Metabolomic GWAS

FIG. 12.22 Metabolomics link the proteomics to phenotypes. The biomolecular organization of information and function is represented on the horizontal axis, from the DNA to phenotype.

Genome-Wide Association Studies With Metabolomics (mGWAS)

Genome-wide association studies (GWAS) analyze a level of association between common genetic variants and phenotypic traits (phenotypes). Initially, GWAS were conducted with disease endpoints, such as diabetes, heart, and kidney diseases. The strength of these found associations, however, is generally very small and is not possible for disease prediction in the case of complex disorders. Moreover, an association of a genetic variant with a disease does not always provide information about the mechanisms leading to the phenotype. To this end, metabolic traits, such as concentrations of blood glucose, amino acids, lipids, and hormones, have been used successfully in GWAS to overcome this drawback.

Metabolomics GWAS (mGWAS) requires the mapping of metabotype quantitative trait loci (mQTL) followed by the genome-wide association framework, which have been enabled by recent advances in bio-analytical technologies that allow accurate and broad metabolite profiles to be obtained in high-throughput. The comprehensive exploration of metabolome perturbations associated with physiological and pathological conditions is a great challenge for accurate disease diagnosis and personalized healthcare. Metabolomics relies on NMR spectroscopy and mass spectrometry to characterize biofluids, such as urine and blood plasma/serum, biopsies and organs, but also cell culture pellets and media. High-resolution (>40 K variables) spectral data are analyzed by univariate and multivariate statistical to identify statistically significant biomarkers associated with the condition under scrutiny.

Chapter 13

Molecular Diagnosis of Chromosomal Disorders

Chapter Outline

CHROMOSOME MORPHOLOGY

The Role of Centromere

Mitotic chromosomes can be distinguished by their relative length and the location of the centromere, which is a region of the chromosome that is responsible for its segregation at mitosis and meiosis. During mitosis, the centromere is moved toward the pole, and the attached chromosome appears to be dragged along behind it (Fig. 13.1). The centromere essentially acts as the luggage handle for the entire chromosome, and its location typically appears as a constricted region connecting all four chromosome arms. The connection is made between microtubules of the spindles and a protein complex known as kinetochore that assembles at the centromere sequences. The kinetochore is a darkly staining fibrous object, and it provides a microtubule attachment point on the chromosome.

Diagnostic Molecular Biology. https://doi.org/10.1016/B978-0-12-802823-0.00013-4

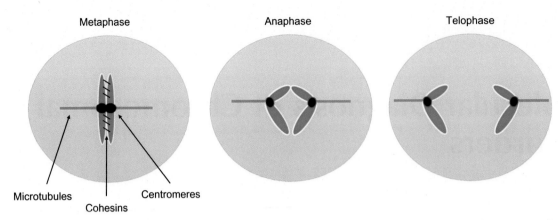

FIG. 13.1 Microtubules attach to centromeres and can pull chromosome to the pole. The sister chromatids are held together by glue protein cohesins.

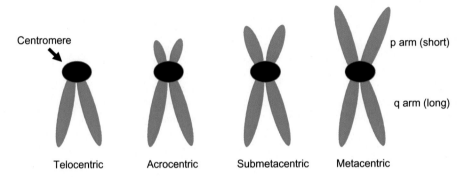

FIG. 13.2 Categorization of chromosomes by the location of centromere in the chromosome.

Location and Classification of Centromere

The centromere region contains a set of highly repetitive alpha satellite sequences, which interfere with chromosome compaction so that the centromere appears as a constriction in the metaphase chromosome. Depending on the location of centromere, the chromosome can be classified as metacentric, submetacentric, acrocentric, or telocentric chromosome (Fig. 13.2). The metacentric chromosome's arms are relatively equal in length, whereas one arm of the submetacentric chromosome is longer than the other. In acrocentric chromosome, one arm is extremely small and the other is extremely long. In the telocentric chromosome, one arm is missing completely.

The chromatin structure of centromere region is epigenetically unique. Centromere-specific histone H3 is a primary determinant in establishing functional centromeres and kinetochore assembly sites. In humans, the centromere-specific histone H3 variant is CENP-A. It replaces the normal H3 histone at sites where centromeres reside and kinetochores attach chromosomes to spindle fibers. This specialized centromeric chromatin is the foundation for binding of other centromere-associated proteins.

CYTOGENETIC METHODS FOR CHROMOSOME

The Karyotype

A karyotype can be defined as the accurate organization (matching and alignment) of the chromosomal content of any given cell type. In a karyotype, chromosomes are arranged and numbered by size, from the largest to the smallest. Karyotype is the normal nomenclature used to describe the normal or abnormal, constitutional or acquired chromosomal complement of an individual, tissue, or cell line. Numerical chromosome aberrations occur in all dividing cells because of errors during mitosis. In most cases extra or missing chromosomes compromise viability, and it can lead to cell death. In rare cases where extra chromosomes result in an adaptive advantage, however, aneuploidy cells can increase in number and replace the normal cell population. Acquisition of chromosome aberrations is the main mechanism by which cancer cells evolve because chromosome changes can lead to abnormal expression of many genes.

The Methods to Prepare the Karyotype

The most common karyotype preparation method used for years is the conventional or classic cytogenetic techniques, in which the cells are cultured, metaphases are obtained, and the chromosomes with an appropriate stain are studied. This technique, although first described some time ago, is still valid and informative, although newer methods of molecular cytogenetics could resolve the main problem of the conventional technique—that is not always possible to obtain an adequate number of metaphases in the process or that the quality of these metaphases does not permit a detailed study of the chromosomes.

CONVENTIONAL CYTOGENETICS

Conventional cytogenetic techniques are used routinely to determine numerical chromosomal abnormalities or structural rearrangements, mainly translocations, in any given cell type. For example, it can be used for detecting genomic aberrations, including both gains and losses of segments of the genome and rearrangements within and between chromosomes. The resolution of standard cytogenetics techniques remained limited, however, with a count of approximately 400–500 bands per haploid genome.

The Conventional Approach to Prepare the Karyotype

Typically, cells are arrested at the metaphase stage by colcemid. Subsequently, cells are fixed and placed on a slide followed by staining with the chemical dye Giemsa, after proteolytic enzyme (trypsin) treatment. The Giemsa dye generates distinct chromosome-specific patterns called G-bands. Fig. 13.3 represents an example of the human set. The Giemsa dye can map structural aberrations clearly and is a commonly used staining method for analyzing chromosome.

Banding Patterns

G-banding allows each chromosome to be identified by its characteristic banding pattern. The banding pattern can distinguish chromosomal abnormalities or structural rearrangements, such as translocations, deletions, insertions, and inversions. G-banding has been divided into regions, bands, and subbands. Fig. 13.4 shows a typical diagram of the bands of the human X chromosome. In general, the bands have a lower G-C content than the interbands, where genes tend to be located. Harsher treatment of chromosome (87°C for 10 min) before Giemsa staining can produce a pattern called R banding, which is opposite to the G-banding pattern. The R banding can stain the euchromatin region.

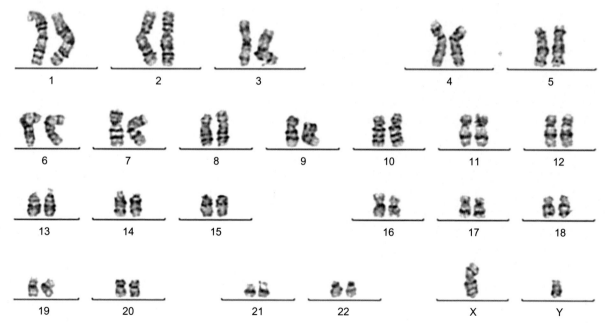

FIG. 13.3   An example of a male karyotype with G-bands.

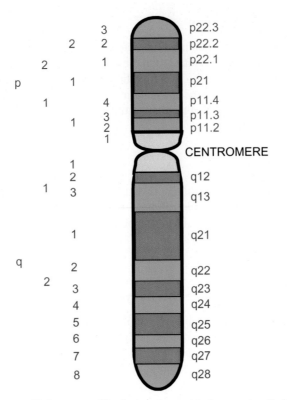

FIG. 13.4 An example of G-banding pattern on X-chromosome. The short arm is p, and the long arm is q. Each arm is divided into larger regions that are further subdivided into smaller bands and interbands.

The region flanking the centromere often displays a considerable amount of heterochromatin. When the entire chromosome is condensed, the centromeric heterochromosome is not visible in a mitotic chromosome. The centromere, however, can be seen by a staining that generates C-bands (Fig. 13.5). The C banding or centromere banding results from the alkali treatment of chromosome. Centromere staining is absent in G-band patterns. C bands are associated with heterochromatin along the chromosomes and around the centromeres.

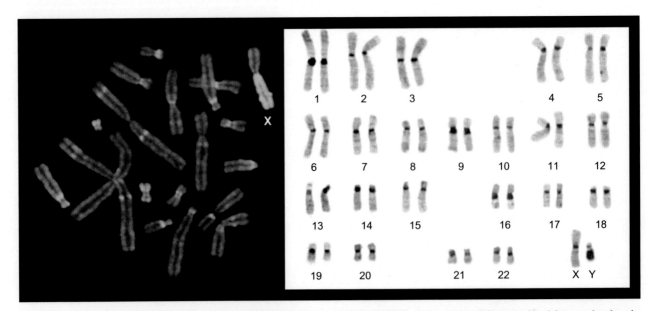

FIG. 13.5 C-banding in CHO and human chromosomes. *Left*: C-banded metaphase of CHO9 cell line. This cell line contains eight normal and twelve rearranged autosomes with only one X chromosome. Giemsa-stained C-band regions are visualized in *yellow* (reflected light microscopy). *Right*: C-banded karyotype of a human peripheral lymphocyte metaphase showing centromeric, pericentric (Chromosomes 1, 9, and 16), and distal Yq heterochromatic blocks.

Detection of Genome and Chromosomal Mutations

Genome mutations, or aneuploidy, can be detected by karyotyping, which is the direct observation of a complete set of metaphase chromosomes structure arranged by size in a cell. During the staining process, the chromosomes in dividing cells that arrest in metaphase will yield a chromosome spread when the cell nuclei are disrupted with hypotonic solution. The 23 pairs of chromosomes then can be assembled into an organized display according to their size and centromeric placement. A schematic diagram of the karyotyping protocol is shown in Fig. 13.6. Because the karyotype can directly observe the metaphase chromosome, it is useful to visualize all types of chromosomal mutations (Fig. 13.7). A list of terms used in karyotyping for chromosomal mutations is shown in Table 13.1.

MOLECULAR CYTOGENETICS: FLUORESCENCE IN SITU HYBRIDIZATION (FISH)

The In Situ Hybridization (ISH)

The considerable gap between the limited resolution for observing chromosome structure through banding techniques (> 5 Mb, depending on the banding resolution applied) at light microscopy and gene levels is bridged by the introduction and application of several molecular cytogenetic approaches. The first applications of molecular techniques to chromosome slide preparations, called in situ hybridization (ISH), were attempts to identify and locate specific nucleic acid sequences inside cells or on chromosomes. The ISH technique was based on the discovery that radioactively labeled ribosomal RNA hybridized to acrocentric chromosomes. The hybridization was visualized using autoradiography, which had been applied to human chromosomes since the early 1960s. The use of ISH technology provided another dimension to the study of chromosomes, facilitating the visualization of DNA or complementary RNA sequences on chromosomes and in cells at the molecular level. The use of this method, however, was limited because of the use of radioactive isotopes, highly repetitive DNA sequences, and corresponding RNA in the satellite regions of chromosomes and centromeres.

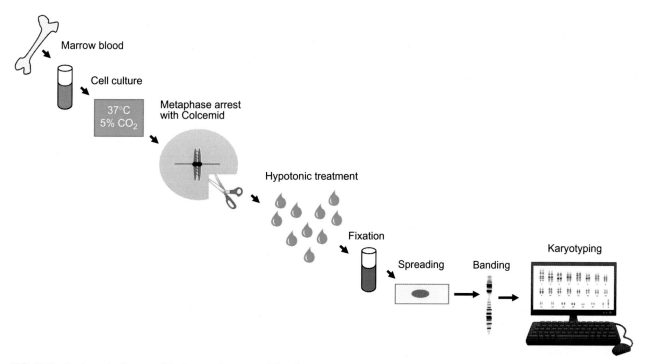

FIG. 13.6 A schematic diagram of the karyotyping protocol from bone marrow.

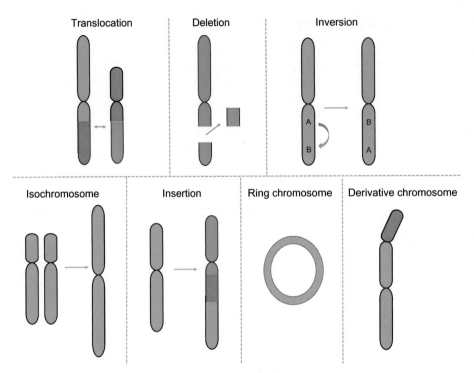

FIG. 13.7 Different types of chromosomal mutations that involved alterations in chromosome structure.

TABLE 13.1 Genetic Abbreviations

Abbreviation	Indication
+	Gain
−	Loss
del	Deletion
der	Derivative chromosome
dup	Duplication
ins	Insertion
inv	Inversion
i, iso	Isochromosome
mat	Maternal origin
pat	Paternal origin
r	Ring chromosome
t	Translocation
tel	Telomere

The Fluorescence In Situ Hybridization (FISH)

The ISH technique was improved with the development of a technique involving the use of a nonradioactive probe (such as biotin) for indirect labeling through nick translation. The hybridization between DNA probe and target sequence could be visualized through avidin or streptavidin fluorescent labeling. The development of fluorescent molecules led to direct (combined with a fluorochrome) or indirect (through an intermediate molecule incorporated into a probe) binding to DNA bases, which eventually evolved into fluorescence in situ hybridization (FISH). FISH is based on fluorescently labeled probes that hybridize to unique DNA sequences along the chromosomes. FISH can be performed on either metaphase preparations or interphase cells. FISH increased the resolution at which chromosome rearrangements could be identified at submicroscopic levels, making this technique applicable for both clinical diagnosis and research. As such, there are wide applications of FISH, mainly in cancer research and molecular diagnosis.

The Metaphase FISH

The FISH technique that is widely used in laboratories involves the hybridization of labeled DNA probe in situ chromosomal target. The standard FISH protocol, illustrated in Fig. 13.8, has five steps: sample pretreatment, denaturation of probe and sample, hybridization of probe to target cells or metaphase spreads (annealing), posthybridization washing, and detection using a simple epifluorescence microscope with appropriate filter sets. Preparation of metaphase FISH samples begins with the culture of cells for 72 h. About 4 min before harvesting, colcemid is added to the culture to arrest dividing cells in metaphase. The cells are then suspended in a hypotonic media, for example 0.075 M KCl, and fixed with methanol/acetic acid. The fixed-cell suspension is applied to a slide and allowed for dry. Subsequently, a target DNA in cells, nuclei, or metaphase chromosomes is fixed and denatured on the surface of the slide. The probe DNA must be labeled with a nucleotide that is either conjugated to fluorescein (direct labeling) and/or a nonfluorescent hapten (indirect labeling). Probe and target DNAs are denatured using high-temperature incubation in a formamide or salt solution.

Prior to hybridization, the metaphase chromosome suspension and/or interphase nuclei are enzymatically pretreated to enhance accessibility to the probe and reduce the amount of cytoplasm. The pretreated slide containing the target and probe DNA is heated to denature the DNA. The prepared probe is subsequently applied to the slide for ~ 16–$48\,h$ at $37°C$ for hybridization. The probe is applied in excess, so the kinetics ensures that the probe anneals to the target DNA.

Detection Techniques

Probe detection is accomplished by ultraviolet light excitement of a fluorochrome, such as fluorescein-5-thiocynate (FITC) or tetramethyl rhodamine isothiocyanate (TRITC), which is attached directly to probe DNA, or by incubation of hapten-labeled probe with a fluorescent conjugate. After washing, an antifade solution containing DAPI (4',6-diamidino-2-phenylindole) is applied to the slide, and a coverslip must be added. DAPI is a fluorescent stain used extensively in fluorescence microscopy. FISH signals are typically observed using epifluorescence microscopes with specific filters for identifying fluorochromes, a CCD camera captures the image, and the fluorescent signals are subsequently quantified. The resulting images can be analyzed using commercially available softwares. Therefore, FISH can be used to detect DNA, RNA, and protein location in the cell that is fixed into a slide. FISH allows the study of chromosome exchanges and gene rearrangements, amplification, and deletions at the single-cell level. The majority of probes used for laboratory purposes are commercially available.

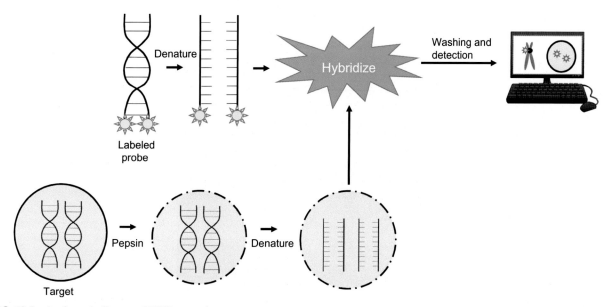

FIG. 13.8 A schematic diagram of FISH protocol.

Interphase FISH

Cell culture is not required in interphase FISH, which is commonly used in the study of the prenatal samples, tumors, and hematological malignancies. Because interphase FISH does not need to culture cells and stimulate the division, it is valuable for analysis of cells that do not divide well in culture. Metaphase FISH is useful in detecting small or complex rearrangements that are not apparent by regular chromosome banding.

Preparation of interphase FISH samples is critical in permeabilizing the cells for optimal probe-target interaction and maintaining cell morphology. Optimal results are obtained if fresh interphase cells are incubated overnight after deposition on slide. After aging overnight, cells are treated with protease to minimize interference from cytoplasmic proteins and fixed with 1% formaldehyde to stabilize nuclear morphology.

Before DNA denaturation, the cells are dehydrated in graded concentrations of ethanol. The quality of the probe also needs to be checked and validated before use. Normally, the bound probe is visualized under a fluorescent microscope as a point of fluorescent light in the nucleus of the cell. Probes are designed to be complementary to a particular chromosome or chromosomal locus, so that the fluorescent image can correlate the state of that chromosome or locus.

FISH PROBES

Three broad types of FISH probes are used in clinical genetics laboratories, each with a different application: Whole-chromosome painting (WCP) probes for deciphering cytogenetic aberrations; repetitive sequence probes for chromosome enumeration; and locus-specific identifier (LSI) probes for gene fusions, gene deletions or duplications. These probes also can be used in various combinations when investigating complex chromosome abnormalities. Telomeric and subtelomeric probes are used for detecting the chromosome telomeres, which is in the very ends of the chromosome arms and thus are not used as widely as other kinds of probes.

WCP Probes

WCP probes are designed to mark the entire chromosome of interest. Fig. 13.9 shows a typical whole painting of chromosome 6. WCP probes are useful for deciphering cytogenetic aberrations that are difficult to resolve on morphological grounds, such as marker chromosomes of uncertain nature or complex changes. For example, WCP can be used in the analysis of sperm chromosome segregation in a man who is heterozygous for an uncommon Robertsonian translocation. This translocation can result in reproductive failure because of imbalances in chromosome meiotic segregation. In Fig. 13.10, two WCP probes, including a chromosome 13 painting probe (WCP 13) labeled in *green*, and a chromosome 22 painting probe (WCP 22) labeled in *orange*, are used to label sperm nuclei, and various segregation patterns (alternate and adjacent) are observed according to the fluorescent. Normal and balanced spermatozoa resulting from alternate segregation are found when both signals are present, whereas unbalanced fluorescent patterns, resulting from adjacent

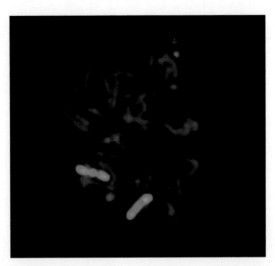

FIG. 13.9 An example of whole chromosome painting of Chromosome 6 (shown in *green*). The centromere probe of Chromosome 10 is labeled *red*.

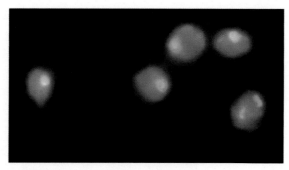

FIG. 13.10 Labeling of sperm nuclei with WCP probes to determine the segregation patterns.

segregation modes, are observed when one signal is present. This result also demonstrates that the use of WCP probes in interphase cells is limited, because the chromosomes are dispersed in the cell nucleus during interphase and do not form discrete units.

Centromeric Enumeration Probes

Centromeric enumeration probes (CEPs) hybridize to alpha (or beta) satellite repeat sequences within the specific centromeric regions of each chromosome and are used for chromosomal enumeration. CEPs detect aneuploidy of any chromosome. Combination of CEPs and region-specific probes often is used to confirm deletions or additions in specific chromosomes. Interphase FISH analysis using dual color CEPs for chromosomes X and Y is an efficient method to predict aneuploidy of sex chromosome. For example, Fig. 13.11 demonstrates an Interphase FISH study of monolayer cultured bone marrow stromal cells. Both probes DXZ1 (chromosome X alpha satellite DNA) and DYZ1 (chromosome Y satellite DNA) are used to show an XXY hybridization pattern that suggests an aneuploidy of sex chromosome of 47, XXY.

Locus-Specific Identifier (LSI) Probes

Locus-specific identifier probe can be used to identify the aneuploidy of chromosome. For example, the 2q11.1-specific probe has been used in the prenatal diagnosis of mosaic trisomy 2. Fig. 13.12 shows that Interphase FISH analysis of uncultured amniocytes using the RP11-468G5 (2q11.1) probe (spectrum *green*) shows (A) two *green* signals in an amniocyte with disomy 2 and (B) three *green* signals in an amniocyte with trisomy 2.

In the dual-color system, two main LSI FISH probe systems are used to detect gene rearrangements, dual-color translocation probes and dual-color break-apart probe. The dual-color translocation probes are designed to detect chromosomal translocations involving known partner genes, such as BCR-ABL1 that result from t(9;22)(q34;q11.2) in chronic myeloid leukemia (CML) (Fig. 13.13A). Dual-color break-apart probes are useful for detecting chromosomal translocations that involve genes with unknown or multiple translocation partners, such as myeloid lymphoid leukemia (MLL), anaplastic lymphoma kinase (ALK), and retinoic acid receptor alpha (RARα) (Fig. 13.13B). Break-apart translocation FISH probes conveniently provide important information about the presence of gene rearrangements, although they are unable to identify the specific partner gene. With the use of long-distance inverse-polymerase chain reaction (LDI-PCR), it has

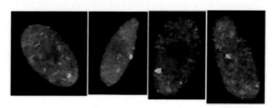

DXZ1 Xp11.1-q11.1
DYZ1 Yq12

FIG. 13.11 Interphase FISH study of monolayer cultured bone marrow stromal cells using probes for DXZ1 (chromosome X alpha satellite DNA) and DYZ1 (chromosome Y satellite DNA) to show an XXY hybridization pattern.

FIG. 13.12 Interphase FISH analysis of uncultured amniocytes using the BAC (bacterial artificial chromosome) probe RP11-468G5 (2q11.1) (spectrum *green*) at the second amniocentesis. (A) Two *green* signals in an amniocyte with disomy 2 and (B) three *green* signals in an amniocyte with trisomy 2.

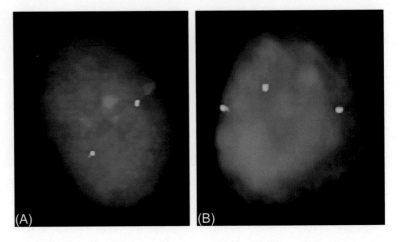

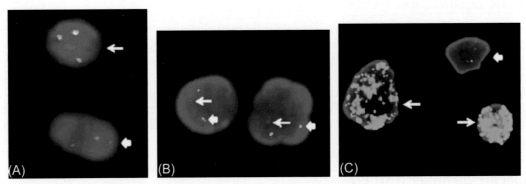

FIG. 13.13 Interphase dual-color FISH images to detect gene rearrangement. (A) Interphase FISH using a dual color dual fusion BCR-ABL1 translocation probe, showing a 2G2R pattern in a normal cell (*block arrow*) and 1G1R2F in a Philadelphia-positive cell (arrow). (B) Interphase FISH using a dual color break apart MLL translocation probe, showing an MLL split signal (distal MLL region, *arrow*; proximal MLL region, *block arrow*), and indicating MLL gene rearrangement. (C) MYC amplification in a neuroblastoma (*arrow*) and a normal cell with a 2G2R pattern (*block arrow*). The MYCN gene is labeled with a *green* fluorochrome, whereas the centromeric probe for Chromosome 2 is labeled with a *red* fluorochrome.

become possible to identify unknown translocation partners and to map breakpoints at the base-pair level. LDI-PCR requires only approximate sequence information for one partner, and thus it is ideal for use in combination with FISH to extend and refine cytogenetic breakpoint data. LSI FISH probes also can be used to identify the origin of the amplified DNA that constitutes homogeneously staining regions (HSRs) and double minutes (DMs), which occur in a variety of tumor cells (Fig. 13.13C).

Telomere is a unique set of repeat sequences located just before the end of each chromosome. Telomeric probes are useful for detecting chromosome structural abnormalities, including cryptic translocations or small deletions. In Fig. 13.14, FISH use a red RP11-25F24 (21q21.2), a green RP11-323F14 (21q22.3) BAC probes, and a red D22S105 probe (22q telomere) showing the inversion of the inserted chromosome 21 segment. *Yellow arrow*: Chromosome 21 inserted in chromosome 22.

ADVANCES IN FISH TECHNOLOGY

Numerous methodological advances in molecular cytogenetic technology have been developed, including comparative genomic hybridization (CGH), spectral karyotyping (SKY), multicolor FISH (MFISH), multicolor banding (MCB), and array CGH (aCGH). All of these cytogenetic techniques add colors to the monotonous world of conventional banding.

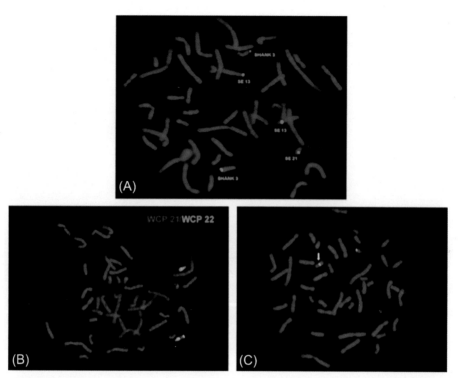

FIG. 13.14 The use of telomeric probes for detecting chromosome structural abnormality. (A) D21S1446 (21q telomere), SE13/21 (centromeres of chromosomes 13 and 21), and SHANK3 (22q telomere) FISH probes showing the deletion of a 21q subtelomeric region and the integrity of both 22q subtelomeric regions. (B) Painting wcp 21 (Chromosome 21) and wcp 22 (Chromosome 22) FISH probes showing the insertion of Chromosome 21 into Chromosome 22. (C) FISH using a *red* RP11-25F24 (21q21.2), a *green* RP11-323F14 (21q22.3) BAC probe, and a *red* D22S105 probe (22q telomere) showing the inversion of the inserted Chromosome 21 segment. *Yellow arrow*: Chromosome 21 inserted in Chromosome 22.

MFISH and SKY

MFISH and SKY, the two multicolor fluorescence technologies, are spectral karyotyping methods that use fluorescent dyes that bind to specific regions of chromosomes. By using a series of specific probes each with varying amounts of dyes, different pairs of chromosomes show unique spectral characteristics. Slight variations in color, undetectable by the human eye, are detected by a computer program and then reassigns an easy-to-distinguish color to each pair of chromosomes. The result is a digital image rather than film, in full color. Pairing of the chromosomes is simpler because homologous pairs are the same color, and aberrations and cross-overs are more easily recognizable. In addition, the spectral karyotype has been used to detect translocations not recognizable by traditional banding analysis (Fig. 13.15).

The main difference between these two techniques is the way to catch the colored images. MFISH combines binary ratio labeling and color-changing karyotyping. These methods offer the capacity to hybridize simultaneously 24 or more DNA probes, based on simultaneous hybridization of 24 chromosome specific painting probes labeled with different combinations of five fluorochromes. In MFISH, a highly sensitive monochrome camera captures a series of images that have passed through different filters. Specific computer software analyzes the acquired data from the probes and pseudocolors of the chromosomes for analysis. The SKY is based on the principles of spectral imaging and Fourier spectroscopy. Five pure dyes that are spectrally distinct are used in combination to create the unique chromosome cocktail of the probe. Image acquisition is accomplished by the use of a specially designed triple filter. Both the SKY and MFISH methods require the use of human WCP that are differentially labeled, so that each chromosome has a unique combination of colors emitted following hybridization for identification purpose. Each method can be performed as a laboratory procedure and is capable of identifying the cytogenetic origins of all chromosomes in the complement in one image acquisition step. These two techniques are suitable to identify subtle chromosomal aberrations that include an unidentified chromosome (marker chromosome) and an unbalanced chromosomal translocation.

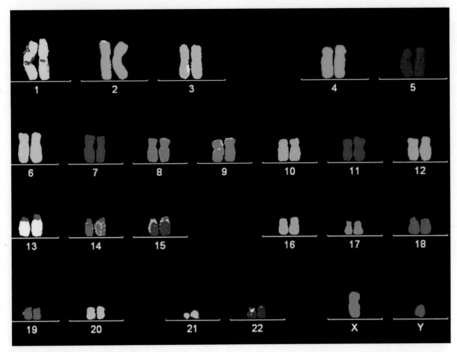

FIG. 13.15   An example of MFISH karyotype image.

Multicolor karyotyping is limited by the inability to detect intrachromosomal duplications and deletions and does not provide landmarks or bands to precisely describe chromosome breakpoints and rearrangements. In addition, multicolor karyotyping can lead to misinterpretations when, for example, overlapping chromosomes distort the fluorescent color of an adjacent chromosome, suggesting that three chromosomes are involved in a translocation when, in fact, there are only two.

Multicolor Banding (MCB)

A high-resolution molecular cytogenetic technique for the analysis of metaphase chromosomes is known as multicolor banding (MCB), and this technique involves the microdissection of chromosomal loci to obtain a set of probes that produce multicolor pseudo-G-banding. Therefore, the MCB DNA probe contains a mixture of region-specific partial chromosome paint probes (PCPs) that are labeled with three to five different fluorophores. The neighboring PCPs partially overlap each other, which decreases the fluorescent intensity toward the margins of the signals and leads to a consistent variation of fluorescent intensity ratios along the longitudinal axis of the chromosomes. This technique provides precise information about intra-chromosomal rearrangements and exact breakpoint mapping (Fig. 13.16)

Comparative Genomic Hybridization (CGH)

The comparative genomic hybridization (CGH) technique is an efficient approach to genome-wide screening for chromosomal copy number changes (gains/duplications and losses/deletions). CGH has been introduced to study chromosomal abnormalities that occur in solid tumors and other malignancies. Chromosomal CGH is based on quantitative two-color FISH and overcomes the problems of tissue culture failure and artifacts because this method is based on using tumor DNA extracted directly from either fresh or archival tumor tissue. The major advantage of CGH over standard FISH techniques is that only the DNA from the tumor cells is needed for analysis, avoiding the difficulties of obtaining metaphase chromosomes with good morphology and resolution for the analysis. As such, CGH can be applied as a molecular cytogenetic technique to search for potential anomalies in genomic DNA by comparing the target DNA (either DNA from any stem cell or patient-specific DNA) with a reference control DNA from either a healthy volunteer or commercially available

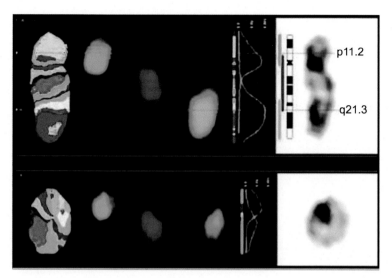

FIG. 13.16 An example of mBAND image and profiles showing the normal 18 (*top*) and the ring Chromosome 18 (*bottom*).

standard DNA. This technique allows a genome screening of chromosomal imbalances without prior knowledge of genomic regions of interest.

In CGH, total genomic DNA obtained from control cells and test samples is labeled differentially using green (fluorescein isothiocyanate, FITC) and red (Texas red) fluorescent dyes, denatured, coprecipitated in the presence of blocking DNA to suppress repetitive sequences and subsequently cohybridized to normal metaphase chromosomes. These are mixed together in equal proportions and hybridized to normal metaphase preparations. Because of the simultaneous hybridization to normal denatured metaphase chromosome spreads, there is competition for DNA hybridization to homologous sites. After hybridization and washing, the metaphase spreads are observed under a fluorescent microscope, and image analysis is performed using image analysis software. The resulting fluorescence intensities of the test and reference hybridizations are quantified digitally along the length of each chromosome. The green:red ratio generated by the two samples of DNA is analyzed by a computer software program that detects gains and losses of material from the DNA test. Fluorescence intensity ratio profiles are generated, and peaks and valleys indicating possible gains and losses of DNA tests are calculated and recorded. Normally, if they exceed ratios lower than 0.75 and higher than 1.25, they are considered as losses and gains of DNA, respectively (Fig. 13.17). If there is a normal amount of genetic material, the equal hybridization of green DNA and red control DNA will appear yellow. If there is a loss of genetic material, the segment will appear in red and the gain of material will produce a green signal as shown in Fig. 13.17. The presence and localization of chromosomal imbalances then can be detected and quantified by analyzing the green:red ratio using digital image analysis. These ratios show which chromosomal regions in the test genome are over- or under-represented relative to the reference genome. As ratio profiles are calculated along all chromosomes in the normal metaphase, an overview of the chromosomal imbalances throughout the total reference genome is created.

Array CGH

Although chromosomal CGH has increased the potential for identifying new chromosomal abnormalities, this technique is time consuming and does not significantly improve resolution (>3 Mb) compared with routine G-banding chromosome analysis. More recently, the development of array-based CGH (a-CGH) approaches involving the substitution of the hybridization targets, the metaphase chromosomes, with genomic DNA sequences adhered onto glass slides has increased the resolution for detecting copy number changes in the human genome, and this has led to more detailed information about genomic gains and losses (Fig. 13.18). Therefore, aCGH is an advanced technology for "molecular karyotyping," and its resolution is 100 kb to 1 Mb.

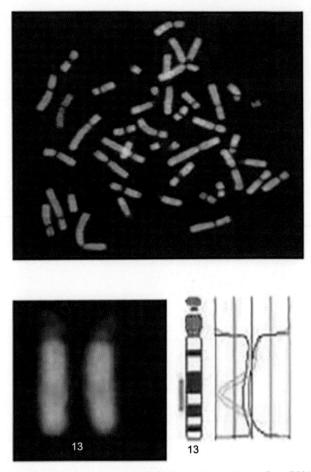

FIG. 13.17 A typical comparative genomic hybridization. Conventional CGH analysis: a mixture of test DNA from a patient and a normal reference DNA labeled with different fluorochromes are hybridized to normal chromosome spreads (*top panel*). The *left panel* illustrates the hybridization pattern of Chromosome 13. The interstitial segment of q-arm appears *red*, which indicates a loss of the region (13q21q31). The *right panel* shows a graph of the ratio profiles of Chromosome 13. The *black line* represents the balanced fluorescence intensities, the *red line* is the threshold for loss, and the *green line* is the threshold for a gain of material.

The fundamental principle of aCGH is essentially the same as that in CGH. The actual array comprises thousands of spots of reference DNA sequences applied in a precisely gridded manner on the slide. The number, size, and distribution of the DNA segments on the glass slide determine the array resolution, but commonly, the higher the number of DNA fragments, the higher the resolution. The initial arrayed DNA segments could be larger (~150 kb) human DNA segments inserted into a bacterial artificial chromosome (BAC clones) or bacterial/P1-derived artificial chromosomes (PAC clones). Shorter sequences, including smaller cDNA fragments, PCR products and synthesized oligonucleotide products, can be used as targets to improve resolution. aCGH also provides resolution at the nucleotide level. Single-nucleotide polymorphism arrays (SNP arrays) have the highest resolution (5–10 kb) of all of the available array-based platforms. The cohybridization of the test and reference DNAs is not required because the test DNA can hybridize directly to the SNP array. In addition to copy number variations (CNVs), the genotype information obtained from SNP arrays enables the detection of stretches of homozygosity and thus the identification of recessive disease genes, mosaic aneuploidy, or uniparental disomy (UPD). Although only SNP arrays enable the detection of copy number-neutral regions in the absence of heterozygosity, these arrays have limited ability to detect single-exon copy CNVs. This is because of the distribution of SNPs across the genome. Combining both aCGH and SNP genotyping in a single platform optimizes the clinical diagnostic capability, offering the simultaneous detection of copy number neutral and small intragenic copy number changes. In addition, these methods can reveal additional important information about the genetics of specific disorders, such as leukemia, with normal cytogenetic and FISH analyses.

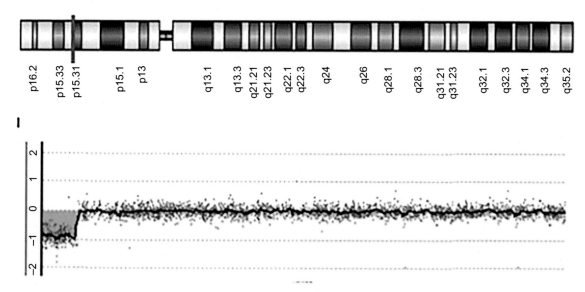

FIG. 13.18 Chromosome 4 array-CGH profile of a test DNA and a reference DNA. The figure shows a copy number loss corresponding to the segment of 4p16.3-p15.33 in a genomic segment with the median log2 ratio shifted to −1.0.

A modified array protocol, called translocation CGH (tCGH), has been developed to address recurrent translocation breakpoints in hematological neoplasms. Before the hybridization step in the array procedure, a linear PCR amplification is performed across the known recurrent translocation breakpoints in hematological neoplasms. Thus, it is possible to detect copy number changes and known recurrent translocations near or at the breakpoints.

FISH APPLICATIONS IN PRENATAL AND POSTNATAL DIAGNOSTICS AND RESEARCH

Molecular cytogenetic approaches have become indispensable for a range of research purposes, and the increasing use of molecular techniques has become a crucial tool in clinical diagnosis. Since the introduction of FISH in the late 1980s, there has been a tremendous increase in the use of this technique to detect chromosomal abnormalities and evaluate CNVs in the human genome in preimplantation genetic, prenatal, and postnatal diagnoses and cancer cytogenetics.

The Advantage of Using FISH in Clinical Diagnosis

A significant advantage of FISH is that it can be applied in nondividing cells, thereby facilitating the direct investigation of chromosomes in cytological preparations and tissue sections. This is because labeled DNA probes in the size of 100–800 Kb have enhanced analytical resolution and allowed rapid screening of locus-specific numerical aberrations. For example, interphase FISH on uncultured amnion cells has become a useful method for the rapid and early diagnosis of the most common chromosome disorders, such as trisomies 21, 13, 18, and sex chromosome aneuploidies, in fetal cells. Thus, interphase FISH in prenatal diagnosis is a quick, accurate, sensitive, and relatively specific method to detect aneuploidies in samples of uncultured chorionic villus and amniotic fluid cells. For prenatal aneuploidy screening using uncultured amniocytes, no time-consuming cell culture is required, and the results can be obtained within 24–48 h. Three satellite centromeric probes for chromosomes X, Y, and 18 and two locus-specific probes for the 13q14 and 21q22.13 regions are the most commonly applied. Using site-specific DNA probes (YACs, BACs, PACs, and cosmids), FISH is applied typically for mapping chromosomal regions with located breakpoints. In addition, using locus-specific probes, FISH has been used to confirm clinical diagnoses of known microdeletion and microduplication syndromes. Furthermore, mFISH approaches have been valuable for cancer cytogenetics.

The Limitation of Using FISH in Diagnosis

Occasionally, patients with small and unusual deletions might escape detection, depending on the specificity of the fluorescent probe. Moreover, FISH has limitations in the detection with gene or imprinting mutations, particularly occurring in some microdeletion syndromes, including Angelman syndrome, Prader-Willi syndrome, Sotos syndrome, Miller-Diecker syndrome, Smith-Magenis syndrome, and Rubinstein-Taybi syndrome. As such, these disorders cannot be detected through FISH. The analysis of telomeres using FISH techniques has been conducted in cancer and aging research (telomere biology), however, because of the lack of specificity of the DNA probes (TTAGGG repetitive sequence motifs), this technique is poorly applicable for cancer and aging diagnosis.

APPLICATIONS OF CGH ANALYSIS

CGH has been applied to study solid tumors, leukemia, and lymphoma. Because CNVs are associated with many conditions, ranging from cancer to developmental abnormalities, CGH has been applied to identify constitutional chromosomal abnormalities in clinical samples. Standard CGH or array-CGH can also be used to detect chromosomal abnormalities in single cells of preimplantation embryos.

aCGH in Detecting Chromosomal Imbalance

Array-CGH can be applied to identify chromosomal imbalances through the detection of CNVs in tumors to distinguish candidate genes involved in the pathogenesis of cancer. In clinical diagnostics, both oligonucleotide aCGH and SNP genotyping are powerful genomic technologies for evaluating idiopathic mental retardation (also referred to as developmental delay, intellectual disability, or learning difficulty), associated congenital abnormalities, autistic spectrum disorders, schizophrenia, and other neuropsychiatric disorders. The introduction of genome-wide array platforms can facilitate the detection of chromosomal abnormalities. It also provides a greater diagnostic sensitivity than conventional cytogenetics. aCGH increases the diagnostic yield for detecting additional genomic imbalances 1%–5% compared with normal karyotyping, depending on the reference source. However, aCGH analysis also has its limitations in the detection of polyploidy, balanced chromosomal rearrangements, inversions and low level mosaicism.

aCGH in Cancer Detection

The implementation of array-based chromosome analysis has been critical in hematologic and oncologic disorders. The complexity of cancer cells requires a sensitive technique that facilitates the detection of small genomic changes in a mixed-cell population and segmental regions of homozygosity. However, recurrent balanced genomic aberrations with important prognostic value in cancer might be not detected through array-based analyses. Because aCGH is based on the principle of CNV detection, this technique is limited by an inability to identify balanced translocations and inversions. Nevertheless, arrays are clinically essential for identifying novel genomic abnormalities that escape detection using current diagnostic methodologies in a number of hematological diseases, such as chronic lymphocytic leukemia, myelodysplastic syndrome, multiple myeloma, acute lymphoblastic leukemia, acute myeloid leukemia, and chronic myelomonocytic leukemia.

Prenatal Genetic Diagnosis Using aCGH

Prenatal genetic counseling for detected cytogenomic abnormalities involves the assessment of fetal viability and the prediction of severity of phenotype from the family history and the clinical evidence in the literature. Prenatal genetic diagnosis has been driven by innovative technologies and accumulated clinical knowledge. Recent advances in maternal serum screening and ultrasonography have resulted in significant decline in invasive amniocentesis of amniotic fluid (AF) and chorionic villus sampling (CVS) procedures. Although the detection rate of diagnostic yield for chromosomal abnormalities has increased significantly for numerical chromosomal abnormalities and for structural chromosomal abnormalities, routine prenatal cytogenetic analysis still has two obvious limitations: the low analytical resolution of about 5–10 megabase (Mb) by the Giemsa banding and the long turn-around time because of the in vitro cell culture procedures. Simultaneous aCGH analysis on uncultured amniotic fluid in addition to routine prenatal cytogenetic analysis is helpful to identify undetected familial diseases.

Comparison of Different Rapid Prenatal Screening and Diagnosis

Various DNA-based molecular approaches have been introduced to provide rapid prenatal screening and diagnosis. For example, FISH and quantitative fluorescence-PCR (QF-PCR) have been used for rapid screening of prenatal aneuploidy but these methods were limited to a few chromosomal loci. Although microarray approaches are used increasingly in prenatal settings in pregnancies, challenges in interpreting the results, quality control, and ethical issues have delayed the use of microarray approaches in prenatal care compared with postnatal diagnoses. Multiplex ligation-dependent probe amplification (MLPA) has been validated as a rapid, cost-effective, and multiallelic approach for prenatal screening of common aneuploids, recurrent genomic disorders, and subtelomeric imbalances. MLPA uses aneuploidy probe mix, subtelomeric probe mixes, and microdeletion syndromes probe mix and is a robust method for rapid prenatal screening of aneuploidy, unbalanced subtelomeric rearrangements, and recurrent microdeletions. The turn-around-time of MLPA screening is usually one to three days. MLPA screening, however, lacks genome-wide coverage and cannot detect copy number changes outside the represented regions. Therefore, MLPA should be used as a rapid and cost-effective screening for common aneuploids, subtelomeric rearrangements, and known genomic disorders.

For pediatric patients with intellectual and developmental disabilities, genome-wide array comparative genomic hybridization using synthesized oligonucleotides has been validated and recommended as first-tier genetic testing. For prenatal diagnosis, the aCGH analysis is useful in defining gene content of chromosomal imbalances and in detecting genomic disorders and cryptic aberrations. There are, however, technical challenges and clinical concerns for rapid replacement of prenatal cytogenetics with genomic analysis. For example, proper DNA extraction avoiding maternal cell contamination, cautious on confined-placental mosaicism, and timely confirmatory and followup parental analyses need to be considered. In order to improve this situation, an approach coupling MLPA and aCGH analyses for prenatal detection of cytogenomic abnormalities has been developed and is a cost-effective approach to detect cytogenomic abnormalities prenatally. The workflow of MLPA, karyotyping, and aCGH analysis is shown in Fig. 13.19. This integrated approach represents a robust and cost-effective approach for prenatal diagnosis of chromosomal aneuploids, subtelomeric rearrangements, and genomic imbalances. The MLPA screening results guide focused chromosome analysis of aneuploids and subtelomeric rearrangements, and aCGH analysis for fine mapping of breakpoints and gene content of imbalanced regions. The aCGH analysis provided genomic maps of breakpoints, gene content of imbalanced regions, and better inference of related phenotypes for genetic counseling.

aCGH in Preimplantation Genetic Diagnosis

Preimplantation genetic diagnosis (PGD) is a procedure that involves the removal prior to the implantation of one or more nuclei from oocytes [polar bodies (PBs)] or embryos (blastomeres or trophectoderm cells) to test for mutations in the genome sequence or for chromosomal aneuploidy. Because the embryonic cells are still at a very early stage of development, conventional chromosome testing (karyotyping) is not yet possible.

PGD techniques are widely carried out by many in vitro fertilization (IVF) centers and provide patients with new hope of having unaffected children, as well as avoiding the necessity of terminating an affected pregnancy for couples carrying affected genes or chromosome problems. Therefore, it is a test screening for aneuploidy in specific groups of patients.

Two main types of array-based technology combined with whole genome amplification have been developed for use in PGD: CGH and SNP-based arrays. Both allow the analysis of all chromosomes, and the latter also allows the haplotype of the sample to be determined. The procedures of PGD require the removal of one or more cells from embryos in order to have sufficient genetic material for diagnosis (Table 13.2). May different methods can be performed for biopsies, including PB biopsy from an oocyte or zygote, cleavage stage biopsy, and blastocyst biopsy. Cleavage stage biopsy is the most common type of biopsy.

DNA from the test sample and normal control DNA are amplified using whole genome amplification (WGA). Different WGA approaches can be used, including degenerate oligonucleotide primer PCR and primer extension preamplification (PEP) PCR, as well as newer methods such as iPEP (improved PEP), multiple displacement amplification, and GenomePlex. The amplified DNAs then are differentially labeled with one of two fluorochromes. For example, red is for the test DNA and green is for the control DNA. After labeling, both DNAs are mixed together in equal proportions and are allowed to competitively hybridize to either the metaphase spreads from a normal male control cell line (mCGH) or onto an array platform containing small pieces of chromosome (aCGH). Although both mCGH and aCGH have been applied in clinical practice, the time-consuming nature of mCGH has hindered the wide application of this approach to PGS. On the other hand, aCGH is increasingly used in clinical practice because of shorten time and easy to use. The CGH approach is reliable to detect aneuploidy in patients with recurrent implantation failure and recurrent

FIG. 13.19 Workflow of integrated MLPA (multiplex ligation-dependent probe amplification) screening, karyotyping, and aCGH analysis for prenatal diagnostics.

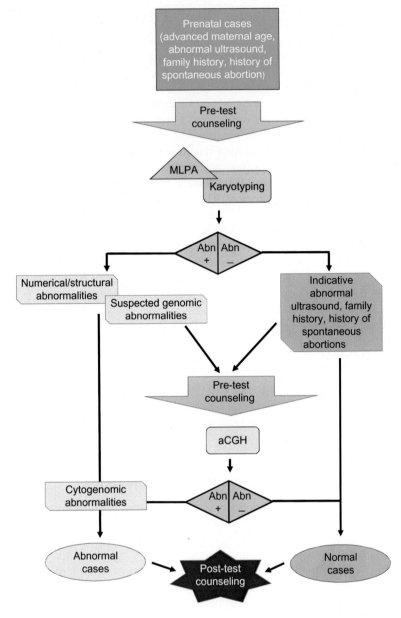

miscarriage. The CGH approach still exhibit some inadequacies, including the inability to differentiate between balanced translocations and inversions, and the inability to detect some types of specific ploidy status, such as polyploidy or monoploidy. The systems also cannot detect changes in DNA sequences, such as point mutations, intragenic insertions or deletions, and triplet repeat expansion, and they are unable to find gains or losses in regions of the genome not covered by the array.

SNPs are the most common type of variation in the genome. SNP microarray analysis is used to detect polymorphisms that exist at a frequency greater than 1% within a population. SNPs not only can be applied for single-gene disorder detection and the detection of aneuploidy, but also can determine which parental haplotype has been transmitted to the embryos. The use of SNPs in clinical practice has several limitations, including the fact that the association of CNVs with clinical information remains unknown and that parental DNA is required for linkage analysis. The accuracy of SNPs can be limited by contamination and allele dropout. For the aneuploidy screening, there was a false detection rate of 2.1% using SNP methods compared to 1% by traditional metaphase karyotyping. The SNP-based array approach with optimized

TABLE 13.2 Preimplantation Genetic Diagnosis Biopsy Methods

Stage of Biopsy	Advantages	Disadvantages and Limitations
Oocyte/zygote (Polar Body)	• No effect on subsequent embryo development • Enough time to perform analysis prior to transfer	• Only 1 or 2 cells available for analysis • Only maternal message obtained DNA liable to degenerate
Cleavage Stage	• Diagnosis of maternal and paternal inherited disease • Possibility of sex determination • Up to two cells available for analysis	• Limited time for analysis • High incidence of chromosomal mosaicism
Blastocyst	• Fewer number of embryos to be biopsied and fewer specimens to process • Three or more cells per embryo available to overcome allele dropout • Less problem of mosaicism	• Occasions when embryos failing to blastocysts • Need for cryopreservation most of the time

protocols and parental support algorithms could be particularly suited to certain applications such as the PGD of single-gene defects and translocation chromosome imbalance combined with comprehensive detection of aneuploidy. In translocation cases, normal embryos can be differentiated from balanced ones.

CHROMOSOMAL DISORDERS: NUMERICAL CHROMOSOME ABNORMALITIES

Chromosomal disorders are caused by abnormalities in the number or the structure of chromosomes and cytogenetic karyotyping is a standard practice to identify all chromosomal disorders. Since an individual's phenotypes result from the expression of genes, the phenotype of a person with a chromosomal disorder can vary with the type of chromosomal defect.

An abnormality in which the chromosome number is an exact multiple of the haploid number ($n = 23$) and is larger than the diploid number ($n = 46$) is called polyploidy. Polyploidy arises from fertilization of an egg by two sperm (total number of chromosomes increases to 69) or the failure in one of the divisions of either the egg or the sperm so that a diploid gamete is produced. The survival of such a fetus to full-term pregnancy is rare.

Aneuploidy occurs when the chromosome number is not an exact multiple of the haploid number and results from the failure of paired chromosomes (at first meiosis) or sister chromatids (at second meiosis) to separate at anaphase. Thus, two cells are produced, one with a missing copy of a chromosome and one with an extra copy of that chromosome (Fig. 13.20).

Trisomy 21

For the trisomy of autosomal chromosomes, the first human chromosomal disorder discovered and the most important one is the full trisomy of chromosome 21 (+21), which is an abnormality that displays an extra copy (total of three copies) of chromosome 21. Trisomy 21 causes Down syndrome. Chromosome 21 is the smallest human autosome with 48 million nucleotides and depicts almost 1%–1.5% of the human genome; more than 400 genes are estimated to be on chromosome 21 (Table 13.3). The isochromosome of the long arm (chromosome with loss of short arm and duplication of the long arm) of chromosome 21 is a less common cause of Down syndrome (Fig. 13.21).

Down syndrome is one of the most frequent congenital birth defects, and the most common genetic cause of mental retardation. It affects between 1 in 400–1500 babies born in different populations, depending on maternal age and prenatal screening schedules. Although the syndrome occurs due to trisomy of the whole or part of chromosome 21 in all or some cells of the body and the genes on all three copies of chromosome 21 are normal, not all individuals with Down syndrome show the same physical characteristics. This suggests that the phenotypes can vary because of gene dosage of the trisomic genes. It is believed that only some genes on chromosome 21 might be involved in the phenotypes of Down syndrome, and some gene products might be more sensitive to gene dosage imbalance than others.

Fig. 13.22 shows various clinical conditions associated with Down syndrome. About half of all affected children are born with a heart defect. Digestive abnormalities, such as a blockage of the intestine, are less common. This syndrome is also associated with intellectual disability, a characteristic facial appearance, and weak muscle tone (hypotonia) in infancy. People with Down syndrome have a typical facial appearance (the face is flat and broad) that includes an abnormally small chin, skin folds on the inner corners of the eyes, poor muscle tone, a flat nasal bridge, a protruding tongue due to a small oral

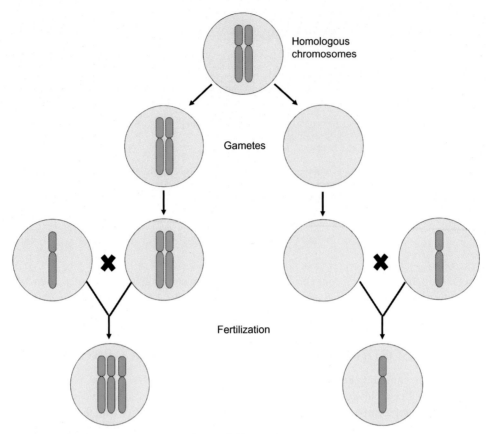

FIG. 13.20 Nondisjunction during gamete formation results in aneuploidy.

cavity, an enlarged tongue near the tonsils, a short neck, and white spots on the iris. Growth parameters such as height, weight, and head circumference are smaller in children with Down syndrome than in typical individuals of the same age. All Down syndrome patients have some degree of mental retardation. Despite this condition, many persons with Down syndrome can be educated and live with minimal daily assistance, while others require much attention and care. Those individuals who survive into their 50s and 60s, however, are at very high risk of developing Alzheimer's disease.

Cytogenetic analysis of metaphase karyotype remains the standard practice to identify not only trisomy 21, but also all other aneuploidies and balanced translocations. Details about diagnostic methods with advantages and disadvantages are listed in Table 13.4. Some rapid molecular assays-FISH, QF-PCR, and MLPA also are used for prenatal diagnosis.

More recent advances in genomics and related technologies have resulted in the development of a noninvasive prenatal screening test (NIPT), which uses cell-free fetal DNA sequences isolated from a maternal blood sample. This is because about 4%–10% of DNA in maternal serum is of fetal origin. Fetal trisomy detection by cell-free DNA (cfDNA) from maternal blood can be done using massively parallel shotgun sequencing (MPSS). By NGS platforms, millions of amplified genetic fragments can be sequenced in parallel. MPSS detects higher relative amounts of DNA in maternal plasma from the fetal trisomic chromosome compared with reference chromosomes. In digital PCR, individual fetal and maternal circulating cfDNA fragments are amplified under limiting-dilution conditions and the total number of chromosome 21 amplifications (representing maternal plus fetal contributions) divided by the number of reference chromosome amplifications should yield a ratio indicating an over- or underrepresentation of chromosome 21. Although the digital PCR approach is conceptually solid, the low percentage of cell-free fetal DNA in the maternal plasma sample requires the performance of thousands of PCRs to generate a ratio with statistical confidence.

Trisomy 18

Trisomy 18, the presence of three copies of chromosome 18, elicits Edward syndrome. Most cases of trisomy 18 are not inherited but occur as random events during the formation of eggs and sperm. An error in cell division called nondisjunction

TABLE 13.3 Candidate Dosage Sensitive Genes on Chromosome 21 Leading to Down Syndrome Phenotype

Gene Symbol	Full Name	Location
APP	Amyloid beta (A4) precursor protein	21q21.2/21q21.3
OLIG1	Oligodendrocyte transcription factor 1	21q22.11
OLIG2	Oligodendrocyte lineage transcription factor 2	21q22.11
DYRK1A	Dual-specificity tyrosine (Y) phosphorylation regulated kinase 1A	21q22.13
DSCAM	Down syndrome cell adhesion molecule	21q22.2
SYNJ1	Synaptojanin 1	21q22.2
JAM2	Junctional adhesion molecule 2	21q22.2
SIM2	Single-minded homolog 2 (Drosophila)	21q22.2/21q22.13
ERG	V-ets avian erythroblastosis virus E26 oncogene homolog	21q22.3
PTTG1IP	Pituitary tumor-transforming 1 interacting protein	21q22.3
ADAMTS1	ADAM metallopeptidase with thrombospondin type 1 motif 1	21q21.3
ITSN1	Intersectin 1	21q22.1-q22.2
SYNJ11	Synaptojanin 1	21q22.2
ERG	V-ets avian erythroblastosis virus E26 oncogene homolog	21q22.3
ETS2	ETS proto-oncogene 2, transcription factor	21q22.3
SLC19A1	Solute carrier family 19 member 1	21q22.3
COL6A1	Collagen type VI alpha 1	21q22.3

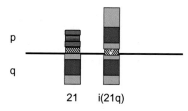

p

q

21   i(21q)

FIG. 13.21   The isochromosome of the long arm (chromosome with loss of short arm and duplication of the long arm) of Chromosome 21.

results in a reproductive cell with an abnormal number of chromosomes. For example, an egg or sperm cell might gain an extra copy of chromosome 18. If one of these atypical reproductive cells contributes to the genetic makeup of a child, the child will have an extra chromosome 18 in each of the body's cells. Only 10% of babies with this syndrome survive past their first year. Children with trisomy 18 usually have problems with breathing and eating, and many have seizures or serious heart conditions. All individuals with trisomy 18 have severe mental retardation. Most babies with trisomy 18 are very small and have certain recognizable facial features. They also tend to overlap their fingers in a very distinct pattern.

Trisomy 13

Trisomy 13, the presence of three copies of chromosome 13, causes Patau syndrome. Most pregnancies involving trisomy 13 end in miscarriage, and only 5% of infants with this disorder survive past their first year. The incidence of this rare syndrome has been estimated at between 1 in 5000 and 1 in 20,000 live births, and it is the third most viable trisomy syndrome after trisomy 21 and trisomy 18 (Edward's syndrome). The clinical spectrum of this syndrome exhibits a variety of

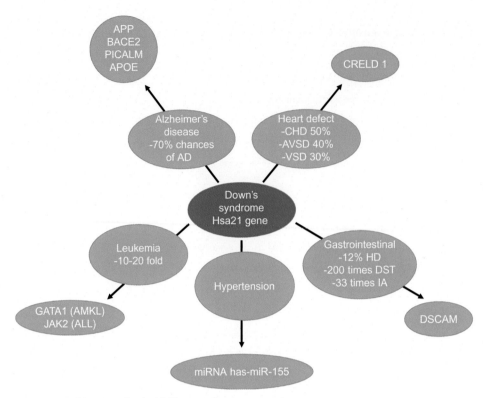

FIG. 13.22   Various clinical conditions associated with Down syndrome.

anomalies, including central nervous malformation, facial, ocular, limb, finger, lung-cardiac system, and urogenital involvement. Children with trisomy 13 usually have a lot of trouble breathing, especially when they sleep, and many have seizures. All individuals with Patau syndrome have severe mental retardation, and other common characteristics include a small head, extra fingers and/or toes, and cleft lip or cleft palate.

Prader-Willi Syndrome

Prader-Willi syndrome results from the absence or nonexpression of a group of genes on chromosome 15. Most cases of Prader-Willi syndrome (about 70%) occur when a segment of the paternal chromosome 15 is deleted in each cell. People with this chromosomal change lose certain critical genes in this region. This is because the genes on the paternal copy have been deleted, and the genes on the maternal copy are turned off (inactive). In another 25% of cases, a person with Prader-Willi syndrome has two copies of chromosome 15 inherited from his or her mother (maternal copies) instead of one copy from each parent. This phenomenon is called maternal uniparental disomy. Rarely, Prader-Willi syndrome can be caused by a chromosomal rearrangement called a translocation, or by a mutation or other defect that abnormally turns off (inactivates) genes on the paternal chromosome 15. A specific form of blood cancer, CML, can be caused by a chromosomal translocation, in which portions of two chromosomes (chromosomes 9 and 22) are exchanged. No chromosomal material is gained or lost, but a new, abnormal gene that leads to the development of cancer is formed.

CHROMOSOMAL DISORDERS: SEX CHROMOSOME ABNORMALITIES

The phenotype of chromosomal disorders can vary depending on whether the chromosomal abnormality occurs on the maternally or paternally derived chromosomes. Sex chromosomal aneuploidy (SCA) is the most common disorder of sex chromosomes in humans.

TABLE 13.4 Advantages and Disadvantages of Common Diagnostic Techniques of Down Syndrome

Method	Description	Advantages	Disadvantages
Cytogenetics analysis	Giemsa banding performed on fetal cells at metaphase stage on amniocytes or CVS	• Allows for high level of diagnostic skill at low cost/absence of laboratory services	• Takes a lot of time • Resolution can be low because of spontaneous dividing cells • Occurrence of confined placental mosaicism and occurrence of aberrant cells that do not represent fetus status • False results possible
FISH	Hybridization of selected chromosome specific DNA sequences that have been labeled with fluorescence. Labeled sequences anneal to corresponding chromosomal DNA and are visualized with microscopy	• Smaller probes yield more distinct dot signals • Because of use of higher number of interphase nuclei, suspected mosaicism is resolved	• Diffused signals possible because of interphase chromosomes • Time consuming • Maternal and fetal XX is not distinguished by FISH
QF-PCR	Amplification and detection of STR with fluorescent primers, followed by visualization and quantification of peak areas using a DNA sequencer and Gene Scan software	• Highly reliable and reproducible • False results are rare • Maternal contamination easily detected • Quick results	• Mosaicism creates a challenge • Normal XX female abnormalities can be indistinguishable from single X Turner issues
PSQ Paralogous Sequence Quantificaiton	PCR detection of targeted chromosome abnormalities based on the use of paralogous genes. These sequences have high degree of sequence identity but accumulate nucleotide substitution in a locus specific manner and thus can be quantified via pyrosequencing	• Reduces sample throughput and decreases the probability of handling errors • Handles bulk samples and yields quick results	• Expensive
MLPA	Based on hybridization and PCR. DNA denaturation, probe hybridization, probe ligation, and PCR amplification then are analyzed through capillary electrophoresis	• Short time for diagnosis • Relatively low cost	• Unable to exclude low level placental and mosaicism
NGS	Amplified sequences in massively parallel fashion that provides digital quantitative information where each sequence is read as a countable "sequence tag" representing an individual clonal DNA template or single DNA molecule	• Time for sample processing, sequencing, and data interpretation is around 5–8 days	• Expensive • Complex data analysis

Turner Syndrome

Turner syndrome (45,X) is an example of monosomy, in which a girl is born with only one sex chromosome, an X chromosome. Although a rare disease with an incidence of 1 in 2500 female births, it is nevertheless the most common sex chromosome abnormality of human females. Turner syndrome is characterized by short stature and ovarian dysgenesis, together with a broad range of other phenotypic characteristics, including an increased risk for heart and renal defects.

Karyotyping is considered to be the best cytogenetic technique for diagnosing Turner syndrome (Fig. 13.23). Karyotyping for neonatal screening has important limitations, including running time, cost, and need for specialized personnel. Therefore, various molecular methods have been proposed for diagnosis or neonatal screening of Turner syndrome, including Southern blotting, PCR-RFLP, fluorescent PCR genotyping, genotyping, pyrosequencing coupled with the analysis of single-nucleotide polymorphisms regions on the X chromosome, and real-time PCR.

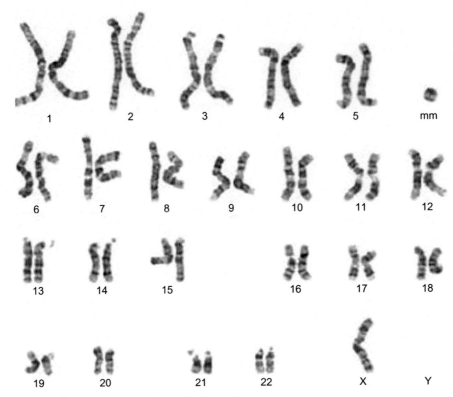

FIG. 13.23 Turner syndrome karyotype (female with only 1 X) of the patient.

Klinefelter Syndrome

Klinefelter syndrome (KS) is resulted from an additional X chromosome. About 80% of KS patients show a 47,XXY karyotype; the remaining 20% have other numeric sex chromosome abnormalities (48,XXXY, 48,XXYY, 49,XXXXY), 46,XY/47,XXY mosaicism, or structurally abnormal sex chromosomes. Males with the genotype 47,XXY (Fig. 13.24) are classified as having a sex chromosome trisomy (three sex chromosomes), because they carry two X chromosomes and one Y chromosome. This syndrome affects about 1 in 1000 males. Most affected males are taller than average, and they might have more body fat in the hips or chest as well as little facial and body hair. Although some KS males are mentally retarded, many others have normal intelligence. The most common feature of this syndrome is infertility; about 2% of infertile men have KS. The XXY also has an increased risk for developing malignancy, including breast cancer, germ cell tumors, non-Hodgkin lymphomas, and lung cancer. The XXYY syndrome is a variant of KS with hypogenitalism disorders, mental retardation and psychiatric problems (Fig. 13.25). The XYY males had normal phenotype, but they can exhibit variable abnormalities, including skeletal, cardiovascular, and genital systems and behavioral problems. The addition of more than one extra X and/or Y chromosome to a normal male karyotype is less frequent and has its own distinctive physical and behavioral profile. Fig. 13.26 demonstrates the aneuploidy of a very rare chromosome abnormality involving pentasomy X chromosome. The exact prevalence is unknown. Pentasomy X is associated with developmental delays, short stature, craniofacial anomalies (microcephaly, micrognathia, plagiocephaly, hypertelorism, up-slanting palpebral fissures, a flat nasal bridge, and ear malformations), musculoskeletal abnormalities, and cardiovascular malformations. The hands and feet are generally small with common findings of camptodactyly, clinodactyly, and radioulnar synostosis.

CHROMOSOMAL DISORDERS: STRUCTURAL CHROMOSOME ABNORMALITIES

Geometric rearrangements can alter genome architecture and yield clinical consequences. Several genomic disorders caused by structural variation of chromosomes were discovered by cytogenetics. The development of molecular cytogenetic techniques have enabled the rapid and precise detection of structural rearrangements on a whole-genome scale. This

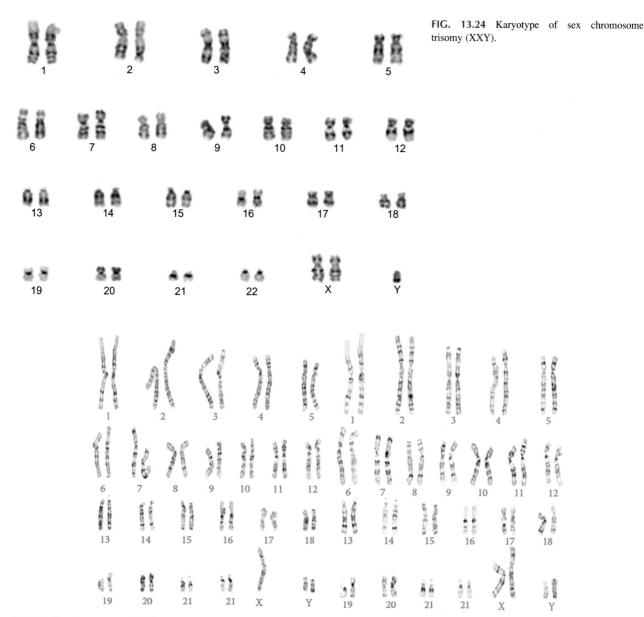

FIG. 13.24 Karyotype of sex chromosome trisomy (XXY).

FIG. 13.25 Karyotypes of a Klinefelter patient with XXYY.

high resolution illustrates the role of structural variants (SVs) and single nucleotide polymorphisms (SNPs) in normal genetic variation.

Structural Variants

In analyzing the role of structural variants in cell functions, it is important to consider the two types—gain-of-function variants and loss-of-function variants. A gain-of-function variant results from a mutation that confers new or enhanced activity on a protein. Most mutations of this type are not heritable (germ line) but are somatic mutations. A somatic mutation is a change in the structure of a gene that can arise during DNA replication. It is not inherited from a parent, nor can it be passed to an offspring. The mutation that results in a single base substitution in DNA is known as a somatic point mutation. A loss-of-function variant results from a point mutation that leads to reduced or abolished protein function. Most lost-of-function mutations are recessive, indicating that clinical signs typically are observed only when both chromosomal copies of a gene carry the same mutation.

FIG. 13.26 G-banding karyotype showing aneuploidy involving pentasomy X chromosome.

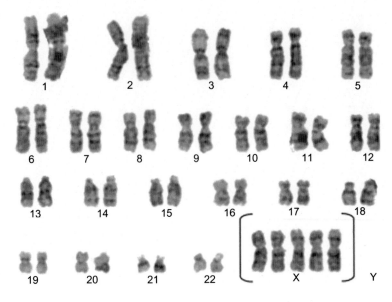

SNP

At a given chromosomal site, one individual might have the A nucleotide and the other individual might have the G nucleotide. This type of site in DNA is known as an SNP (Fig. 13.27). If each of the two different DNA sequences at this site can be inherited, this site is part of a gene, and the alternate forms of the DNA sequence is called an allele.

Types of Chromosomal Rearrangements

Various genomic rearrangements can arise by several mechanisms, including deletions, amplifications, translocations, and inversions of DNA fragments (Fig. 13.28). Deletions occur when a portion of a chromosome is missing or removed. Duplications result from the copying of a portion of a chromosome more than once, producing extra genetic material. Deletions and duplications are known as CNVs.

Chromosomal translocation is a common chromosomal mutation. Burkitt's lymphoma (BL) is a heterogeneous group of highly aggressive mature B-cell malignancies. It is characterized by a high rate of turnover of malignant cells and deregulation of the *c-myc* gene. The histologic hallmark of BL is the presence of apoptotic cells within scattered macrophages, a feature responsible for a "starry sky" microscopic appearance. A characteristic chromosome translocation associated with this disease typically takes place between chromosome 8q24 (*c-myc*) and one of the Ig gene-containing chromosomes,

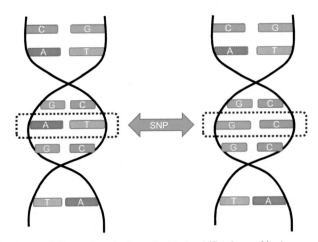

FIG. 13.27 An example of SNP containing a difference in a single nucleotide (an A/G polymorphism).

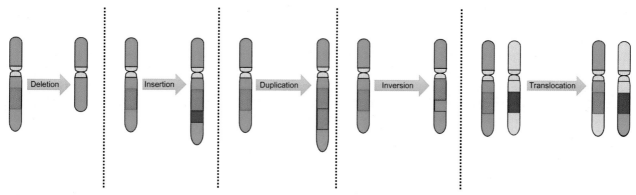

FIG. 13.28 Examples of chromosomal rearrangement.

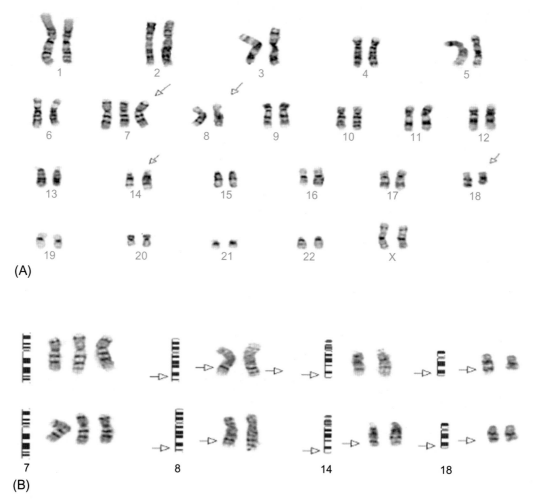

(A)

(B)

FIG. 13.29 An example of chromosomal translocation. (A) Karyotype 47, XX, +7, t(8;14;18) (q24.1;q32;q22). *Arrows* indicate abnormal chromosomes. (B) Diagramatic representation of +7 and 3-way chromosome translocations. *Arrows* indicated abnormal chromosome regions involved in the 3-way translocation.

14q32 (*Ig H* gene). However, some cases of BL have a novel three-way chromosome translocation, t(8;14;18) (Fig. 13.29). In a reciprocal translocation, DNA segments from two different chromosomes are exchanged. A DNA inversion results when a portion of a chromosome is broken off, turned upside down, and reattached. When a portion of a chromosome is disrupted, the chromosome can form a circle, or ring, without any loss of DNA. As such, rearrangements are elicited by multiple events, including external factors such as cellular stress and incorrect DNA repair or recombinations.

THE CAUSES OF CHROMOSOME ABNORMALITIES

Chromosome abnormalities arise from genomic variants. Because most chromosome abnormalities occur accidentally in the ovum or sperm, the abnormality is present in every cell of the body. Some abnormalities arise after birth, however, resulting in a condition in which a few cells have the abnormality and others do not. Chromosomes abnormalities can be either inherited from a parent (e.g., translocation) or develop spontaneously for the first time. This is why chromosome studies are performed on parents when a child is found to have an abnormality. Advances in molecular biology and cyto-genetic techniques permit the identification of many diverse types of SVs, which contribute to human disease, phenotypic variation, and karyotypic evolution.

Errors in Cell Division

Chromosome abnormalities usually occur when there is an error in cell division (Fig. 13.30). There are two kinds of cell division. Meiosis results in cells with half the number of usual chromosomes, 23 instead of the normal 46. These are the eggs and sperm. Mitosis produces two cells that are duplicates (46 chromosomes each) of the original cell. This kind of cell division occurs throughout the body, except in the reproductive organs. In both processes, the correct number of chromosomes appears in the daughter cells. Errors in cell division, however, can result in cells with too few or too many copies of a chromosome.

Maternal Age

Additional factors, such as maternal age, can increase the risk of chromosome abnormalities. Women are born with all the eggs they will ever have. Therefore, when a woman is at certain age, so are her eggs. Chromosomal errors can appear in eggs as they age. Thus, older women are at greater risk of giving birth to babies with chromosome abnormalities than younger women. A second factor might be environmental conditions, although conclusive evidence is currently lacking.

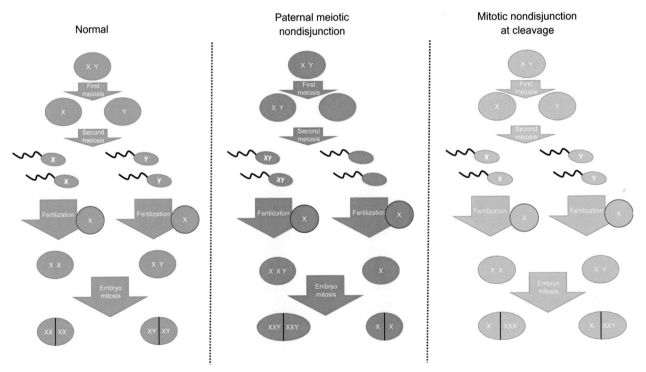

FIG. 13.30 Effects of nondisjunction of sex chromosomes during cell division.

Chapter 14

Molecular Diagnosis of Mutation and Inherited Diseases

Chapter Outline

VARIATIONS THAT CAUSE HUMAN DISEASE

Diagnostic molecular genetics is one of the fastest-growing and one of the most powerful diagnostic and screening tools of the 21st century. With the rapidly accelerating pace of identification of new disease genes from the help of various techniques such as NGS, GWSS, and GWAS, the field of genetic diagnosis can only continue to expand. Its unique capability to diagnose disease both prenatally and presymptomatically should confer on it a primary role in preventive medicine and personalized medicine. With knowledge about many genetic disease genes, loci, and mutational mechanisms, diagnostic molecular genetics must take advantage of the entire spectrum of modern molecular biology techniques available. The US National Library of Medicine has published Genome Home Reference, which provides consumer-friendly information about the effects of genetic variation on human health (https://ghr.nlm.nih.gov/). Since we have discussed chromosomal disorders in Chapter 13, we will focus on genetic diseases at the molecular level (DNA sequences) in this chapter.

Single-Nucleotide Variants

The human genome is composed of two copies of a 3.08-billion sequence code of nucleic acid bases spread across 46 chromosomes. Any sequence change is called a sequence variant, variation, or mutation. The most common sequence variations are single-base changes, also known as single-nucleotide variants (SNVs) (Fig. 14.1). The vast majority of SNVs occur in

Diagnostic Molecular Biology. https://doi.org/10.1016/B978-0-12-802823-0.00014-6

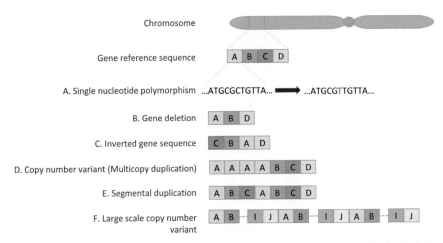

FIG. 14.1 Different types of genomic variation. (A) A single nucleotide polymorphism (SNP) occurs as a result of a single base substitution at an individual site in the DNA sequence along the chromosome, and this can result in the coding of a different amino acid. (B) A deletion is the loss/absence of DNA sequence. (C) An inversion is a rearrangement causing a segment of DNA to be present in reverse orientation. (D) A copy number variant (CNV) is a segment of DNA that is 1 kb or larger and is present at a variable copy number in comparison with a reference genome. CNVs can be either deletion variants where there is loss of copy number relative to the reference sequence or multicopy duplications where there is gain of copy number relative to the reference sample. (E) A segmental duplication is a segment of DNA at least 1 kb in size that occurs in 2 or more copies per haploid genome, with the different copies sharing at least 90% sequence identity. (F) A large-scale copy number variant is a CNV that involves a segment of DNA more than 50 kb in size.

noncoding regions, and only 3% of SNVs are associated with exons. Most SNVs that cause disease are missense which result in an amino acid substitution. Some are nonsense variants that result in a termination codon and premature polypeptide chain termination.

Small Insertions and Deletions

An insertion refers to the presence of extra bases; deletion implies the absence of certain bases in comparison with a reference sequence. Insertions and deletions often cause a shift of the codon reading frame, resulting in altered amino acid sequence downstream of the variation.

Structural Variants

Structural variants include duplications or deletions of entire exons or genes, chromosomal translocations, inversion, simple sequence repeat expansion, gene rearrangement, and copy number variants (CNVs). Although SNVs are the most common sequence variant, CNVs cover more of the genome than SNVs. These CNVs occur in stretches of DNA that can range from 100 bases up to several megabases in size. CNVs might be duplicated in tandem or might involve complex gains or losses of homologous sequences at multiple sites in the genome. Most CNVs are inherited. CNVs can involve genes or contiguous sets of genes. When the normal complement of the gene is two, but more than two functional copies of a gene are present, then the gene is amplified. Amplification of a dosage-sensitive gene, such as HER-2-neu (ERBB2), usually leads to overexpression of messenger RNA and protein, resulting in cellular abnormalities and possible progression to a disease such as cancer. When the normal gene dosage is two, and one of the functional copies of the gene is lost, disorders such as mental retardation and developmental delay can occur. Structural variants can be determined by cytogenetic techniques, including karyotyping, FISH, CGH, and SNV microarrays.

Alterations in Mitochondrial DNA and Hemizygous Genes

All human chromosomes are in pairs, except the unpaired sex chromosome in male and mitochondrial DNA. The presence of only a single copy of genes on the X and Y chromosome in a male leads to well-known sex-linked disorders. On the other hand, the 16,500 bp mitochondrial genome is present in more than 1000 copies per cell, consisting about 0.3% of human DNA. Thus, mitochondrial DNA demonstrates heteroplasmy, which means the ratio of the wild-type allele to a variant allele within a cell can vary almost continuously, sometimes resulting in a wide range of symptoms even when only one sequence variant is involved.

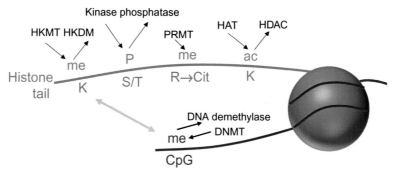

FIG. 14.2 Histone modification patterns at the histone tail.

Human Epigenetic Alterations

Epigenetic modifications refer to the covalent modification of DNA or histones associated with DNA (Fig. 14.2). For example, specific bases in DNA can be methylated, and acetyl, ubiquitin, methyl, phosphate, or sumoyl groups can be added to specific amino acids in histone proteins. Epigenetic changes are heritable, but potentially reversible changes in gene expression that do not represent a change in the sequence of the cell's genomic DNA. The most striking examples of epigenetic inheritance are genomic imprinting and mammalian X-chromosome inactivation, both involving transcriptional silencing of genes through methylation of cytosines at CpG dinucleotides. The methylation status of DNA is maintained following DNA replication by methylases, which act on hemi-methylated double-stranded DNA to methylate the newly synthesized DNA strand as well. The process is perpetuated indefinitely in succeeding cell division. De novo methylation of CpG dinucleotides might take place in the promoters of some tumor suppressor genes. This CpG methylation can silence these genes and in essence constitute one or both of the "hits" in a tumor-suppressor gene that leads to tumor development. Another category of epigenetic inheritance that is less clearly understood involves the property of some proteins to influence the conformation of newly synthesized or assembled proteins in a self-perpetuating manner. The most notable examples in mammals are the prion diseases, responsible for scrapie and bovine spongiform encephalopathy in animals and kuru and Creutzfeldt-Jakob disease in humans. The disease-producing prion protein, which can result from a coding sequence mutation in familial cases, is folded into an abnormal conformation and exerts an effect on newly synthesized prions in such a way that it perpetuates and proliferates the conformational abnormality that produces disease.

CHOICE OF TECHNIQUES

The choice of which technique to use in a particular case will depend, in a large extent, on two factors: the current knowledge of the gene(s) associated with the disease in question and its degree of molecular heterogeneity. The first criterion roughly divides all genetic diseases into two categories: those for which the causative gene has been isolated and those for which it has not. Those in the first category often can be approached by direct gene/mutation analysis; those in the latter can be approached only by linkage analysis using polymorphic DNA markers on the same chromosome, as long as the disease has been mapped to crude level. The second concept, heterogeneity, refers to the number of different genes, or the variety of mutations within a single gene, that cause the same disease. The greater the heterogeneity, the more difficult, labor-intensive, and expensive the DNA test becomes, and the more complex the results reporting and genetic counseling become.

Linkage techniques usually are less favored than the direct mutation detection approach because of the need to test multiple family members and because meiotic recombination between the gene and the marker can disrupt the apparent phase of linkage between parent and offspring, leading to false-positive or false-negative interpretation of results. Since the completion of the human genome sequence, which is saturated with polymorphic markers (both short tandem repeats and single nucleotide polymorphisms), however, there is no longer any need to rely on linked marker that is so far away from the disease gene.

The Application of PCR in Direct Mutation Tests

Direct mutation tests can be achieved by PCR through the judicious choice of primers. PCR allows the laboratory to hone in on the precise mutation of interest, or a "hot spot," within a gene containing several possible mutation sites, using minute

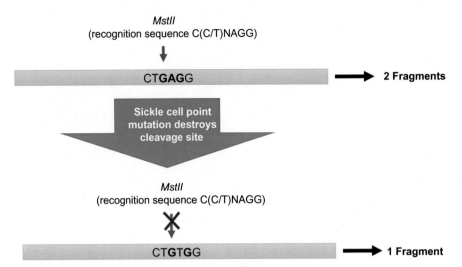

MstII
(recognition sequence C(C/T)NAGG)

CTG**GA**GG → 2 Fragments

Sickle cell point mutation destroys cleavage site

MstII
(recognition sequence C(C/T)NAGG)

CT**GT**GG → 1 Fragment

FIG. 14.3 Detection of a point mutation by differential cleavage with a restriction enzyme.

amounts of starting material. After the region containing the suspected mutation is amplified, it can be analyzed by gel or capillary electrophoresis, sequencing, or DNA probe hybridization.

For a deletion, one would expect to see the change of the amplicon length, accurate molecular sizing of the PCR products by electrophoresis will be sufficient. Alternatively, if the deletion or point mutation disrupts (or creates) a restriction endonuclease cleavage site, it can be detected by electrophoretic analysis of PCR products digested with that enzyme (Fig. 14.3).

Another option is the dot blot hybridization between the PCR products and allele-specific oligonucleotide (ASO) probes, short DNA fragments that are precisely complementary to either the normal or mutant target sequence. If the hybridization is performed under sufficiently stringent conditions, target DNA containing the mutation will hybridize only with the mutant probe, and vice versa for wild-type target DNA (Fig. 14.4). Several mutation hot spots in a gene can be

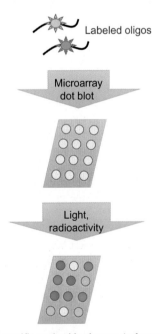

Labeled oligos

Microarray dot blot

Light, radioactivity

FIG. 14.4 Allele specific hybridization. Detection of specific nucleotide changes (polymorphisms) by hybridization of oligonucleotides of known sequences to target DNA. The target DNA is generally obtained using the polymerase chain reaction and specific primers. Allele-specific oligonucleotides are then used to detect single base changes in the DNA samples. Typically, target DNA is immobilized on a solid support and denatured. Labeled (radioactive or fluorescent) oligonucleotides are then allowed to anneal. Complementary sequences bind while noncomplementary sequences do not. Sequences that match the oligonucleotide are detected by fluorescence or when the oligonucleotide is radiolabeled by exposure to X-ray film.

amplified together by multiplex PCR. As a variation on this approach, any number of allelic probes can be spread out on a solid support for subsequent hybridization with the specimen DNA (or amplicons) in the form of a microarray. Several commercial reagents and instruments are available that detect point mutations by differential probe/quencher hybridization or by capillary electrophoresis and other sophisticated techniques.

Techniques Used in Unknown Mutations

To screen a disease gene for unknown mutations that might lie anywhere within it, several mutation scanning techniques can be used, including single-strand conformation polymorphism (SSCP) (Fig. 14.5), denaturing gradient gel electrophoresis (DGGE) (Fig. 14.6), denaturing high-performance liquid chromatography (Fig. 14.7), and other variations of the same principle. These techniques can detect point mutations anywhere within a gene, and these approaches can be performed only on limited PCR-amplified stretches of the gene (usually single exons or parts of exons) at one time. However, they obviate the need for specific ASO probes for every possible mutation. The only technique that should be 100% sensitive in detecting all possible point mutations is complete DNA sequencing of the gene, although even it will miss mutations that lie outside the usual coding region targets.

Techniques Used in Detecting Gene Structural Changes

Disorders caused by gene expansion by a trinucleotide repeat mutation can be diagnosed by Southern blot or PCR. The principle is to examine the size of DNA fragments by observing a larger-than-normal target DNA fragment. Disorders caused by large deletions can be diagnosed by Southern blot which shows loss or decrease in size of a target fragment, by PCR which show loss of a product normally amplified from that site or appearance of a new junction fragment, or by multiplex ligation-dependent probe amplification (MLPA). The multiplexing ability of MLPA is a powerful tool for targeted DNA copy number analyses, but it has its own shortcomings. For example, MLPA depends on the length-based discrimination of the ligation products provided by capillary electrophoresis, therefore, "stuffer" sequence elements of different sizes are required. The potential problems for long stuffer elements are nonspecific hybridization and nonuniform

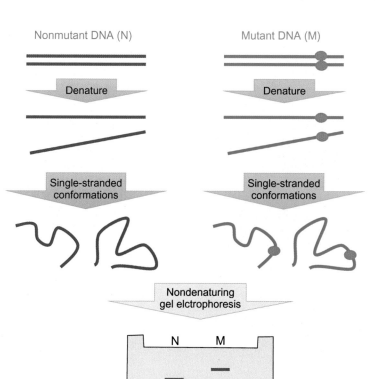

FIG. 14.5 Single-strand conformation polymorphism is a rapid screening for DNA variations. This relies on detecting conformational changes in secondary structure caused by the nucleotide sequence alteration. The change in structure can be detected in a number of ways including denaturing gradient electrophoresis and denaturing gradient high-performance liquid chromatography.

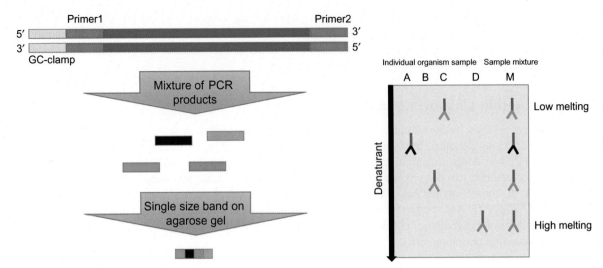

FIG. 14.6 Principle of denaturing gradient gel electrophoresis (DGGE). DNA fragments from a sample containing multiple organisms are amplified using the PCR. These amplified products generally entail sequences that are well conserved from organism to organism. This collection of fragments is then subjected to the DGGE component of the procedure. DGGE is a particular type of gel electrophoresis in which a constant heat (about 60°C), and an increasing concentration of denaturing chemicals are used to force DNA molecules to unwind. This is a separation technique based on the electrical charge, shape, and molecular weight of DNA. In DGGE, DNA reaches the concentration of denaturing reagents at which it unwinds, it is said to have melted. This determines the melting domains, which are defined as stretches of base pairs with an identical melting temperature. In other words, base pairs formed by nucleotides A (adenine) and T (thymine), and those formed by C (cytosine) and G (guanine) are chemically melted apart. Hydrogen bonding between the base pairs is broken by the temperature and the increasing gradient of denaturing chemicals (urea and formamide). Any variation of DNA sequences within these domains will result in different melting temperatures, causing different sequences to migrate at different positions in the gel. This provides DGGE with the power to distinguish between mutated and wild type sequences without prior knowledge of what these sequences are.

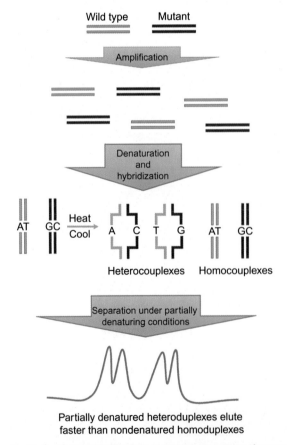

FIG. 14.7 Denaturing high-performance liquid chromatography (DHPLC) compares two or more chromosomes as a mixture of denatured and reannealed PCR amplicons, revealing the presence of a mutation by the differential retention of homo- and heteroduplex DNA (wild type-mutant) on reversed-phase chromatography supports under partial denaturation. Temperature determines sensitivity, and its optimum can be predicted by computation.

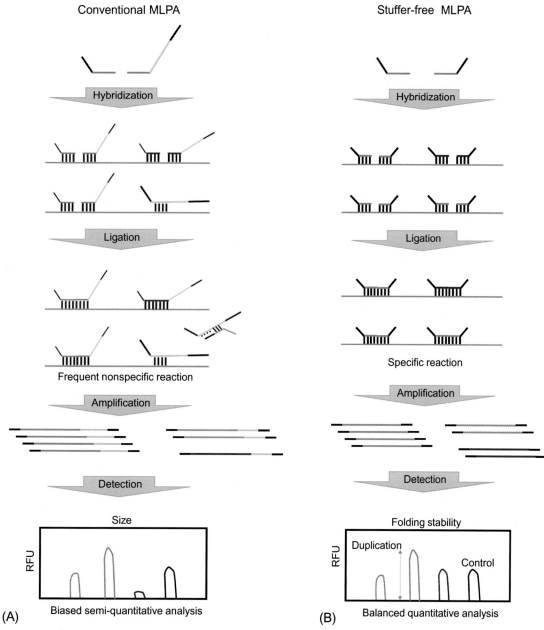

FIG. 14.8 Multiplex ligation-dependent probe amplification (MLPA) technology for detecting copy number variations. (A) The amplified products are separated by conventional MLPA based on length differences in the "stuffer sequences." The stuffer sequence, which is required to separate multiple amplicons using length-based capillary electrophoresis analysis, has potential problems such as nonspecific hybridization and nonuniform amplification. (B) Multiple targets can be specifically amplified with equal efficiency using stuffer-free MLPA by eliminating the stuffer sequence, because the ligated probes are of similar size and amplification uses common primers. *RFU*, relative fluorescence unit.

amplification. Because of these potential problems, it is limited for the use of this technology. A stuffer-free multiplex CNV detection method that combines the advantages of MLPA and capillary electrophoresis-strand conformation polymorphism can be used to overcome these limitations (Fig. 14.8).

Techniques Used for Many Unknown Mutations or an Unknown Gene

Linkage analysis can be used to detect disorders with too many unknown mutations, or an unknown gene. This analysis requires comparative testing of other affected and unaffected siblings and parents, and it requires knowledge of closely linked, flanking or intragenic polymorphic, DNA markers that can be observed to co-segregate consistently with either the normal or disease phenotype within the family. Commonly used detection techniques include the restriction fragment length polymorphisms (RFLPs) detected by Southern blot, microsatellite polymorphisms (tandem oligonucleotide sequences of variable repeat length) detected by PCR and gel or capillary electrophoresis. The later techniques have become favored because of abundance microsatellite throughout the genome, the multiallelic nature of their polymorphisms, and the relative ease of the testing method. Very large genes, such as those for neurofibromatosis and Duchenne muscular dystrophy (DMD), usually have intra genic microsatellites that can be accessed, minimizing the chances of recombination between the mutation and the marker.

Whole-Genome Analysis Techniques

Whole-genome analysis can be done using either high-density oligonucleotide microarray platforms or NGS. Only the former is used routinely in current clinical genetics, in the form of aCGH. Briefly, this technique employs a high number (500,000 to more than 1,000,000) DNA probes on the array, chosen so that their complementary genomic sequences span the entire genome at regular intervals. Total genomic DNA from the patients competes for hybridization to the array against a "normal" total genomic control sample, and each labeled with a different fluorescent dye. If the patient's DNA contains a deletion on a certain chromosome, it will be indicated by an excess of the control DNA hybridizing to the probes on the array that span the deletion site. Conversely, if the patient's DNA contains a duplication of a chromosomal region, an excess of patient DNA will compete for hybridization to the encompassed probes on the array. The resolution for detecting these CNVs is much higher than can be achieved by classic karyotype analysis under the light microscope. It is also far more cost-effective and comprehensive than running many different FISH probes in series. As such, it rapidly has become a predominant technique for detecting deletions/duplications in patients with congenital problems that do not readily suggest a particular known genetic syndrome, including those with nonspecific physical malformations/dysmorphisms, developmental delay, and/or autism.

There are many disadvantages for aCGH to detect unknown mutation. First, because all it does is examine differences in hybridization intensity across the total genomic DNA, balanced translocations with no gain or loss of genetic material will not be detected. Second, all human genomes contain many CNVs that are nonpathologic. Some of these are well-known polymorphisms and thus can be discounted as etiologic if observed in a patient. However, many others are not yet known or studied extensively and this produce the challenging CNV of uncertain clinical significance.

The ultimate gold standard for identifying all possible mutations would be microarray analysis and whole genome sequencing. Microarray analysis can detect mutations at multiple loci associated with complex genetic diseases simultaneously. The NGS automated DNA sequencers is an attractive alternative to array-based assays in molecular cytogenetics. Information obtained by sequence analysis permits the detection of SVs of all types and sizes, including chromosomal gains and losses, and all types of rearrangements. In addition, breakpoints can be mapped with high resolution down to the base pair level, and complex rearrangements can be characterized to analyze multiple breakpoints in a single run.

Single-Cell Sequencing

The new generation of molecular diagnostics based on single-cell sequencing (SCS) has been accessible to clinicians only recently (Fig. 14.9). SCS can measure genome-wide messages or any specific set of targets in individual cells. SCS is uniquely powerful because of its high-sensitivity, high-resolution, and comprehensive analysis of heterogeneous, rare and thus valuable, biopsies. This method has revealed previously unknown molecular characteristics of cancer cells, neurons, immune cells, stem cells, and diseased cells, that is, cells associated with gene-related disease, and is a promising tool for the development of increasingly precise and personalized medicine.

Single-cell whole genome sequencing (scWGS) and single-cell whole exome sequencing (scWES) already have been applied successfully in the clinic, particularly for the genetic evaluation of preimplantation embryos, fetuses, and cancer patients (guiding precise or personalized therapy). In addition, scRNA-seq has also been applied in clinics recently, for example, scRNA-seq of circulating tumor cells (CTC) is used to analyze patients with chemo-resistant prostate cancer. This strategy will be of immense benefit to clinical practice in the future. Single-cell methylome sequencing (scM-seq) is still in the development phase, but the clinical evaluation of CpG methylation in bulk cell samples is emerging. Given the vast and complex role of epigenetics in embryonic development and many diseases, it is anticipated that after the technology is sufficiently developed, rapid clinical use of scM-seq will become powerful.

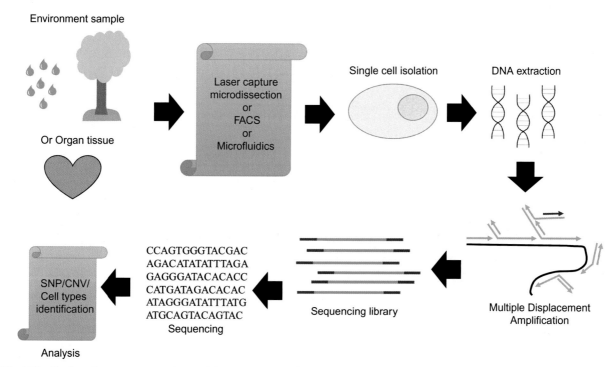

FIG. 14.9 Single cell genome sequencing workflow includes the following steps. 1. Single cell preparation. 2. Whole genome amplification. 3. Amplified product purification. 4. Concentration and size measurement. 5. Estimation of amplification uniformity. 6. Construction of libraries for NGS. 7. NGS.

SINGLE-GENE DISORDERS

Single-gene disorders are caused from mutations in a single gene occurring in all cells of the body. The inheritance of these types of disorders follows a Mendelian segregation pattern, in which the alleles of a gene undergo random segregation during the formation of gametes, so that each gamete receives one or the other with equal likelihood. These single-gene diseases can be classified into three main categories: autosomal dominant, autosomal recessive, and X-linked (Fig. 14.10). In autosomal dominant inheritance, an affected person has at least one affected parent, and an affected person has at least a 50% chance of passing the disorder on to their children. Autosomal recessive disorders occur only when both alleles are mutated. Therefore, affected children can be born to unaffected parents, which mean they usually do not have disease symptoms but each carries a single copy of the mutated gene. The X-linked disorders are linked to mutations in genes on the X chromosome. The X-linked alleles can also be dominant or recessive, but these alleles are expressed more in men because they carry only one copy of the X chromosome.

Achondroplasia

Achondroplasia (ACH) is the most frequently encountered type of genetic dwarfism, with a prevalence of 1 in 20,000–30,000 live-born infants. It is a skeletal disorder with an autosomal dominant inheritance. ACH is characterized by a disproportionately short stature and other skeletal abnormalities resulting from a defective gene encoding for fibroblast growth factor receptor 3 (*FGFR3*). The vast majority of ACH cases are caused by the unique amino acid substitution of G380R as the result of G to A transition (97%) and G to C transition (3%) at cDNA position 1138. The *FGFR3* gene makes a protein called fibroblast growth factor receptor 3 that is involved in converting cartilage to bone. *FGFR3* is the only gene known to be associated with achondroplasia. This is an autosomal dominat disorder. All people who have only a single copy of the normal *FGFR3* gene and a single copy of the *FGFR3* gene mutation have achondroplasia.

Achondroplasia is generally detected by abnormal prenatal ultrasound findings in the third trimester of pregnancy and confirmed by molecular genetic testing of fetal genomic DNA obtained by aspiration of amniotic fluid. Traditionally, this can be confirmed by PCR analysis to determine whether the fetus carries the de novo missense genetic mutation at nucleotide 1138 in *FGFR3* gene. A noninvasive prenatal diagnosis combines digital droplet PCR with minisequencing, which

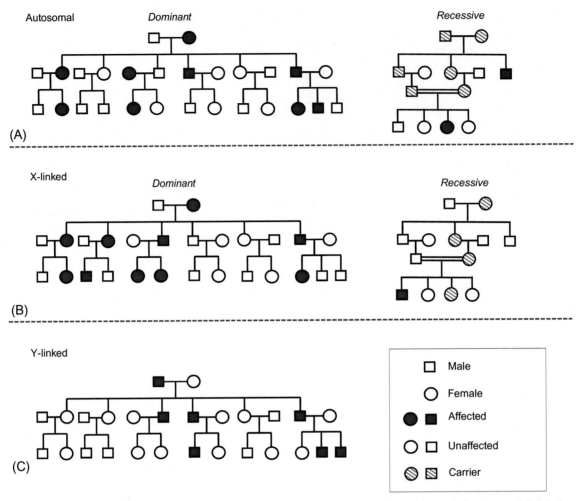

FIG. 14.10 Patterns of Mendelian inheritance. (A) Autosomal dominant and autosomal recessive; (B) X-linked dominant and X-linked recessive; (C) Y-linked.

can analyze fetal DNA from maternal blood. Although free fetal DNA accounts for about 10% of cell-free DNA in maternal plasma, the presence of ambient maternal DNA might make it difficult to detect fetal alleles of paternal origin.

Alpha-1 Antitrypsin Deficiency

Alpha-1 antitrypsin deficiency (AATD) is an inherited condition that causes low levels of, or no, alpha-1 antitrypsin (AAT) in the blood. AATD occurs in approximately 1 in 2500 individuals. This condition is found in all ethnic groups, however, it occurs most often in whites of European descent. AATD is an autosomal codominantly inherited disease that predominantly affects the lungs and liver. This is because AAT is a protein that is made in the liver and is released into the bloodstream. AAT protects the lungs so they can work normally. Without enough AAT, the lungs can be damaged, and this damage can make breathing difficult.

Since it is autosomal dominant disorder, individuals with AATD have one normal copy and one damaged copy or two damaged copies. Most individuals who have one normal gene can produce enough AAT to live healthy lives, especially if they do not smoke. People who have two damaged copies of the gene are not able to produce enough AAT, which leads them to have more severe symptoms.

AAT genotyping is a common test done on the blood sample, and it examines a person's genes and determines genotype. For example, DNA can be extracted from dried blood spot or serum/plasma that provides sufficient quantity and quality of DNA and effectively eliminates any natural PCR inhibitors, allowing for successful AAT genotyping by real-time PCR and direct sequencing.

Autosomal Dominant Polycystic Kidney Disease

Autosomal dominant polycystic kidney disease (PKD) is one of the most common forms of polycystic kidney disease. It is present at birth in 1 in 400–1000 babies, and it affects approximately 400,000 people in the United States. PKD occurs in individuals and families worldwide and in all races.

PKD is a genetic disorder characterized by the growth of numerous fluid-filled cysts in both kidneys. The progressive expansion of PKD cysts slowly replaces much of the normal mass of the kidneys, can reduce kidney function, and lead to kidney failure. When PKD causes kidneys to fail, usually after many years, the patient requires dialysis or kidney transplantation. About one-half of people with the major type of PKD progress to kidney failure, also called end-stage kidney disease. PKD also can cause cysts in the liver and problems in other organs, such as the heart and blood vessels in the brain.

Two genes are known to be associated with PKD. *PKD1*, which is on chromosome 16, is associated in approximately 85% of individuals who have PKD. *PKD2*, which is on chromosome 4, is associated in about 15% of patients (Fig. 14.11). A genetic test can detect mutations in the *PKD1* and *PKD2* genes, which cause autosomal dominant PKD. Although this test can detect the presence of the autosomal dominant *PKD* mutations before cysts develop, its usefulness is limited by two factors: it cannot predict the onset or ultimate severity of the disease and no cure is available to prevent the onset of the disease. A young person who knows about a *PKD* gene mutation, however, might be able to forestall the disease through diet and blood pressure control.

Genetic testing for PDK1 and PDK2 is available for prenatal diagnosis and preimplantation genetic diagnosis but is not usually requested. This is because PKD is usually an adult-onset condition. Since PKD is an autosomal dominant disease with high penetrance, it is highly unlikely to clinically skip a generation.

Locating *PKD1* and *PKD2* allows PKD to be diagnosed by genetic linkage analysis. This study is indirect and is based on the analysis of genetic markers located in the regions of the *PKD1* and *PKD2* genes. It can determine whether it is the *PKD1* or *PKD2* haplotype that segregates with the disease in a given family and, consequently, which gene is responsible for the illness in the family (Fig. 14.12). Databases are available that locate a minimum of 15 microsatellite markers useful for genetic linkage to *PKD1* and eight markers for *PKD2*. The major drawback of this type of diagnosis is that it can be used only in family cases. Also, because of the genetic heterogeneity of the disease, several other affected and unaffected family members must be available for radiological study for an accurate ascertainment of the disease. It is also imperative that one member of the family is diagnosed with PKD with absolute certainty (this person is called the testing or index case). Only certain families are large enough for the study to confirm the linkage to one of the genes and discard the linkage to the other

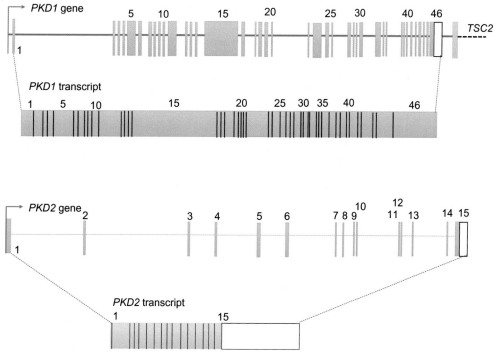

FIG. 14.11   Genes and transcripts of PKD1 and PKD2.

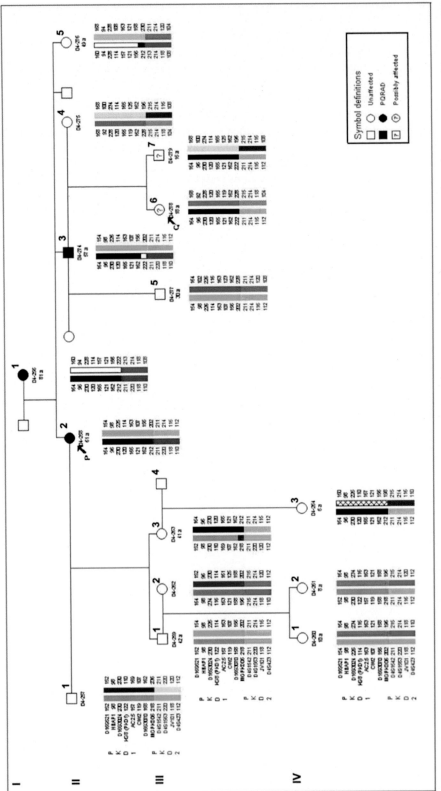

FIG. 14.12 Genetic linkage analysis for PKD1 and PKD2. This family's genetic linkage analysis illustrates why it is necessary to have some affected family members and some who are not affected for the study to provide information. Eight microsatellite markers linked to PKD1 locus (*black*) were used in this study and four markers linked to PKD2 locus (*blue*). The testing (T) or index case was the II-2 family member with a 100% sure diagnosis of PQRAD and the patient was III-6 who was 18 years old. He had come to the hospital to find out whether he had the disease. At the start, only the I-1, II-2, II-3, II-4, II-5, III-5, III6, and III-7 family members took part in the linkage study. As can be seen, the three affected family members (I-1, II-2, II-2) shared the PKD1 haplotype (represented by the *black bar*) as well as the PKD2 (represented by the *red bar*). The only family member that was clearly not affected (II-5) did not have either of these haplotypes. With only this branch of the family the study did, therefore, not provide any information, that is to say that we could not reach a conclusion on which of the genes, PKD1 or PKD2, was responsible for the disease in this family. The study provided information when another branch of the family (II-1, III-1, III-2, III-3) participated a posteriori in the study, given that III-3 also had the at-risk PKD1 haplotype but was not affected. We were, therefore, able to come to the conclusion that PKD2 was the gene that was causing the disease in this family.

gene. In some cases, all affected family members share both the *PKD1* and the *PKD2* haplotype, and none of the unaffected family members carries them. Therefore, the linkage study does not always provide any useful information.

Genetic testing is cumbersome because of the size and structure of major PKD genes, as well as marked clinical and genetic heterogeneity, with most mutations only occuring in single families. Consequently, molecular diagnosis plays an increasingly important role in the diagnosis and management of patients with PKD. Analysis of the *PKD1* gene, however, is complicated by the presence of six pseudogenes (PKD1P1–P6) on chromosome 16 with 97.7% sequence identity as a consequence of segmental duplications of the main part of the 5′ genomic regions (exons 1–33) (Fig. 14.11). The key step in PKD genotyping procedures is amplification of the *PKD1* gene region, while excluding the pseudogenes. This step can be achieved traditionally by using long-range PCR (LR-PCR) with primers located to the rare mismatch sites that distinguish *PKD1* and the pseudogenes, followed by nested PCR of the individual exons, whereas the single-copy regions of *PKD1* and *PKD2* are amplified directly from genomic DNA. Subsequently, amplicons are analyzed directly by sequencing or by sequencing coupled with a mutation screening step to lower the testing cost. Since genetic analysis of PKD is challenging due to the large size, complex genomic structure, and allelic heterogeneity of *PKD1* and *PKD2*, NGS has revolutionized the genetic test and is slowly replacing the traditional genetic testing. For example, one of the NGS-based PKD genetic analysis is an amplicon-based NGS restricted to *PKD1* and *PKD2* and dependent on laborious LR-PCR. In summary, NGS-based PKD genetic analysis is a highly accurate and reliable approach for mutation analysis, achieving high sensitivity and improved intronic coverage with a faster turnaround time and lower cost. It will become an appropriate new standard for clinical genetic testing of PKD.

Cystic Fibrosis

Cystic fibrosis (CF) is the most common fatal autosomal recessive genetic disease in the United States. The incidence is approximately 1 in 2000 of the general population. About 30,000 people in the United States have the disease.

CF is caused by mutations in the cystic fibrosis transmembrane conductance regulator (*CFTR*) gene, whose protein product is involved in the transport of chloride ions across the apical membrane of epithelial and blood cells. Loss of CFTR protein function causes thickened extracellular mucus to accumulate, which impairs mucociliary clearance in the airways. CF prominently leads to microbial pathogen colonization in the lung, followed by recurrent pulmonary infection and chronic inflammation.

About 1900 different mutations have been discovered in the *CFTR* genes of patients with CF, the most-common mutation being an in-frame deletion of three nucleotides in exon 10 of the *CFTR* gene that leads to loss of the amino acid phenylalanine at codon 508 (ΔF508). Allele-specific hybridization is commonly used to screen for CF. With this technique, an individual's *CFTR* gene is amplified by PCR and then hybridized to labeled oligonucleotide probes for the mutant and wild-type genes, respectively (Fig. 14.13). In this way, it is possible to distinguish whether the individual is healthy with two wild-type alleles, a carrier with one wild-type and one mutant alleles, or CF with two mutant alleles.

Duane's Retraction Syndrome

Duane's retraction syndrome (DRS) is a congenital anomaly of the sixth cranial nerve nuclei with aberrant innervation supplied from the third cranial nerve. It is the most common of the congenital cranial dysinnervation disorders. It is a rare condition, with an incidence of approximately 1 in 1000 of the general population.

Individuals with DRS typically present with limitation or absence of globe abduction, variable limitation of adduction, palpebral fissure narrowing on attempted adduction secondary to globe retraction, absence of abducens motor neurons and nerve, and aberrant innervation of the lateral rectus muscle by a branch of the oculomotor nerve.

DRS patients carry an autosomal dominant trait in which unique heterozygous missense mutations are found in the *CHN1* gene. Direct sequencing of the *CHN1* gene is available as a clinical test. The *CHN1* mutations, however, have not been found to be a common cause of simplex Duane retraction syndrome.

Duchenne Muscular Dystrophy

Duchenne muscular dystrophy (DMD) is an X-linked recessive disorder. It occurs in boys with an incidence rate of 1 in 3500, and it is caused by an alteration in dystrophin gene, called the *DMD* gene, that can be inherited in families in an X-linked recessive fashion. Individuals who have *DMD* have progressive loss of muscle function and weakness, which begins in the lower limbs. Boys with Duchenne muscular dystrophy do not make the dystrophin protein in their muscles.

FIG. 14.13 An example of allele-specific hybridization application to screen for cystic fibrosis. The CFTR gene is amplified by PCR. Subsequently, the PCR products are denatured and hybridized to wild-type or mutant alleles. Hybridization is detected by measuring fluorescence emitted by the fluorophore attached to the probe. The healthy individuals are homozygous for the wild-type allele and heterozygous for the CFTR alleles, whereas those affected by the disease carry two mutant alleles.

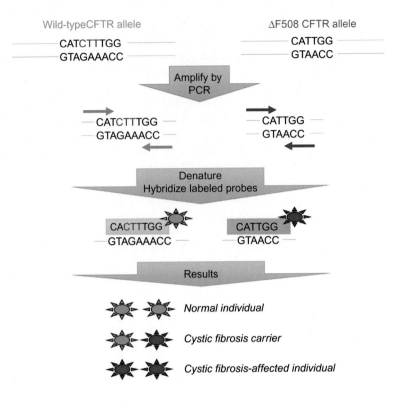

Dystrophin is the largest gene in human genome, spanning 2.4 Mb and containing 79 exons. The full-length transcript expressed in human skeletal muscle encodes a protein of 3685 amino acids, which gives rise to a 427 kDa dystrophin protein that links cytoskeletal actin to the extracellular matrix via the sarcolemmal dystrophin-associated glycoprotein complex. The well-known types of DMD-causing mutations include large mutations (deletions larger than 1 exon), large duplications (larger than 1 exon), small mutations (deletions less than 1 exon), small insertions (less than 1 exon), splice site mutations (less than 10 bp from exon), point mutations (nonsense, missense), and midintronic mutations. Large deletions, such as deletion of exon 45 whose length is 176 bp resulting in frameshift, are the most commonly observed and account for about 68% of the total mutations.

Multiplex ligation-dependent probe amplification (MLPA) is a widely used method for DMD diagnosis, and it is the initial diagnostic test of choice in many hospitals. Although MLPA can diagnose only patients with deletions/duplications, another 30% of patients with point mutations need direct sequencing of all coding regions. Therefore, diagnostic methods have been conducted separately, which eventually increase medical costs and delay diagnosis. NGS approaches have become major tools for the genetic diagnosis of DMD. Furthermore, targeted NGS on HiSeq2000 can perform a mutational search for DMD and female carriers.

Fragile X Syndrome

Fragile X syndrome (FXS) is the most common monogenic cause of intellectual disability and autism. It affects about 1 in 4000 boys and 1 in 8000 girls, and occurs in all racial and ethnic groups. It is an X chromosome-linked recessive trait with variable expression.

Partial or complete loss-of-function mutations in the fragile X mental retardation 1 (*FMR1*) gene cause FXS. Most FXS-affected individuals (more than 99%) harbor a hypermethylated CGG-repeat stretch in the 5′-untranslated region of the *FMR1* exon 1. These allelic variants containing more than 200 CGGs—referred to as full-mutation (FM)—trigger aberrant methylation and heterochromatization of the *FMR1* promoter region, the epigenetic consequences that ultimately block *FMR1* gene expression. Absence of the FMR1-encoded fragile X mental retardation protein (FMRP), which is critical for neuronal development and synaptic plasticity, results in the manifestation of FXS. Rare deletions and point mutations that result in FMRP deficiency account for fewer than 1% of FXS cases.

Comprehensive molecular characterization of *FMR1* alleles requires information about CGG-repeat size, length of uninterrupted CGG repeats, number of AGG interruptions at the 3′ and 5′ ends of *FMR1*, respectively, and methylation status and mosaicism for CGG-repeat size. Generally, a combination of methods might be required to characterize these different facets of the *FMR1* expansion mutation.

The gold standard fragile X test is Southern blot analysis which allows concurrent detection of large CGG-repeat expansions and determination of its methylation status. However, precise resolution of *FMR1* allelic variants and AGG-interruption mapping cannot be achieved using Southern blot analysis. Alternatively, PCR-based approaches can be used to characterize CGG-repeat size and/or AGG-interruption patterns. Therefore, many diagnostic laboratories employ a first-tier, high-throughput PCR-based approach to exclude nonexpansion carriers and reflex only samples with an expansion for confirmatory Southern blot analysis, as testing all samples by Southern blot is both labor- and time-intensive.

Hemophilia

Hemophilia is a bleeding disorder that slows down the blood clotting process. People who have hemophilia often have longer bleeding after an injury or surgery, and people with severe hemophilia have spontaneous bleeding into the joints and muscles.

Hemophilia is caused by permanent gene changes. Mutations in the *FVIII* gene cause hemophilia A, and mutations in the *FIX* gene cause hemophilia B. One in 5000–10,000 males worldwide have hemophilia A. Hemophilia B is less common, affecting 1 in 20,000–34,500 males worldwide. Being an X-linked recessive disorder, females are generally not affected, although they can be carriers of this disorder. A classical female hemophilia is possible only when a carrier female marries a hemophilic male and this is very rare. Hemophilia A and B are diagnosed by measuring factor clotting activity. Individuals who have hemophilia A have low factor VIII clotting activity. Individuals who have hemophilia B have low factor IX clotting activity.

Genetic testing is available for both the factor VIII gene and the factor IX gene. Genetic testing of the FVIII gene finds a disease-causing mutation in up to 98% of individuals who have hemophilia A. Genetic testing of the FIX gene finds disease-causing mutations in more than 99% of individuals who have hemophilia B.

Genetic testing usually is used to identify women who are carriers of a FVIII or FIX gene mutation, and to diagnose hemophilia in a fetus during a pregnancy (prenatal diagnosis). It sometimes is used to diagnose individuals who have mild symptoms of hemophilia A or B.

Huntington's Disease

Huntington's disease (HD) is an autosomal dominant genetic neurodegenerative disorder that can affect motor, cognitive, and psychiatric functioning. Mutant huntingtin protein (mHTT) gains toxic function, primarily in medium spiny neurons, with aberrant interactions in several pathways, including gene expression, oxidative stress, mitochondrial function, energy metabolism, Ca^{2+} homeostasis, and signaling. The mHTT is ubiquitously expressed and evidence is growing that it causes cytotoxicity not only in the central nervous system but also in the peripheral tissue of patients with HD. HD is a progressive neurodegenerative disorder that presents with motor symptoms, cognitive impairment, and psychiatric disturbances. The first symptoms usually manifest between 35 and 50 years of age and the duration of the disease is between 15 and 20 years. The incidence of HD is 3–10 per 100,000 in populations of Western European descent. It is much less frequent (0.1–0.4 per 100,000) in other populations.

HD is an inherited single-gene dominant disorder caused by the abnormal expansion of CAG repeats in the *HTT* gene (Table 14.1). The number of CAG repeats can be established by PCR analysis of the region encompassing the CAG repeat, followed by fragment sizing through capillary or gel electrophoresis at sufficient resolution to allow separation of alleles with one repeat difference. PCR using flanking primers allows amplification up to approximately 100 CAG repeats but is unreliable above this size. The diagnostic threshold for this series of diseases is between 40 to 50 repeats with alleles above this size being in the affected range. The myotonic dystrophy CAG repeat can expand to give alleles of greater than 5 kb. PCR amplification of DNA from an affected person carrying such an expansion gives a single normal allele and fails to pick up a larger allele. For people apparently homozygous for small common repeat size, the presence of a large unamplifiable allele can be excluded by performing a Southern blot and probing with a locus specific fragment flanking or containing the CAG repeat.

Another method to diagnose HD is the triplet repeat primed PCR (TP PCR). TP-PCR is a fluorescently labeled locus-specific primer flanking the CAG repeat together with paired primers amplifying from multiple priming sites within the

TABLE 14.1 Implications of Varying Repeat Ranges in Huntington's disease

# of Repeats	Implications for Patient		Implications for Family
	Diagnostic Test	Predictive Test	
6–26 normal allele	Diagnosis not confirmed	Will not develop HD	No increased risk for HD
27–35 intermediate allele	Diagnosis not confirmed	Will not develop HD	Increased risk for HD (<10%)
36–39 incomplete penetrance allele	Diagnosis of HD confirmed	Might or might not develop HD	Increased risk for HD
40 and more complete penetrance allele	Diagnosis of HD confirmed	Will develop HD	Increased risk for HD

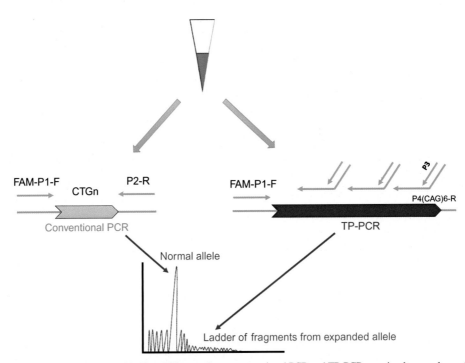

FIG. 14.14 The principle of the modified TP-PCR. Two PCR procedures, conventional PCR and TP-PCR, are simultaneously carried out. The conventional PCR primers, P1-F and P2-R, bind to the gene specific regions that closely flank the CTG repeat, generating the smaller alleles. Primer P4(CTG)$_6$-R has two parts: 6 CAG trinucleotides at the 3'-end sequence and an artificial sequence with no homology to the human genome at the 5'-end sequence. By the triplet priming from its 3'-end sequence, a ladder of PCR fragments differing by 1 CTG unit is produced.

CAG repeat (Fig. 14.14). TP PCR gives a characteristic ladder on the fluorescence trace enabling the rapid identification of large pathogenetic CAG repeats that cannot be amplified using flanking primers.

Marfan Syndrome

Marfan syndrome is one of the most common inherited disorders of connective tissue. It is an autosomal dominant condition occurring once in every 10,000 to 20,000 individuals. Marfan syndrome has a wide variability of clinical symptoms, with the most notable occurring in eyes, skeleton, connective tissue, and cardiovascular systems. The most common symptom of Marfan syndrome is myopia which is nearsightedness from the increased curve of the retina because of connective tissue changes in the globe of the eye. About 60% of individuals who have Marfan syndrome have lens displacement from the center of the pupil. Individuals who have Marfan syndrome also have an increased risk for retinal detachment, glaucoma, and early cataract formation.

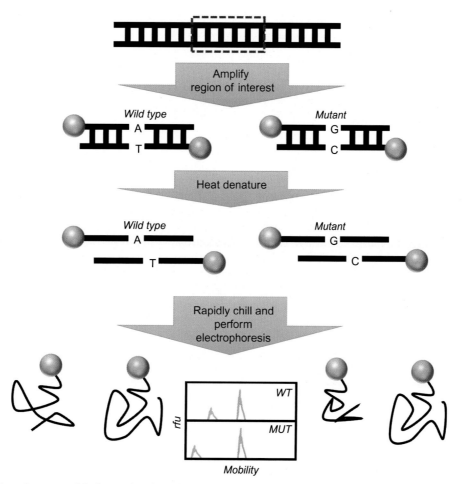

FIG. 14.15 The schematic process of single-strand conformation polymorphism (SSCP) analysis. SSCP analysis involves the following four steps: (1) polymerase chain reaction (PCR) amplification of DNA sequence of interest; (2) denaturation of double-stranded PCR products; (3) cooling of the denatured DNA (single-stranded) to maximize self-annealing; (4) detection of mobility difference of the single-stranded DNAs by electrophoresis under nondenaturing conditions. Several methods have been developed to visualize the SSCP mobility shifts. These include the incorporation of radioisotope labeling, silver staining, fluorescent dye-labeled PCR primers, and more recently, capillary-based electrophoresis.

Marfan syndrome is caused by mutations in the *FBN1* gene, which is a large gene with 65 exons. *FBN1* mutations are associated with a broad continuum of physical features, ranging from isolated features of Marfan syndrome to a severe and rapidly progressive form in newborns. The diagnosis of Marfan syndrome is commonly based on family history and the presence of characteristic clinical findings in ocular, skeletal, and cardiovascular systems. This is because the *FBN1* gene is a large gene, screening and mutation detection for Marfan syndrome are not cost-effective. Genetic testing of the *FBN1* gene identifies 70%–93% of the mutations and is available in clinical laboratories. Patients negative for the test for gene mutation, however, should be considered for evaluation for other conditions. Therefore, for prenatal diagnosis or presymptomatic detection of the disease in families with Marfan syndrome linkage analysis can be useful for detecting the affected allele. Mutational analysis using techniques such as single-strand conformation polymorphism (Fig. 14.15), denaturating high performance liquid chromatography, and direct sequencing can be employed for the genetic diagnosis of Marfan syndrome.

Sickle Cell Disease

Sickle cell disease, an autosomal recessive disease, is the most common inherited blood disorder in the United States, and it is most prevalent among African Americans. About one in 12 African Americans and about 1 in 100 Hispanic Americans carries the sickle cell trait, which means they are carriers of the disease.

Sickle cell disease is caused by a mutation in the hemoglobin-β gene found on chromosome 11. Hemoglobin transports oxygen from the lungs to other parts of the body. Red blood cells with normal hemoglobin (hemoglobin-A) are smooth and round and glide through blood vessels. In people with sickle cell disease, abnormal hemoglobin molecules (Hbs) stick to one another and form long, rod-like structures. These structures cause red blood cells to become stiff, assuming a sickle shape that causes them to pile up, causing blockages and damaging vital organs and tissue. Sickle cells are destroyed rapidly in the bodies of people with the disease, causing anemia. This anemia is what gives the disease its commonly known name—sickle cell anemia.

The sickle cells also block the flow of blood through vessels, resulting in lung tissue damage that causes acute chest syndrome, pain episodes, stroke, and priapism (painful, prolonged erection). It also causes damage to the spleen, kidneys, and liver. The damage to the spleen makes patients—especially young children—easily overwhelmed by bacterial infections.

Sickle cell disease can be diagnosed by a simple blood test. Currently, the most common screening techniques include sickle solubility testing, hemoglobin electrophoresis, high-performance liquid chromatography (HPLC), and isoelectric focusing (IEF), each with its own advantages and limitations. The sickle solubility test is a low-cost assay that relies on the relative insolubility of Hbs in the presence of a reducing agent, such as sodium dithionite, by detecting turbidity or crystal formation from lysis of Hbs-containing erythrocytes. Because it detects only the presence or absence of sickle hemoglobin, the solubility test cannot differentiate individuals with sickle cell disease and sickle cell trait and can be falsely negative, making confirmatory testing essential. Hemoglobin electrophoresis, HPLC, and IEF are used either for primary identification of sickle cell trait or as confirmatory tests. These techniques can provide discrimination and relative quantification of hemoglobin, allowing for differentiation of sickle cell trait from sickle cell disease syndromes. For example, hemoglobin electrophoresis, an inexpensive and frequently used technique, uses the principles of gel electrophoresis to separate hemoglobin molecules by size and charge. Comigration of certain rare hemoglobin variants with Hbs can obscure the diagnosis with standard electrophoresis; therefore, the use of different gels such as citrate agar or cellulose acetate or IEF methods often are required for further hemoglobin discrimination.

Molecular protocols for hemoglobinopathies often are used in the research setting to identify sickle cell trait carriers, especially in studies where hemoglobin electrophoresis samples have not been collected. Rapid advances in NGS have allowed for detection of sickle cell trait from DNA through exome sequencing, direct genotyping for the SNP that encodes the sickle mutation (rs334), and even GWAS.

Thalassemia

Thalassemia is actually a group of autosomal recessive inherited diseases of the blood that affect a person's ability to produce hemoglobin, resulting in anemia. About 100,000 babies worldwide are born with severe forms of thalassemia each year. Thalassemia occurs most frequently in people with Italian, Greek, Middle Eastern, Southern Asian, and African ancestry.

The two main types of thalassemia are called "alpha" and "beta," depending on which part of an oxygen-carrying protein in the red blood cells is lacking. Alpha-thalassemia is characterized by reduced or absence of α-globin chain synthesis and is caused mainly by deletions in the α-globin gene complex. Beta-thalassemia is characterized by reduced or absence of β-globin chains. Both types of thalassemia are inherited in the same manner. The disease is passed to children by parents who carry the mutated thalassemia gene. A child who inherits one mutated gene is a carrier, which is sometimes called thalassemia trait. Most carriers lead completely normal, healthy lives.

Blood tests and family genetic studies can show whether an individual has thalassemia or is a carrier. The PCR is a common method to detect the deletional thalassemias. For example, gap-PCR amplifies the deleted DNA sequence using the primers flanking the deleted region. Three primers are designed for each deletion to amplify the normal (undeleted) and deleted gene sequences. Because conventional gap-PCR requires post-PCR handling and is time-consuming, it is not suitable for large-scale screening. High resolution melting (HRM) analysis is a high-throughput mutation scanning method that is based on melting temperature (T_m) profiles. The changes in T_m of the DNA duplexes are detected during dissociation of the dsDNA to single-stranded DNA. Differences in the T_m profile enable the identification of different genotypes. HRM has various additional advantages compared with other mutation scanning methods. For example, it not only can detect multiple known and unknown mutations but also offers straightforward and rapid analysis. HRM is a sensitive and specific high-performance platform in a closed-tube system.

Tay-Sachs Disease

Tay-Sachs disease (TSD) is a fatal autosomal recessive genetic disorder, most commonly occurring in children. TSD is caused by mutations in the *HEXA* (hexosaminidase-A) gene localized on chromosome 15. Without normal *HEXA*, a fatty substance, or lipid, called GM2 ganglioside, accumulates abnormally in cells, especially in the nerve cells of the brain. This ongoing accumulation causes progressive damage to the cells and leads to a neurodegenerative disease. The overall prevalence of TSD is estimated to be about 1 in 200,000 live births in the general population. A high prevalence is found in the Ashkenazi Jewish population, with about one in 4000 live births. Furthermore, one in 31 individuals of Ashkenazi Jewish (AJ) ancestry are likely a carrier, with an incidence of one in 2500–3900 live births, although TSD does occur in other populations, including the French Canadian, Cajun, Irish, and Italian populations.

TSD has three different forms: classic infantile, juvenile, and adult late-onset. The classic infantile form is the most common and is characterized by onset at 4–8 months and progressive neurological deterioration with macular cherry-red spots, blindness, intractable seizures, and paralysis. Affected children rarely survive beyond 5 years of age. Juvenile and adult forms are very rare with a later onset and slower course. Juvenile or subacute GM2 gangliosidosis is characterized by gait disturbances, incoordination, speech problems, and intellectual impairment. Behavior or psychiatric disturbances, proximal and distal weakness, and muscle wasting tend to appear later in the disease course.

Nearly 130 mutations have been reported in the *HEXA* gene to cause TSD and its variants, including single base substitutions, small deletion, duplications and insertions splicing alterations, complex gene rearrangement, and partial large duplications. Most of these mutations are private mutations and have been present in a single or few families. Only few mutations have been found commonly in particular ethnicities or geographically isolated populations. Carrier screening usually measures HEXA activity because below-normal levels indicate carrier status. Because of the variations, different targeted mutation analyses can be used in different populations. In the Ashkenazi Jews, 94%–98% patients are caused by one of the three common mutations c.1277_1278insTATC, c.1421+1 G>C and c.805 G>A (p.G269S). A 7.6 kb deletion that includes the entire exon 1 and parts of the flanking sequence is the major mutation causing TSD in the French-Canadian population. The mutation c.571–1 G>T accounts for 80% of mutant alleles among Japanese patients with TSD. The c.1277_1278insTATC and the c.805 G>A(p.G269S) mutations also are found in non-Ashkenazi Jewish populations, along with an intron 9 splice site mutation (c.1073+1 G>A) and the 7.6 kb French-Canadian deletion. A technique such as amplification refractory mutation system-polymerase chain reaction (ARMS PCR)—can be used to detect common mutations by using normal and mutant primers as reverse primers and common forward primers. Detection rate by sequence analysis has been evaluated and reported at 92%–100% in individuals from many ethnic groups.

POLYGENIC DISORDERS AND GENE CLUSTERING

Genetic disorders are polygenic when they are associated with the effects of many genes. Such disorders also can be multifactorial because they do not follow simple Mendelian pattern of inheritance and can be influenced by several different lifestyle and environmental factors. An individual's genetic background usually is not sufficient to cause the disorder in most polygenic disorders, but it might render an individual more susceptible to the disorder. It is common, therefore, to find clustering of such disease in certain families because of similarities in genome sequence and because genome architecture promotes gene-gene interactions and gene-environment interactions in family members. An individual might not be born with a disease but might be at high risk of acquiring it through lifestyle habits or exposure to chemicals. This indicates a role for gene-environment interactions in the development of a specific disease. This condition is usually referred to as genetic predisposition or genetic susceptibility.

The genome-wide strategies currently used to map polygenic disease loci include linkage analysis, association studies, and high-throughput direct DNA sequencing. Many common disease variants are mapped using GWAS. This approach can identify loci associated with complex traits without a dependence on prior hypotheses and becomes feasible because of the completion of the Human Genome Project and subsequent rapid advancements in genotyping technology. In contrast to linkage studies, GWAS can be performed with unrelated subjects, who can be recruited more readily than families, and identifies specific risk alleles.

The simple principle of GWAS entails the genotyping of hundreds of thousands of genetic variants and comparison of allele distributions between cases with a disease and controls without the disease (Fig. 14.16). Such studies became feasible with the development of highly multiplexed genotyping arrays, sequencing of the human genome, and the recognition that the most common form of genetic variation is SNPs, which are relatively correlated over sizable genetic distances in the general population. The majority of common variation, therefore, can be assayed by genotyping a few hundred thousand SNPs.

The foundation of GWAS lies in the "common disease, common variant" hypothesis, that is, that relatively common variants in the population with low penetrance contribute to genetic susceptibility to common diseases. It is now recognized that the risk loci identified could explain only a small portion of the variance in the traits. Thus, the focus has been broadened from use of a standard genome-wide significance threshold to the use of other indicators of association. The

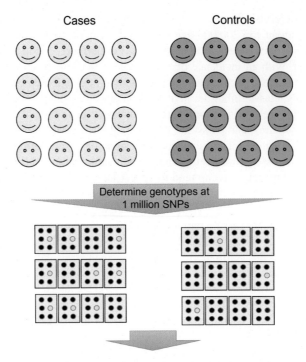

FIG. 14.16 The basic principle of GWAS. Large-scale genomic studies often have high-dimensional data consisting of tens of thousands of individuals, genotypes data on a million (or more!) SNPs for all individuals in the study, phenotype or trait values of interest such as height, BMI, HDL cholesterol, blood pressure, diabetes, etc. Subsequently, it is necessary to identify SNPs where one allele is significantly more common in cases than controls and test for association between SNPs and a quantitative trait that underlies the interest.

development of large meta-analysis and mega-analysis based collaborative efforts have identified replicable risk loci for a variety of complex traits.

More than 2000 common genetic variants have been associated with disease and been deposited in GWAS catalogs, such as those available at http://www.genome.gov/gwastudies. Most genetic effects for common variants are typically small and inversely related to allele frequency. Many loci with small impact contribute to the genetic basis of most complex phenotypes with >100 loci identified for traits with studies of hundreds of thousands of cases. Estimates indicate that many more additional loci of similar or smaller effects have yet to be identified, collectively accounting for substantial proportions of disease risk.

Alcohol Dependence

Alcohol dependence (AD) is a common, complex psychiatric disorder characterized by the excessive and compulsive use of alcohol, which often results in physical and social consequences. The National Epidemiologic Survey on Alcohol and Related Conditions estimated the lifetime prevalence of AD in the general American population to be 12.5% and the 12-month prevalence to be 3.8%. Overconsumption of alcohol is known to be a contributing factor to more than 60 diseases, including several types of cancer, and accounts for approximately 2.5 million deaths each year. Thus, AD represents a significant public health problem.

Although environmental factors play a significant role in AD risk, twin and family-based studies consistently have demonstrated a heritability of approximately 50%. Identifying genes that influence risk for AD can add to our knowledge about the mechanisms underlying the disorder, potentially improving its diagnosis, treatment, and prevention. GWAS contribute significantly in the identification of risk loci for AD in the past decade. The most robust associations for AD have been with the alcohol metabolizing enzyme genes, especially *ALDH2* in East Asian populations, and *ADH1B* in European-American and African-American populations. The individual loci that have been identified, however, explain just a fraction of the variance in AD risk. This has led to efforts to exploit GWAS data using different analytic approaches, including studies of

TABLE 14.2 Examples of Alcohol Dependent Genome-Wide Association Studies (GWAS)

Phenotype	Study	Discovery Sample Size	GWS Findings ($P < 5 \times 10^{-8}$)
AD case-control status	German GWAS	1024 German ancestry cases; 996 German ancestry controls (male)	rs7590720, rs1344694 (PECR)
AD case-control status	German GWAS	1333 German ancestry cases; 2168 German ancestry controls (male)	rs1789891 (*ADH1C*)
AD case-control status	Korean GWAS	117 Korean cases; 279 Korean controls	rs1442492 and rs10516441 (ADH7), rs10516441 (ALDH2)
AD case-control status	Chinese GWAS	102 Chinese cases; 212 Chinese controls (male)	rs3782886 and rs671 (ALDH2)
DSM-IV AD criteria count; AD case-control status	Yale-Penn and SAGE	2379 Euro-Americans; 3318 Afr-Americans (Yale-Penn) 2752 Euro-Americans; 1311 Afr-Americans (SAGE)	14 SNPs in chromosome 4 in AAs including rs28542574 (LOC100507053) And rs2066702 (ADH1B) and rs1229984 (ADH1B) in Eas; rs1437396 on chromosome 2 (intergenic)

the heritability, risk prediction, and pathways involved in AD (Table 14.2). Samples in the tens of thousands likely will be necessary to detect additional loci beyond these identified to date. Post-GWAS approaches to studying the genetic influences on AD are increasingly common and could greatly increase our knowledge of both the genetic architecture of AD and the specific genes and pathways that influence risk (Table 14.3).

Analyses of GWAS data beyond traditional SNP association testing have increased our understanding about the genetic architecture of AD. Common genetic variation can regulate or encode proteins found within important biological pathways for addiction. In line with this notion, pathway analysis of GWAS data has led to new discoveries in the genetics of AD and likely will become more important as pathway annotations and other resources continue to improve. The integration of data from GWAS, gene expression, and whole-genome sequencing, which is becoming increasingly available, will present new opportunities and challenges in the discovery of novel loci and pathways associated with AD.

Charcot-Marie-Tooth Disease

Charcot-Marie-Tooth disease (CMT) is the most common inherited neuromuscular disorder with an incidence of 1 in every 2500 individuals worldwide. The classical symptoms include slowly progressive distal muscle weakness, muscle atrophy, and sensory loss of the lower and then upper limbs. As the motor and sensory peripheral nerves are affected, it is also called hereditary sensory and motor neuropathy.

On the basis of electrophysiological results, CMT is usually subdivided into two main groups: demyelinating type (CMT1) and axonal type (CMT2). The majority of patients with CMT have autosomal dominant inheritance, although many will have forms with X-linked (CMTX) or autosomal recessive (AR) (CMT4) inheritance (Table 14.4). X-linked forms are distinguished by a lack of male-to-male transmission on pedigree analysis and have intermediate-to-demyelinating range conduction velocities. Autosomal recessive inheritance is less common, but should be considered in consanguineous families or those with multiple affected siblings and asymptomatic parents, or in simplex cases. Autosomal recessive forms are frequently characterized by a more severe phenotype with an earlier onset of symptoms. More than 80 disease-causing genes have been identified as being associated with CMT; the duplication/deletion of *PMP22* is the most common cause of CMT1 (Fig. 14.17). Genetic testing for CMT first became available after the discovery of the *PMP22* duplication in 1991. *PMP22* duplication accounts for the majority of CMT1 cases and close to 40% of all

TABLE 14.3 Post-GWAS Analyses of AD GWAS Datasets

Post-GWAS Method	Methods and Findings
SNP heritability	Using Genome-wide Complex Trait Analysis (GCTA), estimated a SNP heritability of 16% to 22% for AD
Risk profile scoring	Study 1: Used a polygenic risk score to predict AD
	Study 2: Used k-fold cross-validation and observed polygenicity; as the P-value threshold increased, the percent of variation increased
	Study 3: Used the Collaborative Study on the Genetics of Alcoholism (COGA) and SAGE samples to conduct risk profile scoring for candidate gene SNPs as well as with SNPs from GWAS. It is possibe to predict AD case-control status in the target sample using results from the discovery sample
	Study 4: Used a Convergent Functional Genomic approach to integrate multiple sources of information and generated a list of 713 nominally significant SNPs in 135 candidate genes.
Gene set analysis	Used a network-based gene set analysis approach to identify human protein interaction network (HPIN) enriched for AD-associated genes. Identified seven HPIN that were enriched for AD-associated genes
Pathway analysis	Study 1: Conducted pathway analysis in MGS2 controls GWAS sample using ALIGATOR and found significant enrichment of Gene Ontology (GO) categories and Kyoto Encyclopedia of Genes and Genomes (KEGG) pathways, including those related to lipid and cholesterol metabolism and cell adhesion
	Study 2: Evaluated enrichment of pathways/ontologies enriched among SNPs identified in risk profile scoring analysis using the COGA discovery sample and SAGE target sample. Identified four ontologies relevant to brain development and inhibitory neurotransmission with significant enrichment
	Study 3: Conducted pathway analysis in AD cases and controls using annotations from KEGG, Reactome, Gene Ontology, Biocarta, microRNA targets, transcription factor targets, and positional information. Found that XRCC5 is critical for AD development. In a human laboratory-based self-administration study, found significant association between maximum blood alcohol concentration and the top XRCC5 SNP from the GWAS, rs828701

TABLE 14.4 Classification and Genetics of CMT Disease

Inheritance	Pathophysiology	Type	Example Gene Associations
Autosomal dominant	Demyelinating	CMT1	PMP22, MPZ, LITAF/SIMPLE, EGR2, NEFL, FBLN5
	Axonal	CMT2	KIFIB, MFN2, RAB7, TRPV4, GARS, NEFL, HSPBI, MPZ, GDAPI, HSPB8, DNM2, AARS, DYNCIHI, LRSAMI, DHTKDI, DNAJB2, HARS, MARS, MT-ATP6, TFG
	Intermediate	CMTDI	DNM2, YARS, MPZ, IFN2, GNB4
Autosomal recessive	Demyelinating	CMT4	GDAPI, MTMR2, MTMRI3 (SBF2), SBFI, SH3TC2, NDRGI, EGR2, PRX, HKI, FGD4, FIG4, SURFI
	Axonal	CMT2	LMNA, MED25, GDAPI, MFN2, NEFL, HINTI, TRIM2, IGHMBP2, GAN
	Intermediate	CMTRI	GDAPI, KARS, PLEKHG5, COX6AI
X-linked	Intermediate or axonal	CMTX	GJBI, AIFMI, PRPSI, PDK3

CMT cases. The three next most common genetic forms of CMT are those caused by mutations in *GJB1*, *MPZ*, and *MFN2*, accounting for the majority of the additional yield of current genetic testing (Table 14.5). With conventional testing methods, it has been estimated that approximately 60% of CMT cases can be genetically confirmed with higher rates of genetic confirmation for CMT1. The yield is lower for CMT2, where, in approximately one-third of cases, a molecular diagnosis is obtained. Since CMT is a highly heterogeneous disorder, MLPA followed by the combined targeted NGS and *PMP22* duplication/deletion analysis is a diagnostic strategy for CMT patients.

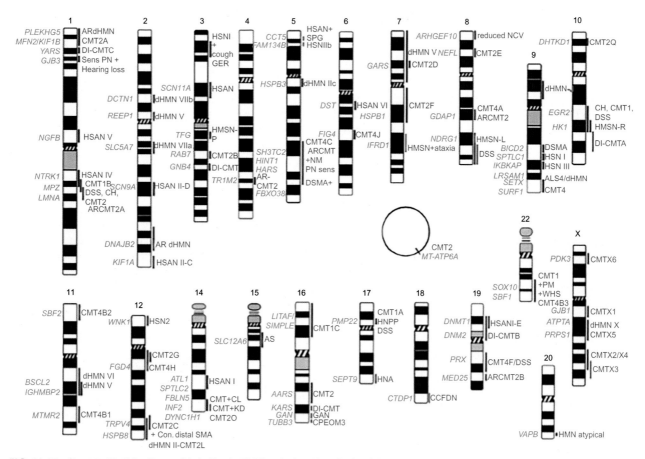

FIG. 14.17 Genes and loci for Charcot-Marie-Tooth (CMT) and related inherited peripheral neuropathies. The figure shows 80 currently known genes *(orange symbols)* and their corresponding chromosomal loci *(vertical bars)*. The corresponding phenotypes are indicated by *blue symbols* and according to the disease nomenclature. Note that the disease names may not always correspond to information available in OMIM, GeneReviews, or other publicly available databases.

TABLE 14.5 Relative Frequencies of Known Mutations

Known Mutation	Relative Frequency
PMP22 duplication	61.1%
GJBI	10.7%
MFN2	7.0%
MPZ	6.7%
Others	Less than 2% each

For molecular diagnostic, the strategy of NGS can be applied to a set of genes (panels), all protein-coding genetic material (the exome), or all genetic material (the genome). Of these options, targeted panel testing is currently the least expensive, has the highest per-gene technical performance, and offers the most straightforward result interpretation with fewer variants of unknown significance. NGS panel testing for CMT is considered as the first line for clinical testing, with the exception of initially testing for the PMP22 duplication in CMT1. NGS technology does not directly ascertain the presence of pathogenic duplications or deletions, but indirect methods such as relative read depth are increasingly being applied to identify duplications and deletions from NGS data. As such, a single NGS panel test has the potential to ascertain all known major mutation types, including PMP22 duplications and deletions. Whole exome or genome sequencing can be considered if panel testing is negative to further increase the yield of diagnosis, but clinical caution is necessary. A significant limitation of whole exome and genome testing currently is that it does not detect small tandem repeats, other copy number variations, mutations in regulatory sequences distant from a gene, epigenetic changes, or mutations in mitochondrial DNA. As such, this technology does not yet have the capability to provide genetic diagnosis in all cases of CMT. With large multigene panels, whole exome, and whole genome testing for CMT, it is critical that clinicians are able to determine the relevance of molecular findings in the setting of the patient's clinical phenotype. The targeted NGS mainly focuses on *GJB1*, *MFN2*, *HSPB8*, *BSCL2*, and *YARS*. More than 20 genes (*COX6A1*, *C12*, *f65*, *EGR2*, *FGD4*, *FIG4*, *GALC*, *GAN*, *GDAP1*, *HINT1*, *HK1*, *LMNA*, *MED25*, *MTMR2*, *MTMR5*, *NDRG1*, *PRX*, *SACS*, *SBF1*, *SBF2*, *SH3TC2*, *SURF1*, and *TRIM2*) are involved in AR-CMT, and can be evaluated through whole-exome sequencing analysis.

Cri du chat Syndrome

Cri du chat syndrome (CdCS) also is known as 5p-syndrome and cat cry syndrome. It is a rare genetic condition caused by the deletion of genetic material on the small arm (the p arm) of chromosome 5, and is among the most common deletion syndromes. The incidence of CdCS ranges from approximately one in 15,000–50,000 in live-born infants, and about 1 in 350 among people with intellectual disabilities. The cause of this rare chromosomal deletion is unknown.

The hallmark features of CdCS include a high-pitched monotonous cry, low birth weight, microcephaly, hypotonia, poor growth, and developmental delay. Specific craniofacial findings include round face, bilateral epicanthal folds, hypertelorism with a broad nasal bridge, down-slanting palpebral fissures, and short philtrum. CdCS also is associated with variable intellectual disabilities and neuropsychiatric manifestations, including autistic behaviors. Some individuals with CdCS also can present with retinal vessel abnormalities, opticatrophy, or congenital renal and heart defects. Most patients with CdCS, however, do not have other major congenital malformations.

Terminal 5p deletions and the mode of inheritance can be characterized and diagnosed by a combination of microarray and FISH analyses. Loss of several genes in the 5p region contributes to the phenotype. For example, five genes (*TERT*, *SEMA5A*, *MARCH6*, *CTNND2*, *NPR3*) are classified as dosage sensitive leading to haploinsufficiency; six genes (*SLC6A3*, *CDH18*, *CDH12*, *CDH10*, *CDH9*, *CDH6*) have effects that are conditionally haploinsufficient and depend on another genetic or environmental factor to cause an abnormal phenotype; and two genes (*RICTOR*, *DAB2*) have effects that are haplolethal (Table 14.6). FISH analysis can be complemented with other techniques, such as aCGH and quantitative PCR. MLPA combined with aCGH analyses also can be used on prenatal analysis of CdCs.

Cardiovascular Disease (CVD)

Cardiovascular disease (CVD), a leading cause of morbidity and mortality in developed countries, has a significant genetic background. A small portion of CVD follows monogenic Mendelian pattern. Hypercholesterolemia and Tangier disease generally are well recognized and their genetic backgrounds have been explained. The vast majority of CVD is of a polygenic character, meaning that the disease has a late onset, a mild genotype, and inheritance based on the presence of risky alleles of many common polymorphisms with relatively minor effects.

With increasing age, arterial walls typically undergo gradual stiffening, thickening, accumulation of atherosclerotic plaques, and deposition of calcium. Structural changes, such as vascular thickness, calcification, plaque composition, and pulse wave velocity, can be measured using ultrasound or computed tomography. Functional assessment of vasodilator capacity can be assessed by flow-mediated dilatation. Several such markers of cardiovascular aging have been examined in GWAS. Through the central concepts of GWAS, the GWAS of aging-related cardiovascular phenotypes and diseases have identified genetic loci that are important to cardiovascular disease. To date, meta-analysis of GWASs of coronary disease, including more than 60,000 cases and 130,000 controls, have identified 46 genetic loci (Table 14.7). Of these, 16 loci also are consistently associated with known risk factors in a consistent direction: 10 also have been associated with increased LDL, five with increased blood pressure (of which one is also associated with lower LDL and HDL), four with increased triglycerides, and seven with decreased HDL. Other SNPs have not been associated with traditional risk factors. Interestingly, none of the SNPs associated with type 2 diabetes or other glucometabolic traits such as fasting glucose and insulin, HOMA-B and HOMA-IR, has been associated with coronary disease. Although the exact genes have not been identified, the majority of SNPs act through novel pathways and thus have the potential to point to novel mechanisms in the pathophysiology of coronary artery disease.

TABLE 14.6 Chromosome 5p Genes With Haploinsufficient and Conditionally Haploinsufficient Effects

	Locus	Cytogenetic Location	Gene Symbol	Gene Name	OMIM
Haploinsufficient	1,253,281-1,295,177	5p15.33	TERT	Telomerase reverse transcriptase	187270
	9,035,137-9,546,23	5p15.31	SEMA5A	Semaphorin 5A	609297
	10,353,750-10,440,499	5p15.2	MARCH6	Membrane-Associated Ring Ch finger protein	613297
	10,971,951-11,904,154	5p15.2	CTNND2	Catenin, delta-2	604275
	32,710,283-32,791,829	5p13.3	NPR3	Natriuretic peptide receptor C	108962
Conditionally Haploinsufficient	1,392,790-1,445,428	5p15.33	SLC6A3	Solute carrier family 6 (neurotransmitter transporter, dopamine), member 3	126455
	19,473,139-20,575,971	5p14.3	CDH18	Cadherin 18	603019
	21,750,869-22,853,730	5p14.3	CDH12	Cadherin 12	600562
	24,487,208-24,645,084	5p14.2-p14.1	CDH10	Cadherin 10	604555
	26,880,708-27,038,688	5p14.1	CDH9	Cadherin 9	609974
	31,193,761-31,329,252	5p13.3	CDH6	Cadherin 6	603007

MITOCHONDRIA DISORDERS

Mitochondria are versatile organelles in continuous fusion and fission processes in response to various cellular signals. Mitochondrial dynamics, including mitochondrial fission/fusion, movements and turnover, are essential for the mitochondrial network quality control. Because of the important functions of mitochondria in cell homeostasis, mitochondrial dysfunctions cause a great variety of diseases that can affect almost all the tissues and organs in the body.

All mitochondrial fission and fusion proteins are nuclear-encoded. Generally, fission allows segregation of damaged mitochondria and fusion grants material exchange between functional mitochondria and damaged mitochondria, establishing the integrity of the entire mitochondrial network. Therefore, alterations in mitochondrial dynamics often play a crucial role in mitochondrial disorders (Fig. 14.18).

Primary mitochondrial disorders represent a diverse group of diseases associated with dysfunction of the mitochondrial respiratory chain because of mutations in nuclear DNA or mitochondria DNA. The presentation usually depends on generalized or tissue-specific decreases in ATP production. In addition, the vast majority of mitochondrial diseases are characterized by an increased ROS production leading to oxidative stress. Increased mitochondrial ROS production includes damage of mitochondrial and/or nuclear DNA and RNA, protein and lipid oxidation. Some mitochondrial disorders affect a single organ, but many affect most of them. The range of symptoms of mitochondrial diseases is broad and includes neuropathies, myopathies, cardiovascular disorders, and gastrointestinal, endocrine, and hematological alterations.

TABLE 14.7 Coronary Artery Disease Genetic Loci With Genome-Wide Significant Association

Clinical Coronary Artery Disease

1p13 (SORT1) rs602633[a,d] $P=1\times10^{-25}$	3q22 (MRAS) rs9818870 $P=3\times10^{-9}$	6q26 (PLG) rs4252120[f] $P=5\times10^{-10}$	10q23 (LIPA) rs2246833 $P=9\times10^{-6}$	17p11 (PEMT) rs12936587 $P=1\times10^{-9}$
1p32 (PPAP2B) rs17114036 $P=6\times10^{-12}$	4q31 (EDNRA) rs1878406 $P=3\times10^{-8}$	7p21 (HDAC9) rs2023938 $P=5\times10^{-8}$	10q24 (CYP17A1) rs12413409[e] $P=6\times10^{-8}$	17p13 (SMG6) rs2281727 $P=8\times10^{-9}$
1p32 (PCSK9) rs11206510a $P=2\times10^{-5}$	4q32 (GUCY1A3) rs7692387 $P=3\times10^{-11}$	7q22 (COG5) rs12539895 $P=5\times10^{-4}$	11q22 (PDGFD) rs974819 $P=4\times10^{-11}$	17q21 (GIP) rs15563[f] $P=9\times10^{-6}$
1q21 (IL6R) rs4845625 $P=4\times10^{-10}$	5q31 (SLC22A4) r rs273909 $P=1\times10^{-9}$	7q32 (ZC3HC1) rs11556924[e,f] $P=7\times10^{-17}$	11q23 (APOA1) rs9326246[a,c,d] $P=2\times10^{-7}$	19p13 (LDLR) rs1122608[a] $P=6\times10^{-14}$
1q41 (MIA3) rs17464857 $P=6\times10^{-5}$	6p21 (ANKS1A) rs12205331[d] $P=4\times10^{-5}$	8p21 (LPL) rs264[c,d] $P=3\times10^{-9}$	12q24 (SH2B3) rs3184504[b,d,e,f] $P=5\times10^{-11}$	19q13 (APOE) rs2075650[a,c,d] $P=6\times10^{-11}$
2p11 (VAMP8) rs1561198 $P=1\times10^{-10}$	6p21 (KCNK5) rs10947789 $P=1\times10^{-8}$	8q24 (TRIB1) rs2954029[a,c,d] $P=5\times10^{-9}$	13q12 (FLT1) rs9319428 $P=7\times10^{-11}$	21q22 (KCNE2) rs9982601 $P=8\times10^{-17}$
2p21 (ABCG5) rs6544713[a] $P=2\times10^{-9}$	6p24 (ADTRP) rs6903956 $P=5\times10^{-12}$	9p21 (CDKN2A/ CDKN2B) rs1333049 $P=1\times10^{-52}$	13q34 (COL4A1– COL4A2) rs4773144 p−1×10^{-11}	
2p24 (APOB) rs515135[a] $P=310^{-10}$	6p24 (PHACTR1) rs9369640 $P=8\times10^{-22}$	9q34 (ABO) rs579459[a] $P=3\times10^{-8}$	14q32 (HHIPL1) rs2895811 $P=4\times10^{-10}$	
2q22 (ZEB2) rs2252641 $P=5\times10^{-8}$	6q23 (TCF21) rs12190287 $P=5\times10^{-13}$	10p11 (KIAA1462) rs2505083 $P=1\times10^{-11}$	15q25 (ADAMTS7) rs7173743 $P=7\times10^{-13}$	
2q33 (WDR12) rs6725887[f] $P=1\times10^{-15}$		10q11 (CXCL12) rs501120 $P=2\times10^{-9}$	15q26 (FURIN) rs17514846[e] $P=9\times10^{-11}$	

Coronary Artery Calcium

Chromosome	Candidate Gene	SNP	P-Value
6p24	PHACTR1	rs9349379	$P=3\times10^{-12}$
9p21	CDKN2A/CDKN2B	rs1333049	$P=8\times10^{-19}$

[a]10 loci associated with higher LDL for CAD risk alleles.
[b]1 locus with lower LDL.
[c]4 loci with higher TG.
[d]7 loci with lower HDL.
[e]5 loci with higher blood pressure.
[f]5 loci with coding variants at r^2 > 0.9 to index SNP in 1000 Genomes CEU.

Chromosomal loci, candidate genes, SNP, and p-values at the 46 genetic loci associated with coronary artery disease in genome-wide association studies ($P<5\times10^{-8}$). Adapted from the CARDIoGRAMplusC4D study (up to >50,000 cases and >130,000 controls) and a study of coronary artery calcium on computed tomography (nearly 16,000 subjects). One locus (6p24 in ADTRP) was reported separately in a Chinese study.

DISEASES CAUSED BY MUTATIONS IN MITOCHONDRIAL DYNAMICS MACHINERY

Dominant Optic Atrophy (DOA)

DOA is believed to be the most common hereditary optic neuropathy that severely impairs vision. DOA is caused by a loss of retinal ganglion cells located only in the inner retina and projecting their axons via the optic nerve to the brain. Retinal ganglion cell loss and atrophy of the optic nerve are accompanied by weakening of the optic nerve and the characteristic fundus with pallor of the optic disc.

Mutations in two genes—*OPA1* and *OPA3*—and three loci—*OPA4, OPA5, OPA8*—are known for DOA. OPA1 is a multifunctional protein located within the mitochondrial inner membrane, which regulates a number of critical cellular

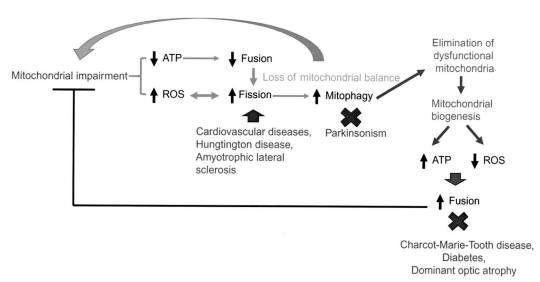

FIG. 14.18 Schematic diagram demonstrating the role of mitochondrial dynamics in mitochondrial diseases. Mitochondrial dysfunction and reactive oxygen species (ROS) production induces decreased mitochondrial fusion and mitochondrial fragmentation and mitophagy. Elimination of dysfunctional mitochondria, associated with increased mitochondrial biogenesis, restores ATP levels, decreases ROS production, and increases mitochondrial fusion and function.

functions, including mitochondrial fusion, mitochondrial DNA maintenance and apoptosis. The OPA1 mutations can affect mitochondrial fusion, energy metabolism, apoptosis, calcium levels, and maintenance of mitochondrial genome integrity. Mitochondrial cristae remodeling and fusion are functionally distinct from each other, however, they can be correlated with apoptotic cytochrome c release, which can be prevented by OPA1 overexpression. The lack of OPA1 function results in a misbalance toward mitochondrial fission that leads to mitochondrial network fragmentation, impaired mitochondrial oxidative phosphorylation, increases in reactive oxygen species, and altered calcium homeostasis.

The analysis of *OPA1* gene can be performed on the DNA sample of the index patient by amplifying and sequencing all the 31 coding exons and their flanking intronic regions. If a mutation is identified, its segregation in the family must be analyzed, and its identity has to be compared to the database to find out whether the mutation already is recognized as pathogenic or not. If not, the possible consequence of the mutation on *OPA1* transcript and protein integrity can be analyzed in silico, and by assessing the expression of the mutated allele by RT-PCR amplification and sequencing. If no significant mutation is found in *OPA1*, then the presence of a deletion in the gene can be tested with the MLPA. Otherwise either *OPA3* gene or the full-length mitochondrial genome should be considered for the test. If results are still negative, then when the family is large and many members are affected, genetic analysis of chromosome markers can be performed to identify the causative locus and eventually a novel pathogenic gene. Nevertheless, the identification of a morbid mutation greatly helps the genetic counseling.

Mitochondrial Dynamics in Neurodegenerative Diseases: Parkinson's Disease

Neuronal development requires both mitochondrial dynamic proteins Drp1 and mitofusin-2. In addition, most neurodegenerative diseases, such as Parkinson's disease, show abnormal mitochondrial dynamics.

Some cases of parkinsonism are caused by autosomal recessive mutations in mitophagy related proteins, such as PINK1 and Parkin. Mutations in one or both proteins impair mitophagy and cause neurons to accumulate Drp1, resulting in an excessive mitochondrial fission, oxidative stress and reduced ATP production.

Parkinson's disease (PD) is a common, age-related progressive neurodegenerative disease, affecting more than 1% of the population older than 60 years of age. Diagnosis of PD relies on clinical history, physical examination, and the response to dopaminergic drugs, but misdiagnosis is common in early phases of the disease. In PD, degeneration of dopaminergic neurons between the substantianigra and the striatum occurs. As a result, a great majority of dopamine-producing cells in the substantianigra is lost in patients with PD. Therefore, the symptoms of PD are trembling in hands, arms, legs, and face; stiffness of the limbs and trunk; slowness of movement; and impaired balance and coordination. As these neurons are progressively destroyed, patients can have difficulty walking, talking, and completing simple tasks. PD usually affects people older than 60.

PD is inherited in a Mendelian autosomal dominant or autosomal recessive fashion in a small number of families. No standard diagnostic test exists for Parkinson's; the best objective test for PD currently consists of specialized brain scanning techniques to measure the dopamine system and brain metabolism. With the help of GWAS and SNP analysis, some Parkinson's disease susceptibility genes have been identified. Mutations in α-synuclein (*SNCA*) and leucine-rich repeat kinase2 (*LRRK2*) genes are responsible for late-onset disease. Parkin (*PARK2*), ubiquitin carboxy-terminal hydrolase L1 (*UCH-L1*), PTEN Induced Putative Kinase1 (*PINK1*), oncogene DJ1 (*DJ1*) are responsible for early onset. Point mutations, duplications, and triplication in *SNCA*, located on chromosome 4, are characteristic of PD. They occur in most forms, including the rare early onset familial form of PD. More than 70 mutations on the large *PARK2* have been associated with the early-onset form of Parkinson's. Mutations in *PARK2* might account for PD in as many as 50% of familial cases of autosomal recessive juvenile PD. *UCH-L1* is located on chromosome 4 encodes a protein that belongs to the family of deubiquitinating enzymes. Protein UCH-L1 constitutes 1% of brain protein, and its function is presumed to recycle ubiquitin by hydrolyzing the ubiquitinated peptides. This enzyme plays a role in modifying the damaged proteins that otherwise might accumulate to toxic levels in the neuron. Two homozygous mutations in *PINK1* gene associated with PD found in Spanish and Italian families. As such, molecular diagnosis can focus on these genes' mutations through PCR, MLPA or NGS sequencing.

Chapter 15

Molecular Diagnosis of Infectious Diseases

Chapter Outline

DETECTION OF BACTERIAL PATHOGENS, FUNGAL PATHOGENS, AND VIRUSES

The detection of bacteria traditionally has been based upon the phenotypic properties of the microorganism grown in pure culture under different specific conditions. Although they are useful in most circumstances, the physiological characteristics of bacteria are mutable and not always consistent within a given species. DNA-DNA hybridization techniques have long been used to classify prokaryotes. However, this method is time-consuming and impractical in the clinical laboratory. With recent advancement in molecular diagnostic techniques, nucleic acids sequence-based diagnosis, also known as genotyping assays, including the various nucleic acid amplification techniques (NAATs), microarray, NGS, and FISH, can provide simultaneous detection and differentiation of numerous microbial pathogens. DNA sequence-based identification can replace the performance of many time-consuming and labor-intensive biochemical reactions. The average time spent per specimen is also dramatically reduced, allowing medical laboratory scientists more time to accomplish other necessary tasks. This is especially important in the current environment, in which there is a growing shortage of well-trained medical laboratory professionals. Furthermore, 16S rDNA PCR with sequencing is a useful supportive tool for the diagnosis of the causing agents, especially for cases of culture-negative samples. PCR and sequencing assays are particularly useful in the diagnosis of microorganisms in patients who have had prior antimicrobial therapy, had tested positive by blood culture owing to suspected contaminants, or are predisposed towards rare infections such as *Bartonella* endocarditis owing to homelessness and alcoholism. As such, current DNA-based molecular diagnosis has become powerful and more accurate and reliable than before.

Diagnosis Using NAATs

The molecular detection of pathogens directly in clinical specimen such as blood, tissue, bronchoalveolar lavage fluid, and other body fluids most often is accomplished through PCR-based amplification. This technique can overcome low sensitivity to improve the specificity of conventional phenotypic tests. As such, they have been expected to provide early diagnosis for proper treatment with potential reduction in mortality and morbidity. Technical advances are making these diagnostic tests faster and less expensive.

Diagnostic Molecular Biology. https://doi.org/10.1016/B978-0-12-802823-0.00015-8

All PCR assays have been designed to detect a low load of a given microorganism among a huge number of human cells, but many factors still can affect the performance of NAATs. For example, preanalytical factors, such as specimen collection, transport, and nucleic acid extraction, can influence the ability of all amplification techniques to detect bacterial nucleic acid. Conventional manual nucleic acid extraction for the isolation of pathogen DNA or RNA from clinical samples is the most labor-intensive and critical part in current molecular diagnosis. With care, the quality of nucleic acids for NAATs can be improved dramatically. Furthermore, many different types of PCR formats have been developed for analysis and they can have different goals to differentiate different species. For example, the small amount of fungal DNA in clinical specimens can use nested PCR for identification of *Aspergillus* and *Candida*.

Because amplification of a gene with several hundred copies improves PCR assay sensitivity compared to that of tests amplifying single-copy genes, most novel PCR assays have targeted multicopy rRNA genes. For example, 16S rRNA and 23S rRNA are commonly used for detection of bacterial pathogens, whereas 18S rRNA and 28S rRNA can be used to detect fungal pathogens (Table 15.1)

Although the significant advantages and strengths that PCR assays offer in terms of rapid detection, sensitivity and quality are two major concerns common to all assays, regardless of the target detected. It is important that the type of specimens acceptable for assaying must be delineated, as well as the physical criteria for the specimen's suitability for analysis, including optimum sources and volume, appropriate collection method, transport and storage conditions, and specimen longevity. The choice of specimens plays a key role in the performance and interpretation of test results because if any of the aforementioned criteria are not fulfilled, the sensitivity and specificity of the assay will vary accordingly.

Decreased sensitivity with clinical specimens is one of the main concerns about using NAATs. During the laboratory procedure, it is very common that the inhibitory substances in the specimen are carried through the extraction process and

TABLE 15.1 Commonly Used Bacterial 16S and 23S rRNA Primers for Broad-Range PCR to Identify Bacterial Phyla

rRNA	Identifier	Sequence 5'–3'	Location	Specificity
16S	27 forward	AGAGTTTGATYMTGGCTCAG	8–27	Most Bacteria
	355 forward	ACTCCTACGGGAGGCAGC	338–355	Most Bacteria
	338 reverse	GCTGCCTCCCGTAGGAGT	338–355	Most Bacteria
	533 forward	GTGCCAGCMGCCGCGGTAA	515–533	Universal
	515 reverse	TTACCGCGGCKGCTGGCAC	515–533	Universal
	556 reverse	CTTTACGCCCARTRAWTCCG	556–575	Most Bacteria
	806 forward	ATTAGATACCCTGGTAGTCC	787–806	Most Bacteria
	787 reverse	GGACTACCAGGGTATCTAAT	787–806	Most Bacteria
	926 forward	AAACTYAAAKGAATTGACGG	907–926	Most Bacteria
	907 reverse	CCGTCAATTCMTTTRAGTTT	907–926	Most Bacteria
	930 forward	TCAAAKGAATTGACGGGGGC	911–930	Most Bacteria
	911 reverse	CCGTCAATTCMTTTRAGTTT	911–930	Most Bacteria
	1194 forward	GAGGAAGGTGGGGATGACGT	1175–1194	Most Bacteria
	1175 reverse	ACGTCATCCCCACCTTCCTC	1175–1194	Most Bacteria
	1371 reverse	AGGCCCGGGAACGTATTCAC	1390–1371	Most Bacteria
	1391 reverse	TGACGGGCGGTGWGTRCA	1391–1408	Universal
	1492 reverse	GGTTACCTTGTTACGACTT	1510–1492	Universal
	1525 reverse	AAGGAGGTGWTCCARCC	1525–1541	Most Bacteria
23S	1077 forward	AGGATGTTGGCTTAGAAGCAGCCA	1054–1077	Most Bacteria
	1950 reverse	CCCGACAAGGAATTTCGCTACCTTA	1950–1926	Most Bacteria

interfere with amplification. Another laboratory performance that can affect the sensitivity and quality is that organisms taken directly from humans by means of a clinical sample are more resistant to lysis or are more buoyant and thus are more difficult to extract nucleic acid from or concentrate by centrifugation than broth-grown organisms. In addition to monitoring for inhibition, it is imperative that nucleic acid extraction protocols be proven efficient and optimized for the particular specimen type being assayed.

The selection and use of appropriate controls is also important for PCR analysis. First, an appropriate number of negative controls should be included, to be processed along with patient specimens to monitor for the contamination. Other controls to monitor the efficacy of nucleic acid extraction are important. Furthermore, in light of amplification inhibitors, inhibition must be monitored for each sample either by use of internal control or by spiking the native specimen with target nucleic acid. Also, measures such as the design of the workspaces, direction of workflow, and other contamination control measures should be addressed prior to PCR analysis.

Diagnosis With Sequencing

DNA-DNA hybridization techniques have long been used to classify prokaryotes, but this method is time-consuming and might be impractical in the clinical microbiology laboratory. Nucleic acid sequencing of various bacterial genes and other DNA targets, with advances in technology in the past decade, this approach has become popular in the molecular diagnosis. DNA target sequencing often is faster and more accurate in identifying bacteria, with the added benefit of being independent of a microorganism's growth characteristics.

The 16S rRNA molecule contains alternating regions of sequence conservation and heterogeneity (Fig. 15.1). This alternating sequence feature makes it well suited as a target for sequence analysis. The conserved regions are ideal primer targets for amplification from all bacterial species. Regions of DNA sequence diversity between these conserved regions provide sequence polymorphisms that serve as "signatures" unique to a genus or species. The signature sequence obtained is compared to a database containing sequences of known microorganisms. The number of similar nucleotide bases between sequences is used to calculate the percent identity and ascertain the identification of the microorganism. While this strategy is adequate for identification of many bacterial species, the degree of divergence observed within the 16S rRNA molecule is not sufficient to distinguish some closely related species.

The rRNA genes and their intergenic regions found in bacteria are commonly used for prokaryotic phylogenetic studies. The small-subunit rRNA molecule is a fragment with a sedimentation coefficient of 16S and is encoded by a roughly 1542-bp gene. The large-subunit rRNA contains 23S and 5S molecules. Partial sequencing of the 16S rRNA gene, particullarly the first 500 bp, is most frequently used for bacterial identification in the clinical laboratory, including anaerobes and mycobacteria. Most sequences that have been deposited in publicly available databases correspond to this region of the 16S rRNA gene.

Other DNA targets have been used to better separate closed related species include *rpoB* (beta subunit of RNA polymerase), *sodA* (manganese-dependent superoxide dismutase), *gyrA* or *gyrB*, tuf (elongation factor Tu), *recA*, *secA*, and heat shock proteins. The utility of each target varies depending on the microorganism. Similar to 16S rRNA gene, these alternative DNA targets have conserved regions flanking variable regions that can be used to differentiate closely related bacterial species.

For the hepatitis B virus, the Siemens Trugene HBV genotyping assay uses direct sequencing for HBV genotyping. It derives the sequence from domains B through E of the viral polymerase from a single amplicon. The advantage for this

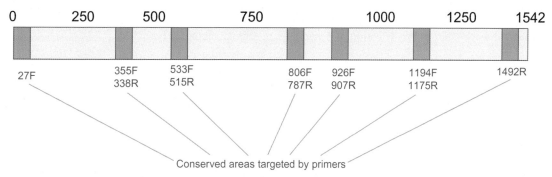

FIG. 15.1 Sequence structure of 16S rRNA. Highlighted region indicates the conserved regions that serve as primer target for PCR amplification and DNA sequencing of bacteria.

approach is its ability to detect mutations relevant to drug resistance, phylogenetic genotype, and HBsAg (surface antigen) alteration in a single assay. Assay-chemistry allows simultaneous bidirectional sequencing of a PCR amplicon in the same tube using different end-labeled primers for each sequencing reaction. The sequence ladder can be resolved using either slab gel or capillary electrophoresis. In the case of the hepatitis C virus, the Siemens Trugene HCV 5′NC assay is based on nucleotide sequence analysis of 5′ UTR of the genome. Other regions that are used for sequencing methods are NS5B, core, and core-E1 regions of the genome.

DETECTION OF BACTERIA

Bartonellae Species

Bartonellae are fastidious, facultative, intracellular zoonotic, and arthropod vector-borne bacteria distributed among mammalian reservoirs worldwide. Because of their hemotropic lifestyle, *Bartonellae* typically are transmitted by blood-sucking arthropods within mammalian reservoir communities. The genus *Bartonella* contains more than 30 species of bacteria. During the past 20 years, the number of *Bartonella* species associated with human disease has risen from two to at least 14. Sand flies, lice, fleas, biting flies (e.g., *Hippoboscidae*, *Muscidae*), and ticks are among the arthropods associated with *Bartonella* transmission and/or infection. Special attention has been given to this bacterial genus because of the pathogenicity exhibited by many of its species, including the two anthropogenic *Bartonellae* (*B. bacilliformis* and *B. quintana*) and the zoonotic species (*B. henselae*, *B. grahamii*, *B. elizabethae*, *B. koehlerae*, and *B. rochalimae*).

The range of symptoms associated with *Bartonella* infections includes fever, malaise, headache, rash, lymphadenopathy, psychiatric disorder, bone and joint pain, anemia, and various inflammatory and vascular proliferative disorders. Although a single species can cause a wide range of clinical presentations, the diseases caused by different species also can overlap. For example, *B. quintana* has been known to cause trench fever since the early twentieth century, but it was identified as a cause of culture negative endocarditis in 1993. Subsequently, human endocarditis has been linked to six additional *Bartonella* species and subspecies: *B. henselae*, *B. elizabethae*, *B. vinsonii* subsp. *berkhoffii*, *B. vinsonii* subsp. *arupensis*, *B. koehlerae*, and *B. alsatica*.

The pathogenic potential of many *Bartonella* species has increased the interest in these bacteria. Isolation of *Bartonella* using classical bacteriological methods is laborious and requires specific conditions and prolonged incubation periods. Serology has been most commonly used to diagnose *Bartonella* infections. Serological assays for *B. henselae* and *B. quintana* are widely available, and serodiagnostics for a few other *Bartonella* species have been developed, but serology cannot reliably discriminate among *Bartonella* species.

Molecular methods for detection of *Bartonella* DNA have been considered as more practical and sensitive than classical and serological analysis. This because *Bartonella* bacteria emerge constantly and their importance in public health increases, rapid diagnostic tools are required. The use of real-time PCR assays for the initial screening of *Bartonella* spp., followed by several molecular confirmatory assays targeting several loci (preferably of longer fragments than those amplified by the realtime assay), has become more and more popular. Genus- and/or species-specific PCRs directed against *gltA*, *rpoB*, *nuoG*, *htrA*, *ribC*, *pap31*, the 16S rRNA gene, *ssrA*, and the internal transcribed spacer can be used to distinguish *Bartonella* species (Table 15.2). To differentiate between *Bartonella* species, the use of real-time PCR assays for primary screening followed by several molecular confirmatory technologies, including restriction fragment length polymorphism analysis, PCR-based enzyme immunoassay, dot blot, DNA fingerprinting, amplicon size analysis, high resolution melt realtime PCR, and sequencing. While direct sequencing of a PCR product is likely the most straightforward and specific of these technologies, conventional sequencing is time consuming and prone to contamination issues. Alternatively, NGS sequencing, such as pyrosequencing, provides a rapid and clinically relevant sequencing technology. Fig. 15.2 illustrates the recommended work-chart in the detection of *Bartonella* in animal samples.

Borrelia Species

The infectious disease most frequently transmitted by ticks in North America, Europe, and Asia is Lyme borreliosis. The *Borrelia* are transferred to the skin during the blood-sucking process of the hard-bodied tick. There, the *Borrelia* either are killed off by the (unspecific, innate) immune system, or a localized infection occurs, leading to illness in a small percentage of those infected. Most often, there is an inflammation of the skin, typically in the form of an erythema migrans or, seldom, as borrelial lymphocytoma. In the course of the infection, the *Borrelia* can disseminate and attack various organs, primarily the skin, joints, and nervous system. Acrodermatitis chronica atrophicans can develop as a chronic or late-form of skin manifestation.

TABLE 15.2 Commonly Used PCR Targets for Detection of *Bartonella spp.*

Targets	*Bartonella* Locus	Assay Type
Molecular characterization of isolates from cultures	Citrate synthase *gltA*	Conventional PCR
	β-subunit of RNA polymerase *rpoB*	Conventional PCR
	16S-23S Internal transcribed spacer ITS	Conventional PCR
	16S ribosomal RNA gene *16S*	Conventional PCR
	Riboflavin synthase *ribC*	Conventional PCR
	NADH dehydrogenase gamma subunit *nuoG*	Conventional PCR/real-time PCR
	Cell division protein *ftsZ*	Conventional PCR
	Heat shock protein *groEL*	Conventional PCR
Direct molecular detection (from blood/tissues/arthropods)	mtRNA *ssrA*	Real-time PCR
	16S-23S Internal transcribed spacer ITS	Real-time PCR
	β-subunit of RNA polymerase *rpoB*	Real-time PCR

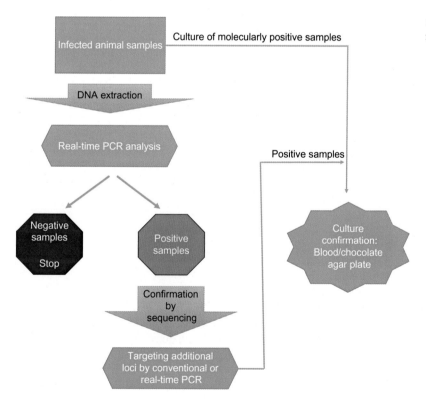

FIG. 15.2 A schematic diagram of the flow chart for molecular diagnosis of *Bartonella* in animal samples.

Five human-pathogenic genospecies from the *Borrelia burgdorferi sensu lato* complex have been found in Europe. *B. afzelii* is the most frequent, followed by *B. garinii*, *B. bavariensis*, *B. burgdorferi sensu stricto*, and *B. spielmanii*. The human pathogenicity is still unclear for *B. valaisiana*, *B. lusitaniae*, and *B. bissetii*. All of the species that have been ascertained to be human-pathogenic are found in Europe. Only *B. burgdorferi sensu stricto* is present in the United States, and all of the species, except for *B. burgdorferi sensu stricto*, are found in Asia. The various genospecies of the *Borrelia burgdorferi sensu lato* complex are genetically heterogenic and exhibit an organotropism in human infections. Erythema migrans is triggered by all five genospecies. *B. afzelii* is detected with acrodermatitis chronica atrophicans, *B. garinii* and *B. bavariensis* often are present in neurological manifestations, and *B. burgdorferi sensu stricto* mainly affects the joints. *B. spielmanii* has been isolated only from erythema migrans.

B. burgdorferi is transmitted to birds, mammals, and humans from hard-bodied ticks of the *I. ricinus/I. persulcatus* species complex during the blood meal. In Europe, this transmission is primarily from *I. ricinus*, in Asia from *I. persulcatus*, and in the United States predominantly from *I. scapularis*. Ticks suck blood in the course of their cycle of development from larva to nymph to adult tick, and before they lay eggs. It is at this time that they can acquire and/ or transmit *Borrelia*. Small rodents, particularly mice, and birds are the main reservoirs. Birds contribute to the geographical propagation of infected ticks. In Germany, ticks are ubiquitously infected with *Borrelia*, however percentages can vary heavily from region to region, even between areas close in proximity.

Various methods can be applied to detect the presence of *Borrelia* in vectors. Widely used approaches that demonstrate significant sensitivity, specificity, and reliability. For example, nested PCR targets different genomic loci, selection of which depends on the sample origin (template). Reverse-line blotting is based on hybridization of amplified selected *Borrelia* genes with spirochete-specific probes. Multilocus sequences analysis and multilocus sequence typing is based on the sequence analysis of amplified fragments of spirochete genome or microscopy with stained spirochetes in tick midgut or salivary glands. PCR provides a valuable diagnostic approach in acutely ill patients. This overcomes the poor sensitivity of microscopy and can be used either to diagnose relapsing fever borreliosis or to further characterize the infecting spirochete. The absence of GlpQ, a glycerophosphodiester phosphodiesterase identified previously in *E. coli*, *Haemophilus influenzae*, and *Bacillus subtilis*, in Lyme *borreliosis* species makes it a specific target for detection of relapsing fever spirochetes. Other assays either can spirochete specific relapsing fever *borreliae* or be designed to detect a single member of the relapsing fever clade such as *B. miyamotoi*.

The most recently applied techniques include NGS and proteomic approaches. Although NGS offers huge potential, the challenge is which diagnostic sample type to investigate for Lyme *borreliosis* in the absence of focal lesions. Sensitivity can be improved, especially among high levels of host DNA. Care should be taken to avoid bias when using target enhancement strategies to amplify low copy-number targets. Data analysis represents an additional computational challenge.

Chlamydia trachomatis

Chlamydia, with its *Chlamydia trachomatis* etiology, is the most common bacterial sexually transmitted infection (STI) in the United States. In contrast to *Neisseria gonorrhoeae* infection in which most patients develop symptoms and seek care promptly, most females and males with *C. trachomatis* infection were asymptomatic or mildly symptomatic upon clinical presentation. Diagnosis was largely on the basis of screening or having a symptomatic contact. Chlamydia is a prevalent disease rather than an incident disease comes from extrapolations of STI agent acquisition rates.

Sexually active asymptomatic populations have been implicated in widespread transmission of the *C. trachomatis* etiology. Selective screening of sexually active women has yielded infection rates ranging from 8% to 40% (typical mean of 15%), while approximately 10% of sexually active asymptomatic males are infected. Data from the CDC have revealed that chlamydia rates are highest in late adolescents and young adults. African Americans and Native Americans normally have higher rates of chlamydia compared to other races or ethnicities.

For molecular diagnosis, three main targets have been used for detection of *C. trachomatis*: the cryptic plasmid, genomic sequences, or rRNA (16S rRNA and 23S rRNA), with sequences on the cryptic plasmid being the most common. The advantage of using target sequences present on the plasmid is the intrinsic amplification caused by its presence in multiple (7–10) copies. The advantage of using rRNA as the target is the presence of high numbers, leading to an intrinsic amplification, and being indispensable for the organism. Thus, the likelihood for genetic changes leading to false-negative results is minimal.

There are several reasons and applications for typing *C. trachomatis* to examine association between types and clinical manifestations and pathogenicity. For example, it can be used in investigation of sexual assaults, for analysis of transmission patterns in sexual networks, and for examination of persons with repeatedly positive Chlamydia tests to examine for persistence versus reinfection.

Typing of *C. trachomatis* has been based mainly on discrimination of the 15 different serovars encoded by the *ompA* (*omp1*) gene. Molecular genotyping of *ompA* can be performed by using restriction fragment length polymorphism on PCR products from culture isolates, but also directly from clinical specimens.

Increased possibilities for genotyping can be achieved by multiplex real time PCR and hybridization arrays. All these methods can discriminate mixed infections and provide sensitive detection and typing of *C. trachomatis* directly from clinical specimens. The nine polymorphic membrane protein genes, *pmpA* to *pmpI*, can be used for typing but the discriminating capacity is limited. The highly conserved genome of *C. trachomatis* has given difficulties in developing discriminating typing methods based on single genes, but improved resolution can be achieved using multilocus target systems. Analysis of variable numbers of tandem repeats in three loci combined with *ompA* sequencing can bring a significantly higher diversity index than by using *ompA* alone.

Clostridium difficile

C. difficile is the major causative agent of antibiotic-associated diarrhea, colitis, and pseudomembranous colitis and is the major recognized cause of nosocomial diarrhea. Concomitant with the increase of *C. difficile* infection (CDI) in adults in the past decade, the rates of hospitalized children with the infection also have been increasing. The number of cases per patient-days in the hospital has increased, and overall there has been a rise in hospitalizations attributable to *C. difficile* in children. While the rates of CDI have increased, the severity of disease (e.g., colectomy and in-hospital deaths) has not increased in pediatric patients as it has in adults.

C. difficile is an anaerobic, Gram-positive, spore-forming bacillus that can be found in the environment and the gastrointestinal tract of animals and humans. The pathogenicity of *C. difficile* is related closely to the production of toxins A and B. Outbreaks of *C. difficile* infections with increased severity, high relapse rate, and significant mortality have been related to the emergence of a new hypervirulent *C. difficile* strain (PCR ribotype 027). Toxinogenic *C. difficile* detection by tissue culture cytotoxin assay often is considered the best method. Several real-time PCRs targeting the *C. difficile* toxin genes including *tcdA*, *tcdB* and *tcdC117* have been developed.

A new multiplex PCR for the detection of *tcdA*, *tcdB*, and the binary toxin (*cdtA/cdtB*) gene is a one-step, rapid, and specific screening method for *C. difficile* toxin genes. This toxin gene profiling can allow an evaluation of the pathogenic potential of *C. difficile*. Isothermal helicase-dependent amplification for the detection of toxinogenic *C. difficile* also has been developed. Although the NAAT is the most sensitive analytically, questions remain about its use as a first-line, standalone test. The best test or testing algorithms for the diagnosis of CDI and the ability to reliably differentiate disease from colonization are still in debate.

Mycoplasma pneumoniae

Mycoplasma pneumoniae is a common cause of upper and lower respiratory tract infections in children and adults. The organism is spread easily through respiratory droplets and can cause a variety of clinical manifestations, including pharyngitis, tracheobronchitis, and pneumonia. Extrapulmonary manifestations sometimes occur after primary respiratory tract infection, either by direct spread or autoimmune effects. *M. pneumoniae* is recognized as one of the most common pathogens causing community-acquired pneumonia in certain populations, especially in children among whom infection can lead to severe pneumonia requiring hospitalization.

PCR is the most widely applied NAAT for detection of *M. pneumoniae*. PCR also has been adapted to detect antimicrobial resistance determinants and to analyze genetic relatedness of clinical isolates. Gene targets for PCR assays have included 16S rRNA, *P1*, *tuf*, *parE*, *dnak*, *pdhA*, ATPase operon, CARDS toxin gene (mpn372), and the noncoding repetitive element *repMp1*. Analytical sensitivity is generally high, with some assays being capable of detecting a single organism when purified DNA is used.

Both conventional and real-time NAATs have been used to detect *M. pneumoniae* RNA. NAATs can provide rapid results with sensitivity at least as good as PCR and this assay can be used in monoplex and multiplex format. The LAMP assay also has been applied to detect *M. pneumoniae* in clinical specimens using *P1* sequences for primers in direct comparison to real-time PCR. The assay was specific, with a detection limit of 200 copies, and had 100% concordance with PCR. The development of commercial assays using this type of technology for *M. pneumoniae* is anticipated because instrumentation and reagents are undergoing clinical trials.

A multiplex real-time PCR assay can be used to detect point mutations in all three positions of the 23S rRNA gene (2063, 2064, and 2617) that are related to the macrolide resistance on all clinical samples that are positive for *M. pneumoniae* in the repMp1 real-time PCR assay. This assay uses fluorescence resonant energy transfer hybridization probes and this method is based on the fact that nucleic acid will melt at a precise temperature that is related to the nucleotide base composition. The presence of one or more point mutations in 23S rRNA that impair drug attachment to the bacterial ribosome will be detected by this extremely sensitive method, which can be completed in just a few hours. The detection limit is as low as seven mutant molecules per microliter in the PCR mixture.

Rickettsia Species

The genus *Rickettsia* (*Rickettsiales*: *Rickettsiaceae*) includes Gram-negative, small, obligate intracellular, nonmotile, pleomorphic coccobacilli bacteria transmitted by arthropods. Some of them cause human and probably also animal disease that is life-threatening in some patients. The main clinical manifestations of a rickettsial syndrome in humans are fever, rash, and eschar with different combinations, but they are not pathognomonic. Whenever a rickettsiosis is suspected, an early antimicrobial treatment must be started before confirming the infection. Confirmatory assays provide information to retrospectively validate the accuracy of the clinical diagnosis and contribute to the epidemiological knowledge of the pathogenic circulating species.

Molecular methods based on PCR have enabled the development of sensitive, specific, and rapid tools for the detection and identification of *Rickettsia* species in human and animal specimens, including ticks and other arthropods. Different types of specimens can be used for molecular diagnosis of rickettsioses, such as whole blood, buffy coat, skin or eschar biopsies, and eschar swabs. Other samples, such as like organ tissues, cerebrospinal fluid, or pleural fluid, also could be used (Table 15.3). Collection of patient specimens should be performed early in the course of the infection and before the patient initiates specific treatment. PCR appears to be more useful for detecting *rickettsiae* in eschars and skin or organ biopsies than in acute blood because typically low numbers of *rickettsiae* circulate in the blood in the absence of advanced disease or fulminant infection. In specific situations in which there are no other options, molecular detection also can be performed from plasma, serum, paraffin-embedded tissues, or even fixed-slide specimens.

Molecular detection strategies have been based mainly on recognition of sequences from different targets such as 16S rRNA gene and other protein-coding genes: 17-kDa protein (htr), citrate synthase (gltA), and surface cell antigen (sca) autotransporter family, including the outer membrane proteins, ompA and ompB, and the surface cell antigens, sca4 and sca1. The guidelines for taxonomic classification and identification of a new rickettsial species suggest the characterization by at least these five genes: *rrs*, *gltA*, *ompA*, *ompB*, and *sca4*.

Regular PCR assays are used frequently for the characterization or detection of DNA of *Rickettsia* species from culture, arthropods, or eschar biopsies. The use of nested PCR technique for human specimens, such as blood, buffy coat, or plasma with a low level of rickettsiemia, however, is advisable to increase the analytical sensitivity. Real-time PCR for *Rickettsia* species and species-specific detection has been developed. This type of assays offers the advantages of speed, reproducibility, quantitative capability, and low risk of contamination compared to conventional PCR.

Staphylococcus aureus

Staphylococcus aureus is a leading cause of bone and joint infections and one of the most common causative pathogens of bacterial pneumonia in children. An increased risk of mortality from invasive staphylococcal infections has been attributed

TABLE 15.3 Samples Handling Methods for Detection of *Rickettsia* spp.

Specimen	Collection/Storage Method	Transportation	Preservation	Molecular Assay
Whole blood/Buffy coat	EDTA or citrate tube	−20°C, more than 24 hours	−20°C, more than 24 hours	PCR
Serum/plasma	Serum separator tube/ Anticoagulant tube	−20°C, more than 24 hours	−20°C, more than 24 hours	PCR
Skin or eschar biopsy and autopsy organ tissue	Sterile tube (Tissue dry)	−20°C, more than 24 hours	−20°C, more than 24 hours or Process immediately	PCR Culture
Eschar swab	Sterile tube (Dry swab)	2–8°C, more than 24 hours	2–8°C more than 24 hours	PCR/Culture
Formalin-fixed tissue paraffin-embedded tissue	Tube/cassette	Room Temperature	Room temperature	PCR/Immunohistochemistry

to methicillin-resistant *S. aureus* (MRSA) in comparison to methicillin-susceptible *S. aureus* (MSSA) strains. *MecA* and *mecC* are the main genes responsible for the resistance of MRSA to most of the β-lactam antibiotics. The financial burden of MRSA infections is considerable, and this microorganism has been associated with prolonged hospital stays and the occupation of isolation rooms, resulting in extra in-hospital costs.

Because *S. aureus* causes life-threatening invasive disease worldwide and mortality rates for bacteraemia remain high, at 30%, prompt effective antibiotic treatment is important to optimize clinical outcome. Clinicians, therefore, need reliable rapid tests to determine susceptibility. Increasing awareness of MRSA as an agent of nosocomial infection has made disease prevention and control as key targets in many countries. However, the control and management of *S. aureus* infections are hindered by incomplete understanding of pathogenesis and virulence, and by limitations in our ability to characterize highly related strains with sufficient resolution to identify routes of transmission.

S. aureus strains can be assigned to lineages (or clonal complexes) that share characteristics because of common descent. Conventional typing techniques such as multilocus sequence typing and spa typing rely on variation in one or more genes to allocate strains to lineages. The short sequences (400–4000 bp) characterized by conventional typing methods account for only a fraction of the whole genome (c. 2.8 million nucleotides). In suitably equipped laboratories, spa typing can be turned around rapidly, but such methods must be performed as an adjunct to conventional frontline methods of identification and characterization. It can take weeks before informative results can be obtained from a reference laboratory (Fig. 15.3).

PCR is considered to be the best molecular diagnostic tool for MRSA detection. PCR amplification and detection of the methicillin resistance determinant *mecA* confirm the results obtained by conventional culture methods. PCR assays that detect a single target (*mecA*) are both robust and easy to perform, but amplification inhibition can lead to false-negative results. Duplex PCR assay detecting *mecA* and *femB/nuc* genes can be very useful. A triplex PCR targeting 16SrRNA, *mecA*, and *nuc* genes can reach 98% accuracy within 6h of visible growth detection.

The latest development in (direct) MRSA detection and identification is that of real-time PCR. Real-time PCR enables the detection of its generated products even during the amplification process, resulting in a significantly shorter time than conventional PCR, which requires a post amplification PCR product identification step. Moreover, real-time PCR is less sensitive to contamination.

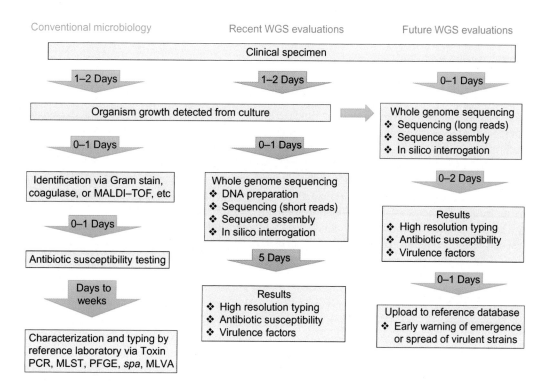

FIG. 15.3 A comparison of methods used to identify and characterize *Staphylococcus aureus*, including (i) existing microbiological methods, (ii) recent evaluated whole genome sequencing (WGS) techniques, and (iii) potential future applications of WGS. *MLST*, multilocus sequence typing; *MLVA*, multiple-locus variable-number tandem-repeat analysis; *PFGE*, pulsed-field gel electrophoresis.

As a frontline tool, whole genome sequencing (WGS) offers the prospect of a method that could be used to predict resistance, assess virulence, and type isolates at the highest possible resolution. The results could be used to guide clinical management and infection control practice. Therefore, WGS offers the prospect of a portable high-resolution typing method for surveillance. Coupled with syndromic surveillance, this would allow precise monitoring of strains and assessment of their clinical significance, and could provide early warning of emerging pathogenic clones. The variety of conventional molecular methods has resulted in differing strain nomenclature and difficulties in communication of data. Standardized WGS nomenclature, data transfer, and interpretative algorithms could provide an enhanced global surveillance system. Early identification of newly emerging strains could allow the prompt implementation of targeted infection control resources.

Streptococcus pneumonia

Streptococcus pneumoniae is a Gram-positive coccus, surrounded by a polysaccharide capsule that allows the microorganism to avoid phagocytosis, representing a major virulence factor. Capsular variability also permits different subtypes to avoid immune detection by antibodies previously generated by infection or administration of vaccine. Some capsular types are less immunogenic and therefore associated with prolonged asymptomatic carrier status. The ability to adhere to mucosal linings is another important virulence factor.

S. pneumoniae is one of the major contributors to mortality and morbidity worldwide. It causes a variety of diseases ranging from simple respiratory infections, such as otitis media and pneumococcal pneumonia, to reach threatening invasive infections, such as meningitis and septicemia. Even though proper antimicrobial treatments are available in the market, pneumococcal diseases kill 1.6 million people in developing countries each year, of which one million are children younger than 5 years.

Early and accurate diagnosis of pneumococcal pneumonia remains difficult because of the limitations of conventional diagnostic methods. Blood and sputum cultures are time-consuming and lack sensitivity, particularly for patients with antecedent antibiotic exposure. Sputum culture is also difficult to interpret because of oropharyngeal contamination. Several PCRs have been employed with varying degrees of success, using primers specific to repetitive regions and genes encoding rRNA, pneumococcal surface adhesion A molecule (*psaA*), pneumolysin (*ply*), penicillin binding protein and autolysin (*lytA*). In recent years, real-rime PCR has improved diagnostics. As such, *S. pneumonia* can be diagnosed rapidly by analyzing sputum through real-time PCR with the targets at *lytA*.

Streptococcus pyogenes

Streptococcus pyogenes or Lancefield group A Streptococcus (GAS) is responsible for a diverse range of clinical manifestations. Acute infection causes pharyngitis, impetigo, scarlet fever, cellulitis, acute bacterial endocarditis, and necrotizing fasciitis. Ineffective or delayed treatment, and infection with drug-resistant bacteria eventuates in chronic immune-mediated disorders, such as acute rheumatic fever (ARF) and acute glomerulonephritis (AGN) with attended morbidity and mortality worldwide.

The fast emergence of antimicrobial resistance amongst *S. pyogenes* has caused a need for molecular characterization of the newly variant strains. The conventional M serotyping is cumbersome, less effective, and demands preparatory requirements. A large number of *S. pyogenes* isolates remains M untypable by the available present range of high titer antisera.

One of the molecular diagnosis—the GASDirect test identifies specific rRNA sequences of *S. pyogenes* in pharyngeal specimens by a single-stranded chemiluminescent nucleic acid probe, and the sensitivity and specificity is 98%–99%. The GASDirect test can be applied for primary testing, has been used as a backup test to negative antigen tests, and is suitable for batch screening of throat cultures.

Streptococcus agalactiae

Streptococcus agalactiae (beta-hemolytic group B Streptococcus, GBS) is the leading cause of neonatal sepsis and meningitis and is responsible for high mortality and morbidity, particularly in invasive infections of neonates. It is also an important pathogen in elderly patients and those with underlying diseases that impair immunological defense mechanisms against pathogenic microbes. Some cases of neonatal GBS infections are caused by vertical GBS transmission to the neonate during accouchement. Colonization occurs in 15%–40% of pregnant women, contributing to a neonatal infection rate of 0.73 per 1000 live births. Approximately 5 percent of GBS-infected infants die, and those who survive often develop severe neurological sequelae, such as mental retardation and visual or auditory disabilities. Because of this, CDC recommends that all pregnant women be screened for vaginal and rectal *S. agalactiae* colonization at 35–37 weeks of gestation. The CDC recommendations include the culture of GBS from vaginal/rectal specimens of pregnant women to arrive at a diagnosis of GBS colonization. Some clinical laboratories perform the Christie-Atkins-Munch-Peterson (CAMP) test to discriminate GBS from other streptococci. This test detects the CAMP factor, a pore-forming cytotoxin that generates a cohemolytic reaction with *Staphylococcus aureus* β-toxin; however, it requires two consecutive overnight cultures. Therefore, it is important to develop a more rapid, highly sensitive, and specific method to facilitate prompt diagnosis.

Conventional and real-time PCR assays can be for the rapid detection of GBS. Real-time PCR for GBS from vaginal-rectal swabs is commercially available and has the ability to both decrease turnaround time and increase sensitivity. Aside from vaginal-rectal swabs, GBS-PCR tests also can be performed with alternative specimen types, including amniotic fluid, neonatal screening swabs, blood, breast milk, urine, and serum. Universal 16S PCR also has been used to identify GBS from blood, bone, and joint infections. Targets for GBS-PCR include the *sip*, *cfb*, *scpB*, and *ptsI* genes. The *sip* gene codes for the Sip immunogenic protein. The *cfb* gene codes for the Christie-Atkins-Munch-Petersen factor. The *scpB* gene codes for the C5a peptidase and the *ptsI* gene, which codes for phosphotransferase. The LAMP method can amplify target nucleotide sequences at isothermal conditions (usually 60–65°C) within 90 min by using four or six primers. Amplification can be detected by macroscopic observation of turbidity or color change of a fluorescent dye. The use of LAMP for detecting GBS has been developed successfully by using probes targeting cfb gene.

MOLECULAR DETECTION AND CHARACTERIZATION OF FUNGAL PATHOGENS

Fungi are eukaryotic organisms and represent a specific kingdom between plants and animals. Because fungi are widely present, humans are exposed to fungi on a daily basis, either by contact with the skin, or by inhalation. Most fungi are opportunistic pathogens, so humans are naturally resistant to fungal pathogenesis. The development of fungal diseases often is seen in immunocompromised patients, or when natural barriers are damaged. As such, fungal infections represent a significant cause of morbidity and mortality in immunocompromised patients and those undergoing invasive procedures.

Identification and classification of fungi from clinical samples are important for antifungal susceptibility testing and epidemiological investigation. The immediate primary consequence for the clinical laboratory is to isolate fungus from a clinical specimen and identify the fungi. Sequence-based molecular techniques increasingly being used in the identification and taxonomy characterization of fungal infections, but it often is not necessarily true for a causal relationship between the fungi and the disease. When culture from mucosal specimens is positive, the result is questionable because it is difficult to differentiate among true infection, colonization, or bystander recovery. Likewise, results of PCR can be interpreted only in conjunction with the overall clinical picture.

PCR Probes and PCR Formats

DNA sequence-based approaches have been developed in recent years to provide rapid and accurate methods to identify medically important fungi as a complement to traditional phenotypically based methods. The molecular detection of pathogenic fungi directly in clinical specimens (blood, tissue, bronchoalveolar lavage fluid, and other body fluids) most often is accomplished through PCR-based DNA amplification or isothermal amplification. Because amplification of a gene with several hundred copies improves PCR assay sensitivity compared to that of tests amplifying single-copy genes. As such, most novel PCR assays have targeted multicopy rRNA genes. The choice of the primers and probes depends on the overall detection strategy.

DNA targets that have been used for fungal identification and classification include rDNA, cytochrome b, β-tubulin, calmodulin, enolase, chitin synthase, heat shock protein, and other housekeeping and functional genes (Table 15.4). The nuclear rDNA genes and intergenic regions are the most commonly used.

Among eukaryotic organisms, the nuclear rDNA array—small (18S-like), large (26S- or 28S-1ike), and 5.8S rDNA-subunit genes (Fig. 15.4)—provides an ideal target for PCR amplification, also because of their homogenization of multiple array copies by molecular processes such as gene conversion. The 26S and 28S fragments are associated with yeasts and plants and with animals, respectively. Within the rDNA array, the ITS1 and ITS2 and the embedded expansion segment (ES) regions of the 26S-like gene are the most variable. Therefore, these sequences are commonly serve as diagnostic, target sequences (Figs. 15.4 and 15.5). For differentiation between strains of the same fungal species, multilocus sequence typing of housekeeping genes can be used. Using genus or species-specific primers and probes, several methods, including Southern blotting, slot blotting, PCR-enzyme immunoassay, microarray, nested PCR, PCR-restriction fragment length polymorphism, and microsatellites, can successfully identify a variety of specific fungal pathogens.

TABLE 15.4 Commonly Used Target Genes for Sequence Identification of fungi

Target Gene	Target Fungi
Actin	• *Aspergillus*
β-Tubulin	• Phaeacremonium • Aspergillus • Pseudallescheria
Calmodulin	• Aspergillus • Pseudallescheria info
Chitin synthase 2	• Lacazia loboi
Cytochrome b	• Aspergillus, Trichosporon, and Rhodotorula
D1-D2	• Most fungi
Elongation factor 1a	• Fusarium species
ITS	• Medically significant yeasts • fungi
26S	• Medically relevant Fusarium and Scedosporium

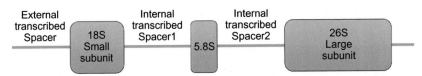

FIG. 15.4 Sequence structure of fungal rDNA. *ETS*, external transcribed spacer; *SSU*, small subunit; *LSU*, large subunit.

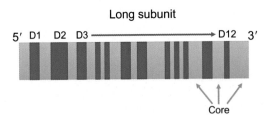

FIG. 15.5 The map of 26S rDNA expansion segments.

A broad range PCR approach, in which assays are designed to detect many different pathogenic fungi with multi- or even panfungal primers, can be used, however, pan-fungal primers can hybridize other eukaryotic DNAs. Therefore, the pan-fungal approach can be limited when tissue samples contain mainly human DNA, little fungal DNA, and potentially mixtures of fungi.

As with every PCR assay designed to detect a low load of a given microorganism among a huge number of human cells, all PCR parameters must be optimized to obtain a positive result with the highest sensitivity possible. By doing this, however, one increases the risk of false-positive results, usually by contamination either with previously amplified products or with environmental microorganisms. Because of the small amount of fungal DNA in clinical specimens, nested PCR formats can be used to detect *Aspergillus* and *Candida*.

Identification of species responsible for lesions in tissue is essential to adapt antifungal treatments. Tissue, however, is processed frequently for histopathology with FFPE tissue, which hampers the molecular identification workflow because of technical issues. Fresh frozen tissue gives rise to a better yield than FFPE tissue in terms of qPCR detection. After the DNA is extracted from biopsies, various strategies can be used, such as testing multiple species/genera by qPCR, so that it can provide accurate diagnosis of fungal pathogens.

The Application of Sequencing Techniques

For sequencing analysis, genomic DNA is extracted from culture or clinical specimens, the target gene fragment is PCR amplified using specific PCR primers and cycling conditions, the PCR amplicon is purified, and the order of the nucleotide bases in the DNA target is determined by NGS. The dye-terminator sequencing method, along with automated DNA sequence analyzers can be used. Another DNA sequencing method is pyrosequencing, which is a unique method for short-read DNA sequencing. For direct sequencing, the 26S and 28S fragments can be used for PCR amplification (Figs. 15.4 and 15.5). For differentiation between strains of the same fungal species, multilocus sequence typing of house-keeping genes should be used.

Nucleic acid sequencing is the best approach for the identification of organisms that are not commonly seen. For example, by using an extensive library containing DNA sequences (many based upon type strains), approximately 1200 nucleotides in length, from the 26S nuclear rDNA gene array, it is possible to identify the organisms (Fig. 15.4). Fungal identification also can be based on the complete DNA sequence of the three largest and most variable ES regions, which also provides information for strain or even phylogenetic species differentiation. The pattern of sequence variability within the fungal ES fragment is intermittent among more conservative invariant regions. Such a pattern of variation is well understood. Benefits of interspersed variability include the ability to design efficient DNA sequencing approaches.

Molecular Detection of *Aspergillus*

Aspergillosis, defined as an infection caused by organisms included in the genus *Aspergillus*, constitutes a spectrum of diseases that range from allergic reactions to disseminated disease in immunocompromised hosts. *Aspergillus fumigatus* is the predominant organism causing invasive aspergillosis (IA), but other species have become increasingly recognized as important opportunistic pathogens. Despite recent advances in antifungal therapy, IA continues to cause significant morbidity and mortality. This is because it is difficult and slow in making a microbiologic diagnosis. Although the use of fungal cell wall biomarkers, such as 1,3-β-D-glucan or *Aspergillus galactomannan* antigen, has improved early diagnosis of IA, these methods also have significant limitations, including poor sensitivity in certain patient groups and issues with nonspecificity.

For the rapid and accurate diagnosis of aspergillosis. NAAT is highly sensitive and specific for the detection of *Aspergillus* DNA or RNA. Amplification of DNA by PCR has been the most widely applied NAAT method. Nested PCR protocols can be used to optimize sensitivity. Multiplex PCR detection methods coupled with fluorescently labeled probes is commonly used. Real-time PCR also can be used to quantify fungal burden, while multiplexed arrays allow for the differentiation of larger numbers of pathogenic species. Isothermal amplification techniques, such as nucleic acid sequence-based amplification (NASBA), also have been used to detect *Aspergillus* nucleic acids. NASBA offers several potential advantages over PCR, including more efficient amplification, the potential to assess cell viability, and lack of concern over contamination with fungal DNA. Additionally, the use of RNA templates can increase sensitivity by capitalizing on the fact that highly expressed genes produce thousands of transcripts within a cell.

In situ hybridization (ISH) probes targeting fungal rDNA have been used for the detection of *Aspergillus* in fresh and FFPE tissues without requiring nucleic acid extraction or amplification. ISH enables direct visualization of organisms with the use of labeled probes that bind to complementary fungal sequences. Genus-specific probes are potentially useful for differentiating *Aspergillus* species from other fungal pathogens that produce septate branching hyphae in vivo. The drawbacks for ISH is that sensitivity is limited compared to that of traditional histopathology stains, and results sometimes have been difficult to interpret because of the autofluorescence of hyphae or high background produced by necrotic tissue.

The rDNA gene cluster is the most common target of assays designed for direct detection in clinical specimens because of the presence of highly conserved (18S rDNA and 28S rDNA) and variable (ITS and D1/D2) regions. In general, genus-level assay designs often target 18S, while *A. fumigatus*-specific assays target ITS-1, mitochondrial DNA, alkaline protease, or Aspergillus collagen-like (*acl*) genes. In addition to robust rDNA sequence databases and the availability of universal primers, another advantage of using the rDNA genes as a diagnostic target is their copy numbers. This is because high copy number increases the sensitivity of detection. A potential limitation of rDNA targets is the close phylogenetic relationship between *Aspergillus*, *Penicillium*, and *Paecilomyces* species. Depending on the primer and probe sequences, cross-reactivity among these species can be predicted and is an important consideration when testing respiratory specimens.

DETECTION OF VIRUSES

Diagnoses of Human Immunodeficiency Virus Infection

Approximately, 36.7 million people worldwide are living with human immunodeficiency virus (HIV) at the end of 2016, and 1.8 million new infections were reported in 2016. HIV is classified as a member of the family *Retroviridae* and genus *Lentivirus* based on the biological, morphological, and genetic properties. Two types of HIV have been isolated and characterized from patients infected with the virus: HIV-1 and HIV-2. HIV-1 is the most virulent and pathogenic, and when people talk about HIV without stating the type of virus they usually are referring to HIV-1. HIV-2 is constrained only to some areas of Central and Western Africa. A detailed investigation about HIV structure and its mechanism of infection have not only allowed to characterize and develop new and effective vaccines and drugs but has also described new approaches for diagnosis of HIV at the laboratory scale.

Group M of HIV-1 is the major cause of worldwide HIV epidemic. There are nine known phylogenetic subtypes, sub-subtypes of Group M, clades (A-K) and an intersubtype circulating recombinant forms (CRFs) among which intersubtype genetic diversity is 25% for the *env* gene and 15% for the *gag* gene. Subtypes and sub-subtypes are a result of founder effects in the past, whereas if two different subtypes co-infect a patient they give rise to intersubtype recombinants. These recombinants are called CRFs if they have a significant epidemic spread. Subtype B of HIV-1 dominates in Australia, the Americas, and Europe, whereas subtype C predominates in India and Africa.

Current techniques of HIV antibody detection can detect the HIV antibody within or after one or 2 weeks of HIV infection. Additional confirmatory tests are available for the accuracy of results. Table 15.5 is a list of diagnostic methods that have been developed. These techniques include the detection of antibodies in patient's serum or plasma representing the presence of viral nucleic acid either by PCR chain reaction, p24 antigen, or rising viruses in cell culture. Most commonly antibody test is used for the detection of HIV infection. Seroconversion can be detected using high sensitive enzyme immunoassay within 2–3 weeks of infection in most of the cases. P24 antigenic protein of virus might become visible or detectable before antibody detection. Therefore, proper diagnosis depends upon the use of accurate testing by laboratories along with the availability of patient's clinical history, including high-risk symptoms associated with seroconversion illness. Reactive results can be confirmed by using alternative assay approaches. Highly sensitive fourth-generation tests can be used to confirm and detect both the p24 antigen and HIV antibodies. Western blot is another sensitive immunoblot used as a confirmatory test.

High sensitivity and specificity of DNA PCR pick and choose a small number of viral particles. Because newborn babies from an HIV-infected mother carry HIV antibodies up to 15 months of age, an HIV antibody test is not reliable. Under this circumstance, detection should proceed through PCR. Taking into account primers specificity, HIV result assessment needs to be cautious. PCR helps solve indeterminate Western blot results and detection of HIV in immunocompromised persons.

Quantitative RNA-PCR and genotyping RNA-PCR are performed to monitor HIV-positive patients. In combination with CD4 count, RNA-PCR is performed to decide when HIV therapy should start in clinical aspects. Genotyping is useful in assessing drug resistance in patients under the process of therapy and to help the physician decide what combination of drug therapy should be used.

Quantitative PCR always produces many false positive and false negative results and is not a good diagnostic test. Because of high genetic variability, HIV-1 is divided into many genotypes or subtypes. Therefore, it is more difficult

TABLE 15.5 HIV Diagnostic Methods

Type of Test	Detection of DNA/RNA	Detection of Antigen	Detection of Antibody
PCR or viral load		✓	✓
P24 test	✓		✓
4th generation antigen/antibody tests	✓		✓
1st/2nd/3rd generation antigen only tests	✓	✓	
Rapid tests (finger prick/oral swab are antibody only)	✓	✓	
Western blot	✓	✓	

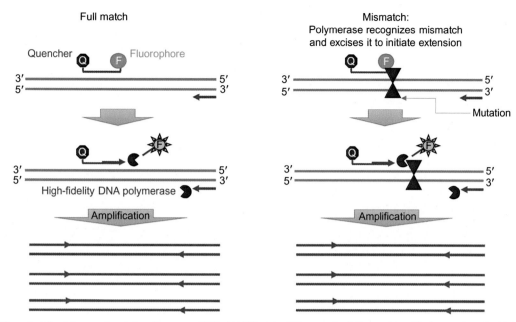

FIG. 15.6 The principle of the novel real-time fluorescent PCR. (A) When the HFman probe fully matches with the template, the high-fidelity DNA polymerase can non-specifically recognize 3′ fluorophore-modified base and remove the 3′-base to initiate primer extension. (B) When the HFman probe forms a mismatch with the template at the 3′-end, the high-fidelity DNA polymerase recognizes the 3′ mismatched base and removes the mismatched base to initiate extension. The removal of 3′ fluorophore-labeled base will emit the fluorescent signal.

to find highly conserved regions shared by all genotypes or subtypes of the HIV-1. It is harder to design and select appropriate qPCR primers and probes for HIV-1 virus detection with high detection sensitivity and specificity than for bacteria, fungi, and other pathogens. For the TaqMan qPCR, since the probe length usually requires more than 25 nt to ensure relatively high Tm value (68–70°C), it is very difficult for primer and probe design. Although MGB (minor groove binder) modification had been demonstrated to increase the stability of duplexes formed by probe and target, and allows shorter probes, it is still a big challenge to develop qPCR methods for high-efficient detection and quantification of all genotypes/subtypes of HIV-1. A novel, simple real-time qPCR use a probe as a primer to initiate PCR amplification under the high fidelity (HF) DNA polymerase and the emitted fluorescence during the amplification can be monitored during a real-time PCR amplification (Fig. 15.6). The new method allows the presence of mismatches between the 3′-ends of the HFman probe/primer and the target sequence. It has better performance in quantification of gene expression and viral load than the conventional TaqMan probe-based qPCR. Tolerance to mismatches between primer/probe and target and the one primer-one HFman probe system simplifies the method, improves the sensitivity and accuracy for the quantification of gene expression, and diagnosis of various pathogens. As such, HF DNA polymerase-mediated qPCR method is a simple, sensitive, and promising approach for the diagnostics for viral infectious diseases.

Molecular Detection of Human Papillomavirus

Human papillomavirus (HPV) is a broad term that refers to a group of related papillomavirus strains that infect humans. HPVs are highly tissue tropic, and infection is limited to stratified epithelium at either cutaneous or mucosal surfaces, usually at specific body sites. The vast majority of HPV types cause benign warts of the skin or genital region (Table 15.6). Some HPV types have the potential to cause lesions that progress to cancer. Fig. 15.7 shows group 1 carcinogens (carcinogenic to humans), group 2A carcinogens (probably carcinogenic to humans), and group 2B carcinogens (possibly carcinogenic to humans). The 12 HPVs group 1 (HPV16, HPV18, HPV31, HPV33, HPV35, HPV39, HPV45, HPV51, HPV52, HPV56, HPV58, and HPV59) are considered high-pressure liquid chromatography risk. There is less evidence for including HPV68 as high risk, so it is classified in group 2A as probably carcinogenic. Ninety-six percent of cervical cancers are attributable to one of these 13 HPV types (groups 1 and 2A). Other alphapapillomaviruses, HPV26, HPV30, HPV34, HPV53, HPV66, HPV67, HPV 69, HPV70, HPV73, HPV82, HPV85, and HPV97, have been associated

TABLE 15.6 Cancers Attributable to HPV in the United States

Disease			Most Frequently Associated HPV Type(s)
Nongenital skin disease	Benign	Common warts	1, 2, 4, 7
		Palmoplantar warts	1, 2, 4
		Flat plane warts	3, 10; less frequently 26–29, 41
		Ungual warts	1, 2, 4, 27, 57
		Butcher's warts	2, 7
		Epidermodysplasia verruciformis	5, 8
	Malignant	Ungual squamous cell carcinoma	16, 18
		Epidermodysplasia verruciformis	5, 8
Oral mucosal disease	Benign	Respiratory papillomatosis	6, 11
		Laryngeal papillomatosis	6, 11
		Oral squamous cell papillomas	6, 11
		Oral focal epithelial hyperplasia of Heck	13, 32
	Potentially malignant	Oral lichen planus	6, 11, 16
		Oral leukoplakia	6, 11, 16
		Oral erythroplakia	6, 11, 16
	Malignant	Oropharyngeal carcinoma, squamous cell	16
Ocular mucosal disease	Benign	Conjunctival papillomas	6, 11
	Malignant	Squamous cell carcinoma of conjunctiva	16
Anogenital disease	Benign	Condyloma acuminate	6, 11
		Anogenital warts	6, 11, 40, 42, 43, 44, 54, 61, 72, 81, 89
		Giant condylomata	6
	Potentially Malignant	Bowenoid papulosis	16, 55
	Malignant	Carcinoma of vulva	16, 18
		Carcinoma of vagina	16, 18
		Squamous cell carcinoma of cervix	Group 1, 16, 18, 31, 33, 35, 39, 45, 51, 52, 56, 58, 59; group 2A, 68; group 2B, 26, 30, 34, 53, 66, 67, 69, 70, 73, 82, 85, 97
		Adenocarcinoma of cervix	16, 18, 45
		Squamous cell carcinoma of anus	16, 18
		Carcinoma of penis	16, 18, 6, 11

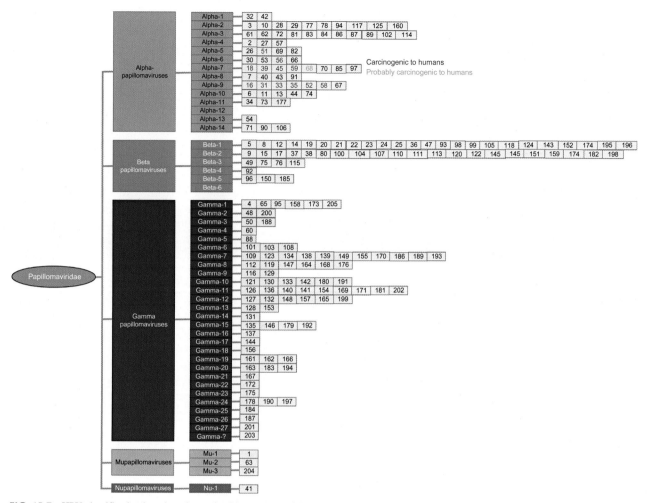

FIG. 15.7 HPV classification based on the nucleotide sequence of the capsid protein L1 gene.

with rare cases of cervical cancer and are considered group 2B probable carcinogens. Because the number of cases attributed to the group 2B HPV types is smaller, carcinogenicity is somewhat more difficult to assess. HPVs in the alpha-papillomavirus genus cause both mucosal and cutaneous lesions. Beta-, gamma-, mu-, and nu-HPVs cause cutaneous lesions.

HPV infection can follow either of two pathways: productive or transforming (nonproductive or abortive). In productive infections, the HPV genome is propagated in episomal form, and this is a polyclonal proliferative process with viral copies increasing about two to four logs. Eventually, most of the virus-containing keratinocytes desquamate and are removed from the body by natural processes. HPV virions can be released as a result of degeneration of the infected exfoliated host cells. In productive infection, HPV DNA can be measured inside as well as outside host cells. HPV can persist in the environment and remain infectious for several days without a host. Productive infections produce huge amounts of virions (with HPV DNA inside) but never lead to cancer because desquamating cells never divide and are eliminated from the body. Productive HPV infections are transient, and HPV becomes undetectable after several months. In the nonproductive (transforming) pathway, the ordered expression of genes does not occur and the normal HPV life cycle cannot be completed, resulting in a nonproductive or abortive infection. The highest risk for cancer development occurs as a result of prolonged persistent infection over many years (Table 15.7).

Head and neck squamous cell carcinomas (HNSCC) include the oral cavity, pharynx, and larynx. Since oral and oropharyngeal cancers often are detected in later stages, substantial efforts have been made toward early detection. A number of oral lesion detection systems have been introduced and are based on autofluorescence or tissue reflectance. Because HPV status has biological significance in oropharyngeal squamous cell carcinomas (OPSCC), tumors often are described by the presence of HPV and/or molecular changes associated with HPV instead of relying on specific histologic features.

TABLE 15.7 HPV Attributable Cancers in the United States

| Anatomic Site | Cases Attributable to HPV | | Annual No. of Cases Attributable to HPV | |
	%	Range (%)	Avg	Range
Cervix	96	95–97	11,500	11,400–11,600
Vulva	51	37–65	1600	1200–2000
Vagina	64	43–82	500	300–600
Penis	36	26–47	400	300–500
Anus	93	86–97	2900 women 1600 men	2700–3000 1400–1600
Oropharynx	63	50–75	1500 women 5900 men	1200–1800 4700–7000

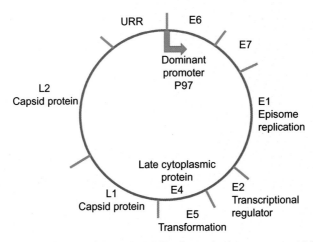

FIG. 15.8 Schematic diagram of genomic organization of HPV16 ORF. P97 refers to dominant promoter. All PVs are similar in their genetic organization and appearance under electron microscope. *URR*, upstream regulatory region.

Detection of HPV16 DNA in saliva rinses and plasma after primary treatment by PCR can allow for early detection of recurrence in patients with known HPV16-positive OPSCC.

Detection of E6/E7 mRNA generally is considered to be the standard to indicate transcriptionally active, clinically relevant oncologic HPV. Extraction of RNA and amplification of E6/E7 mRNA by PCR can be applied to fresh or frozen tissues. It remains technically challenging and is unreliable for FFPE tissues (Fig. 15.8). Reliable immunohistochemical probes for E6 and E7 proteins are not available. On the other hand, HPV DNA ISH and p16 IHC have emerged as the most useful tests. DNA ISH assays for high-risk HPV allow for direct visualization of the virus in the nuclei of tumor cells in the tissue, which better demonstrates HPV as a causal agent.

All cervical cancers are caused by high-risk HPV infections. HPV infections are common, and it is estimated that approximately 75%–80% of sexually active individuals will become infected in their lifetime. Most women become infected shortly after becoming sexually active, and the highest prevalence is seen in women younger than age 25. Common risk factors for HPV infection in young women include earlier age of onset of sexual activity, two or more sexual partners in the previous year, and coinfection with *Chlamydia trachomatis*, herpes simplex virus, or bacterial vaginosis. Of the 12 class 1 carcinogenic HPV types, HPV16 is the most frequently detected in cervical cancer, followed by HPV18. Both HPV16 and HPV18 are responsible for about 70% of cervical squamous cell carcinomas worldwide. The next most frequently detected HPV types in cervical cancer are HPV31, -33, -35, -45, -52, and -58. HPV infection with a high-risk type is necessary for development of cervical cancer, but cofactors such as smoking, immune suppression, long-term oral contraceptive use, chronic inflammation, and having given birth two or more times can increase the risk. It generally is estimated that it takes several years to decades from the time of an initial HPV infection until development of a high-grade lesion and eventually formation of a tumor.

The majority of molecular diagnoses are designed to detect nucleic acids of the 12 IARC HPV group 1 carcinogens (HPV16, -18, -31, -33, -35, -39, -45, -51, -52, -56, -58, and 59). For example, multiplex real-time PCR and nucleic acid hybridization with four different fluorescent reporter probes can be used to concurrently detect the L1 gene of HPV16 and HPV18 as individual reactions and HPV31, -33, -35, -39, -45, -51, -52, -56, -58, -59, -66, and -68 as a pooled result. Detection of HPV E6/E7 oncogene mRNA in cervical cells is an alternative to detection of HPV DNA. HPV tests based on nucleic acid sequence-based amplification (NASBA) can detect very little expression of E6/E7 mRNA for transient infection. In persistent infections, there is overexpression of E6/E7 mRNA. Detection of overexpression of E6/E7 mRNA might be more directly associated with disease progression. In the United States, although DNA testing is more sensitive than cytology for identifying women with cervical precancer, cervical cytology serves as the primary screening test, and HPV DNA or RNA testing is used as a cotest or reflex test to triage patients with normal cytology. Internationally, the trend is toward reversing this sequence. Some countries, including China, India, Sweden, the Netherlands, and Australia, are beginning to establish programs with HPV testing as the primary screen and cervical cytology used to triage patients who are HPV positive.

Molecular Detection of Hepatitis Viruses

Viral hepatitis is a serious public health problem requiring early diagnosis and timely treatment. A number of hepatitis viruses already have been characterized based on their molecular structure and named alphabetically as hepatitis viruses A, B, C, D, E, and G (HAV, HBV, HCV, HDV, HEV, and HGV, respectively). These are hepatotropic and noncytopathic in nature and cause liver damage by immune mediated cell lysis. HAV infects mainly the pediatric age group, occurs both sporadically as well as in epidemics, and accounts for an estimated 1.4 million cases annually. HBV is responsible for approximately 350 million chronic infections worldwide and more than one million annual deaths, mostly as a result of end-stage liver disease and liver cancer. About 25% of adults infected with HBV during childhood are reported to die from hepatocellular carcinoma (HCC) or liver cirrhosis. In addition, 3 million to 4 million people are infected with HCV each year, and a high proportion of them develop chronic HCV infection. Many people infected with HCV die from serious liver diseases annually. Co-infections of Hepatitis A and E infections usually run a benign course of disease and resolve in time without developing chronic diseases. In contrast, hepatitis B and C infections cause severe liver diseases, developing chronicity in a significant number of patients.

Although the diagnosis of hepatitis viral infections usually is done with serological markers in blood, the nucleic acid-based test have been developed to detect the viral genome in serum for the diagnosis of viral hepatitis. Real time PCR is one of the latest techniques frequently used to diagnose viral hepatitis. Several PCR-based assays, coupled with oligonucleotide microarray technology, allow for the simultaneous detection and genotyping of several viruses, including blood borne pathogens, respiratory viruses, and adenoviruses. Multiplex qPCR assays use the most conserved region representing all the strains/variants for detection of the hepatitis viral genome in body fluid. Table 15.8 shows the list of target conserved

TABLE 15.8 Conserved Hepatitis Virus Regions Used as qPCR Templates

Virus	Conserved Region
HAV	5' UTR
HBV	S-gene
	X-gene
HCV	5' UTR
HDV	Ribozyme-1
HEV	ORF2
	ORF3
HGV	5' UTR

regions used in various multiplex qPCR assays developed for hepatitis viruses. The 5′ untranslated region (UTR) is the main target template in HAV, HCV, and HGV. It is based on the availability of most conserved sequence in the 5′ UTR for amplification purpose. Similarly, S-gene or X-gene is used for HBV, ribozyme-1 gene for HDV, and open reading frame (ORF)-2 or ORF-3 region for HEV.

The multiplex qPCR assays used both for comparison and in combination with other molecular technologies to improve the sensitivity for detection of the viral genome. Various other assay systems also were developed for simultaneous detection of HBV, HCV, and HIV, in addition to multiplex qPCR. The other assay systems included flowcytometric microsphere-based hybridization assay, transcription-mediated amplification (TMA), and nucleic acid sequence-based amplification (NASBA). Comparatively, TMA was reported to be an equally sensitive technique. When comparing qPCR with NASBA and TMA for the detection of hepatitis viruses, however, the level of sensitivity of TMA was found to be associated closely with qPCR. The qPCR assay was reported to be faster, more economic, and easier to perform compared to all other assays that were evaluated. Multiplex qPCR has been employed successfully for the simultaneous detection of hepatitis virus A, B, C, D, and E and genotyping of their strains. It appears to be a good tool for screening blood donor samples in blood banks for hepatitis viruses. Moreover, a single-step multiplex qPCR assay allows for an early diagnosis and timely treatment of patients with viral hepatitis. Several studies in this field are in progress, with more important information likely to be available until the next update is necessary.

The Detection of HBV

HBV infection of hepatocytes initiates with concentrated attachment of the HBV virion to cell surface glycosaminoglycans (GAGs) followed by a high-affinity binding of the myristoylated N-terminal PreS1 of the HBV large-envelope protein to the sodium taurocholate cotransporting polypeptide (NTCP) receptor on the hepatocytes. Following virion entry into the cell (likely by endocytosis), the envelope is removed, and the HBV nucleocapsid releases the enclosed relaxed circular DNA (rcDNA), which is transported to the nucleus. In the nucleus, rcDNA is converted into cccDNA (Fig. 15.9), which then acts as a template for the transcription of all viral RNA molecules, including the major mRNAs encoding the viral proteins: hepatitis B core antigen (HBcAg), hepatitis B e antigen (HBeAg), HBV polymerase (pol), hepatitis B surface antigen (HBsAg), and HBV X protein (HBx). HBV replicates its genomic rcDNA via reverse transcription of pregenomic RNA (pgRNA), one of the RNA forms, and the resulting rcDNA in the nucleocapsid can be enveloped and secreted or recycled to the nucleus to amplify the cccDNA pool. In addition to the rcDNA from incoming virions and the intracellular amplification pathway, cccDNA can be generated from double-stranded linear DNA (dslDNA), which is the major precursor for HBV integration by nonhomologous recombination. In this process, the generated cccDNA usually is defective because of a loss of sequence fragments.

Covalently closed circular DNA (cccDNA) in the nucleus of infected cells cannot be eliminated by current therapeutics and can result in persistence and relapse. Southern blotting is regarded as the best test for quantitative cccDNA detection (Fig. 15.10). Real-time PCR, competitive PCR, semi-nested/nested PCR, droplet-digital PCR, and rolling circle amplification qPCR have been developed for better specificity and accuracy. In the case of real-time PCR, because of the partial homology between rcDNA and cccDNA, specificity is compromised in the presence of abundant rcDNA. To reduce rcDNA contamination, a chimeric sequence composed of two segments to increase specificity can be designed: segment A is complementary to the HBV DNA plus strand from nt 1615 to nt 1604, and segment B is consensual to the HIV long terminal repeat (LTR) region, which differs from the HBV DNA sequence. This can effectively distinguish cccDNA from other HBV DNA.

EMERGING TECHNOLOGY: MALDI-TOF MS FOR MICROBIAL SPECIES IDENTIFICATION

The integration of molecular diagnosis into microbial identification supported critical advances in analytical sensitivity, allowing clinical microbiologists to explore options other than routine culture and to begin testing patient specimens directly for the presence or absence of particular organisms. The ability to rapidly and uniformly test both direct patient specimens and cultured organisms in near real time by molecular methods has improved clinical analysis significantly. Nucleic acid-based methods, such as FISH and PCR-based strategies, drastically decreased the time to result and provided significant improvements to both laboratory workflow and patient prognoses. A major limitation of molecular diagnosis is that these assays require advance knowledge about the characteristics of the microorganisms or a likelihood of that particular organism being present, in order to select the correct assay to fit the testing application. In the case of polymicrobial infections, multiple molecular assays, preliminary culture and separation, or additional downstream testing is sometimes required for full characterization of the clinical specimen, adding to the result turnaround time and the overall financial cost.

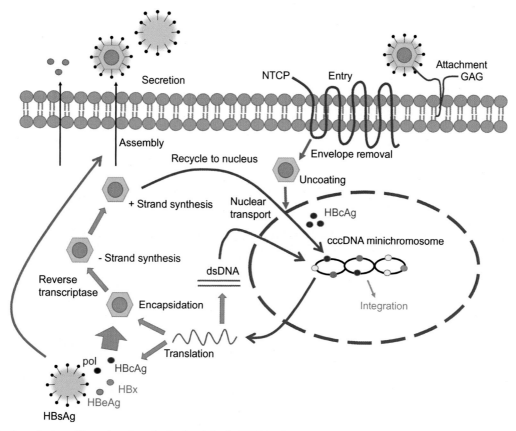

FIG. 15.9 A schematic steps of formation of covalently closed circular DNA (cccDNA) in the hepatitis B virus (HBV) life cycle. 1. Attachment and entry of HBV via glycosaminoglycan (GAG) and sodium taurocholate cotransporting polypeptide (NTCP). 2. Envelope is removed and the inner nucleocapsid is released into the cytoplasm. 3. Transport of nucleocapsid into the nucleus pore and release of the polymerase (pol)-linked relaxed circular DNA (rcDNA). 4. Conversion of cccDNA from rcDNA and double-stranded linear DNA (dslDNA). rcDNA is the main source of cccDNA. The incoming or intracellular amplified rcDNA fills in the gap in plus strand and removes the pol and RNA primer from the 5'-terminus of minus and plus strand to form the protein-free (PF)-rcDNA, which is also called deproteinized-rcDNA (DP-rcDNA). The end of both strands of PF-rcDNA is further ligated to form cccDNA. In addition, a fraction of cccDNA is synthesized from dslDNA by non-homologous end-joining (NHEJ). The episomal cccDNA, once formed, is assembled into mini-chromosomes including loose or spread and condensed forms bound to histone and non-histone proteins. 5. Transcription of viral mRNAs (3.5 kb prege-nomic RNA (pgRNA) and precoreRNA, 2.4 kb and 2.1 kb preS/S mRNAs and 0.7 kb X mRNA) from cccDNA minichromosomes. 6. Translation of HBV proteins: hepatitis B surface antigen (HBsAg), hepatitis B e antigen (HBeAg), hepatitis B core antigen (HBcAg), HBV X protein (HBx), and polymerase (pol). 7. Encapsidation involving pgRNA, pol, and HBcAg. 8. Synthesis of HBV minus strand via reverse transcription of pgRNA. 9. HBV plus strand synthesis. 10. Assembly of capsid containing rcDNA with HBsAg to form virion, which secretes out of the cell.

NGS has provided an attractive option for universal identification of fungi and bacteria, It has been considered among the most definitive of all molecular microbiological analyses since its implementation. Although 16S rRNA and 18S rRNA gene sequencing (for bacterial and fungal identifications, respectively) are powerful diagnostic tools with high discriminatory power for species and strain-level determinations, the NGS is employed primarily by large, high-complexity clinical and reference laboratories and it often requires specialized instrumentation and dedicated laboratory space and staff. These constraints often render rRNA gene sequencing impractical for most laboratories; therefore, the use of automated instruments for the phenotypic analysis of bacterial isolates still predominates as the basis for routine microbial identification, despite imperfections in accuracy, robustness, and time to identification. Clearly, the development and validation of alternative rapid and universal identification methods are warranted.

Matrix-assisted laser desorption ionization time-of-flight mass spectrometry (MALDI-TOF MS) has revolutionized the identification of microbial species in clinical microbiology laboratories. This next-generation microbial identification tool has key advantages of simplicity and robustness, making it the new best method to use. Although the technology is refined and based on physical parameters, the practical aspects of the method are really simple, and it requires limited processing, time, and disposables. The technician needs to put a small amount of microbial biomass or a protein extract on a sample holder, add a chemical compound, and the equipment does the rest. A reference database assists in species identification. Methods can be conveniently standardized, although, in some, cases culture conditions affect the nature and sometimes the

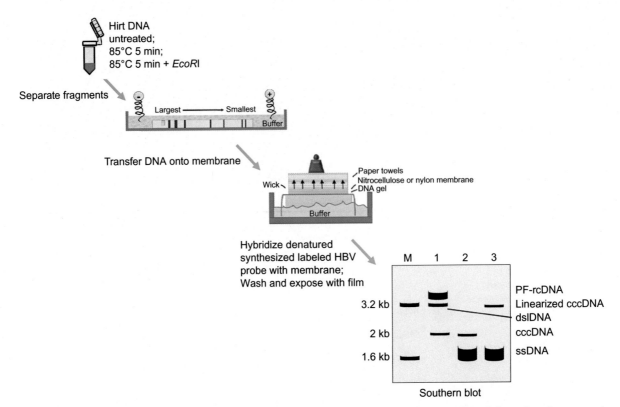

FIG. 15.10 A typical Southern blot assay to detect HBV cccDNA. The typical pattern of Hirt DNA is shown after gel electrophoresis, transmembrane, hybridization and detection. M is an HBV DNA marker composed of 1.6 kb and 3.2 kb HBV fragments. The untreated extract (Lane 1) contains all types of extrachromosomal protein-free DNA: PF-rcDNA, dslDNA and cccDNA. After 5 min 85°C treatment before gel loading, PF-rcDNA and dslDNA turn into ssDNA, and only cccDNA remains undenatured and immobile (Lane 2). When it is further digested with *Eco*RI, cccDNA is linearized to dslDNA (Lane 3). *cccDNA*, covalently closed circular DNA; *rcDNA*, relaxed circular DNA; *dslDNA*, double-stranded linear DNA.

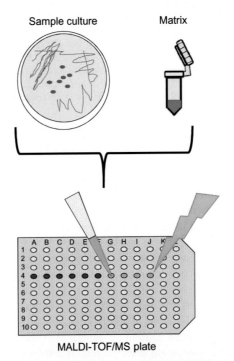

FIG. 15.11 The general procedure for the identification of bacteria and yeast by MALDI-TOF MS using the intact-cell method. Bacterial or fungal cells are isolated from culture media and applied directly onto the MALDI test plate. Samples are then overlaid with matrix and dried. The plate is subsequently loaded into the MALDI-TOF MS instrument and analyzed by software associated with the respective system for rapid identification of the organism.

overall quality of the MALDI-TOF MS spectra. This variable did not limit the development of spectral databases and assay development. Results can be obtained in a few minutes, and identifications are very reproducible and therefore reliable. The technology has been broadly adopted and many laboratories already use MALDI-TOF MS as the key component of their microbial identification strategy.

The Principle of MALDI-TOF MS in Clinical Analysis

In an MS analysis using MALDI as a soft ionization mechanism, a saturated solution of a low-mass organic compound, called a matrix, is added to the sample, and the mixture is spotted onto a metal target plate for analysis (Fig. 15.11). In the case of bacterial or fungal identification, a microbial colony is analyzed, or in some cases, direct blood culture material, urine, CSF, or protein extract is used. Upon drying, the clinical material and the matrix cocrystallize and form a solid deposit of sample embedded into the matrix. The matrix acts both as a scaffold by which ionization can occur and as a supplier of protons for the ionization of the clinical material. It is critical for the successful ionization of the clinical samples. This sample-matrix crystal, now present on the surface of the metal plate, is irradiated with a UV laser beam or an N_2 laser beam with a wavelength of 337 nm. Irradiation occurs for a short time to avoid damage or degradation of the sample embedded in the matrix, which could be caused by excess heating.

The laser beam is focused on a small spot on the matrix-clinical sample crystalline surface, and a beam attenuator is employed in the laser optics to adjust the irradiance which is defined as the intensity per unit of surface. This laser attenuation can be adjusted individually for each measurement, depending upon the sample type, but usually is standardized by the manufacturer for routine applications. The interaction among the photons from the laser and matrix molecules caused by uptake of energy from the beam triggers a sublimation of the matrix into a gas phase, forming a plume, which is directly followed by the ionization of the clinical sample (Fig. 15.12). These generated ions are separated on their mass-to-charge ratio and a profile of the spectral representation is created.

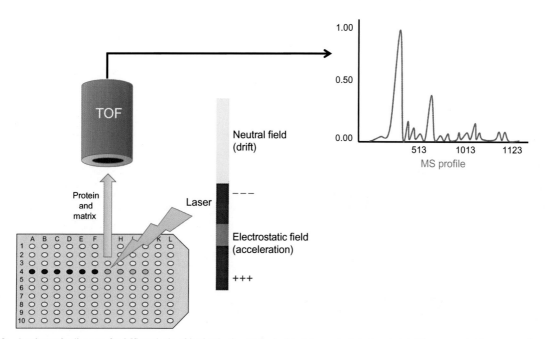

FIG. 15.12 A schematic diagram for MS analysis of ionized microbiological isolates and clinical material. The appropriately processed samples are added to the MALDI plate, overlaid with matrix, and dried. Subsequently, the sample is bombarded by the laser. This bombardment results in the sublimation and ionization of both the sample and matrix. These generated ions are separated based on their mass-to-charge ratio via a TOF tube, and a spectral representation of these ions is generated and analyzed by the MS software, generating an MS profile. This profile is subsequently compared to a database of reference MS spectra and matched to either identical or the most related spectra contained in the database, generating an identification for bacteria or yeast contained within the sample.

Sample Preparation

A key element of MALDI-TOF MS is sample processing, which might involve the selective enrichment of various biomolecules. In most current clinical microbiology applications, focus is on the evaluation of protein rather than nucleic acid, lipid, or sugar spectra. For this, the only requirement is to have a pure culture of a single species of bacteria, mycobacteria, or fungi, from liquid or solid growth medium. Therefore, the primary application of MALDI-TOF MS in a clinical laboratory is species identification of cultured isolates of bacteria and fungi. Typically, the pure culture grown on a solid agar medium are prepared by direct deposit of a small amount of cell material onto the target plate (Fig. 15.13). For yeasts, and

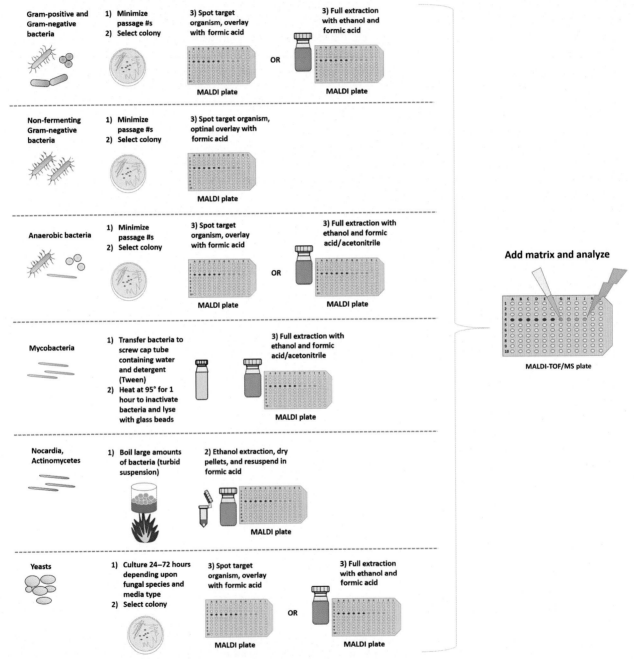

FIG. 15.13 The sample preparation procedures for the application of MALDI-TOF MS in identifying different classes of microbes. Different groups of microorganisms vary fundamentally in their cellular composition and architecture. These differences can affect the quality and accuracy of spectra generated during MS experiments. Different methods for sample preparation of different groups of microorganisms have been established. Proper biological safety precautions should be followed with respect to dangerous members of these groups of organisms.

Gram-positive cocci in some cases, on-target extraction with formic acid is applied to improve extraction. For molds and acid-fast bacteria, a multistep extraction procedure in a reaction tube generally improves the spectrum quality and identification performance. Most importantly, the various extraction procedures are not competitive. That is, in the case where a direct deposit does not yield an identification result, an on-target or tube extraction method likely improves the spectrum quality without increasing the risk of obtaining false-positive results.

In general, and for most microbial species, approximately 10^4 cells per sample is the minimum biomass required to yield a spectrum of sufficient quality for identification. Therefore, only urine samples from urethritis patients are amenable to direct MALDI-TOF MS without an incubation step. This is because infected urine samples contain sufficient numbers of bacteria ($>10^3$ cells/mL) that can be concentrated from larger volumes (>10 mL). Direct identification from urine is not always easy, however, because of the presence of potentially obscuring defensin peaks that can interfere with identification. Working directly with clinical samples also can be impaired by the fact that samples might be contaminated or contain more than one microorganism, which will compromise the identification results.

Without a doubt, MALDI-TOF MS provides a very strong methodology for microbial species identification with accuracies and specificities depending upon the microbial species involved or the spectrum of clinical samples analyzed. Identification of isolates that are considered to be difficult-to-identify microbes with conventional procedures generally is not problematic with MALDI-TOF MS and depends primarily on the integrity of the database. The reason is simply that all microorganisms produce housekeeping proteins, such as ribosomal proteins, that allow for mass spectral identification. The use of MALDI-TOF MS-based identification is the ease and relatively low cost of a single identification result, which permits the selection of multiple colonies from agar plates. These advantages will lead to a widened view of infectious and commensal microorganisms and, possibly, to the recognition of new pathogens or specific disease-related microbial consortia.

Chapter 16

Guidance for Molecular Clinical Laboratory

Chapter Outline

THE NEED FOR MOLECULAR ANALYSIS

The Request of Molecular Testing

The vast majority of medical disciplines can request molecular analysis to define a treatment strategy for individual patients. In many cases, a multidisciplinary team will decide to request a test. The process should ensure that the request is made appropriately, and that every patient who needs a test is offered one in a timely manner. This is one of the most difficult aspects of operating a molecular diagnosis laboratory in hospital. The goal of the molecular diagnosis procedure is to select the right test for each patient, generating accurate, clinically relevant results to optimally influence clinical care and to conserve health care resources. The process of diagnostic analysis begins with evaluation, selection, and implementation of appropriate diagnostic tests for the clinical setting. The second step is to incorporate guidance for health care providers regarding the use of testing for appropriate patients. The final process is to ensure timely sample collection, transport, and processing, and timely reporting of results. Molecular diagnostic testing is still expensive, although costs are decreasing. The decision to implement molecular diagnosis should be based on a business plan, taking into account the costs in time and money of specimen collection from the patients or retrieval from pathology archives when the result becomes clinically necessary. Additionally, the amount of tissue that can be obtained versus that required for molecular diagnosis should be included in considerations of the sampling strategy. Flexibility also is needed because technology and requirements are changing rapidly and frequent reassessment of clinical need is essential. Good communication is required among the laboratory, the physicians, and administrators to ensure that necessary tests are requested timely.

Diagnostic Molecular Biology. https://doi.org/10.1016/B978-0-12-802823-0.00016-X

Workflow of Molecular Test

All molecular diagnosis workflows can be divided into preanalytical, analytical and postanalytical process steps, where different standards are applicable. Preanalytical standards aim at quality, format, and amount of specimen/sample, which should be documented in detail. The analytical standards consider the proof of these aspects, concerning DNA/RNA extraction, quantification, and purity determination via fluorometry or spectrophotometry. If a sample does not meet the defined minimum requirements, the laboratory should refuse to process the sample and order a new one. The laboratory must know the original source of the sample (blood, tissue, etc.), request which methods were used to isolate the sample, and which quantification was performed. Postanalytical analysis mainly focuses on the documentation and bioinformatics process.

PREANALYTICAL STANDARDS

Reduction of Identification Errors

Since the molecular diagnosis laboratory has to be able to handle multiple sample types, this raises a number of challenges when the needs of molecular analysis are included within a diagnostic pathway. Though many of the issues are common to any diagnostic test, there is potential for samples to be misidentified, and careful attention should be given to this danger.

It is the responsibility of the person taking the specimen to identify the patient and the sample required. He or she must ensure that the sample is correctly labeled, fixed in formalin (where necessary), and/or dispatched to the laboratory. At least two patient identifiers should be used before collecting a specimen. All identifying labels must be attached to specimen containers at the time of collection, not later. Identification (ID) errors related to laboratory specimens can involve misidentification of a patient or a patient specimen. Patient ID errors and sample mislabeling at the time of collection can cause a serious problem in clinical laboratories. Misidentified specimens not only adversely affect patient care but also increase the cost of health care delivery. ID errors can have serious consequences for patients, including missed or delayed diagnosis, incorrect or unnecessary treatment, patient injury, and severe transfusion reactions. Patient and specimen ID errors have been reported at rates of 0.005%–1.12% among various institutions. Normally, 55.5% of ID errors were reported to be associated with primary specimen labeling errors. It is estimated that approximately 1 in 18 ID errors result in adverse events, and the cost of misidentified specimens is estimated to be around $280,000 (US) per million specimens.

Most ID errors occur in wards or in the emergency department (ED). These departments are labor intensive and not under the control of the clinical laboratory. Reducing patient misidentification could be approached using nontechnical methods (patient safety guidelines and procedures) and/or technical solutions (ID wristbands containing a barcode). Barcoding is effective for reducing patient ID errors in diverse hospital settings.

Sample Receipt and Handling

A dedicated process should be in place for the transport and the receipt of samples, it should be suitably staffed and allow accession numbers to be allocated to samples as they enter the laboratory (as per ISO 15189 and ISO 9001). Most laboratory information management systems (LIMS) allow this, but few have as yet acknowledged the need for integration of molecular diagnosis needs. Barcoding of samples is good practice and is encouraged. All samples should be tracked from patient to report.

Tissue, blood, and fluid processing within the laboratory should be performed to defined standard operation procedures (SOPs). While most tissue biopsy samples are received fixed, fresh tissue also can be sent to the laboratory for intraoperative frozen sectioning or molecular analysis. This requires separate SOPs and dedicated portering, nursing, and laboratory staff. Adequate processing of tissue samples is essential in a reliable molecular diagnosis. SOPs should be established for secure and, where possible, controlled collection, handling, and storage of tissue and blood samples, and their fractions (including extracted nucleic acids). The SOPs should follow ISO 15189 and any national guidelines. Few laboratories maintain their stores of tissue blocks and slides in temperature-controlled environments. Blood samples usually are discarded within days of receipt. Blood and blood-derived products are considered biohazards and should be handled accordingly. Fresh tissue poses similar risk, but FFPE tissue is regarded as safe for handling, though chemicals involved in tissue processing need careful storage and handling.

Tissue samples go through a macroscopic examination and cut-up stage, when blocks are taken and placed in bar-coded and/or numbered plastic cassettes for processing. Spare tissue in formalin can be retained at this stage. Vacuum or microwave-based processors are temperature-controlled devices that ensure excellent penetration of graded alcohols and xylene substitutes, allowing paraffin embedding. The temperature for different machines and protocols can vary substantially, and this can affect nucleic acid recovery. The use of fixed tissue that has been frozen previously is not advised, and decalcification may reduce DNA and particularly RNA recovery. Close collaboration between clinical teams and molecular diagnosis laboratories is advised, and it is critical that changes in processing are communicated to the clinical laboratory and adapted where necessary.

Minimization of Cross-Contamination

For molecular testing, it is essential to take strict precautions to prevent cross-contamination between samples. It is best practice to replace the knife (blades) regularly especially for FFPE block. Additionally, disposable plasticware should be used to transfer sections to glass slides (if sections are used). The use of water baths to stretch sections can be avoided by using a droplet of PCR-quality water on glass slides. Care should be taken that decontamination procedures do not reduce DNA or RNA levels required for assay when sections are cut. The use of substances likely to inhibit PCR should be avoided.

Blood samples usually are spun down at low speed to pellet red cells and retrieve the buffy coat and plasma fractions. Processing should begin as soon as possible (ideally, within 30 min) after sampling. Collection time should be indicated on the tube, and the elapsed period between sampling and processing should be recorded. Cells usually are removed from plasma by centrifugation. Again, a protocol should be observed strictly. It is feasible to use density centrifugation to obtain white cells, exosomes, and plasma from the same sample. Specimens can be flagged on many LIMS for molecular analysis. Aliquoted plasma samples can be stored at $-80°C$.

Quality Control Metrics

The implementation of molecular diagnostic technology in the clinical diagnostic environment is complex and requires significant changes in the infrastructure of the laboratory. For example, because of the significant improvement and changes of the sequencing technology, the technical aspects of quality management for test and system validation, quality control, quality assurance, and the measurement of quality characteristics is more challenging and with degrees of complexity that vary depending on the platform and comprehensiveness of the NGS assays. Quality control measures must be established to the specific NGS assay and must be applied to every run.

Specimen Requirements

Specimen requirements for molecular diagnosis can be highly variable, depending on the disease setting, individual practices of acquisition, tissue handling, processing protocols, and testing method. Laboratories must validate all potential specimen types (i.e., FFPE, fresh tissue, blood/bone marrow, cell-free DNA) and establish criteria for specimen acceptability. Nonvalidated specimens should be considered as inappropriate for NGS testing and rejected. Laboratories also must establish individual requirements of minimum nucleic acids content to match the sensitivity of the assay and qualification criteria for quality and quantity of nucleic acids based on the assay validation to ensure maximal performance and accurate results. Before testing, samples should be reviewed by an appropriately trained laboratory personnel for specimen suitability, including specimen type, quality and quantity, and selection of areas for identifications. Minor deviations from the established validation criteria can be warranted in rare cases. Furthermore, the acceptability and reporting of unique cases are at the discretion of the laboratory director or designee.

DNA Requirements

Nucleic acids quantification is a critical component in molecular diagnosis. This is because it affects not only the sensitivity of the assay but also the number and the accuracy of the variants reported. As quality and quantity of nucleic acids can vary widely based on the source and extraction method, and requirements can differ depending on the diagnostic method, each laboratory must establish a standardized and cost-effective workflow for nucleic acid quantitation to guarantee accurate and reproducible results. Quality control of the nucleic acid samples should focus on the quality and quantity. To this end, the quality control methods must be well developed to clearly define the preanalytical sample quality, accurately assess the minimum amount of DNA/RNA required to detect a variant depending on the sensitivity of the assay, and must establish a way to adjust DNA/RNA inputs.

There are multiple methods of DNA and RNA quantification as discussed in Chapter 7. Spectrophotometric methods, although the most commonly used across laboratories, should be avoided. These measure total nucleic acids (including double-stranded DNA, single-stranded DNA, oligo, and free nucleotides) as well as impurities that markedly overestimate amplifiable DNA or RNA content. Double-stranded DNA-specific fluorometric quantitation methods are recommended to evaluate the DNA quantity. Fluorescence-based DNA measurements are far lower than those quantified by spectrophotometry, but results are more accurate and precise, particularly at lower concentration ranges.

Library Qualification and Quantification Requirements for NGS

Accurate library qualification and quantification are critical to obtaining optimal NGS data. Underloading or using libraries with inappropriate fragment size ranges can lead to reduced coverage and read depth. Conversely, overloading would cause saturation of the flow cell or beads and lead to read problems.

Library qualification should be performed to detect potential problems such as high percentage of short DNA fragments or adapter dimers. Electrophoretic methods generally are used to assess the overall range of DNA fragment sizes constituting the library. The laboratory must standardize protocols to obtain fragment sizes of the expected molecular weight in narrow range. Specific measures should be taken to minimize primer dimers, adapter dimers, and broader bands of higher molecular weight. Primer dimers that usually are minimized by the use of magnetic beads do not constitute a significant problem unless they dominate the reaction.

Choice of Analytical Method

Selecting the right test for the clinical setting involves the evaluation of test performance, laboratory feasibility, and cost versus value. Evaluation of basic intrinsic test characteristics, such as sensitivity and specificity, is essential, but findings can be challenging to interpret when there is no clear gold standard. This is a particular challenge for molecular methods when conventional culture-based or serological techniques are unable to confirm the presence of infection and reliance on clinical findings is necessary. For instance, a multiplex PCR assay detecting pneumococcal DNA in a spinal fluid sample from a patient with suspected meningitis who was pretreated with antibiotics and has negative culture results could indicate a true-positive result or a false-positive result, depending on the clinical scenario. The consideration of positive and negative predictive values of a test, specific to the local incidence of disease, is more important than sensitivity and specificity. Clinically, these statistics assess whether a clinician can trust a positive test result and whether the test can be used to rule out an infection. Multiplex panels, however, challenge traditional notions of positive and negative predictive values. This is because these values can be calculated for individual pathogens, for all pathogens in the panel, or for a clinical syndrome. While a rapid multiplex PCR assay might have a high negative predictive value for an individual pathogen or for all pathogens included in the panel, it might have a low overall negative predictive value for the clinical condition of meningitis if bacteria that commonly cause meningitis are not included in the panel, limiting the clinical utility of the assay for ruling out bacterial meningitis.

To ensure that the diagnostic technology selected is appropriate for the clinical setting, it is important to consider testing volumes, diagnostic yield, and the feasibility of performing the test in the laboratory setting. Clinician guidance can be useful to determine the potential utility of newly emerging diagnostic tests with regard to medical decision-making in a particular setting. Retrospective data can be collected to predict testing volumes and diagnostic yield in that setting. Normally, a single-pathogen, point-of-care, influenza PCR test might be appropriate to triage and to treat patients with respiratory symptoms in an outpatient setting. On the other hand, a more extensive rapid multiplex PCR assay for respiratory pathogens might be appropriate to diagnose ventilator-associated pneumonia in the critical care unit of a tertiary care referral center. Factors determining the laboratory feasibility of implementing a new rapid diagnostic test include the technologist training required, the time to perform the test, and the way in which testing would fit into the laboratory workflow. It is important to consider whether the testing would supplant or be conducted in addition to existing diagnostic tests. For example, with multiplex PCR platforms for cerebrospinal fluid testing, culture is still needed for detection of bacterial pathogens not detected by the panel, confirmation of positive results, and susceptibility testing. Multiplex PCR assays for stool testing, however, might supplant monoplex PCR assays for viral and parasitic pathogens and save the time and expense of conducting multiple monoplex tests with a single sample.

Cost-effectiveness and potential clinical impact are important factors for deciding whether to adopt an emerging diagnostic technology in the clinical microbiology laboratory. Cost analysis is often a priority of hospital administrators, but it can be challenging, given the fact that sometimes it is urgent to obtain the results. Although a simple analysis can be used to compare costs and charges for existing diagnostic tests versus new rapid diagnostic tests, a true cost-value analysis should include back-end cost savings of decreases in resource utilization, as well as effects on morbidity and mortality rates. Because of the fast turnaround times for the rapid diagnostic technology, the potential clinical impact can be estimated to determine what changes in medical decision-making could occur if results were available at that time point. For instance, a rapid multiplex PCR test of spinal fluid for which results are available in 2 h is predicted to decrease acyclovir exposure by 2 to 3 days for nearly one-half of children undergoing lumbar punctures.

Many laboratories are moving away from single-gene assays toward panel testing. Other considerations in making the choice of analytical method are the number of samples that require testing, the number of genes being tested, and the percentage of mutations tested for in each gene. Most clinical laboratories now have real-time PCR machines that are perfectly capable of running well-established kits that cover 95% of the actionable mutations in a particular gene. Sanger and pyrosequencing methods also are available widely. In high throughput units, it can be economical to use NGS platforms, which have the advantage of a more extensive coverage of an increasing number of genes of interest. The amount of tissue available is often the limiting factor as to the extent of the testing performed and might influence which technology is chosen. The degree of staff expertise, equipment, and infrastructure required also should be considered. Many workflows in molecular diagnosis require separation of stages in different compartments (ideally rooms) to prevent DNA and RNA contamination. Separation of pre-PCR and post-PCR steps is particularly important.

ISO 15189 requires thorough validation of in-house tests compared to certified in vitro diagnostic (CE-IVD) tests. Because most tests are only part of the whole process, the complete procedure, including preanalytic, analytic and postanalytic phases, should be validated in-depth. These considerations typically determine the choice of technology. In practice, the options are based on real-time PCR or sequencing techniques. The pace of change in technology is considerable and no recommendations are made here, other than to state that instruments and kits usually should be IVD CE-marked, and assays validated against existing performance specifications. Validation and verification of tests is important, both within consortia of companies, academic and hospital laboratories, and by individual laboratories offering tests for clinical use. Batch-to-batch variation is a particular problem for complex reagents, and several mutation detection kits have shown problematic fluctuations in performance in recent years. ISO 15189 requires checks on new batches of reagents on the premise that nothing should be assumed to work without rigorous checking.

It is likely that panel testing, including array and NGS, will become widespread within a few years. This mainly is because these techniques provide information on mutations in multiple genes without requiring unattainable amounts of DNA or RNA. It should be noted that NGS is sensitive to low-level genetic alteration during PCR steps in the process, and the incorporation of bioinformatics rules in data analysis are necessary to allow correct and appropriate interpretation.

After the appropriate rapid diagnostic test has been selected for implementation, the next step is directing testing toward the right patients, for whom the test results will affect clinical care. It is very important to realize that overuse of rapid diagnostic tests can add to health care costs without having a significant impact on patient care, whereas underuse can lead to suboptimal clinical outcomes. Inappropriate use of these tests in cases with low pretest probability can lead to misinformation from false-positive results potentially misleading clinicians.

ANALYTICAL PHASE

DNA and RNA Extraction

Ensuring good DNA/RNA quality begins during isolation and extraction. The majority of incubation steps must be kept at lower temperatures to inhibit possible nuclease activity while storing DNA/RNA samples permanently on ice, and to avoid repeated freeze-thaw cycles. It also has to be taken into account that there are some special sample quality requisitions for NGS, such as Pacific Biosciences RS II, because of omission of DNA amplification. These requisitions include double-stranded format of DNA, prevention of pH extremes (< 6 or > 9), and absence of chelating agents, detergents, divalent metal cations, denaturants and RNA. The quality assessment should yield an $OD_{260/280}$ ratio of 1.8 to 2.0 and an $OD_{260/230}$ ratio of 2.0 to 2.2, with the latter being an additional value for purity determination. It also is recommended that an initial DNA damage repair for genomic DNA sequencing applications be performed, and the quality of DNA should be always assessed before library preparation. This will eliminate any possible uncertain factors and ensure that the DNA input amount and following amplification steps are major bias-related factors. Depending on the application, using 30–50 ng of DNA input is recommended.

Many companies produce kits, though few are sold for clinical use, and it is the responsibility of the laboratory to validate and verify any methods performed against existing standards. Consideration about which method to use should consider throughput, required quality and quantity of DNA or RNA, and intended methods of analysis.

Manual methods are commonly used. The majority of manual methods use precipitation-wash steps with centrifugation of spin columns or filters. Even kits require training and considerable skill, but they can provide excellent results. Special care should be taken to avoid sample misidentification and contamination. Careful adherence to SOPs is important; the use of checklists and careful staff training can help.

Automated methods are useful because they can save on staff time and can help to prevent sample misidentification. They tend to use magnetic bead extraction methods, which are suited to robotic systems. It is important to try such systems before their implementation.

Storage of DNA and RNA

Storage of DNA and RNA should be controlled carefully. Accession logs or barcoded vials can be used to prevent sample misidentification. Temperature logs should be maintained. In general, it is good practice to

- Store extracted DNA and RNA samples, clearly labeled, at −20°C or lower
- Store PCR products in a separate freezer at −20°C or lower
- Store sequencing libraries in a separate freezer at −20°C or lower.

There is no standard protocol about the effects of long storage time on tissue and extracted DNA or RNA for different analytical methods. In general, DNA, or cDNA, is more stable than RNA and is likely to survive many years under the recommended conditions.

Standard Devices for Sample Quality Assessment

Capillary gel electrophoresis is used almost exclusively for quality control in sequencing projects because it can be used to investigate fragment size distribution and final library quality assessment. A fluorometer, which offers fluorometric quantitation, is used commonly to determine the input DNA/RNA amount and to check sample quality at appropriate steps of the sequencing or PCR workflow. Most frequently used devices for this purpose are the Fluorometer and the NanoDrop. For accurate quantification of DNA/RNA at certain workflow steps and to determine the final library quantity during NGS, established systems such as qPCR or digital droplet PCR are recommended.

Validation of Comprehensive Molecular Assays

Before testing patient samples, clinical laboratories must establish the analytical validity of the intended tests because the quantitative analytical performance of a laboratory test does not necessarily predict performance at a clinical level. The intrinsic biological variability of disease has the greatest impact on the clinical sensitivity and specificity of molecular diagnosis. The intrinsic biological variability results in a scenario in which, generally, only a subset of cases of a specific disease harbors a characteristic mutation; more than one characteristic genetic abnormality is associated with a specific disease; more than one disease can share the same mutation and the clinical significance of the same mutation will be different for different disease; and a mutation that is characteristic of a disease can be identified in a subset of healthy individuals. Thus, even an NGS assay with perfect analytical performance (i.e., 100% analytical sensitivity and 100% analytical specificity) could have a lower clinical sensitivity and specificity, and the differences between analytical performance and clinical performance need to be addressed. The clinical laboratory must know the intended use of the panel, and whether it is going to be used for tumor diagnosis, prognostication, or making therapeutic decisions. Some NGS assays will sequence a more limited target region (e.g., a hotspot test), which might be sufficient if intended use is limited to specific genetic alterations. The panel might have a lower clinical sensitivity than a more comprehensive test if its intended use is tumor diagnosis and prognostication. This is because other variants outside the hotspot region will not be detected. Therefore, the clinical validity (i.e., clinical sensitivity and specificity) and clinical utility of the molecular diagnosis, especially the NGS, needs to be determined during the assay design and evaluated during the validation process. The validation protocol has to include the NGS panel content, the intended use of the test, potential clinical validity and utility, and references. Currently, validation occurs according to the criteria of the local laboratory quality control management system, with no global consensus for validation of molecular diagnosis.

Implementation of new technology in the laboratory needs determination of test-conditions including DNA-input, setup of SOPs, determination of coverage needed and testing software applications. The desired sensitivity of the test in diagnostic samples determines the read-depth, which reflects how often a genomic region has been sequenced. A read-depth per amplicon of $500\times$ at minimum, which enables accurate detection of low frequency allelic variants, is necessary. Validation of broad-spectrum mutation detection requires essentially the same process as for other conventional mutational analysis. For this purpose, it is necessary to address technology specificity by analysis of different known mutations in parallel with a golden standard, such as Sanger sequencing, and to assess sensitivity and reproducibility by replicating the analyses of samples with different mutant allele frequencies. Commercially available predesigned reference standards might be used especially for detection of low abundance mutations. During the validation process the laboratory must not only demonstrate that the test works well but also that reliable results are provided within the desired turnaround time.

An essential difference with standard Sanger sequencing is that, during targeted NGS, tens to hundreds of amplicons are amplified in a multiplex PCR that can be analyzed simultaneously by massive parallel sequencing. As a consequence, NGS of 50 amplicons at a read depth of $500\times$ yields 25,000 independent sequence reads of one sample. The complexity of one NGS analysis multiplies when simultaneously running multiple samples on the same chip. In this strategy in which samples are pooled, data identification is achieved by incorporating a sample-specific DNA barcode in the amplified DNA fragments. Because of the enormous complexity of these data, one has to rely on bioinformatic analysis using appropriate software much more than with Sanger sequencing. Validation of software in the NGS-pipeline is required. In general, exchange of raw datasets between laboratories that use different software packages should be part of software validation in order to establish the accuracy. In addition, new software updates need to be validated, by analysis of prior NGS-datasets covering various simple and complex mutations. Finally, the use of raw NGS datasets should be part of in silico external assessment schemes.

The validation process should assess the entire workflow, including the molecular report. In addition, a molecular diagnostic report should include which genes or regions of the genes are investigated, information about the gene coverage, the sensitivity of the detection, and the frequency of the detected mutation. Evaluation of the mutant allele frequency in the context of the percentage of neoplastic cells should be estimated.

Performance of Molecular Tests

It is essential to monitor the performance of all tests within clinical laboratories to ensure that all measurable parameters remain within acceptable limits, as defined in the validation report for the test. In most cases, these are defined statistically based on known variations. Using simple statistics, it is possible to establish confidence intervals (CIs) within which the assay is performing correctly. For example, commercial kits often provide internal controls or reference genes, the levels of which should not vary for the same DNA input. While variation between samples is common, it should not exceed certain parameters, definable from the standard operating procedure. In-house tests should use externally sourced controls wherever possible.

Turnaround time often is used by health care managers to judge services. The timeliness of reports is critical to patient care and depends on patient pathways in practice in individual hospitals. Acceptability of longer turnaround time can permit greater control of testing or batching of samples to improve the cost effectiveness of testing. For all laboratories, regular participation in external quality assessment is needed to verify and improve the quality of testing.

Continuity planning is another issue. The solution to problems with equipment because of breakdowns or limited staff availability should not be allowed to interrupt a service. Sending samples away to another laboratory might be necessary, but ensuring that alternative equipment or staff is available can prevent cessation of service and avoid patients having to wait for treatment decisions. The referring laboratory should check the service standards of laboratories used for referral.

POSTANALYTICAL ANALYSIS

Data Analysis

For a rapid diagnostic test to have rapid impact, it is essential that providers be notified the results in real time. Active readback reporting ensures that results have been received in a timely manner and is superior to passive reporting that requires the clinician to continually check the medical record for results. A rapid diagnostic test result that remains unchecked in the medical record wastes expensive health care resources and potentially leads to delayed optimization of care.

Understanding of Test Results

As the complexity of molecular diagnostic testing increases, the communication of clinically relevant results becomes more important to ensure proper interpretation by clinicians. The reporting of increasingly specific, species-level identification of pathogens not widely familiar to general clinicians can create confusion. This confusion can be avoided by grouping species and using more familiar, clinically relevant nomenclature or providing interpretive decision support. Clinicians also can require decision support to navigate some of the pitfalls in the interpretation of rapid molecular infectious disease diagnostics. The detection of nucleic acids by molecular diagnostics does not always equate to detection of a viable organism that is the cause of the patient's disease process. Pathogens could represent colonization, asymptomatic infection, reactivation in the setting of acute infection, chromosomal integration, or prolonged shedding after an unrelated prior infection. For instance, detection of a rhinovirus/enterovirus in a respiratory pathogen panel for a febrile infant without respiratory symptoms could reflect shedding from a prior respiratory infection, an asymptomatic infection, or a true source of fever. With more broadly based syndromic testing panels and NGS, pathogens that are unlikely to be the cause of disease might be detected, creating clinical confusion. For example, gastrointestinal panels might detect *Escherichia coli* species that are asymptomatically carried in a large portion of the population or *Clostridium difficile* in children younger than 1 year of age, for whom this result more likely represents carriage than infection requiring treatment. Selective reporting of relevant pathogens is one approach to deal with these issues, but it requires full disclosure of the pathogens being reported versus withheld from reporting for the panel. The detection of multiple pathogens within a single specimen suggest that some of which might be causing disease and others of which might be contaminants or innocent bystanders. As such, it can be challenging for clinicians to interpret. The decision support strategies can assist clinicians in dealing with these new challenges posed by rapid molecular diagnostics.

Decision-Making Process

After a test result is delivered and interpreted properly, prompt and appropriate clinical action must be taken to adopt the suitable care optimally. The first step in this process is creating evidence-based clinical practice guideline recommendations based on test results. National guidelines are useful in creating consensus of empirical treatment recommendations to standardize care. Providing decision support tools with rapid diagnostic results is more effective than traditional methods of reporting results alone in the electronic medical record or having clinical laboratory technicians call providers with results.

As new molecular diagnostic tests for infectious diseases are implemented, real-time feedback from end users and troubleshooting mechanisms are needed to optimize practices. During the implementation process, one should design protocols and algorithms according to the diverse perspectives of the microbiology laboratory, infectious diseases, infection control, and, most importantly, end user clinicians. Everyone involved in the process can help gather ongoing feedback during the implementation process. Provider surveys are another method to obtain feedback regarding the accuracy, timeliness, impact, and acceptability of the new test. During the rollout period, telephone and email hotlines can help troubleshoot unforeseen issues and provide education and support for end users who are not familiar with the new technology.

Molecular Diagnostic Report

A molecular diagnostic report should convey accurately the information the clinician needs to treat the patient on whom the test was performed, with sufficient information to allow correct interpretation of the result. The key, therefore, is accurate communication. This is often a challenge, given the different groups of clinicians involved and the variable capabilities of laboratory information management systems. There is an urgent need for greater communication among scientists, laboratory personnel, and physicians, so that all groups are educated to use and interpret tests, their results, and associated treatments appropriately.

The following are the three key areas that should be included in reports. A typical NGS clinical report is shown in Table 16.1.

1. Patient identification
   The patient must be identified correctly. Laboratories require a minimum of two unique patient identifiers plus a unique sample identifier on the request form and report. These include the patient name (family name), date of birth, hospital, and national identification number. The ward or service, date of biopsy, and referring clinician name also should be stated.
2. Reporting style and content
   Long reports are rarely read in full. One page, or better still, single-screen reports are preferred, provided that they are legible. If reports are more than one page long, then each page should have appropriate patient identifiers in order to be linked to the correct patient, and the pages should be numbered, that is, page 1 of 3, so the reader of the report is aware if any pages are missing (refer to ISO 15189 for guidance). Clear presentation of the results, the tests performed, and any limitations of the tests. For some tests, the percentage of the sample occupied by the target of the test (e.g., neoplastic cells) is important. Ideally, the report should contain basic information about fixation (fixative, time of fixation) for quality control, though this might be held within the departmental records rather than reported. The name of the

TABLE 16.1 Clinical NGS Report Information

Element	Example
Patient Identification	Registration number, age, gender, and ordering physician
Specimen Type	Formalin-fixed paraffin embedded, fresh, or frozen
Pathological Diagnosis	Name of disease
Tissue Sample Identification	Specimen number, block number
Important Dates	Date on reception or on report
NGS Method	Amplicon-based or hybridization capture-based
	Version of Reference Genome Used
	Sequencing technique used
	Key Quality Control Metrics including mean target coverage, percentage of selected bases, duplication rate
	Variant details according to Human Genome Variation Society mutation nomenclature
	Genes or Genomic Regions such as exonic regions of gene A, B, C, etc...
Interpretation and Summary	

individual taking responsibility for the test and contact details is important, and some form of authorization with date. Because of legal regulations, the laboratory personnel who is finally responsible for the report must be clearly identifiable. Spell checkers are useful to prevent spelling and typographical errors.

3. Interpretation

The results of the test must be reported correctly and interpretation provided, particularly where this involves a treatment decision. Genotyping results should be given according to Human Genome Variation Society nomenclature (http://www.hgvs.org/content/guidelines) at the DNA level and the protein level, with appropriate reference sequences provided. External quality assurance schemes require laboratories to provide accurate interpretation of the results obtained and include any relevant advice. Care should be taken to ensure this advice is as up to date as possible. If conflicting data are published, this should be included in the report. In view of the growth of molecular diagnostic tests in the hospital, it is important that the clinical interpretation needs to be very accurate and precise. The type and extent of molecular analysis used should be described clearly to allow oncologists to request further testing if warranted by the clinical situation.

GENERAL ADMINISTRATIVE GUIDELINES

Quality Assurance for Diagnostic Laboratories

Given the implications of molecular analyses on treatment of patients, high quality is required of tests and laboratory performance. Procedures for continuous measurement and improvement of laboratory performance should be integrated fully in the lab's internal quality assurance (IQA) system to ensure a consistently high standard of performance. External quality assurance (EQA) programs, also known as proficiency testing (PT), are inevitable for monitoring of performance. It is necessary to periodically assess the analytical performance of molecular tests by inter-laboratory comparison, which will assist laboratories in monitoring their assays and improve assay performance and evaluation of results. One of the requirements to become accredited according to ISO 15189 is to participate in EQA programs.

IQA is necessary to ensure high assay reproducibility and performance and enable detection and correction of errors in daily practice. Assays should be performed according to SOPs using appropriate positive and negative controls. Implementation of new tests requires a validation procedure using predefined parameters. Dedicated trained personnel (laboratory technicians and members of the medical staff) as well as a manageable, easy-to-use quality management system are necessary to maintain high level of performance and improve test results or logistics. A laboratory quality manager is essential to take charge of participation and performance in quality assurance schemes and to organize internal and external audits. The wide variety of tests in a molecular diagnosis laboratory, the rapid technological advances, and high complexity of the tests require experienced personnel. Both technicians and supervisors should have an adequate theoretical and technical training in molecular biology techniques and should remain up-to-date through peer-reviewed literature and by regularly attending meetings and symposia.

Standardization and Quality Management

Standards is the critical guidelines to ensure comparability and exchange of experimental data to accelerate the innovation process and aid improvement of transferability, transparency, and reproducibility of results. Standardization can potentially decrease costs, therefore enabling improved financial planning and scheduling and a possible expansion of services. Standardization, however, is a complex topic characterized by several problems and challenges such as failure of initializing standards, missing consensus and deadlocks, and incompatible implementations of finished standards. In particular, the formulation of NGS standards also requires an extensive collection and evaluation of appropriate platform-dependent and independent information as well as comparative analysis of different sequencing systems. The results gained from successful standardization of NGS might transfer to other fields in life sciences. An overview of the general NGS workflow annotated with single steps and appropriate quality control (QC) checkpoints is shown in Fig. 16.1.

Quality Management (QM) and Quality Assurance (QA)

A good starting point in standardization measures is the introduction of quality documentation through a system of quality management (QM). The QA program, an important component of QM, should contain predetermined QC checkpoints for monitoring QA and extensive documentation, including used devices, reagent lot numbers, reagent expiration date, and any deviation from standard procedures. The QA program also should contain QC methods for contamination identification at several stages within the diagnosis workflow. The detailed documentation of SOPs is critical for quality assurance of a complex, multistep molecular diagnosis process. All SOPs of each step of the wet bench process must be documented so that each step can be traced. This includes documentation of all methods, reagents, instruments, and controls (if applicable). Most of the documentations should be similar to those of current standard single-gene testing, but those specific to NGS testing include detailed information regarding captured regions, such as genomic coordinates of captured probes and lists of genes, and target enrichment protocols. Clinical laboratories that process different types of samples, such as FFPE samples or blood, should establish SOPs for each validated sample type. Metrics for quality control to assess run status also must be documented. Examples include mean target coverage, percentage of reads that map to target regions, and the fraction of bases meeting specified quality and coverage thresholds. Laboratories must define and document acceptance or rejection criteria for each step of the wet bench process, such as DNA/RNA extraction, library preparation, and sequencing. Frequently, there is an obvious lack of such SOPs documentation within sequencing experiments. Thus, it is crucial to develop and establish procedures, operating and inspection instructions, and quality records. Verification documents, particularly for providing a string of documents for the verifiable origin of sequencing data, are essential, especially the quality records that could act as a certificate for customers and the general documentation that would improve the traceability and transparency with the aim to prove the reliability of results.

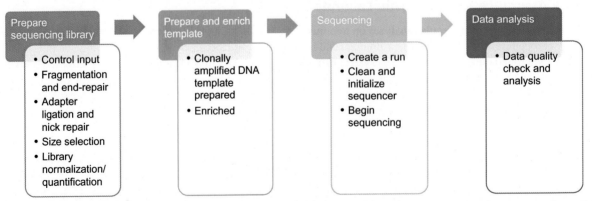

FIG. 16.1 Overview of the general NGS workflow. Each main step, including library and template preparation, enrichment, sequencing, and data analysis are divided into sub-steps.

Equipment Maintenance

Laboratory personnel must have regular access to all equipment required to perform all the analyses within the scope of the laboratory. SOPs and supporting documentation, such as maintenance logs, must exist to demonstrate and provide evidence that all instrumentation and equipment are validated, operated, inspected, maintained, tested, and standardized to ensure optimal quality of assay results. All preventive maintenance and calibrations must be scheduled and performed at least as frequently as suggested by the equipment manufacturers to ensure continued accuracy, precision, and extended usable life of the equipment.

Equipment performance and maintenance must be consistently and routinely documented and reviewed by the laboratory personnel. It is important for a clinical laboratory to maintain all equipment in a working order, not only because the equipment is expensive, but also to ensure accurate results. The laboratory must keep documentation of all scheduled preventive maintenance, unscheduled maintenance, service records, and calibrations for all equipment, as defined by the laboratory or institution. This documentation should be readily accessible. Laboratory personnel should retain preventive maintenance and service logs until otherwise instructed. A supervisory staff member must review, sign, and date all documentation of equipment maintenance at least monthly. All laboratory equipment should be listed on an inventory document.

Staff must conduct all preventive maintenance and service per manufacturer specifications following these guidelines:

- Staff must keep all equipment clean, avoiding any buildup of dust, dirt, and spills that might adversely affect personnel safety or equipment performance.
- The laboratory must employ and adhere to standard daily, weekly, and/or monthly routine maintenance plans for all equipment, and record completion of these tasks on the appropriate logs in a timely fashion.
- A written policy/procedure should be in place, explaining how temperatures are monitored during the absence of laboratory staff.
- Backup instruments/correlation testing: If a laboratory uses more than one instrument to perform the same test, the primary and backup instruments should be compared to each other to determine the consistency of results. It is preferable that these comparisons be performed using study-participant samples. However, where availability or stability precludes the use of study-participant samples, the use of QC, EQA, or other reference materials might be necessary.

In addition to these general guidelines, the laboratory must observe and document the specific service guidelines described below. Trained laboratory staff or certified contractors should perform tests or check on the following items:

- Adjustable and fixed-volume automatic pipettors: Check for volumetric accuracy and reproducibility, and recalibrate as necessary before placing in service initially and at specific defined intervals, at least once every 6 months. Pipettor inaccuracy is one of the most common sources of laboratory error.
- Thermometers: Test against a standardized thermometric device annually.
- Refrigerators and freezers: Establish tolerance limits for temperatures. The refrigerator's temperature tolerance limits must reflect the most stringent needs of all reagents, supplies, and specimens stored within it. Maintain daily records of temperatures and appropriate documentation of corrective action for out-of-range temperatures.
- Incubators and water bath: Establish tolerance limits for temperatures, carbon dioxide level, and humidity as applicable. Maintain daily record of temperatures and appropriate documentation of corrective action for out-of-range temperatures.
- Centrifuges: Measure operating speeds periodically with a tachometer. Document the readings. Maintain daily record of temperatures for refrigerated centrifuges. Verify performance of centrifuge timers by comparing to a known standard every 6 months.
- Autoclaves: Verify content processing using heat sensitive tape with each autoclave batch. Verify effective sterilization with an appropriate biological indicator weekly. Perform autoclave maintenance annually or as per manufacturer, including a pressure check and calibration of temperature device. Check autoclave mechanical timing device periodically. Maintain records of autoclave operation and maintenance in the equipment log.
- Analytical balances: Check accuracy with standard weights and class at a predetermined frequency. Document the results with an evaluation of their capacity. Service and calibrate periodically using qualified personnel. Maintain records of service and calibrations. Place the analytical balance in a stable location so that vibration does not affect the readings.
- Biosafety cabinets/laminar air flow hoods: Verify air intake grills are not obstructed. Certify cabinets/hoods annually by a trained service technician, certified maintenance department or company. Check daily for air flow as instructed by

manufacturer and document the results to verify the effectiveness of the hood's personnel and environmental protective functions. Clean the ultraviolet lamp, if used, weekly with 70% alcohol. Document daily and weekly cleaning.

- Generators: Maintain and keep readily accessible records to verify that the backup generator system is in place and operational. Follow SOPs that detail generator maintenance. Procedure should list the frequency and procedural steps of maintenance and testing. Supporting logs should document monthly checks on critical generator components, including fluid levels (oil, coolant, and fuel), belts, battery, testing startup, and operation.
- Out of service equipment: Any equipment that is out of service for any reason should be clearly identified and removed from the laboratory.

Safety

Clinical laboratories expose staff, and potentially the public, to a variety of hazards, including infectious patients, infectious patient specimens, and potentially hazardous chemicals and radiation hazards, as well as ergonomic/environmental hazards, fire safety, and epidemic emergency preparedness plans. Laboratories are obligated to identify hazards, implement safety strategies to contain the hazards, and continually audit existing practices to determine whether new ones are needed. Plans are required to meet staffing shortages and to manage the infected patient population in case of outbreak. Frequent safety reviews and disaster drills and general employee awareness help maintain a safe work environment.

Good safety practices benefit patients and employees and the laboratory's bottom line. Injuries and harmful exposures can negatively affect the laboratory financially, by reputation because of bad press, and through potential lawsuits, lost workdays and wages, damage to equipment, and poor staff morale. An injured person might be absent for an indefinite period and often cannot work at peak efficiency upon return. During this time off, the workload has to be absorbed by existing staff or through additional temporary services. Careful planning and compliance with the law will minimize undesired outcomes. Although inexperience might be the cause for some accidents, others result from ignoring known risks, pressure to increase workload, carelessness, fatigue, or mental preoccupation. These factors should be under control and should not lead to any major problem.

Biological Hazards

Biological hazards expose an unprotected individual to bacteria, viruses, parasites, or other biological entities that can result in injury. Exposure occurs from ingestion, inoculation, tactile contamination, contaminated needles, aerosol dispersion, or inhalation of infectious material from patients or their body fluids/tissues, supplies or materials with which they have been in contact. The potential also exists for inadvertent exposure to the public through direct contact with aerosolized infectious materials, improperly processed blood products, and inappropriately disposed of waste products.

The spread of HBV, HCV, HIV, and tuberculosis has focused the responsibility on each health care organization to protect its employees, patients, and the general public from infection. The CDC and the Occupational Safety and Health Administration (OSHA) have provided guidelines that recommend universal precautions in handling body fluids and human tissues for all patients, regardless of their blood-borne infection status. OSHA defines occupational exposure as reasonable anticipated skin, eye, mucous membrane, or percutaneous contact with blood or other potentially infectious materials that can result from the performance of an employee's duties. Blood, all other body fluids, and any unfixed tissue samples are considered potentially infectious for various blood-borne pathogens. In the laboratory, individuals should avoid mouth pipetting, consumption of food, smoking, applying cosmetics, potential needlestick situations, and leaving unprotected any skin, membranes, or open cuts. Aerosol contamination can result from inoculating loops, spills on laboratory counter, expelling a spray from needles, and centrifugation of infected fluids.

OSHA strongly recommends that gloves be used routinely as a barrier protection, especially when the health care worker has cuts or other open wounds on the skin, or during phlebotomy procedure. Employees must wash their hands after removal of gloves, after any contact with blood or body fluids, and between patients. Gloves should not be washed and reused because microorganisms that adhere to gloves are difficult to remove. Masks, protective eyewear, or face shields must be worn to prevent exposure from splashes to the mouth, eyes, or nose. All protective equipment that has the potential for coming into contact with infectious material, including laboratory coats, must be removed before leaving the laboratory area and must never be taken home or outside the laboratory. Laboratory coats must be cleaned onsite or by a professional. It is helpful for all employees to know what areas (offices, conference rooms, lounges, etc.) and equipment (telephones, keyboards, copy machines, etc.) are designated for laboratory work because of potential contamination. Avoid contamination by not wearing soiled gloves when in these areas or when using nonlaboratory equipment. Furthermore, the work area should be decontaminated on the regular basis (Table 16.2).

TABLE 16.2 Decontamination Options

Common Decontamination Agents
Heat
Ethylene oxide
2% Glutaraldehyde
10% Hydrogen peroxide
10% Formalin
5.25% Hypochlorite
Formaldehyde
Detergents
Phenols
Ultraviolet radiation
Ionizing radiation
Photo-oxidation

Chemical Hazards

Chemicals should be stored in certain sections of the laboratory, with every chemical having a specific storage place where it should be returned after use. Large containers should be stored near the floor. Chemicals that are not in use or have packaging that shows signs of leakage or corrosion should be replaced. Water reactive chemicals should not be stored where they could come into contact with water. Acids must be stored in an acid cabinet, and flammable chemicals must be stored in a flame cabinet. Storage of flammable liquids requires special treatment, such as stored containers of one gallon or less in a solvent cabinet.

Material Safety Data Sheet (MSDS) is a document that gives detailed information about a material and about any hazards associated with the material.

OSHA does not specify the exact format of the MSDS, nor even how the information should be broken into sections, and so MSDSs prepared by different manufacturers tend to look different and contain different information. Even MSDSs for the same chemical can be quite different if they were prepared by different manufacturers. It is the responsibility of the laboratory personnel to read the MSDS carefully before the chemicals are stored. Below is a brief guideline for handling the chemicals.

- Check the label to verify if the substance is correct before using it. Appropriate chemical-resistant gloves should be worn before handling any chemicals because all gloves are not universally protective against all chemicals.
- If transferring chemicals from their original containers, label chemical containers as to the contents, concentration, hazard, date, and user's initials. Use a spatula to remove a solid reagent from a container and avoid touching any chemicals with hands. Never use a metal spatula when working with peroxides because metals will decompose explosively with peroxides.
- Containers should be held away from the body when transferring a chemical or solution from one container to another. Use a hot water bath to heat flammable liquids. Chemicals should never be heated with a direct flame.
- Add concentrated acid to water slowly. Never add water to a concentrated acid.
- Weigh out or remove only the amount of chemical needed. Do not return the excess to its original container, but dispose of it properly in the appropriate waste container.
- Never touch, taste, or smell any reagents. Never place the container directly under nose and inhale the vapors.
- Use the laboratory chemical hood, if available, when there is a possibility of releasing toxic chemical vapors, dust, or gases. When using a hood, the opening should be kept open at a minimum to protect the user and to ensure efficient operation of the hood.
- Clean up all spills properly and promptly and dispose of according to directions.

A chemical hygiene plan must be put into place in order to protect anyone using the chemicals from particular health hazards that might be caused a result of their use. A chemical hygiene plan (CHP) is a written guideline stating the policies, procedures, and responsibilities that serve to protect laboratory personnel from the health hazards associated with the hazardous chemicals used in laboratory. The OSHA specifies the mandatory requirements of a CHP in its Occupational Exposure to Hazardous Chemicals in Laboratories Standard (Title 29, Code of Federal Regulations, Part 1010.1450), which can be viewed at the OSHA website, www.osha.gov. Certain elements are required in a CHP:

- Defined SOPs relevant to safety and health considerations for all activities that involve the use of hazardous chemicals.
- Criteria to determine and implement control measures to reduce exposure to hazardous materials, including engineering controls, the use of personal protective equipment, administrative controls, and hygiene practices, with particular attention given to of the use of control measures for extremely hazardous materials.
- Requirements ensuring that chemical hoods and other protective equipment are installed correctly and functioning properly.
- Information should be made available for persons working with hazardous substances regarding the hazards of the chemicals in the work area, the location of the CHP, signs and symptoms associated with hazardous chemical exposures, the recommended exposure limits of the chemicals, and the location and availability of information about the hazards, safe handling, storage, and disposal of hazardous material not limited to the MSDS.
- People working with hazardous substances should receive training in methods and observations to detect the presence or release of a hazardous chemical, the physical and health hazards of the chemicals used, the measures to be taken to protect against these hazards (i.e., personal protective equipment, appropriate work practices, emergency response actions) and applicable details of the CHP.
- Laboratory operations or procedures require prior approval from the appropriate administrator.
- Requirements for medical consultation and medical examination whenever a person develops signs or symptoms associated with a hazardous chemical, exposure monitoring reveals an exposure level routinely above the action level, or an event takes place in the work area such as a spill, leak, explosion, or other occurrence resulting in the likelihood of a hazardous exposure.
- Personnel should be designated as responsible for the implementation of the CHP, including the assignment of a chemical hygiene officer.

Many factors should be considered in relevant to purchasing chemicals that would contribute to safe handling.

- A centralized system should be used so that one person who knows all the chemicals being stored and used in the facility does the purchasing. Purchasing requests also can be linked into an inventory tracking system to avoid having excess stock.
- Receiving room, storeroom, and stockroom personnel should be well trained about the proper methods of receiving and handling hazardous substances.

Before ordering chemicals, laboratory personnel should:

- Assess the hazards and physical properties of the chemicals described on the MSDS.
- Evaluate short-term and long-term health risks.
- Ensure that there is sufficient ventilation and proper storage is available for the chemicals being ordered.
- Make certain that the proper personal protective equipment is available and safety equipment is on hand before using the chemical.
- Determine if the chemicals being ordered or its end product will need to be disposed of as a hazardous waste product.

A chemical tracking system has all the information regarding the chemicals stored and used in the laboratory. This system should be used to track chemicals from their time of purchase to when they are used and discarded. A good chemical tracking system can reduce procurement costs by eliminating unnecessary purchases and minimizing the expenses required to dispose of these chemicals.

A tracking system can be set up by organizing chemicals by their name and/or the molecular formula and by creating a computerized system with following tracking fields being recommended:

- Chemical name as printed on the container
- Chemical name as it appears on the MSDS, if different from that on the container
- Molecular formula Chemical Abstract Service (CAS) registry number
- Date received, source (chemical manufacturer and, if known, supplier)

- Type of container
- Hazard classification (for storage, handling, and disposal)
- Required storage conditions
- Location within the room (i.e., shelf #1, acid cabinet)
- Expiration or "use by" date
- Amount of the chemical in the container
- Name of the person who ordered or requested the chemical
- Keep accurate, up-to-date records about the use of each chemical in the system
- Each record should represent a single container of a chemical (rather than just the chemical itself)
- Conduct regularly scheduled inventory inspections to purge any inaccurate data in the system and dispose of outdated, unneeded, or deteriorated chemicals following the written CHP.

This above is an outline of all the safety procedures involved in chemical handling at laboratories. If these rules and regulations are implemented, injury and other hazards should be avoided while conducting experiments with various types of chemicals. Table 16.3 shows a typical chemical hygiene plan.

ACCREDITATION

Accreditation means that medical laboratories have been assessed against internationally recognized standards to demonstrate their competence, impartiality and performance capacity. The international standard ISO 15189 "Medical Laboratories—Particular Requirements for Quality and Competence" is the basis for accreditation of Medical Testing Laboratories. This standard is based upon ISO/IEC 17025 and ISO 9001, and this standard requires medical laboratories to demonstrate a quality system, technical competency and to be able to generate technically valid results. All laboratories providing molecular diagnostic services should have laboratory accreditation according to ISO 15189 or their national equivalent.

Accreditation provides patients, staff, service users, and commissioners with evidence of laboratory competence. The standards include the provision of adequate facilities, trained staff, and documentation. Laboratories that want to proceed toward accreditation should contact their national accreditation body (NAB).

Most NABs have formal application documents which require preliminary documentation about the laboratory, and its existing quality management system (QMS). The accreditation body will appoint a lead assessor and a technical assessor, who are experts in the field. Assessment and pre-assessment visits include the following steps; opening briefing, examination of records, sample handling, interview with technicians, demonstrations of tests and techniques, examination of equipment calibration and maintenance, review of quality documentation, written report of assessor's findings, and finally, closure meeting. The team of assessors will conduct an external audit in the laboratory and will report the assessment findings to the NAB. The laboratory responds to any deficiencies indicated by assessor's findings. The laboratory has

TABLE 16.3 Chemical Hazard Procedure
Chemical Hazard Communications Plan
1. Develop written hazard communication program
2. Maintain inventory of all chemicals
3. Manufacturer must assess and supply information about hazards
4. Employers must maintain MSDS in English
5. MSDS must list all ingredients greater than 1% (if carcinogen, greater than 0.1%)
6. Employers must make MSDS available to employees
7. Employers must ensure labels are not defaced and must post warnings
8. Employers must provide information and training
9. Employers must adhere to Occupational Safety and Health Administration limits
10. Designate person responsible for the program

to send a written corrective action response in due time. In case of minor noncompliance, the assessor will later check whether the corrective actions are adequate, through submitted documentation. In case of major noncompliance, a new onsite assessment might be required to assess the effectiveness of the corrective actions. Accreditation can be granted for 3 or 5 years, but the NABs will perform surveillance visits to ensure that the laboratory continues to meet requirements, which is typically every 1 to 2 years. The procedure for renewal of accreditation is usually similar to the one for new applicants, but the laboratory might use that opportunity to widen the scope of the accreditation by adding new tests.

External Quality Assessment (EQA) Scheme

All laboratories performing molecular tests should be part of an external quality assessment (EQA) scheme. Indeed, in most countries, continued accreditation requires EQA and regular audits of test performance. For example, the use of control materials within each run is recommended where the technique allows this. The results of internal control testing should be monitored to assure end-to-end performance of the test. Internal controls should be run with all batches. Internal positive controls can be prepared from samples with sufficient surplus material, or bought from commercial sources. Standardized products should be used to assure the stability and the external validation of the results.

The errors picked up by EQA can be classified into several categories. Misidentification at all stages of the testing process is an issue, resulting in incorrect assignment of mutation or wild-type results and, ultimately, incorrect treatment of the patient. Genotyping errors do occur, often from human error, by not following SOPs and performing methods suboptimally. Careful internal validation of tests is essential and recording lot and batch numbers can save weeks of investigation with manufacturers. Interpretative errors are less common, and likely to be picked up by clinical teams if the data do not match the conclusion. It is therefore essential that the results of the test and the interpretation are stated in the report.

It is essential that those participating in EQA schemes read the reports that detail errors made by all participants and act on recommendations made, as well as ensuring that their own errors are corrected rapidly. Poorly performing laboratories should terminate their service immediately, perform a root cause analysis and review cases that might have been affected by systemic errors. They should send samples to another provider until certain that their testing is safe. In either case, it is incumbent on the heads of laboratories to act on EQA results and ensure patient safety.

Molecular Diagnosis Checklist

The College of American Pathologists (CAP) NGS Work Group works on means of quality documentation and it has developed 18 laboratory accreditation checklist requirements for upstream analytic processes and downstream bioinformatics solutions for NGS in clinical applications. These requirements include new standards for documentation, validation, QA, confirmatory testing, exception logs, monitoring of upgrades, variant interpretation and reporting, incidental findings, data storage, version traceability, and data transfer confidentiality. The wet-bench process is composed of workflow steps, such as handling of patient samples, extraction of nucleic acids, fragmentation, bar coding, optional enrichment of targets, adaptor ligation, amplification, library preparation, flow cell loading, and generation of sequence reads. The checklist topics for the wet-bench process are summarized in the Table 16.4.

TABLE 16.4 Laboratory Checklist Requirements for Wet-Bench Process in NGS Experiments

Topic	Description	Requirements
Documentation	Use of SOP	• All standard operating protocols must be documented in order to trace each step and manipulations • All used methods, reagents, instruments, instrument software and versions have to be documented • Controls need to be described • Targeted NGS assays and target-enrichment protocols, regarding captured regions must be documented • Development of SOP for each validated sample type • Sequence information of barcodes for pooled analysis must be documented • Metrics and QC parameters regarding run performance have to be documented (% of reads mapping to target region, base quality and coverage thresholds, average)

Monitor of the Quality

The successful introduction of modern molecular technologies in diagnosis is regulated by the profession of the clinical scientists itself, namely the requirements for training and continuous education, as well as the accreditation of the clinical laboratory. The advantage is that it guarantees the quality of the clinical scientists, using their knowledge, expertise and skills to come to optimal results and interpretation of the data. It also guarantees that the laboratory techniques and processes are performed in a standard way and yield high-quality results. Clearly, high-quality molecular testing starts within the laboratory, with appropriately educated and dedicated personnel, analyses performed by SOPs, quality assurance schemes, and accreditation. Internal QA and external QA are essential methods to monitor the quality, performance, and competence of a laboratory. To improve the current standard, all molecular diagnostic laboratories are obliged to become accredited according to the international ISO 15189 standard for medical laboratories. It is essential that every country has a national body/standard with a mandate to enforce laboratories to become ISO 15189-accredited and to participate in EQA schemes (Fig. 16.2). In case of poor performance or lack of participation, the national body should have the authority to undertake adequate steps. As such, we can ensure that the molecular diagnosis provides accurate results for proper medical decision and reduce unnecessary errors.

CHALLENGES FOR MOLECULAR DIAGNOSIS IN A CLINICAL LABORATORY

To offer patients the best treatment options, diagnostic tests should be reliable, reproducible, of sufficient high sensitivity, and able to investigate all potential targets with the constrains of limited amount of tissue, time, and budget. These criteria for comprehensive molecular testing require permanent development of new assays, awareness of their potentials and drawbacks, continuous quality assessment to improve testing of diagnostic tissues, consciousness of budget and costs and clinical demands such as turnaround time.

Tissue Challenges

Molecular diagnostics present many challenges. The vast majority of the DNA to be analyzed is retrieved from routine FFPE tissue, which leads to suboptimal DNA quality for the required assays because of fixation and histoprocessing procedures. Any molecular test should take into account that suboptimal DNA samples (isolated from routine FFPE tissues) allow amplification of only small-sized PCR-amplicons (100–200 bp). In addition, the DNA is isolated from (dissected) tissue fragments composed of mixed populations of normal and diseased cells, reducing the mutant allele frequency, and frequently only a limited amount of (biopsied) diseased tissue is available. Therefore, the test developed must be able to accurately detect low levels of mutations.

Technological Challenges: A Multitude of Different Molecular Tests

To detect the various genomic DNA alterations, including point mutations, large insertions and deletions, genomic rearrangements, and promoter methylation, many diagnostic methods have been developed for daily routine molecular diagnostics. It can be anticipated that, in the era of precision medicine, the number of molecular markers that need to be assessed will steadily increase per sample. The challenge for the diagnostic laboratory is to select high-performing technological

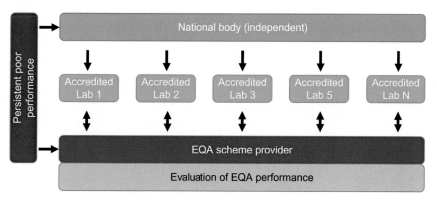

FIG. 16.2   A model for the interaction between the EQA (external quality assurance) providers, the National Body and ISO 15189-accredited laboratories. In case of poor performance, there will be an encouraging interaction between the EQA provider and the laboratory. In case of persistent poor performance, the EQA provider will communicate this to the National Body, who will undertake adequate steps.

methodologies that enable reliable molecular diagnosis, at a high sensitivity, with a limited amount of tissue (biopsies, cytological preparations), within short turnaround times, and at low costs.

Mutation Detection

The majority of diagnostic tests for precision medicine involve a wide range of PCR-based assays that amplify short DNA fragments for analysis of a single gene. Each of these methods has its own unique strengths and challenges. For example, Sanger sequencing allows detection of essentially all diagnostically relevant base substitutions, insertions and deletions, but it has a relatively modest limit of detection. Other commonly used methods, such as mutation- or allele-specific PCR (AS-PCR), pyrosequencing, and high-resolution melting (HRM), are all considerably more sensitive, have a faster turnaround time, and less hands-on time than Sanger sequencing. However, AS-PCR and pyrosequencing identify only mutations at predefined positions and are not suitable for detection of novel mutations. HRM and pyrosequencing require confirmation by another method.

None of the aforementioned tests is suitable for high throughput mutational analysis and might prove insufficient to detect all clinically relevant mutations in a cost-effective fashion in the future. For diagnostic molecular pathology, whole genome sequencing is not yet affordable, requires too much DNA and still has a long turnaround time. NGS-based methods using the Ion Torrent Personal Genome Machine (IT-PGM) and the MiSeq Benchtop Sequencer are now applied for analysis of gene panels for diagnostic purposes. Both platforms use a sequencing-by-synthesis approach and their sensitivity is higher than Sanger sequencing (detection of 2%–10% versus 15%–25% allele frequency). Moreover, the amount of DNA that is needed for the analysis of gene panels is very low, only 10–50 ng for all amplicons versus 10 ng per amplicon needed for Sanger sequencing. The turnaround time and costs can be competitive with respect to low throughput technologies in centers that have sufficient number of samples. The use of small, dedicated gene panels and efficient loading of the chips for IT-PGM also significantly reduce costs per case. Both IT-PGM and MiSeq systems allow the use of commercially established gene panels as well as custom-designed gene panels for amplicon sequencing. The major benefit of targeted sequencing is that it uncovers all kinds of mutations in selected genomic regions instead of mutations only at predefined positions. An additional advantage of NGS is that, in the same assay, mutations and allelic imbalances can be detected and loss of heterozygosity by SNP analyses. Unfortunately, it is not yet possible to detect both mutations and gene rearrangements in one assay by IT-PGM and MiSeq. Currently, a major disadvantage of NGS in implementation in molecular diagnostics is that the data-generating and data-processing technologies are not yet fully developed. Future equipment and software improvements will be required.

Bibliography

Aanensen, D.M., Feil, E.J., Holden, M.T., Dordel, J., Yeats, C.A., Fedosejev, A., Goater, R., Castillo-Ramírez, S., Corander, J., Colijn, C., Chlebowicz, M.A., Schouls, L., Heck, M., Pluister, G., Ruimy, R., Kahlmeter, G., Åhman, J., Matuschek, E., Friedrich, A.W., Parkhill, J., Bentley, S.D., Spratt, B.G., Grundmann, H., 2016. Whole-genome sequencing for routine pathogen surveillance in public health: a population snapshot of invasive *Staphylococcus aureus* in Europe. mBio 7 (3), e00444-16.

Abayasekara, L.M., Perera, J., Chandrasekharan, V., Gnanam, V.S., Udunuwara, N.A., Liyanage, D.S., Bulathsinhala, N.E., Adikary, S., Aluthmuhandiram, J.V.S., Thanaseelan, C.S., Tharmakulasingam, D.P., Karunakaran, T., Ilango, J., 2017. Detection of bacterial pathogens from clinical specimens using conventional microbial culture and 16S metagenomics: a comparative study. BMC Infect. Dis. 17, 631.

Aberg, K.A., Xie, L.Y., Nerella, S., Copeland, W.E., Costello, J., van den Oord, E.J.C.G., 2013. High quality methylome-wide investigations through next-generation sequencing of DNA from a single archived dry blood spot. Epigenetics 8 (5), 542–547.

Adamski, J., Suhre, K., 2013. Metabolomics platforms for genome wide association studies—linking the genome to the metabolome. Curr. Opin. Biotechnol. 24, 39–47.

Agrawal, M., Biswas, A., 2015. Molecular diagnostics of neurodegenerative disorders. Front. Mol. Biosci. 2, 54.

Aksglaede, L., Garn, I.D., Hollegaard, M.V., Hougaard, D.M., Rajpert-De Meyts, E., Juul, A., 2012. Detection of increased gene copy number in DNA from dried blood spot samples allows efficient screening for Klinefelter syndrome. Acta Paediatr. 101, e561–e563.

Alame, M., Lacourt, D., Zenagui, R., Mechin, D., Danton, F., Koenig, M., Claustres, M., Cossée, M., 2016. Implementation of a reliable next-generation sequencing strategy for molecular diagnosis of dystrophinopathies. J. Mol. Diagn. 18 (5), 731–740.

Alanio, A., Bretagne, S., 2017. Challenges in microbiological diagnosis of invasive Aspergillus infections. F1000Res 6 (F1000 Faculty Rev), 157.

Ali, I., Aboul-Enein, H.Y., Singh, P., Singh, R., Sharma, B., 2010. Separation of biological proteins by liquid chromatography. Saudi Pharm. J. 18, 59–73.

Allison, L.A., 2007. Fundamental Molecular Biology, first ed. Wiley-Blackwell. ISBN-10: 1405103795; 13: 978-1405103794.

Anahory, T., Hamamah, S., Andréo, B., Hédon, B., Claustres, M., Sarda, P., Pellestor, F., 2005. Sperm segregation analysis of a (13;22) Robertsonian translocation carrier by FISH: a comparison of locus-specific probe and whole chromosome painting. Hum. Reprod. 20 (7), 1850–1854.

Andersen, C.B., Holst-Jensen, A., Berdal, K.G., Thorstensen, T., Tengs, T., 2006. Equal performance of TaqMan, MGB, molecular beacon, and SYBR green-based detection assays in detection and quantification of roundup ready soybean. J. Agric. Food Chem. 54, 9658–9663.

Andersson, D.I., Jerlström-Hultqvist, J., Näsvall, J., 2015. Evolution of new functions de novo and from preexisting genes. Cold Spring Harb. Perspect. Biol. 7, a017996.

Angelier, N., Penrad-Mobayed, M., Billoud, B., Bonnanfant-Jaïs, M.L., Coumailleau, P., 1996. What role might lampbrush chromosomes play in maternal gene expression? Int. J. Dev. Biol. 40, 645–652.

Ansorge, W.J., 2009. Next-generation DNA sequencing techniques. New Biotechnol. 25 (4), 195–203.

Antonara, S., Leber, A.L., 2016. Diagnosis of *Clostridium difficile* infections in children. J. Clin. Microbiol. 54 (6), 1425–1433.

Aranda, P.S., LaJoie, D.M., Jorcyk, C.L., 2012. Bleach gel: a simple agarose gel for analyzing RNA quality. Electrophoresis 33 (2), 366–369.

Asiello, P.J., Baeumner, A.J., 2011. Miniaturized isothermal nucleic acid amplification, a review. Lab Chip 11, 1420–1430.

Asim, A., Kumar, A., Muthuswamy, S., Jain, S., Agarwal, S., 2015. Down syndrome: an insight of the disease. J. Biomed. Sci. 22 (1), 41.

Avery, O.T., MacLeod, C.M., McCarty, M., 1944. Studies on the chemical nature of the substance inducing transformation of pneumococcal types induction of transformation by a desoxyribonucleic acid fraction isolated from pneumococcus type III. J. Exp. Med. 79 (2), 137–158.

Aziz, N., Zhao, Q., Bry, L., Driscoll, D.K., Funke, B., Gibson, J.S., Grody, W.W., Hegde, M.R., Hoeltge, G.A., Leonard, D.G., Merker, J.D., Nagarajan, R., Palicki, L.A., Robetorye, R.S., Schrijver, I., Weck, K.E., Voelkerding, K.V., 2015. College of American Pathologists' laboratory standards for next-generation sequencing clinical tests. Arch. Pathol. Lab. Med. 139, 481.

Badrinarayanan, A., Le, T.B.K., Laub, M.T., 2015. Bacterial chromosome organization and segregation. Annu. Rev. Cell Dev. Biol. 31, 171–199.

Bailor, M.H., Mustoe, A.M., Brooks 3rd, C.L., Al-Hashimi, H.M., 2011. Topological constraints: using RNA secondary structure to model 3D conformation, folding pathways, and dynamic adaptation. Curr. Opin. Struct. Biol. 21 (3), 296–305.

Bantscheff, M., Schirle, M., Sweetman, G., Rick, J., Kuster, B., 2007. Quantitative mass spectrometry in proteomics: a critical review. Anal. Bioanal. Chem. 389, 1017–1031.

Belyaeva, E.S., Andreyeva, E.N., Belyakin, S.N., Volkova, E.I., Zhimulev, I.F., 2008. Intercalary heterochromatin in polytene chromosomes of *Drosophila melanogaster*. Chromosoma 117, 411–418.

Bentley, D.R., Balasubramanian, S., Swerdlow, H.P., Smith, G.P., Milton, J., Brown, C.G., Hall, K.P., Evers, D.J., Barnes, C.L., Bignell, H.R., Boutell, J.M., Bryant, J., Carter, R.J., Keira Cheetham, R., Cox, A.J., Ellis, D.J., Flatbush, M.R., Gormley, N.A., Humphray, S.J., Irving, L.J., Karbelashvili, M.S., Kirk, S.M., Li, H., Liu, X., Maisinger, K.S., Murray, L.J., Obradovic, B., Ost, T., Parkinson, M.L., Pratt, M.R., Rasolonjatovo, I.M., Reed, M.T., Rigatti, R., Rodighiero, C., Ross, M.T., Sabot, A., Sankar, S.V., Scally, A., Schroth, G.P., Smith, M.E.,

Smith, V.P., Spiridou, A., Torrance, P.E., Tzonev, S.S., Vermaas, E.H., Walter, K., Wu, X., Zhang, L., Alam, M.D., Anastasi, C., Aniebo, I.C., Bailey, D.M., Bancarz, I.R., Banerjee, S., Barbour, S.G., Baybayan, P.A., Benoit, V.A., Benson, K.F., Bevis, C., Black, P.J., Boodhun, A., Brennan, J.S., Bridgham, J.A., Brown, R.C., Brown, A.A., Buermann, D.H., Bundu, A.A., Burrows, J.C., Carter, N.P., Castillo, N., Chiara, E., Catenazzi, M., Chang, S., Neil Cooley, R., Crake, N.R., Dada, O.O., Diakoumakos, K.D., Dominguez-Fernandez, B., Earnshaw, D.J., Egbujor, U.C., Elmore, D.W., Etchin, S.S., Ewan, M.R., Fedurco, M., Fraser, L.J., Fuentes Fajardo, K.V., Scott Furey, W., George, D., Gietzen, K.J., Goddard, C.P., Golda, G.S., Granieri, P.A., Green, D.E., Gustafson, D.L., Hansen, N.F., Harnish, K., Haudenschild, C.D., Heyer, N.I., Hims, M.M., Ho, J.T., Horgan, A.M., Hoschler, K., Hurwitz, S., Ivanov, D.V., Johnson, M.Q., James, T., Huw Jones, T.A., Kang, G.D., Kerelska, T.H., Kersey, A.D., Khrebtukova, I., Kindwall, A.P., Kingsbury, Z., Kokko-Gonzales, P.I., Kumar, A., Laurent, M.A., Lawley, C.T., Lee, S.E., Lee, X., Liao, A.K., Loch, J.A., Lok, M., Luo, S., Mammen, R.M., Martin, J.W., PG, M.C., McNitt, P., Mehta, P., Moon, K.W., Mullens, J.W., Newington, T., Ning, Z., Ling Ng, B., Novo, S.M., O'Neill, M.J., Osborne, M.A., Osnowski, A., Ostadan, O., Paraschos, L.L., Pickering, L., Pike, A.C., Pike, A.C., Chris Pinkard, D., Pliskin, D.P., Podhasky, J., Quijano, V.J., Raczy, C., Rae, V.H., Rawlings, S.R., Chiva Rodriguez, A., Roe, P.M., Rogers, J., Rogert Bacigalupo, M.C., Romanov, N., Romieu, A., Roth, R.K., Rourke, N.J., Ruediger, S.T., Rusman, E., Sanches-Kuiper, R.M., Schenker, M.R., Seoane, J.M., Shaw, R.J., Shiver, M.K., Short, S.W., Sizto, N.L., Sluis, J.P., Smith, M.A., Ernest Sohna Sohna, J., Spence, E.J., Stevens, K., Sutton, N., Szajkowski, L., Tregidgo, C.L., Turcatti, G., Vandevondele, S., Verhovsky, Y., Virk, S.M., Wakelin, S., Walcott, G.C., Wang, J., Worsley, G.J., Yan, J., Yau, L., Zuerlein, M., Rogers, J., Mullikin, J.C., Hurles, M.E., NJ, M.C., West, J.S., Oaks, F.L., Lundberg, P.L., Klenerman, D., Durbin, R., Smith, A.J., 2008. Accurate whole human genome sequencing using reversible terminator chemistry. Nature 456 (7218), 53–59.

Bloch, K.D., Grossmann, B., 1995. Digestion of DNA with restriction endonucleases. Curr. Protoc. Mol. Biol. 31 (1), 3.1.1–3.1.21.

Blokesch, M., 2016. Natural competence for transformation. Curr. Biol. 26, R1126–R1130.

Boesenberg-Smith, K.A., Pessarakli, M.M., Wolk, D.M., 2012. Assessment of DNA yield and purity: an overlooked detail of PCR troubleshooting. Clin. Microbiol. Newsl. 34, 1.

Bonifacino, J., et al. (Eds.), 2003. Short Protocols in Cell Biology. John Wiley & Sons, Hoboken, NJ, 826 pp. ISBN-10 0-471-48339-7; 13 978-0-471-48339-7.

Borst, P., 2005. Ethidium DNA agarose gel electrophoresis: how it started. IUBMB Life 57 (11), 745–747.

Bowie, P., Halley, L., McKay, J., 2014. Laboratory test ordering and results management systems: a qualitative study of safety risks identified by administrators in general practice. BMJ Open. 4. https://doi.org/10.1136/bmjopen-2013-004245.

Branda, K.J., Tomczak, J., Natowicz, M.R., 2004. Heterozygosity for Tay-Sachs and Sandhoff diseases in non-Jewish Americans with ancestry from Ireland, Great Britain, or Italy. Genet. Test. 8 (2), 174–180.

Brown, T.A., 2002. Genomes, second ed. Wiley-Liss, Oxford.

Brown, T.A., 2016. Gene Cloning and DNA Analysis: An Introduction, seventh ed. Wiley-Blackwell. ISBN: 978-1-119-07256-0.

Buchan, B.W., Ledeboer, N.A., 2014. Emerging technologies for the clinical microbiology laboratory. Clin. Microbiol. Rev. 27 (4), 783–822.

Buckingham, L., 2011. Molecular Diagnostics: Fundamentals, Methods and Clinical Applications, second ed. F.A. Davis Company. ISBN-10: 0803626770; 13: 978-0803626775.

Buh Gasparic, M., Tengs, T., La Paz, J.L., Holst-Jensen, A., Pla, M., Esteve, T., Zel, J., Gruden, K., 2010. Comparison of nine different real-time PCR chemistries for qualitative and quantitative applications in GMO detection. Anal. Bioanal. Chem. 396, 2023–2029.

Burd, E.M., 2016. Human papillomavirus laboratory testing: the changing paradigm. Clin. Microbiol. Rev. 29 (2), 291–319.

Burgess, R.R., Deutscher, M.P. (Eds.), 2009. Methods in Enzymology: Guide to Protein Purification, second ed., vol. 463. Academic Press, pp. 1–851.

Buss, S.N., Gebhardt, L.L., Musser, K.A., 2012. Real-time PCR and pyrosequencing for differentiation of medically relevant Bartonella species. J. Microbiol. Methods 91, 252–256.

Butterfield, D.A., Gu, L., Di Domenico, F., Robinson, R.A., 2014. Mass spectrometry and redox proteomics: applications in disease. Mass Spectrom. Rev. 33 (4), 277–301.

Bystricky, K., 2015. Chromosome dynamics and folding in eukaryotes: insights from live cell microscopy. FEBS Lett. 589, 3014–3022.

Campbell, M.K., Farrell, S.O., 2014. Biochemistry, eighth ed. Cengage Learning. ISBN-10: 1285429109; 13: 978-1285429106.

Cariani, E., Bréchot, C., 1988. Detection of DNA sequences by southern blot. Ric. Clin. Lab. 18, 161–170.

Carr, D.W., Scott, J.D., 1992. Blotting and band-shifting: techniques for studying protein–protein interactions. TIBS 17, 246–249.

Catalina, P., Cobo, F., Cortés, J.L., Nieto, A.I., Cabrera, C., Montes, R., Concha, A., Menendez, P., 2007. Conventional and molecular cytogenetic diagnostic methods in stem cell research: a concise review. Cell Biol. Int. 31, 861–869.

Cesca, F., Bregant, E., Peterlin, B., Zadel, M., Dubsky de Wittenau, G., Siciliano, G., Ceravolo, R., Petrozzi, L., Pauletto, G., Verriello, L., Bergonzi, P., Damante, G., Barillari, G., Lucci, B., Curcio, F., Lonigro, I.R., 2015. Evaluating the SERCA2 and VEGF mRNAs as potential molecular biomarkers of the onset and progression in Huntington's disease. PLoS One. 10(4).

Chacon-Cortes, D., Haupt, L.M., Lea, R.A., Griffiths, L.R., 2012. Comparison of genomic DNA extraction techniques from whole blood samples: a time, cost and quality evaluation study. Mol. Biol. Rep. 39, 5961–5966.

Chan, W.-M., Miyake, N., Zhu-Tam, L., Andrews, C., Engle, E.C., 2011. Two novel CHN1 mutations in 2 families with Duane retraction syndrome. Arch. Ophthalmol. 129 (5), 649–652.

Chanderbali, A.S., Berger, B.A., Howarth, D.G., Soltis, D.E., Soltis, P.S., 2017. Evolution of floral diversity: genomics, genes and gamma. Philos. Trans. R. Soc. Lond. B: Biol. Sci. 372(1713).

Chandra De, B., Patra, M.C., Kumar, S., Brahma, B., Goutam, D., Jaiswal, L., Sharma, A., De, S., 2015. Noninvasive method of DNA isolation from fecal epithelial tissue of dairy animals. Anim. Biotechnol. 26, 211–216.

Chang, K., Deng, S., Chen, M., 2015. Novel biosensing methodologies for improving the detection of single nucleotide polymorphism. Biosens. Bioelectron. 66, 297–307.

Che, D., Hasan, M.S., Chen, B., 2014. Identifying pathogenicity islands in bacterial pathogenomics using computational approaches. Pathogens 3, 36–56.

Cheema, M.S., Ausió, J., 2015. The structural determinants behind the epigenetic role of histone variants. Genes 6, 685–713.

Chen, C.P., Su, Y.N., Chern, S.R., Chen, Y.T., Wu, P.S., Su, J.W., Pan, C.W., Wang, W., 2012. Mosaic trisomy 2 at amniocentesis: prenatal diagnosis and molecular genetic analysis. Taiwan. J. Obstet. Gynecol. 51 (4), 603–611.

Chen, Q., Ma, Y., Yang, Y., Chen, Z., Liao, R., Xie, X., Wang, Z., He, P., Tu, Y., Zhang, X., Yang, C., Yang, H., Yu, F., Zheng, Y., Zhang, Z., Wang, Q., Pan, Y., 2013. Genotyping by genome reducing and sequencing for outbred animals. PLoS One. 8(7).

Chen, C.K., Yu, H.T., Soong, Y.K., Lee, C.L., 2014a. New perspectives on preimplantation genetic diagnosis and preimplantation genetic screening. Taiwan. J. Obstet. Gynecol. 53, 146–150.

Chen, Y., Carter, R.L., Cho, I.K., Chan, A.W.S., 2014b. Cell-based therapies for Huntington's disease. Drug Discov. Today 19 (7), 980–984.

Chu, G., Vollrath, D., Davis, R.W., 1986. Separation of large DNA molecules by contour-clamped homogeneous electric fields. Science 234, 1582–1585.

Clark, A.E., Kaleta, E.J., Arora, A., Wolk, D.M., 2013. Matrix-assisted laser desorption ionization-time of flight mass spectrometry: a fundamental shift in the routine practice of clinical microbiology. Clin. Microbiol. Rev. 26 (3), 547–603.

Cole, K.D., Tellez, C.M., 2002. Separation of large circular DNA by electrophoresis in agarose gels. Biotechnol. Prog. 18, 82–87.

Coligan, J.E., Kruisbeek, A.M., Margulies, D.H., Shevach, E.M., Strober, W. (Eds.), 1991. Current Protocols in Immunology. Greene Publishing Associates and Wiley-Interscience, New York. ISBN: 0-471-52276-7.

Coligan, J.E., Dunn, B.M., Ploegh, H.L., Speicher, D.W., Wingfield, P.T., 1995. Current Protocols in Protein Science. John Wiley & Sons, Inc. ISBN: 978-0-471-11184-9

Corrêa, S.C., Rocha, M.N., Richeti, F., Kochi, C., Silva, E., Lima, L.A., Magalhães, M., Longui, C.A., 2014. Neonatal detection of Turner syndrome by real-time PCR gene quantification of the ARSE and MAGEH1 genes. Genet. Mol. Res. 13 (4), 9068–9076.

Coughlin, S.S., 2014. Toward a road map for global -omics: a primer on -omic technologies. Am. J. Epidemiol. 180 (12), 1188–1195.

Craig, N., Cohen-Fix, O., Green, R., Greider, C., Storz, G., Wolberger, C., 2010. Molecular Biology: Principles of Genome Function, first ed. Oxford University Press, 839 pp. ISBN-10: 0199562059; 13: 978-0199562053.

Cree, I.A., Deans, Z., Ligtenberg, M.J., Normanno, N., Edsjö, A., Rouleau, E., Solé, F., Thunnissen, E., Timens, W., Schuuring, E., Dequeker, E., Murray, S., Dietel, M., Groenen, P., Van Krieken, J.H., 2014. Guidance for laboratories performing molecular pathology for cancer patients. J. Clin. Pathol. 67 (11), 923–931.

Culen, C., Ertl, D.-A., Schubert, K., Bartha-Doering, L., Haeusler, G., 2017. Care of girls and women with Turner syndrome: beyond growth and hormones. Endocr. Connect. https://doi.org/10.1530/EC-17-0036. pii: EC-17-0036.

Cunningham, F.C., et al. 2014. Quality and safety issues highlighted by patients in the handling of laboratory test results by general practices—a qualitative study. BMC Health Serv. Res. 14, 206.

Cutler, S.J., Rudenko, N., Golovchenko, M., Cramaro, W.J., Kirpach, J., Savic, S., Christova, I., Amaro, A., 2017. Diagnosing borreliosis. Vector Borne Zoonotic Dis. 17 (1), 2–11.

da Silva, M.A., Arruda, M.A., 2006. Mechanization of the Bradford reaction for the spectrophotometric determination of total proteins. Anal. Biochem. 351, 155–157.

Daban, J.R., 2011. Electron microscopy and atomic force microscopy studies of chromatin and metaphase chromosome structure. Micron 42, 733–750.

Dadkhah, E., Ziaee, M., Davari, M.H., Kazemi, T., Abbaszadegan, M.R., 2012. Informative STR Markers for Marfan syndrome in Birjand, Iran. Iran. J. Basic Med. Sci. 15 (5), 1020–1025.

Dai, Z., Wu, Z., Jia, S., Wu, G., 2014. Analysis of amino acid composition in proteins of animal tissues and foods as pre-column o-phthaldialdehyde derivatives by HPLC with fluorescence detection. J. Chromatogr. B: Analyt. Technol. Biomed. Life Sci. 964, 116–127.

de Hoon, M., Hayashizaki, Y., 2008. Deep cap analysis gene expression (CAGE): genome-wide identification of promoters, quantification of their expression, and network inference. Biotechniques 44 (5), 627–628. 630, 632.

Deforges, J., Locker, N., Sargueil, B., 2015. mRNAs that specifically interact with eukaryotic ribosomal subunits. Biochimie 114, 48e57.

Demirhan, O., Tanriverdi, N., Yilmaz, M.B., Kocaturk-Sel, S., 2015. Report of a new case with pentasomy X and novel clinical findings. Balkan J. Med. Genet. 18 (1), 85–92.

Denden, S., Lakhdar, R., Keskes, N.B., Hamdaoui, M.H., Chibani, J.B., Khelil, A.H., 2013. PCR-based screening for the most prevalent alpha 1 antitrypsin deficiency mutations (PI S, Z, and Mmalton) in COPD patients from Eastern Tunisia. Biochem. Genet. 51 (9–10), 677–685.

DeRespinis, P.A., Caputo, A.R., Wagner, R.S., Guo, S., 1993. Duane's retraction syndrome. Surv. Ophthalmol. 38 (3), 257–288.

Derti, A., Garrett-Engele, P., Macisaac, K.D., Stevens, R.C., Sriram, S., Chen, R., Rohl, C.A., Johnson, J.M., Babak, T., 2012. A quantitative atlas of polyadenylation in five mammals. Genome Res. 22 (6), 1173–1183.

Desch, L., Marle, N., Mosca-Boidron, A.L., Faivre, L., Eliade, M., Payet, M., Ragon, C., Thevenon, J., Aral, B., Ragot, S., Ardalan, A., Dhouibi, N., Bensignor, C., Thauvin-Robinet, C., El Chehadeh, S., Callier, P., 2015. 6q16.3q23.3 duplication associated with Prader-Willi-like syndrome. Mol. Cytogenet. 8, 42.

Desjardins, P., Conklin, D., 2010. NanoDrop microvolume quantitation of nucleic acids. J. Vis. Exp. (45). pii: 2565.

Desjardins, P., Hansen, J.B., Allen, M., 2009. Microvolume protein concentration determination using the NanoDrop 2000c spectrophotometer. J. Vis. Exp. (33). pii: 1610.

Dickerson, J.A., Ramsay, L.M., Dada, O.O., Cermak, N., Dovichi, N.J., 2010. Two-dimensional capillary electrophoresis: capillary isoelectric focusing and capillary zone electrophoresis with laser-induced fluorescence detection. Electrophoresis 31 (15), 2650–2654.

Diegoli, T.M., 2015. Forensic typing of short tandem repeat markers on the X and Y chromosomes. Forensic Sci. Int. Genet. 18, 140–151.

Dogini, D.B., Pascoal, V.D., Avansini, S.H., Vieira, A.S., Pereira, T.C., Lopes-Cendes, I., 2014. The new world of RNAs. Genet. Mol. Biol. 37(1 Suppl.), 285–293.

Downie, A.W., 1972. Pneumococcal transformation—a backward view. Fourth Griffith Memorial Lecture. J. Gen. Microbiol. 73, 1–11.

Dubbink, H.J., Deans, Z.C., Tops, B.B., van Kemenade, F.J., Koljenović, S., van Krieken, H.J., Blokx, W.A., Dinjens, W.N., Groenen, P.J., 2014. Next generation diagnostic molecular pathology: critical appraisal of quality assurance in Europe. Mol. Oncol. 8 (4), 830–839.

Dukic, K., Zoric, M., Pozaic, P., Starcic, J., Culjak, M., Saracevic, A., Miler, M., 2015. How compliant are technicians with universal safety measures in medical laboratories in Croatia?—A pilot study. Biochem. Med. 25 (3), 386–392.

Dumas, M.E., 2012. Metabolome 2.0: quantitative genetics and network biology of metabolic phenotypes. Mol. BioSyst. 8, 2494–2502.

Early, A., Drury, L.S., Diffley, J.F., 2004. Mechanisms involved in regulating DNA replication origins during the cell cycle and in response to DNA damage. Phil. Trans. R. Soc. Lond. B 359, 31–38.

Egger, D., Bienz, K., 1994. Protein (western) blotting. Mol. Biotechnol. 1 (3), 289–305.

Eisenberger, T., Decker, C., Hiersche, M., Hamann, R.C., Decker, E., Neuber, S., Frank, V., Bolz, H.J., Fehrenbach, H., Pape, L., Toenshoff, B., Mache, C., Latta, K., Bergmann, C., 2015. An efficient and comprehensive strategy for genetic diagnostics of polycystic kidney disease. PLoS One. 10(2).

El Ashkar, S., Osman, M., Rafei, R., Mallat, H., Achkar, M., Dabboussi, F., Hamze, M., 2017. Molecular detection of genes responsible for macrolide resistance among *Streptococcus pneumoniae* isolated in North Lebanon. J. Infect. Public Health 10 (6), 745–748.

Elshire, R.J., Glaubitz, J.C., Sun, Q., Poland, J.A., Kawamoto, K., Buckler, E.S., Mitchell, S.E., 2011. A robust, simple genotyping-by-sequencing (GBS) approach for high diversity species. PLoS One. 6(5). https://doi.org/10.1371/journal.pone.0019379.

Endrullat, C., Glökler, J., Franke, P., Frohme, M., 2016. Standardization and quality management in next-generation sequencing. Appl. Transl. Genom. 10, 2–9.

Epstein, C.B., Butow, R.A., 2000. Microarray technology—enhanced versatility, persistent challenge. Curr. Opin. Biotechnol. 11, 36–41.

Fang, C., Otero, H.J., Greenberg, D., Neumann, P.J., 2011. Cost-utility analyses of diagnostic laboratory tests: a systematic review. Value Health 14 (8), 1010–1018.

Feist, P., Hummon, A.B., 2015. Proteomic challenges: sample preparation techniques for microgram-quantity protein analysis from biological samples. Int. J. Mol. Sci. 16, 3537–3563.

Fenselau, C., 2007. A review of quantitative methods for proteomic studies. J. Chromatogr. B 855, 14–20.

Ferretti, J.J., Stevens, D.L., Fischetti, V.A. (Eds.), 2016. *Streptococcus pyogenes*: Basic Biology to Clinical Manifestations [Internet]. University of Oklahoma Health Sciences Center, Oklahoma City, OK.

Fowler, C.B., Cunningham, R.E., O'Leary, T.J., Mason, J.T., 2007. 'Tissue surrogates' as a model for archival formalin-fixed paraffin-embedded tissues. Lab. Invest. 87, 836–846.

Fowler, C.B., Chesnick, I.E., Moore, C.D., O'Leary, T.J., Mason, J.T., 2010. Elevated pressure improves the extraction and identification of proteins recovered from formalin-fixed, paraffin-embedded tissue surrogates. PLoS One. 5(12).

Frühmesser, A., Kotzot, D., 2011. Chromosomal variants in Klinefelter syndrome. Sex Dev. 5 (3), 109–123.

Fullwood, M.J., Wei, C.L., Liu, E.T., Ruan, Y., 2009. Next-generation DNA sequencing of paired-end tags (PET) for transcriptome and genome analyses. Genome Res. 19 (4), 521–532.

Gaginskaya, E., Kulikova, T., Krasikova, A., 2009. Avian lampbrush chromosomes: a powerful tool for exploration of genome expression. Cytogenet. Genome Res. 124, 251–267.

García-Closas, M., Egan, K.M., Abruzzo, J., Newcomb, P.A., Titus-Ernstoff, L., Franklin, T., Bender, P.K., Beck, J.C., Le Marchand, L., Lum, A., Alavanja, M., Hayes, R.B., Rutter, J., Buetow, K., Brinton, L.A., Rothman, N., 2001. Collection of genomic DNA from adults in epidemiological studies by buccal cytobrush and mouthwash. Cancer Epidemiol. Biomarkers Prev. 10 (6), 687–696.

Garrett, R.H., Grisham, C.M., 2016. Biochemistry, sixth ed. Cengage Learning. ISBN-10: 1305577205; 13: 978-1305577206.

Gašparič, M.B., Cankar, K., Žel, J., Gruden, K., 2008. Comparison of different real-time PCR chemistries and their suitability for detection and quantification of genetically modified organisms. BMC Biotechnol. 8, 26.

Georgiou, C.D., Grintzalis, K., Zervoudakis, G., Papapostolou, I., 2008. Mechanism of Coomassie brilliant blue G-250 binding to proteins: a hydrophobic assay for nanogram quantities of proteins. Anal. Bioanal. Chem. 391, 391–403.

Georgiou, T., Christopoulos, G., Anastasiadou, V., Hadjiloizou, S., Cregeen, D., Jackson, M., Mavrikiou, G., Kleanthous, M., Drousiotou, A., 2014. The first family with Tay-Sachs disease in Cyprus: genetic analysis reveals a nonsense (c.78G > A) and a silent (c.1305C > T) mutation and allows pre-implantation genetic diagnosis. Meta Gene 2, 200–205.

Ghatak, S., Muthukumaran, R.B., Nachimuthu, S.K., 2013. A simple method of genomic DNA extraction from human samples for PCR-RFLP analysis. J. Biomol. Tech. 24 (4), 224–231.

Giavarina, D., Lippi, G., 2017. Blood venous sample collection: recommendations overview and a checklist to improve quality. Clin. Biochem. 50 (10–11), 568–573.

Gill, P., Ghaemi, A., 2008. Nucleic acid isothermal amplification technologies: a review. Nucleosides Nucleotides Nucleic Acids 27, 224–243.

Glick, B.R., Patten, C.L., Delovitch, T.L. (Eds.), 2013. Medical Biotechnology. first ed. ASM Press. ISBN-10: 155581705X; 13: 978-1555817053.

Goda, T., Tabata, M., Miyahara, Y., 2015. Electrical and electrochemical monitoring of nucleic acid amplification. Front. Bioeng. Biotechnol. 3, 29.

Golden, R., Valentini, M., 2014. Formaldehyde and methylene glycol equivalence: critical assessment of chemical and toxicological aspects. Regul. Toxicol. Pharmacol. 69 (2), 178–186.

Goldring, J.P., 2012. Protein quantification methods to determine protein concentration prior to electrophoresis. Methods Mol. Biol. 869, 29–35.

Gomez-Escribano, J.P., Alt, S., Bibb, M.J., 2016. Next generation sequencing of actinobacteria for the discovery of novel natural products. Mar. Drugs 14 (4), 78.

Grimberg, J., Nawoschik, S., Belluscio, L., McKee, R., Turck, A., Eisenberg, A., 1989. A simple and efficient non-organic procedure for the isolation of genomic DNA from blood. Nucleic Acids Res. 17 (20), 8390.

Gro, E., Strøm, A., Tellevik, M.G., Hanevik, K., Langeland, N., Blomberg, B., 2014. Comparison of four methods for extracting DNA from dried blood on filter paper for PCR targeting the mitochondrial Plasmodium genome. Trans. R. Soc. Trop. Med. Hyg. 108, 488–494.

Gross, S.J., Pletcher, B.A., Monaghan, K.G., Professional Practice and Guidelines Committee, 2008. Carrier screening in individuals of Ashkenazi Jewish descent. Genet. Med. 10 (1), 54–56.

Guerrera, I.C., Kleiner, O., 2005. Application of mass spectrometry in proteomics. Biosci. Rep. 25 (1–2), 71–93.

Gullett, J.C., Nolte, F.S., 2015. Quantitative nucleic acid amplification methods for viral infections. Clin. Chem. 61 (1), 72–78.

Gustafsson, O.J.R., Arentz, G., Hoffmann, P., 2015. Proteomic developments in the analysis of formalin-fixed tissue. Biochim. Biophys. Acta: Proteins Proteomics 1854 (6), 559–580.

Gutiérrez, R., Vayssier-Taussat, M., Buffet, J.-P., Harrus, S., 2017. Guidelines for the isolation, molecular detection, and characterization of bartonella species. Vector-Borne Zoonotic Dis. 17 (1), 42–50.

Guzel, O., Guner, E.I., 2009. ISO 15189 accreditation: requirements for quality and competence of medical laboratories, experience of a laboratory I. Clin. Biochem. 42 (4–5), 274–278.

Harkins, A.L., Munson, E., 2011. Molecular diagnosis of sexually transmitted chlamydia trachomatis in the United States. ISRN Obstet. Gynecol. 2011, 279149.

Harrington, C.T., Lin, E.I., Olson, M.T., Eshleman, J.R., 2013. Fundamentals of pyrosequencing. Arch. Pathol. Lab. Med. 137, 1296–1303.

Hart, A.B., Kranzler, H.R., 2015. Alcohol dependence genetics: lessons learned from genome-wide association studies (GWAS) and post-GWAS analyses. Alcohol. Clin. Exp. Res. 39 (8), 1312–1327.

Hayes, P.C., Roland Wolf, C., Hayes, J.D., 1989. Blotting techniques for the study of DNA, RNA, and proteins. BMJ 299, 965–968.

Henares, D., Brotons, P., Buyse, X., Latorre, I., de Paz, H.D., Muñoz-Almagro, C., 2017. Evaluation of the eazyplex MRSA assay for the rapid detection of Staphylococcus aureus in pleural and synovial fluid. Int. J. Infect. Dis. 59, 65–68.

Herschleb, J., Ananiev, G., Schwartz, D.C., 2007. Pulsed-field gel electrophoresis. Nat. Protoc. 2 (3), 677–684.

Higuchi, Y., Hashiguchi, A., Yuan, J., Yoshimura, A., Mitsui, J., Ishiura, H., Tanaka, M., Ishihara, S., Tanabe, H., Nozuma, S., Okamoto, Y., Matsuura, E., Ohkubo, R., Inamizu, S., Shiraishi, W., Yamasaki, R., Ohyagi, Y., Kira, J.-i., Oya, Y., Yabe, H., Nishikawa, N., Tobisawa, S., Matsuda, N., Masuda, M., Kugimoto, C., Fukushima, K., Yano, S., Yoshimura, J., Doi, K., Nakagawa, M., Morishita, S., Tsuji, S., Takashima, H., 2016. Mutations in MME cause an autosomal-recessive Charcot–Marie–Tooth disease type 2. Ann. Neurol. 79 (4), 659–672.

Ho, C.S., Lam, C.W., Chan, M.H., Cheung, R.C., Law, L.K., Lit, L.C., Ng, K.F., Suen, M.W., Tai, H.L., 2003. Electrospray ionisation mass spectrometry: principles and clinical applications. Clin. Biochem. Rev. 24 (1), 3–12.

Hofmann, H., Fingerle, V., Hunfeld, K.P., Huppertz, H.I., Krause, A., Rauer, S., Ruf, B., 2017. Cutaneous Lyme borreliosis: guideline of the German Dermatology Society. Ger. Med. Sci. 15, Doc14.

Holden, M.J., Haynes, R.J., Rabb, S.A., Satija, N., Yang, K., Blasic Jr., J.R., 2009. Factors affecting quantification of total DNA by UV spectroscopy and PicoGreen fluorescence. J. Agric. Food Chem. 57, 7221–7226.

Hoyle, J.C., Isfort, M.C., Roggenbuck, J., Arnold, W.D., 2015. The genetics of Charcot–Marie–Tooth disease: current trends and future implications for diagnosis and management. Appl. Clin. Genet. 8, 235–243.

http://www.sigmaaldrich.com/life-science/molecular-biology/pcr/long-and-accurate-pcr.html#sthash.XbyxCSLC.dpuf.

https://commons.wikimedia.org/wiki/File:Single_Cell_Genome_Sequencing_Workflow.pdf.

Hu, J., Sathanoori, M., Kochmar, S.J., Surti, U., 2006. Application of multicolor banding for identification of complex chromosome 18 rearrangements. J. Mol. Diagn. 8 (4), 521–525.

Hubacek, J.A., Adamkova, V., Lanska, V., Dlouha, D., 2017. Polygenic hypercholesterolemia: examples of GWAS results and their replication in the Czech-Slavonic population. Physiol. Res. 66 (Suppl. 1), S101–S111.

Hurd, P.J., Nelson, C.J., 2009. Advantages of next-generation sequencing versus the microarray in epigenetic research. Brief. Funct. Genomic. Proteomic. 8 (3), 174–183.

Ialongo, C., Bernardini, S., 2016. Phlebotomy, a bridge between laboratory and patient. Biochem. Med. 26 (1), 17–33.

Imataka, G., Hagisawa, S., Nitta, A., Hirabayashi, H., Suzumura, H., Arisaka, O., 2016. Long-term survival of full trisomy 13 in a 14 year old male: a case report. Eur. Rev. Med. Pharmacol. Sci. 20 (5), 919–922.

Irshad, M., Gupta, P., Mankotia, D.S., Ansari, M.A., 2016. Multiplex qPCR for serodetection and serotyping of hepatitis viruses: a brief review. World J. Gastroenterol. 22 (20), 4824–4834.

Jalbut, M.M., Sohani, A.R., Dal Cin, P., Hasserjian, R.P., Moran, J.A., Brunner, A.M., Fathi, A.T., 2015. Acute myeloid leukemia in a patient with constitutional 47,XXY karyotype. Leukemia Res. Rep. 4 (1), 28–30.

Jamali, S., Eskandari, N., Aryani, O., Salehpour, S., Zaman, T., Kamalidehghan, B., Houshmand, M., 2014. Three novel mutations in Iranian patients with Tay-Sachs disease. Iran. Biomed. J. 18 (2), 114–119.

Janecka, J., Chowdhary, B., Murphy, W., 2012. Exploring the correlations between sequence evolution rate and phenotypic divergence across the Mammalian tree provides insights into adaptive evolution. J. Biosci. 37 (5), 897–909.

Janecka, A., Adamczyk, A., Gasińska, A., 2015. Comparison of eight commercially available kits for DNA extraction from formalin-fixed paraffin-embedded tissues. Anal. Biochem. 476, 8–10.

Jang, W., Chae, H., Kim, J., Son, J.O., Kim, S.C., Koo, B.K., Kim, M., Kim, Y., Park, I.Y., Sung, I.K., 2016. Identification of small marker chromosomes using microarray comparative genomic hybridization and multicolor fluorescent in situ hybridization. Mol. Cytogenet. 9, 61.

Javadi, A., Shamaei, M., Mohammadi Ziazi, L., Pourabdollah, M., Dorudinia, A., Seyedmehdi, S.M., Karimi, S., 2014. Qualification study of two genomic DNA extraction methods in different clinical samples. Tanaffos 13 (4), 41–47.

Jennings, L.J., Arcila, M.E., Corless, C., Kamel-Reid, S., Lubin, I.M., Pfeifer, J., Temple-Smolkin, R.L., Voelkerding, K.V., Nikiforova, M.N., 2017. Guidelines for validation of next-generation sequencing-based oncology panels: a joint consensus recommendation of the Association for Molecular Pathology and College of American Pathologists. J. Mol. Diagn. 19 (3), 341–365.

Jeon, K. (Ed.), 2012. International Review of Cell and Molecular Biology, first ed., vol. 293. Academic Press, 328 pp. ISBN: 9780123946409; 9780123943040.

Jeong, Y.J., Park, K., Kim, D.E., 2009. Isothermal DNA amplification in vitro: the helicase-dependent amplification system. Cell. Mol. Life Sci. 66, 3325–3336.

Jiang, G., Harrison, D.J., 2000. mRNA isolation in a microfluidic device for eventual integration of cDNA library construction. Analyst 125, 2176–2179.

Jiang, X., Jiang, X., Feng, S., Tian, R., Ye, M., Zou, H., 2007. Development of efficient protein extraction methods for shotgun proteome analysis of formalin-fixed tissues. J. Proteome Res. 6, 1038–1047.

Jiang, Z., Zhou, X., Li, R., Michal, J.J., Zhang, S., Dodson, M.V., Zhang, Z., Harland, R.M., 2015. Whole transcriptome analysis with sequencing: methods, challenges and potential solutions. Cell. Mol. Life Sci. 72, 3425–3439.

Jiang, Z., Wang, H., Michal, J.J., Zhou, X., Liu, B., Woods, L.C., Fuchs, R.A., 2016. Genome wide sampling sequencing for SNP genotyping: methods, challenges and future development. Int. J. Biol. Sci. 12, 100–108.

Jovanovich, S., Bogdan, G., Belcinski, R., Buscaino, J., Burgi, D., ELR, B., Chear, K., Ciopyk, B., Eberhart, D., El-Sissi, O., Franklin, H., Gangano, S., Gass, J., Harris, D., Hennessy, L., Kindwall, A., King, D., Klevenberg, J., Li, Y., Mehendale, N., McIntosh, R., Nielsen, B., Park, C., Pearson, F., Schueren, R., Stainton, N., Troup, C., Vallone, P.M., Vangbo, M., Woudenberg, T., Wyrick, D., Williams, S., 2015. Developmental validation of a fully integrated sample-to-profile rapid human identification system for processing single-source reference buccal samples. Forensic Sci. Int. Genet. 16, 181–194.

Kamata, S., Akahoshi, N., Ishii, I., 2015. 2D DIGE proteomic analysis highlights delayed postnatal repression of α-fetoprotein expression in homocystinuria model mice. FEBS Open Bio 5, 535–541.

Kawamura, T., Nomura, M., Tojo, H., Fujii, K., Hamasaki, H., Mikami, S., Bando, Y., Kato, H., Nishimura, T., 2009. Proteomic analysis of laser-microdissected paraffin-embedded tissues: (1) Stage-related protein candidates upon non-metastatic lung adenocarcinoma. J. Proteomics 73 (6), 1089–1099.

Kazemi, M., Salehi, M., Kheirollahi, M., 2016. Down syndrome: current status, challenges and future perspectives. Int. J. Mol. Cell. Med. 5 (3), 125.

Kho, S.L., Chua, K.H., George, E., Tan, J.A., 2015. A novel gap-PCR with high resolution melting analysis for the detection of α-thalassaemia Southeast Asian and Filipino β-thalassaemia deletion. Sci. Rep. 5, 13937.

Kim, J., Park, W.-Y., Kim, N.K.D., Jang, S.J., Chun, S.-M., Sung, C.-O., Choi, J., Ko, Y.-H., Choi, Y.-L., Shim, H.S., Won, J.-K., The Molecular Pathology Study Group of Korean Society of Pathologists, 2017a. Good laboratory standards for clinical next-generation sequencing cancer panel tests. J. Pathol. Transl. Med. 51 (3), 191–204.

Kim, M.S., Chang, J., Kim, M.N., Choi, S.H., Jung, S.H., Lee, J.W., Sung, H., 2017b. Utility of a direct 16S rDNA PCR and sequencing for etiological diagnosis of infective endocarditis. Ann. Lab. Med. 37 (6), 505–510.

Kimura, K., Yanagisawa, H., Wachino, J.-i., Shibayama, K., Arakawa, Y., 2013. Rapid and reliable loop-mediated isothermal amplification method for detecting *Streptococcus agalactiae*. Jpn. J. Infect. Dis. 66, 546–548.

Kleckner, N., Fisher, J.K., Stouf, M., White, M.A., Bates, D., Witz, G., 2014. The bacterial nucleoid: nature, dynamics and sister segregation. Curr. Opin. Microbiol. 22, 127–137.

Klug, W.S., Cummings, M.R., Spencer, C.A., Palladino, M.A., 2011. Concepts of Genetics, tenth ed. Benjamin Cummings. ISBN-10: 0321724127; 13: 978-0321724120.

Koster, D.A., Crut, A., Shuman, S., Bjornsti, M.-A., Nynke, H., 2010. Cellular strategies for regulating DNA supercoiling: a single-molecule perspective. Cell 142 (4), 519–530.

Krausz, C., Chianese, C., Giachini, C., Guarducci, E., Laface, I., Forti, G., 2011. The Y chromosome-linked copy number variations and male fertility. J. Endocrinol. Invest. 34, 376–382.

Krebs, J. et al. (Eds.), 2014. Lewin's Gene XI, eleventh ed. ISBN-13: 978-1449659851.

Kressler, D., Hurt, E., Bassler, J., 2010. Driving ribosome assembly. Biochim. Biophys. Acta 1803, 673–683.

Kreutzberger, J., 2006. Protein microarrays: a chance to study microorganisms? Appl. Microbiol. Biotechnol. 70, 383–390.

Kroczek, R.A., 1993. Southern and northern analysis. J Chromatogr. 618 (1–2), 133–145.

Kuboki, N., Inoue, N., Sakurai, T., Di Cello, F., Grab, D.J., Suzuki, H., Sugimoto, C., Igarashi, I., 2003. Loop-mediated isothermal amplification for detection of African trypanosomes. J. Clin. Microbiol. 41 (12), 5517–5524.

Kuff, E.L., Mietz, J.A., 1994. Analysis of DNA restriction enzyme digests by two-dimensional electrophoresis in agarose gels. Methods Mol. Biol. 31, 177–186.

Kunkel, T.A., 2004. DNA replication fidelity. J. Biol. Chem. 279 (17), 16895–16898.

Kurien, B.T., Hal Scofield, R. (Eds.), 2012. Protein Electrophoresis: Methods and Protocols, Methods in Molecular Biology. vol. 869. https://doi.org/10.1007/978-1-61779-821-4_3.

Kwok, C.K., Tang, Y., Assmann, S.M., Bevilacqua, P.C., 2015. The RNA structurome: transcriptome-wide structure probing with next-generation sequencing. Trends Biochem. Sci. 40 (4), 221.

Lan, K., Verma, S.C., Murakami, M., Bajaj, B., Robertson, E.S., 2007. Isolation of human peripheral blood mononuclear cells (PBMCs). Curr. Protoc. Microbiol. 6, A.4C.1–A.4C.9.

Landers, J.P., 1995. Clinical capillary electrophoresis. Clin. Chem. 41 (4), 495–509.

Lee, J.T.Y., Cheung, K.M.C., Leung, V.Y.L., 2015. Extraction of RNA from tough tissues with high proteoglycan content by cryosection, second phase separation and high salt precipitation. J. Biol. Methods. 2(2).

Lenaers, G., Hamel, C., Delettre, C., Amati-Bonneau, P., Procaccio, V., Bonneau, D., Reynier, P., Milea, D., 2012. Dominant optic atrophy. Orphanet J. Rare Dis. 7, 46.

Li, X., Wu, Y., Zhang, L., Cao, Y., Li, Y., Li, J., Zhu, L., Wu, G., 2014. Comparison of three common DNA concentration measurement methods. Anal. Biochem. 451, 18–24.

Li, X., Feng, H., Zhang, J., Sun, L., Zhu, P., 2015. Analysis of chromatin fibers in Hela cells with electron tomography. Biophys. Rep. 1 (1), 51–60.

Li, L.X., Zhao, S.Y., Liu, Z.J., Ni, W., Li, H.F., Xiao, B.G., Wu, Z.Y., 2016. Improving molecular diagnosis of Chinese patients with Charcot-Marie-Tooth by targeted next-generation sequencing and functional analysis. Oncotarget 7 (19), 27655–27664.

Li, X., Zhao, J., Yuan, Q., Xia, N., 2017. Detection of HBV covalently closed circular DNA. Viruses 9 (6), 139.

Ligor, M., Olszowy, P., Buszewski, B., 2012. Application of medical and analytical methods in Lyme borreliosis monitoring. Anal. Bioanal. Chem. 402, 2233–2248.

Liu, D., Shimonov, J., Primanneni, S., Lai, Y., Ahmed, T., Seiter, K., 2007. t(8;14;18): a 3-way chromosome translocation in two patients with Burkitt's lymphoma/leukemia. Mol. Cancer 6, 35.

Liu, L., Li, Y., Li, S., Hu, N., He, Y., Pong, R., Lin, D., Lu, L., Law, M., 2012. Comparison of next-generation sequencing systems. J. Biomed. Biotechnol. 2012, 251364.

Long, M., VanKuren, N.W., Chen, S., Vibranovski, M.D., 2013. New gene evolution: little did we know. Annu. Rev. Genet. 47, 307–333.

Losekoot, M., van Belzen, M.J., Seneca, S., Bauer, P., Stenhouse, S.A.R., Barton, D.E., 2013. EMQN/CMGS best practice guidelines for the molecular genetic testing of Huntington disease. Eur. J. Hum. Genet. 21 (5), 480–486.

Luger, K., Mäder, A.W., Richmond, R.K., Sargent, D.F., Richmond, T.J., 1997. Crystal structure of the nucleosome core particle at 2.8 Å resolution. Nature 389 (6648), 251–260.

Lyle, R., Béna, F., Gagos, S., Gehrig, C., Lopez, G., Schinzel, A., Lespinasse, J., Bottani, A., Dahoun, S., Taine, L., Doco-Fenzy, M., Cornillet-Lefèbvre, P., Pelet, A., Lyonnet, S., Toutain, A., Colleaux, L., Horst, J., Kennerknecht, I., Wakamatsu, N., Descartes, M., Franklin, J.C., Florentin-Arar, L., Kitsiou, S., Aït Yahya-Graison, E., Costantine, M., Sinet, P.M., Delabar, J.M., Antonarakis, S.E., 2009. Genotype-phenotype correlations in Down syndrome identified by array CGH in 30 cases of partial trisomy and partial monosomy chromosome 21. Eur. J. Hum. Genet. 17 (4), 454–466.

Maegawa, G.H., Stockley, T., Tropak, M., Banwell, B., Blaser, S., Kok, F., Giugliani, R., Mahuran, D., Clarke, J.T., 2006. The natural history of juvenile or subacute GM2 gangliosidosis: 21 new cases and literature review of 134 previously reported. Pediatrics 118, e1550–e1562.

Maeshima, K., Imai, R., Tamura, S., Nozaki, T., 2014. Chromatin as dynamic 10-nm fibers. Chromosoma 123, 225–237.

Magdeldin, S., Enany, S., Yoshida, Y., Xu, B., Zhang, Y., Zureena, Z., Lokamani, I., Yaoita, E., Yamamoto, T., 2014. Basics and recent advances of two dimensional-polyacrylamide gel electrophoresis. Clin. Proteomics 11 (1), 16.

Malentacchi, F., Pizzamiglio, S., Wyrich, R., Verderio, P., Ciniselli, C., Pazzagli, M., Gelmini, S., 2016. Effects of transport and storage conditions on gene expression in blood samples. Biopreserv. Biobank. 14 (2), 122–128.

Malissovas, N., Griffin, L.B., Antonellis, A., Beis, D., 2016. Dimerization is required for GARS-mediated neurotoxicity in dominant CMT disease. Hum. Mol. Genet. 25 (8), 1528–1542.

Mao, F., Leung, W.-Y., Xin, X., 2007. Characterization of EvaGreen and the implication of its physicochemical properties for qPCR applications. BMC Biotechnol. 7, 76.

Marquet, V., Bourgeois, D., De Mas, P., Bouneau, L., Vigouroux-Castera, A., Molignier, R., Calvas, P., 2015. Double deletion of a chromosome 21 inserted in a chromosome 22 in an azoospermic patient. Clin. Case Rep. 3 (9), 757–761.

Martin, D.F., Martin, B.B. (Eds.), 2013. Column Chromatography. InTech, 209 pp. ISBN-13: 9789535110743.

Martin, D.C., Mark, B.L., Triggs-Raine, B.L., Natowicz, M.R., 2007. Evaluation of the risk for Tay-Sachs disease in individuals of French Canadian ancestry living in new England. Clin. Chem. 53 (3), 392–398.

Martínez-Cano, D.J., Reyes-Prieto, M., Martínez-Romero, E., Partida-Martínez, L.P., Latorre, A., Moya, A., Delaye, L., 2015. Evolution of small prokaryotic genomes. Front. Microbiol. 5, 742.

Maze, I., Noh, K.M., Soshnev, A.A., Allis, C.D., 2014. Every amino acid matters: essential contributions of histone variants to mammalian development and disease. Nat. Rev. Genet. 15 (4), 259–271.

McDowell, G.A., Mules, E.H., Fabacher, P., Shapira, E., Blitzer, M.G., 1992. The presence of two different infantile Tay-Sachs disease mutations in a Cajun population. Am. J. Hum. Genet. 51 (5), 1071–1077.

McKenna, J.P., Cox, C., Fairley, D.J., Burke, R., Shields, M.D., Watt, A., Coyle, P.V., 2017. Loop-mediated isothermal amplification assay for rapid detection of *Streptococcus agalactiae* (group B streptococcus) in vaginal swabs—a proof of concept study. J. Med. Microbiol. 66, 294–300.

Mehrotra, S., Goyal, V., 2014. Repetitive sequences in plant nuclear DNA: types, distribution, evolution and function. Genomics Proteomics Bioinformatics 12, 164–171.

Mehta, N., Lazarin, G.A., Spiegel, E., Berentsen, K., Brennan, K., Giordano, J., Haque, I.S., Wapner, R., 2016. Tay-Sachs carrier screening by enzyme and molecular analyses in the New York City Minority population. Genet. Test. Mol. Biomarkers 20 (9), 504–509.

Meisner, L.F., Johnson, J.A., 2008. Protocols for cytogenetic studies of human embryonic stem cells. Methods 45, 133–141.

Memczak, S., Jens, M., Elefsinioti, A., Torti, F., Krueger, J., Rybak, A., Maier, L., Mackowiak, S.D., Gregersen, L.H., Munschauer, M., Loewer, A., Ziebold, U., Landthaler, M., Kocks, C., le Noble, F., Rajewsky, N., 2013. Circular RNAs are a large class of animal RNAs with regulatory potency. Nature 495, 333–338.

Messacar, K., Parker, S.K., Todd, J.K., Dominguez, S.R., 2017. Implementation of rapid molecular infectious disease diagnostics: the role of diagnostic and antimicrobial stewardship. J. Clin. Microbiol. 55, 715–723.

Miller, J.R., Andre, R., 2014. Quantitative polymerase chain reaction. Br. J. Hosp. Med. 75 (12), c188.

Miller, S.A., Dykes, D.D., Polesky, H.F., 1988. A simple salting out procedure for extracting DNA from human nucleated cells. Nucleic Acids Res. 16 (3), 1215.

Miotto, E., Saccenti, E., Lupini, L., Callegari, E., Negrini, M., Ferracin, M., 2014. Quantification of circulating miRNAs by droplet digital PCR: comparison of EvaGreen- and TaqMan-based chemistries. Cancer Epidemiol. Biomarkers Prev. 23 (12), 2638–2642.

Mirmajlessi, S.M., Destefanis, M., Gottsberger, R.A., Mänd, M., Loit, E., 2015. PCR-based specific techniques used for detecting the most important pathogens on strawberry: a systematic review. Syst. Rev. 4, 9.

Mitra, A., Boroujeni, M.B., 2015. Application of gel-based proteomic technique in female reproductive investigations. J. Hum. Reprod. Sci. 8 (1), 18–24.

Miyake, N., Chilton, J., Psatha, M., Cheng, L., Andrews, C., Chan, W.M., Law, K., Crosier, M., Lindsay, S., Cheung, M., Allen, J., Gutowski, N.J., Ellard, S., Young, E., Iannaccone, A., Appukuttan, B., Stout, J.T., Christiansen, S., Ciccarelli, M.L., Baldi, A., Campioni, M., Zenteno, J.C., Davenport, D., Mariani, L.E., Sahin, M., Guthrie, S., Engle, E.C., 2008. Human CHN1 mutations hyperactivate alpha2-chimaerin and cause Duane's retraction syndrome. Science 321 (5890), 839–843.

Montalvo, A.L., Filocamo, M., Vlahovicek, K., Dardis, A., Lualdi, S., Corsolini, F., Bembi, B., Pittis, M.G., 2005. Molecular analysis of the HEXA gene in Italian patients with infantile and late onset Tay-Sachs disease: detection of fourteen novel alleles. Hum. Mutat. 26 (3), 282.

Mullally, A., Ritz, J., 2007. Beyond HLA: the significance of genomic variation for allogeneic hematopoietic stem cell transplantation. Blood 109, 1355–1362.

Murray, J.R., Rajeevan, M.S., 2013. Evaluation of DNA extraction from granulocytes discarded in the separation medium after isolation of peripheral blood mononuclear cells and plasma from whole blood. BMC Res. Notes 6, 440.

Mutz, K.-O., Heilkenbrinker, A., Lönne, M., Walter, J.-G., Stahl, F., 2013. Transcriptome analysis using next-generation sequencing. Curr. Opin. Biotechnol. 24, 22–30.

Muyzer, G., de Waal, E.C., Uitterlinden, A.G., 1993. Profiling of complex microbial populations by denaturing gradient gel electrophoresis analysis of polymerase chain reaction-amplified genes coding for 16S rRNA. Appl. Environ. Microbiol. 59 (3), 695–700.

Myerowitz, R., Piekarz, R., Neufeld, E.F., Shows, T.B., Suzuki, K., 1985. Human beta-hexosaminidase alpha chain: coding sequence and homology with the beta chain. Proc. Natl. Acad. Sci. U. S. A. 82 (23), 7830–7834.

Nagamine, K., Hase, T., Notomi, T., 2002. Accelerated reaction by loop-mediated isothermal amplification using loop primers. Mol. Cell. Probes 16, 223–229.

Naik, R.P., Haywood Jr., C., 2015. Sickle cell trait diagnosis: clinical and social implications. Hematology Am. Soc. Hematol. Educ. Program 2015 (1), 160–167.

Navarro, E., Serrano-Heras, G., Castaño, M.J., Solera, J., 2015. Real-time PCR detection chemistry. Clin. Chim. Acta 439, 231–250.

Nguyen, J.M., Qualmann, K.J., Okashah, R., Reilly, A., Alexeyev, M.F., Campbell, D.J., 2015 Sep. 5p deletions: current knowledge and future directions. Am. J. Med. Genet. C: Semin. Med. Genet. 169 (3), 224–238.

Nigro, E., Imperlini, E., Scudiero, O., Monaco, M.L., Polito, R., Mazzarella, G., Orrù, S., Bianco, A., Daniele, A., 2015. Differentially expressed and activated proteins associated with non small cell lung cancer tissues. Respir. Res. 16, 74.

Ning, H.C., Lin, C.N., Chiu, D.T., Chang, Y.T., Wen, C.N., Peng, S.Y., Chu, T.L., Yu, H.M., Wu, T.L., 2016. Reduction in hospital-wide clinical laboratory specimen identification errors following process interventions: a 10-year retrospective observational study. PLoS One. 11(8).

Noble, J.E., Bailey, M.J., 2009. Quantitation of protein. Methods Enzymol. 463, 73–95.

Notomi, T., Mori, Y., Tomita, N., Kanda, H., 2015. Loop-mediated isothermal amplification (LAMP): principle, features, and future prospects. J. Microbiol. 53 (1), 1–5.

Nuchtavorn, N., Suntornsuk, W., Lunte, S.M., Suntornsuk, L., 2015. Recent applications of microchip electrophoresis to biomedical analysis. J. Pharm. Biomed. Anal. 113, 72–96.

Okubo, M., Minami, N., Goto, K., Goto, Y., Noguchi, S., Mitsuhashi, S., Nishino, I., 2016. Genetic diagnosis of Duchenne/Becker muscular dystrophy using next-generation sequencing: validation analysis of DMD mutations. J. Hum. Genet. 61 (6), 483–489.

Okutucu, B., Dınçer, A., Habib, Ö., Zıhnıoglu, F., 2007. Comparison of five methods for determination of total plasma protein concentration. J. Biochem. Biophys. Methods 70, 709–711.

Orhant, L., Anselem, O., Fradin, M., Becker, P.H., Beugnet, C., Deburgrave, N., Tafuri, G., Letourneur, F., Goffinet, F., Allach El Khattabi, L., Leturcq, F., Bienvenu, T., Tsatsaris, V., Nectoux, J., 2016 May. Droplet digital PCR combined with minisequencing, a new approach to analyze fetal DNA from maternal blood: application to the non-invasive prenatal diagnosis of achondroplasia. Prenat. Diagn. 36 (5), 397–406.

Paireder, S., Werner, B., Bailer, J., Werther, W., Schmid, E., Patzak, B., Cichna-Markl, M., 2013. Comparison of protocols for DNA extraction from long-term preserved formalin fixed tissues. Anal. Biochem. 439, 152–160.

Pandey, R., Mehrotra, D., Kowtal, P., Mahdi, A.A., Sarin, R., 2014. Mitochondrial DNA from archived tissue samples kept in formalin for forensic odontology studies. J. Oral Biol. Craniofacial Res. 4, 109–113.

Paradisi, F., Corti, G., Cinelli, R., 2001. Streptococcus pneumoniae as an agent of nosocomial infection: treatment in the era of penicillin-resistant strains. Clin. Microbiol. Infect. 7 (Suppl. 4), 34–42.

Parizad, E.G., Parizad, E.G., Valizadeh, A., 2016. The application of pulsed field gel electrophoresis in clinical studies. J. Clin. Diagn. Res. 10 (1), DE01–DE04.

Patterson, A.S., Hsieh, K., Tom Soh, H., Plaxco, K.W., 2013. Electrochemical real-time nucleic acid amplification: towards point-of-care quantification of pathogens. Trends Biotechnol. 31 (12), 704–712.

Pavšič, J., Devonshire, A.S., Parkes, H., Schimmel, H., Foy, C.A., Karczmarczyk, M., Gutiérrez-Aguirre, I., Honeyborne, I., Huggett, J.F., McHugh, T.D., Milavec, M., Zeichhardt, H., Žel, J., 2015. Standardization of nucleic acid tests for clinical measurements of bacteria and viruses. J. Clin. Microbiol. 53, 2008–2014.

Pegadaraju, V., Nipper, R., Hulke, B., Qi, L., Schultz, Q., 2013. De novo sequencing of sunflower genome for SNP discovery using RAD (restriction site associated DNA) approach. BMC Genomics 14, 556.

Peng, X., Yu, K.Q., Deng, G.H., Jiang, Y.X., Wang, Y., Zhang, G.X., Zhou, H.W., 2013. Comparison of direct boiling method with commercial kits for extracting fecal microbiome DNA by Illumina sequencing of 16S rRNA tags. J. Microbiol. Methods 95, 455–462.

Persing, D.H., Tenover, F.C., Tang, Y.-W., Nolte, F.S., Hayden, R.T., Van Belkum, A. (Eds.), 2011. Molecular Microbiology: Diagnostic Principles and Practice, second ed. ASM Press. ISBN-10: 1555814972; 13: 978-1555814977.

Petersen, J.R., Mohammad, A.A. (Eds.), 2013. Clinical and Forensic Applications of Capillary Electrophoresis. Eur. J. Hum. Genet. 21 (5), 480–486 ISBN: 978-0-89603-645-1.

Poater, J., Swart, M., Bickelhaupt, F.M., Guerra, C.F., 2014. B-DNA structure and stability: the role of hydrogen bonding, π–π stacking interactions, twist-angle, and solvation. Org. Biomol. Chem. 12, 4691–4700.

Poland, J.A., Brown, P.J., Sorrells, M.E., Jannink, J.L., 2012. Development of high-density genetic maps for barley and wheat using a novel two-enzyme genotyping-by-sequencing approach. PLoS One. 7(2). https://doi.org/10.1371/journal.pone.0032253.

Portillo, A., de Sousa, R., Santibáñez, S., Duarte, A., Edouard, S., Fonseca, I.P., Marques, C., Novakova, M., Palomar, A.M., Santos, M., Silaghi, C., Tomassone, L., Zúquete, S., Oteo, J.A., 2017 Jan. Guidelines for the detection of rickettsia spp. Vector Borne Zoonotic Dis. 17 (1), 23–32.

Posch, A., 2014. Sample preparation guidelines for two dimensional electrophoresis. Arch. Physiol. Biochem. 120 (5), 192–197.

Powers-Fletcher, M.V., Hanson, K.E., 2016. Molecular diagnostic testing for Aspergillus. J. Clin. Microbiol. 54, 2655–2660.

Pratten, J., 2007. Growing oral biofilms in a constant depth film fermentor (CDFF). In: Current Protocols in Microbiology. Wiley Interscience. https://doi.org/10.1002/9780471729259.

Price, J., Gordon, N.C., Crook, D., Llewelyn, M., Paul, J., 2013. The usefulness of whole genome sequencing in the management of *Staphylococcus aureus* infections. Clin. Microbiol. Infect. 19 (9), 784–789.

Radvanszky, J., Surovy, M., Nagyova, E., Minarik, G., Kadasi, L., 2015. Comparison of different DNA binding fluorescent dyes for applications of high-resolution melting analysis. Clin. Biochem. 48, 609–616.

Rainen, L., Oelmueller, U., Jurgensen, S., Wyrich, R., Ballas, C., Schram, J., Herdman, C., Bankaitis-Davis, D., Nicholls, N., Trollinger, D., Tryon, V., 2002. Stabilization of mRNA expression in whole blood samples. Clin. Chem. 48 (11), 1883–1890.

Rajan-Babu, I.-S., Chong, S.S., 2016. Molecular correlates and recent advancements in the diagnosis and screening of FMR1-related disorders. Genes (Basel) 7 (10), 87.

Rallapalli, G., Kemen, E.M., Robert-Seilaniantz, A., Segonzac, C., Etherington, G.J., Sohn, K.H., MacLean, D., Jones, J.D.G., 2014. EXPRSS: an Illumina based high-throughput expression-profiling method to reveal transcriptional dynamics. BMC Genomics 15, 341.

Rasool, M., Malik, A., Naseer, M.I., Manan, A., Ansari, S., Begum, I., Qazi, M.H., Pushparaj, P., Abuzenadah, A.M., Al-Qahtani, M.H., Kamal, M.A., Gan, S., 2015. The role of epigenetics in personalized medicine: challenges and opportunities. BMC Med. Genomics 8 (Suppl. 1), S5.

Ray-Jones, H., Eyre, S., Barton, A., Warren, R.B., 2016. One SNP at a time: moving beyond GWAS in psoriasis. J. Invest. Dermatol. 136 (3), 567–573.

Raynal, B., Lenormand, P., Baron, B., Hoos, S., England, P., 2014. Quality assessment and optimization of purified protein samples: why and how? Microb. Cell Fact. 13, 180.

Resim, S., Kucukdurmaz, F., Kankılıc, N., Altunoren, O., Efe, E., Benlioglu, C., 2015. Cognitive, affective problems and renal cross ectopy in a patient with 48,XXYY/47,XYY syndrome. Case Rep. Genet. 2015, 4 pages.

Restrepo, C.M., Llanes, A., Lleonart, R., 2017. Use of AFLP for the study of eukaryotic pathogens affecting humans. Infect. Genet. Evol. pii: S1567-1348 (17)30323-4.

Rhoads, A., Au, K.F., 2015. PacBio sequencing and its applications. Genomics Proteomics Bioinformatics 13, 278–289.

Ribeiro, I.P., Barroso, L., Marques, F., Melo, J.B., Carreira, I.M., 2016. Early detection and personalized treatment in oral cancer: the impact of omics approaches. Mol. Cytogenet. 9, 85.

Riedel, S., Junkins, A., Stamper, P.D., Cress, G., Widness, J.A., Doern, G.V., 2009. Comparison of the Bactec 9240 and BacT/Alert blood culture systems for evaluation of placental cord blood for transfusion in neonates. J. Clin. Microbiol. 47, 1645–1649.

Riegel, M., 2014. Human molecular cytogenetics: from cells to nucleotides. Genet. Mol. Biol. 37(1 Suppl.), 194–209.

Rivero, E.R., Neves, A.C., Silva-Valenzuela, M.G., Sousa, S.O., Nunes, F.D., 2006. Simple salting-out method for DNA extraction from formalin-fixed, paraffin-embedded tissues. Pathol. Res. Pract. 202 (7), 523–529.

Robinow, C., Kellenberger, E., 1994. The bacterial nucleoid revisited. Microbiol. Rev. 58 (2), 211–232.

Rodríguez-Caballero, A., Torres-Lagares, D., Rodríguez-Pérez, A., Serrera-Figallo, M.A., Hernández-Guisado, J.M., Machuca-Portillo, G., 2010. Cri du chat syndrome: a critical review. Med. Oral Patol. Oral Cir. Bucal 15 (3), e473–e478.

Romsos, E.L., Vallone, P.M., 2015. Rapid PCR of STR markers: applications to human identification. Forensic Sci. Int. Genet. 18, 90–99.

Rosenberg, B.H., Cavalieri, L.F., 1968. Shear and the melting of DNA: an especially sensitive portion of the *Escherichia coli* genome. Biophys. J. 8, 1138–1145.

Rusu, C., 2016. From quality management to managing quality. Procedia Soc. Behav. Sci. 221, 287–293.

Sakamoto, S., Putalun, W., Vimolmangkang, S., Phoolcharoen, W., Shoyama, Y., Tanaka, H., Morimoto, S., 2018. Enzyme-linked immunosorbent assay for the quantitative/qualitative analysis of plant secondary metabolites. J. Nat. Med. 72 (1), 32–42.

Sanderson, B.A., Araki, N., Lilley, J.L., Guerrero, G., Lewis, L.K., 2014. Modification of gel architecture and TBE/TAE buffer composition to minimize heating during agarose gel electrophoresis. Anal. Biochem. 454, 44–52.

Sapan, C.V., Lundblad, R.L., Price, N.C., 1999. Colorimetric protein assay techniques. Biotechnol. Appl. Biochem. 29, 99–108.

Schaefer, M., Pollex, T., Hanna, K., Lyko, F., 2009. RNA cytosine methylation analysis by bisulfite sequencing. Nucleic Acids Res. 37(2).

Schena, M., Aheller, R., Ptheriault, T., Konrad, K., Lachenmeier, E., Davis, R.W., 1998. Microarrays: biotechnology's discovery platform for functional genomics. Trends Biotechnol. 16, 301–306.

Sedlackova, T., Repiska, G., Celec, P., Szemes, T., Minarik, G., 2013. Fragmentation of DNA affects the accuracy of the DNA quantitation by the commonly used methods. Biol. Proced. Online 15, 5.

Shapiro, B., Hofreiter, M. (Eds.), 2011. Ancient DNA: Methods and Protocols, Methods in Molecular Biology. vol. 840. https://doi.org/10.1007/978-1-61779-516-9_11.

Shen, C.-H., 2012. Histones: Class, Structure and Function, first ed. Nova Science Pub Inc. ISBN-10: 1621002748; 13: 978-1621002741.

Shepard, P.J., Choi, E.A., Lu, J., Flanagan, L.A., Hertel, K.J., Shi, Y., 2011. Complex and dynamic landscape of RNA polyadenylation revealed by PAS-Seq. RNA 17, 761–772.

Shi, S.R., Taylor, C.R., Fowler, C.B., Mason, J.T., 2013. Complete solubilization of formalin-fixed, paraffin-embedded tissue may improve proteomic studies. Proteomics Clin. Appl. 7 (3-4), 264–272.

Shirasu, N., Kuroki, M., 2014. Simultaneous genotyping of single-nucleotide polymorphisms in alcoholism-related genes using duplex and triplex allele-specific PCR with two-step thermal cycles. Anal. Sci. 30 (11), 1093–1096.

Siegel, R.L., Miller, K.D., Jemal, A., 2015. Cancer statistics, 2015. CA Cancer J. Clin. 65, 5–29.

Simonian, M.H., Smith, J.A., 2006. Spectrophotometric and colorimetric determination of protein concentration. In: Current Protocols in Molecular Biology. John Wiley & Sons, Inc., Hoboken, NJ, pp. 10.1A.1–10.1A.9 (10.1.1).

Sloots, T.P., Nissen, M.D., Ginn, A.N., Iredell, J.R., 2015. Rapid identification of pathogens using molecular techniques. Pathology 47 (3), 191–198.

Smith, D.R., 2016. The past, present and future of mitochondrial genomics: have we sequenced enough mtDNAs? Brief. Funct. Genomics 15 (1), 47–54.

Smith, J.G., Newton-Cheh, C., 2015. Genome-wide association studies of late-onset cardiovascular disease. J. Mol. Cell. Cardiol. 83, 131–141.

Steitz, T.A., 1999. DNA polymerases: structural diversity and common mechanisms. J. Biol. Chem. 274, 17395–17398.

Stellwagen, N.C., 2009. Electrophoresis of DNA in agarose gels, polyacrylamide gels and in free solution. Electrophoresis 30 (Suppl. 1), S188–S195.

Stellwagen, N.C., Stellwagen, E., 2009. Effect of the matrix on DNA electrophoretic mobility. J. Chromatogr. A 1216 (10), 1917–1929.

Stuart, D., 2013. The Mechanisms of DNA Replication. InTech. ISBN: 978-953-51-0991-4.

Struniawski, R., Szpechcinski, A., Poplawska, B., Skronski, M., Chorostowska-Wynimko, J., 2013. Rapid DNA extraction protocol for detection of alpha-1 antitrypsin deficiency from dried blood spots by real-time PCR. Adv. Exp. Med. Biol. 756, 29–37.

Suárez-Rivero, J.M., Villanueva-Paz, M., de la Cruz-Ojeda, P., de la Mata, M., Cotán, D., Oropesa-Ávila, M., de Lavera, I., Álvarez-Córdoba, M., Luzón-Hidalgo, R., Sánchez-Alcázar, J.A., 2017. Mitochondrial dynamics in mitochondrial diseases. Diseases 5 (1), 1.

Suwandej, N., 2015. Factors influencing total quality management. Procedia Soc. Behav. Sci. 197, 2215–2222.

Taft, R.J., Pang, K.C., Mercer, T.R., Dinger, M., Mattick, J.S., 2010. Non-coding RNAs: regulators of disease. J. Pathol. 220, 126–139.

Tan, S.C., Yiap, B.C., 2009. DNA, RNA, and protein extraction: the past and the present. J. Biomed. Biotechnol. 2009, 10 pages.

Tan, A.Y., Michaeel, A., Liu, G., Elemento, O., Blumenfeld, J., Donahue, S., Parker, T., Levine, D., Rennert, H., 2014. Molecular diagnosis of autosomal dominant polycystic kidney disease using next-generation sequencing. J. Mol. Diagn. 16 (2), 216–228.

Thompson, J.F., Milos, P.M., 2011. The properties and applications of single-molecule DNA sequencing. Genome Biol. 12 (2), 217.

Thavarajah, R., Mudimbaimannar, V.K., Elizabeth, J., Rao, U.K., Ranganathan, K., 2012. Chemical and physical basics of routine formaldehyde fixation. J. Oral Maxillofac. Pathol. 16 (3), 400–405.

Tortora, G.J., Funke, B.R., 2012. Microbiology: An Introduction, eleventh ed. Benjamin Cummings, 960 pp. ISBN-10: 0321796675; 13: 978-0321796677.

Timmerman, V., Strickland, A.V., Züchner, S., 2014. Genetics of Charcot-Marie-Tooth (CMT) disease within the frame of the human genome project success. Genes (Basel) 5, 13–32. https://doi.org/10.3390/genes5010013.

Torra Balcells, R., Ars, C.E., 2011. Molecular diagnosis of autosomal dominant polycystic kidney disease. Nefrologia 31, 35–43.

Trevisan, P., Rosa, R.F., Koshiyama, D.B., Zen, T.D., Paskulin, G.A., Zen, P.R., 2014. Congenital heart disease and chromossomopathies detected by the karyotype. Rev. Paul. Pediatr. 32 (2), 262–271.

Turan, C., Nanni, I.M., Brunelli, A., Collina, M., 2015. New rapid DNA extraction method with Chelex from Venturia inaequalis spores. J. Microbiol. Methods 115, 139–143.

Ushiki, T., Hoshi, O., 2008. Atomic force microscopy for imaging human metaphase chromosomes. Chromosome Res. 16 (3), 383–396.

van Belkum, A., Chatellier, S., Girard, V., Pincus, D., Deol, P., Dunne Jr., W.M., 2015. Progress in proteomics for clinical microbiology: MALDI-TOF MS for microbial species identification and more. Expert Rev. Proteomics 12 (6), 595–605.

Varga, A., James, D., 2006. Real-time RT-PCR and SYBR Green I melting curve analysis for the identification of Plum pox virus strains C, EA, and W: effect of amplicon size, melt rate, and dye translocation. J. Virol. Methods 132, 146–153.

Vučićević, K.M., Miljković, B.R., Golubović, B.C., Jovanović, M.N., Vezmar Kovačević, S.D., Ćulafić, M.D., Kovačević, M.M., de Gier, J.J., 2015. Expectations, concerns, and needs of patients who start drugs for chronic conditions. A prospective observational study among community pharmacies in Serbia. Eur. J. Gen. Pract. 21 (Suppl. 1), 19–25.

Vernon, S.D., Shukla, S.K., Conradt, J., Unger, E.R., Reeves, W.C., 2002. Analysis of 16S rRNA gene sequences and circulating cell-free DNA from plasma of chronic fatigue syndrome and non-fatigued subjects. BMC Microbiol. 2, 39.

van Belkum, A., Welker, M., Pincus, D., Charrier, J.P., Girard, V., 2017. Matrix-assisted laser desorption ionization time-of-flight mass spectrometry in clinical microbiology: what are the current issues? Ann. Lab. Med. 37, 475–483.

van Vliet, A.H., 2010. Next generation sequencing of microbial transcriptomes: challenges and opportunities. FEMS Microbiol. Lett. 302, 1–7.

Voelkerding, K.V., Dames, S.A., Durtschi, J.D., 2009. Next-generation sequencing: from basic research to diagnostics. Clin. Chem. 55 (4), 641–658.

von Wurmb-Schwark, N., Preusse-Prange, A., Heinrich, A., Simeoni, E., Bosch, T., Schwark, T., 2009. A new multiplex-PCR comprising autosomal and y-specific STRs and mitochondrial DNA to analyze highly degraded material. Forensic Sci. Int. Genet. 3, 96–103.

Walker, J.A., Hedges, D., Perodeau, B.P., Landry, K.E., Stoilova, N., Laborde, M.E., Shewale, J., Sinha, S.K., Batzer, M.A., 2005. Multiplex polymerase chain reaction for simultaneous quantitation of human nuclear, mitochondrial, and male Y-chromosome DNA: application in human identification. Anal. Biochem. 337, 89–97.

Wan, T.S., 2014. Cancer cytogenetics: methodology revisited. Ann. Lab. Med. 34, 413–425.

Wang, R., Ross, C.A., Cai, H., Cong, W.N., Daimon, C.M., Carlson, O.D., Egan, J.M., Siddiqui, S., Maudsley, S., Martin, B., 2014. Metabolic and hormonal signatures in pre-manifest and manifest Huntington's disease patients. Front. Physiol. 5, 231.

Wang, Y., Hayatsu, M., Fujii, T., 2012. Extraction of bacterial RNA from soil: challenges and solutions. Microbes Environ. 27 (2), 111–121.

Wangler, M.F., Hu, Y., Shulman, J.M., 2017. Drosophila and genome-wide association studies: a review and resource for the functional dissection of human complex traits. Dis. Model. Mech. 10 (2), 77–88.

Warner, J.P., Barron, L.H., Goudie, D., Kelly, K., Dow, D., Fitzpatrick, D.R., Brock, D.J., 1996. A general method for the detection of large CAG repeat expansions by fluorescent PCR. J. Med. Genet. 33 (12), 1022–1026.

Watson, A., Mazumder, A., Stewart, M., Balasubramanian, S., 1998. Technology for microarray analysis of gene expression. Curr. Opin. Biotechnol. 9, 609–614.

Weaver, R.F., 2011. Molecular Biology, fifth ed. McGraw-Hill Education. ISBN-10: 0073525324; 13: 978-0073525327.

Westermeier, R., 2014. Looking at proteins from two dimensions: a review on five decades of 2D electrophoresis. Arch. Physiol. Biochem. 120 (5), 168–172.

WHO Guidelines Approved by the Guidelines Review Committee, 2011. Policy Statement: Automated Real-Time Nucleic Acid Amplification Technology for Rapid and Simultaneous Detection of Tuberculosis and Rifampicin Resistance: Xpert MTB/RIF System. ISBN: 978-92-4-150154-5.

Willers, H., Dahm-Daphi, J., Powell, S.N., 2004. Repair of radiation damage to DNA. Br. J. Cancer 90, 1297–1301.

Wisdom, G.B., 1994. Protein blotting. Methods Mol. Biol. 32, 207–213.

Xu, Z., Geng, Q., Luo, F., Xu, F., Li, P., Xie, J., 2014. Multiplex ligation-dependent probe amplification and array comparative genomic hybridization analyses for prenatal diagnosis of cytogenomic abnormalities. Mol. Cytogenet. 7, 84.

Yan, M., Bai, W., Zhu, C., Huang, Y., Yan, J., Chen, A., 2016. Design of nuclease-based target recycling signal amplification in aptasensors. Biosens. Bioelectron. 77, 613–623.

Yim, S.-H., Jung, S.-H., Chung, B., Chung, Y.-J., 2015. Clinical implications of copy number variations in autoimmune disorders. Korean J. Intern. Med. 30, 294–304.

Zaher, H.S., Green, R., 2009. Fidelity at the molecular level: lessons from protein synthesis. Cell 136 (4), 746–762.

Zhang, B., Willing, M., Grange, D.K., Shinawi, M., Manwaring, L., Vineyard, M., Kulkarni, S., Cottrell, C.E., 2016. Multigenerational autosomal dominant inheritance of 5p chromosomal deletions. Am. J. Med. Genet. A 170 (3), 583–593.

Zhang, M., Liu, K., Hu, Y., Lin, Y., Li, Y., Zhong, P., Jin, X., Zhu, X., Zhang, C., 2017. A novel quantitative PCR mediated by high-fidelity DNA polymerase. Sci. Rep. 7 (1), 10365.

Zhang, X., Kuivenhoven, J.A., Groen, A.K., 2015. Forward individualized medicine from personal genomes to interactomes. Front. Physiol. 6, 364.

Zhimulev, I.F., Belyaeva, E.S., Vatolina, T.Y., Demakov, S.A., 2012. Banding patterns in *Drosophila melanogaster* polytene chromosomes correlate with DNA-binding protein occupancy. Bioessays 34, 498–508.

Zhu, W., Zhang, X.Y., Marjani, S.L., Zhang, J., Zhang, W., Wu, S., Pan, X., 2017 Mar. Next-generation molecular diagnosis: single-cell sequencing from bench to bedside. Cell. Mol. Life Sci. 74 (5), 869–880.

Zulfiqar, H.F., Javed, A., Sumbal, A.B., Ali, Q., Akbar, K., Nadeem, T., Rana, M.A., Nazar, Z.A., Nasir, I.A., Husnain, T., 2017. HIV diagnosis and treatment through advanced technologies. Front. Public Health 5, 32.

Index

Note: Page numbers followed by *f* indicate figures and *t* indicate tables.

Printed in the United States
by Baker & Taylor Publisher Services